BATTLE

GLADIATOR

WELLESLEY (LONG RANGE)

MILITARY ENVOY

SPITFIRE

FLAMINGO

HENLEY

QUEEN WASP

D.H. 86 B

SCION SENIOR

PHOENIX

BOMBAY

WELLINGTON

MONARCH

BRISTOL 138

MEW GULL

SUNDERLAND

ALBATROSS

EMPIRE FLYING BOAT

HAMPDEN

MOSSCRAFT

AVIATION
The Complete Book of Aircraft and Flight

AVIATION
The Complete Book of Aircraft and Flight

General Editor
David Mondey
FR Hist S

octopus

**First published 1980 by
Octopus Books Limited,
59 Grosvenor Street,
London W1**

© 1980 Octopus Books Limited

ISBN 0 7064 0962 0

Produced by Mandarin Publishers Limited
22a Westlands Road
Quarry Bay, Hong Kong

Printed in Singapore

CONTENTS

Introduction

Each morning when I go off to work, I see (or sometimes only hear) what is to me a minor miracle. Way above my head, precisely on route, and almost invariably within two or three minutes of its scheduled time, a Boeing 747 'jumbo jet' purrs contentedly overhead, its turbofan engines throttled back, as it makes an approach to London Airport. This is but one of innumerable regular daily flights between the New World and the Old, over the once daunting 'barrier' of the North Atlantic. Why is it a minor miracle? Because of its complete reliability, its superb safety record, and the fact that its pilots are making a routine, accurate, and on-schedule approach to London's Heathrow airport, after almost 3,200 km (2,000 miles) of flight across the grey Atlantic. And the essence of this miracle is that it happens by day or night, and in very nearly all weather conditions.

To you, the reader, the word 'miracle' may seem a little exaggerated, especially if you are in your 'teens or early twenties and so have lived only in a world which regards such travel as an ordinary part of everyday life. But at the risk of being dated I must explain that I was born during World War I when Germany's Gotha bombers made the first mass bombing attack on London. Two years later the first tentative civil airline services were flown across the English Channel, between London and Paris, that is, when the weather permitted such folly. One such crossing was dogged by no less than 22 forced landings, fortunately none of them in the Channel! In that same year, 1919, the North Atlantic was flown non-stop for the first time: an achievement so great that the pilot and navigator of the converted Vickers Vimy bomber which made the crossing were both knighted by HM King George V.

I have been fortunate enough to see, within the short span of my lifetime, the thrilling growth of aviation from the events mentioned above to the establishment of airways which link the nations of the world. There was a time when, soon after World War I, children would run into the street to gaze in wonder as an aircraft 'put-putted' across the sky. Yet throughout the whole of 1978 an aircraft took off or landed at Chicago's O'Hare Field, the world's busiest airport, every 41.8 seconds.

I can recall the visit of Sir Alan Cobham's Flying Circus to my home town, which was one of the first real efforts to make Britain airminded at a time when civil air travel was almost non-existent. Yet today I can see from my office window the Anglo-French Concorde supersonic transport as it sets out across the Atlantic, carrying passengers at speeds quite beyond the research aircraft which flew in the first years after World War II.

I watched the once-majestic passage across the sky of the great civil airships: Britain's R.101 and Germany's *Graf Zeppelin* and *Hindenberg*. However, by World War II, the only lighter-than-air craft were barrage balloons which were flown in order to deter enemy planes from flying too close to their targets.

The exciting roar of the Schneider Trophy contenders, which secured this coveted award for Britain in 1931, led to the reassuring sound of the Rolls-Royce Merlin engine

which made victory possible in the Battle of Britain; and the development of a Fiat engine which, in 1934, enabled the Italians to set a seaplane speed record of 709·20 km/h (440·68 mph).

I remember the first feeble attempts to inaugurate rocket airmail services during the 1930s; I have heard the thunder of a V-2 rocket's approach after the detonation of its warhead, and marvelled as men walked upon the Moon and returned safely to Earth.

The whole of this great canvas of aviation is painted in the pages which follow, from the first stirrings of Man's desire to 'fly like the birds', to the Space Shuttle which, in 1980, will begin to make possible the carriage of men and women astronauts into Earth's orbit as routine. It is not only a story of great technological achievement, but one of courage and endurance.

The hope of aviation's pioneers was that their new creation would bring world peace. Man has, inevitably, used this vehicle for wholesale destruction: but there is, on the other hand, much to record on the credit side.

The internationally-known authors who contributed to this work have written the whole story, and one which is well worth reading if only to understand and appreciate aviation's place in our modern world. It is their sincere wish that you will enjoy their efforts.

David Mondey

Left: a trio of modern hot air ballons almost as brightly decorated as the first Montgolfier balloon which carried Pilâtre de Rozier over the streets of Paris in 1783.

Below: The odd-looking creation of Trajan Vuia made several hops during 1906–1907 but is historically important as the first full size conventionally shaped monoplane.

Endpapers: A series of 48 cigarette cards was issued in 1939 by Gallaher Ltd., of London and Belfast, entitled 'Aeroplanes'. It was the only pre-war collection which contained the Spitfire.

Half title: Jet airliners transformed international travel after World War II, and it is now possible to fly across the Atlantic at twice the speed of sound.

Title page: Rockwell International's B-1 advance strategic bomber which, because of USA commitments to the cruise missile, continues now only as a research project.

Contents page: A McDonnell Douglas F-15 A Eagle. The USAF plan to acquire 749 by 1983.

Introduction: One of the classic fighter aircraft of World War II, North America's P-51 Mustang long-range escort fighter, was evolved to satisfy a British specification.

Early Days

Prehistory

From the earliest times the desire to fly has been a powerful human urge. Although people could walk, swim, or travel by ship or carriage, for centuries they were frustrated by their inability to emulate the birds. In vain attempts to realize their aim, many people strapped themselves to flimsy wings and threw themselves recklessly from roofs and cliff tops, often with disastrous results. However misguided and even comic these experiments may now seem, they had their own rational basis and are best understood in relation to contemporary beliefs about the nature of the world.

Perhaps the most fundamental of all scientific propositions up to about the 17th century, when ideas began to change fairly rapidly, was that all matter was made up of four primary 'elements'– earth, water, air, and fire – each of which had its own natural place in the universe. At the centre of the universe lay the Earth itself, surrounded by a 'sphere' or ring of water. Above the water lay the air, while between the air and the heavens there was supposed to be a region of elemental fire. The elements had various distinguishing characteristics, for instance the degree of heaviness or lightness that each possessed. Lightness was seen not as the absence of weight, but as a positive quality which induced things to fly upwards right away from the centre of the world. Earth was thought to be absolutely heavy and fire to be absolutely light. Water and air contained differing mixtures of heaviness and lightness.

Most things, humans included, were held to be made up of mixtures and combinations of the elements. As man was largely 'earthy' and 'watery,' his proper place in the scheme of things was on or near the surface of the earth, so that any attempts which he might make to fly were literally 'going against nature.' Although birds also contained a great deal of the heavier elements, naturalists noticed the lightness of their bones and the many air spaces within them which reduce the specific gravity of their bodies. Together with the fine comb-like structure of the feathers, these characteristics suggested that birds included in their constitution a high proportion of elemental air which made it possible for them to fly.

If man was ever going to be able to fly, he would need somehow to achieve the quality of 'lightness.' He managed to do just that at the time of the first successful hot-air balloon flights in 1783, and in preceding centuries the techniques tried by many of the aspiring 'birdmen' were directly dependent on this approach to the problem. While to us it may seem self-evident that nothing but disaster can result from any attempt to fly with wings made of feathers, the old scientific beliefs about heaviness and lightness made it quite reasonable to suppose that the more feathers one could attach to oneself the more buoyant one would become.

Since not all birds are extremely efficient fliers, it was necessary to choose one's feathers carefully. In 1507 an Italian charlatan called John Damian, who lived at the court of King James IV of Scotland, fell from the walls of Stirling Castle and broke his thigh when trying to fly. In an attempt to excuse himself for the fiasco, he explained that he had unwisely used the feathers of an earthbound hen instead of the rarer but more effective eagle's feathers.

It was also believed that when birds spread their wings open, exposing the concave underside, they were, among other things, using the curvature of the wings to gather a quantity of air beneath them. This gathered air contributed further lightness and was essential to flight. The air-gathering process is alluded to in several stories of flights using artificial wings, for example, in a 12th-century story, in which a Turk jumped from the top of a tower using cloth wings, whose porous texture imitated the structure of feathers. Before deciding to take the plunge (which resulted in his breaking almost all the bones of his body) he stood for a while with his arms outstretched 'gathering the wind' into the 'pleats and foldings' of his wings.

Some of the very early stories of manned flight emphasize how the urge to become airborne can overcome all prudence. Among the more absurd attempts was one made by a 16th-century French labourer who was fanatical about flying. A solid, thick-set man, he was said to be a great swearer of oaths and to be addicted to drinking fermented milk. One day when the fumes of the milk had gone completely to his head, he hit on the idea of making a flying apparatus. Without saying anything to his wife he went to his barn where he cut his large circular winnowing basket in two and used the halves to make a pair of wings.

First fixing them to his back, he passed his arms through a pair of handles and tried to fly by flapping them. When this failed, he decided that his mistake lay in having forgotten to provide a tail 'which is a great help to birds in flight.' After thinking about the matter for a time, he took his coal shovel and fixed it rather uncomfortably between his legs, tying the handle firmly to his stomach. When he was ready he climbed to the top of a pear tree so as to 'gather the wind. Doubtless still intoxicated by the fumes, he

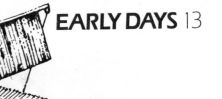

In 1678 at Sablé, in France, a locksmith named Besnier made an attempt to fly with this strange equipment. The four flaps, or wings, functioned like the webbed feet of a water-bird.

threw himself into the air, only to land a second later head first in the drain from his dunghill, breaking his shoulder.

Attempts to fly were quite properly looked on as being dangerous, but the fears expressed were not always the same as those which one might expect today. In addition to the obvious hazards of crashing to the earth from a height, people in medieval and Renaissance times were often aware of dangers associated not so much with the act of falling as with the act of rising.

In the first place, a man might think twice before trying to move above his natural place on earth by soaring to the heavens, because in doing so he might well expose himself to punishment for excessive pride and ambition. In the second place, however attractive a goal the upper regions of the air might seem, they could well prove to be inhospitable.

According to Aristotelian physics and meteorology, the sphere of air connecting men with the heavens and the stars was thought to be divided into three main regions. The lowest, in which we live, was bounded by a higher, middle region where one could find 'the weather.' This middle region was cold, moist, and in a constant state of agitation. It reached as high as the peaks of the highest mountains and was a rather unattractive place. Above it, however, and extending up to the region of fire, was a band of air with much more attractive characteristics. This uppermost part was pure, serene, and warm. In its still and cloudless air man might, if he could fly so high, experience some of the peace and tranquility of the gods. While the beauty of this prospect therefore encouraged men to pursue their age-old dream of flight, there might also be problems, since the nearness of the region of fire led some philosophers to think that man might not be able to survive in such an environment.

Aspects of both dangers – the spiritual and the physical – are expressed in what is probably the best known of all the myths of attempted flight, that of Daedalus and Icarus. While they were trying to escape from the island of Crete, where they had been imprisoned, the famous inventor Daedalus designed a pair of wings for himself and for his son Icarus. Making the appropriate choice of eagle feathers, he and Icarus attached them to their arms with wax and prepared for flight. Before venturing into the air, Daedalus advised his son that, once away from the island, they should both fly at a very low level, close to the sea, in order to remain within the region of air. Although the flight began successfully, Icarus soon grew so enchanted with the delights of this new freedom that, despite his father's warnings, he started to fly higher and higher. Eventually he rose so high that he entered the hot upper region of the atmosphere where the wax melted from his wings, causing him to fall into the sea and drown. Pride and over-ambition were thus punished by the sun god, Apollo, and man was again reminded of the potential physical hazards of unfamiliar parts of the universe.

Many early experimenters, believing that birds fly only in the lower regions of the air, were content to plan quite modest ascents. In doing so they almost always looked to birds for the best example of how to fly, but in this they were seriously misled. Not only did they suppose, quite falsely, that a man has enough muscle-power in his arms to fly, but virtually everyone made erroneous assumptions about the mechanisms of bird-flight. While a bird's wings flap either more or less straight up and down, or downwards and forwards, simultaneously producing lift and propulsion, most writers on flight assumed that a bird moves by a kind of rowing motion, pushing its wings down and backwards against the air. This error, which would in itself have made any machine based on it totally unworkable, is to be found even in the manuscripts of that great early European thinker, Leonardo da Vinci.

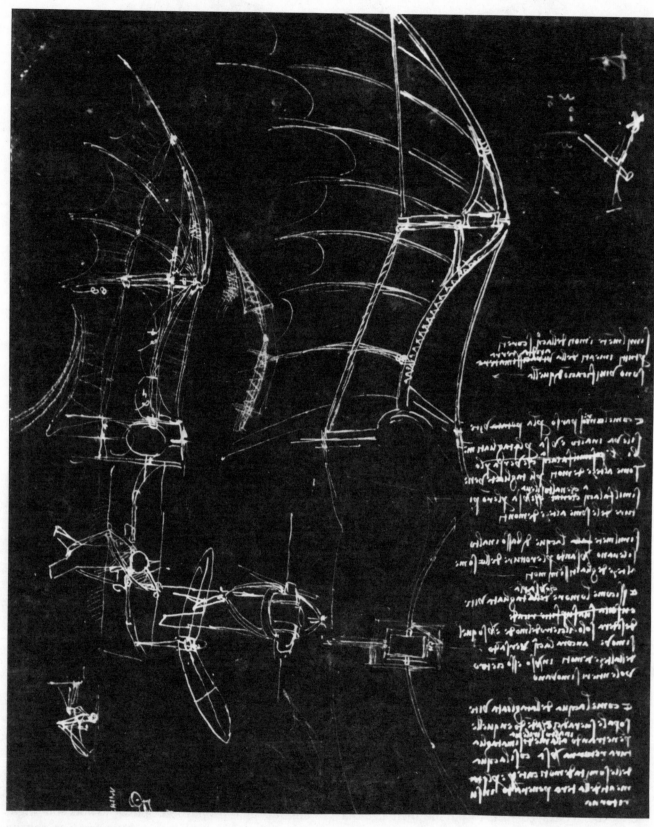

Leonardo da Vinci was
greatly interested in man-
powered flight. Yet it must
be remembered that some of
his work was based on
faulty knowledge. Sadly,
this design was impractical.

Leonardo da Vinci

Leonardo da Vinci (1452–1519) is often described as one of the great forerunners of modern aeronautics. A professional engineer as well as a great painter, he was deeply interested in all scientific and mechanical matters and, as his notebooks show, was especially fascinated – even obsessed – with the problem of discovering how a man might fly. In common with nearly everyone else at the time, he put his faith in the imitation of birds and designed many machines with flappable wings. Some of these were to be operated by the arms alone, some required the use of arms and legs, while a number applied the power of the muscles through various systems of levers and pulleys. Realizing that the muscles of a man's arms are too weak to sustain him for long periods in the air, Leonardo designed his most elegant machines so that the powerful downstroke of the wings would be accomplished by the legs pushing against pedals, while the weaker upstroke would be carried out by the arms. On at least one occasion he thought of using a small motor operated by a powerful bow in which energy could be stored when the string was wound tight and which, after it had run down, could be rewound in flight.

As Leonardo's manuscripts remained virtually unknown until the end of the 19th century, he had no influence on the history of flight. Even if his ideas had been widely circulated, they would not have helped the advance of aerodynamic science. Although his machines were designed with great mechanical ingenuity, they were all based on false assumptions about flapping flight and would have been wholly unstable. In addition, taking an idea from the Roman naturalist Pliny, Leonardo thought that birds 'squeezed' the air with their wingtips, creating an area of high density against which the rowing action of the wing could work with more effect. Some of his wingtips were designed with intricate hinge and pulley arrangements which would have made this imaginary 'squeezing' action possible, but which would also have unbalanced his flying machines still further.

Perhaps the most potentially serious of Leonardo's misconceptions, one which he shared with virtually everyone else until modern times, was his view of the effects of wind. The state of scientific thinking in his day did not allow him to understand that the only effect of a steady horizontal wind is to vary the bird's ground speed. (If it flies downwind it will travel over the ground more rapidly, while if it flies upwind its speed over the earth will be reduced.) The wind does not in any way affect the bird's speed through the air, however, and there can be no question, as Leonardo supposed, of a bird's being able to gain additional lift by turning in flight so as to face upwind, nor is there any danger of its feathers being ruffled the wrong way if it flies downwind. This misconception lies at the bottom of many erroneous calculations about bird flight made by Leonardo and others, and much later it led to innumerable useless designs for sail-driven balloons and airships which, because they were carried along with the wind, could have derived no benefit at all from such an arrangement.

Models

Alongside the history of man's fruitless attempt to fly with wings is a parallel tradition of experiments with flying models. In medieval and Renaissance times the most commonly quoted example was the story of how Archytas of Tarentum (3rd century BC), a Greek mathematician and close friend of Archimedes, built a wooden dove which was reputed to have been powered by some kind of rocket engine or compressed air. If there is any truth at all in the story, Archytas probably succeeded in building a small wooden glider. It is nevertheless wise to be cautious, since the Greeks were skilled at making small working models and had a particular interest in machinery operated by steam power and by air blown through pipes.

Whatever the facts behind the story of Archytas' dove, which inspired many other inventors, it cannot have been powered by gunpowder rockets, as these did not reach Europe from China until medieval times. By 1420, however, at least one man had thought of using rockets to propel a model bird. An interesting manuscript by a Venetian, Giovanni da Fontana, written early in the 15th century, describes a model dove with a rocket embedded in its body. Da Fontana's bird is designed along entirely practicable lines, and there is no reason to doubt that it could have been successfully flown.

Other stories of model aircraft vary greatly in their credibility. Wooden sparrows were said to have been flown around a room to amuse the Emperor Charles V of Spain (1500–58) after his retirement from active political life. Built by a famous and experienced engineer, these may have been small hand-launched gliders. A great eagle, which the mathematician Regiomontanus of Nuremberg (1426–76) flew over the heads of a royal procession, was probably a realistically decorated kite of a kind relatively unfamiliar at the time. On the other hand, it is not so easy to credit a further story concerning

Regiomontanus. In addition to the eagle, he is said to have built an intricate model fly made of iron, and to have astonished his guests by allowing it to fly in a circle around his table. Although sophisticated clockwork machinery was common enough in 15th century Europe, the general ignorance at that time of the true nature of flapping flight makes any such achievement highly unlikely, to say the least. By the mid-17th century, however, at least one flapping model appears to have been successfully flown. The famous English scientist Robert Hooke (1635–1703) mentions in his notes that in 1655 he made a clockwork-powered model bird which managed to lift itself into the air for a short time.

With models as with manned flight, one of the features of early aviation history is the scant attention given to fixed wing gliders, very few of which seem to have been tried. However, although modern experience of work with wind tunnels does make it obvious that much useful information may be drawn from fixed-wing models, men of earlier generations thought it self-evident that the way to success lay not in experiments with gliders but in close imitation of nature, and that meant flapping like a bird.

An imaginative and serious, if also totally misguided attempt to build a flying machine with flapping wings was made in the mid-17th century by an ingenious Italian engineer, Tito Livio Burattini (1615–82). Living at the Polish court, Burattini hoped to put a flying machine at the disposal of the King, who showed a mild interest in his experiments. Burattini's machine, only a model of which was ever built, was designed to look like a fierce flying dragon. It had, in fact, eight wings: four flapped up and down to keep it airborne, two were used to propel it forwards, and a further small pair at the head provided supplementary propulsion. A kind of folded parachute attached by springs to the dragon's back was intended to allow a safe descent in the event of an accident, while directional control was made possible by a tail arranged to move both vertically and horizontally. Burattini's flying machine, designed with great enthusiasm and conviction, though without true aerodynamic insight, caught the imagination of many 17th century scientists and did much to maintain an interest in the practical possibilities of flight.

Some attempts to fly by means of untried apparatus were ludicrous and pathetic. For example, in 1772, a misguided but devoted churchman, the Abbé Pierre Desforges (who had already brought unwelcome attention to himself by writing in favour of the marriage of priests, and who had spent a period in gaol as a result) built a wickerwork gondola about 2m × 1m (7ft × 3½ft.) Made with loving care, the gondola was covered in feathers and surmounted by a kind of feather parasol. Two feather-covered wings were fitted, arranged so as to move like oars and attached to the gondola by hinges which would need, so Desforges said, to be replaced every 36,000 leagues, or about 150,000km (93,205 miles.) The pilot was protected by a large sheet of pasteboard which he could extend across his stomach as a wind shield, and also by a pasteboard flying helmet (carefully painted so as to resemble a bird's head) and fitted with goggles.

While the gondola was made without the slightest thought of aerodynamic principles, Desforges did attempt to estimate its probable performance by reference to experiments which the great French scientist Réaumur (1683–1757) had carried out to determine the speed of swallows. Basing his calculations on these findings, Desforges decided that his machine would be capable of speeds of around 30 leagues per hour (145km/h, or 90mph.) So, having found four peasants to help him, Desforges had the gondola carried to the top of a cliff. After boarding the machine and taking control of the handles which operated the wings, he gave the peasants the signal to drop him over the edge. To everyone's surprise, he escaped from the resulting heavy crash with nothing more serious than a bruised elbow. This brought his aeronautical experimenting to an end, and as one contemporary writer wryly remarked: 'No one will ever need to burn Desforges at the stake because of his achievements as a flying sorcerer.'

First hops and the birth of aerodynamic science
Compared with many other sciences, aerodynamics was slow to develop. Despite a good deal of empirical work in the mid-19th century, a correct formulation of basic aerodynamic principles did not emerge until the 1890s. The theory of flight is still developing, and even today there are aspects of the subject which are not at all well understood. For example, the way certain insects manage to hover is still far from being properly explained and continues to provide a difficult and fruitful line of inquiry.

The first fully thought-out aerodynamic experiments were undertaken early in the 19th century by Sir George Cayley (1773–1857), a meticulous worker whose interesting notebooks have survived and who has rightly been called 'the father of aerial navigation.' Pioneering a technique which was to be useful long after his death, he attached wing surfaces to the end of a rod which could be made to whirl horizontally and which enabled him to measure their lifting force at various angles of attack (the angle between the plane of the surface and the oncoming air stream.) From 1804 to about 1809 he did useful work on some of the most crucial properties of aerofoils (wing surfaces). As well as the angle of attack he studied the different effects produced by altering the camber (curvature), the best shapes for streamlining, the way the lifting forces are distributed over the whole area of the wing, and the effects of angling the wings so that the tips are higher than the centre (dihedral). Although the flying machines which he built were all fixed-wing gliders, he studied the flapping flight of birds and developed some genuine insights into a subject which until then had been little understood. It was Cayley who first showed that on the downstroke the large feathers at the tip of the bird's wing twist so that their trailing edges are higher than the leading edges, causing them to function like small propeller blades, so pulling the bird along.

Cayley had been interested in flight since childhood. When he was only 11, two Frenchmen, Launoy and Bienvenu, made a small toy helicopter which was a development of a plaything common in medieval times. Two contra-rotating airscrews (made from feathers stuck into corks) were driven by a small bow-drill motor, powerful enough to make the toy fly into the air when the bow was released. Having encountered this little helicopter not long

after its invention, Cayley built his own version in 1796. He kept thinking about the possibilities of rotary lift until eventually, in 1809, he published a modified design of his own together with a commentary on its function. This model flying machine became quite well known in England, inspiring a number of other inventors and being rightly looked on as the origin of all subsequent rotorcraft development.

Cayley did not stop there. In 1834, having been stimulated by the ideas of Robert Taylor, he returned to the idea of the helicopter and he published a highly imaginative and ambitious design for a 'convertiplane.' This design used twin full-circle contra-rotating rotors which were placed on either side of a boat-shaped hull. For forward flight the inclined blades (eight on each rotor) could be made to lie flat, forming smooth, circular, biplane wings.

Cayley's 'convertiplane' was never built, but he did undertake practical experiments both with fixed-wing models and with full-sized gliders. In 1804 he built a now famous little model glider, a replica of which may be seen in the Science Museum, London. About 1·5m (5ft) long, it consists of a small arch-topped kite attached to a pole, with a cross-shaped adjustable tail-unit at the rear. A sliding weight at the nose allowed Cayley to vary the position of the centre of gravity. The graceful descent of this glider when launched from the top of a steep hill inspired him to further work, with the emphasis sensibly

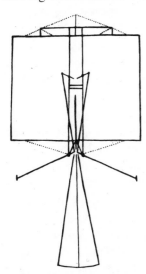

Sir George Cayley, so rightly considered to be the 'father of aerial navigation', was very successful in designing many types of plane. The design *above* of 1799 is for a fixed-wing aircraft, and the one on the *right* is for a fixed-wing man-carrying glider.

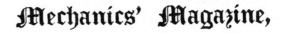

Mechanics' Magazine,

MUSEUM, REGISTER, JOURNAL, AND GAZETTE.

No. 1520.]   SATURDAY, SEPTEMBER 25, 1852.   [Price 3*d*., Stamped 4*d*.

Edited by J. C. Robertson, 166, Fleet-street.

SIR GEORGE CAYLEY'S GOVERNABLE PARACHUTES.

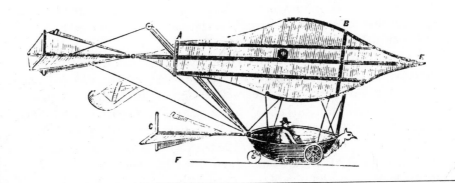

placed on stability and control. Aware of the need to equip any future flying machine with an engine, but also realizing that nothing satisfactory was likely to be produced for some time, Cayley determined to discover as much as he could from unpowered flight. Several other gliders were sketched and constructed and in 1809 he built a full-sized machine which he tested for a time in short hops.

Although he remained interested in flight, it was not until towards the end of his life that Cayley returned to practical experiments. Having thought about multiplane structures since 1843, he built a triplane glider in 1849 and twice tested it in very short flights with a boy on board. In 1852 he returned to his early concern for monoplanes, designing (but not building) a very graceful glider with large wide wings and incorporating many refinements of construction and flight control. Finally, in 1853, he built his most famous machine, a glider in which his coachman was reluctantly flown across a shallow valley. Although the flight – probably the first of its kind – ended without injury to the coachman, the poor man was so frightened by his experience that he immediately gave in his notice. Being used to the sensitive handling of horses. the coachman must have felt doubly vulnerable since he was merely a passenger in the glider, having no means of control at his disposal.

During the middle part of the 19th century, the development of fixed-wing aircraft progressed sporadically. After Cayley the most serious work was carried out by William Henson (1805–1868), who designed an Aerial Steam Carriage. The drawing for this intelligently conceived aeroplane became very famous, being frequently reproduced during the 19th century and even later. Although it was never built, it contained many prophetic features, including braced monoplane wings of 45·75m (150ft) span, twin pusher propellers (that is, propellers operating behind the wings), an enclosed cabin, and tricycle landing gear. Despite the inadequacy of the control system, which consisted merely of a rudder and an 'all-moving tailplane' (the whole surface moving as an elevator), such an aircraft could probably have flown after a fashion if a sufficiently light and powerful engine could have been found. Henson's only practical experiments were with a model of the Aerial Steam Carriage having a wingspan of 6m (20ft). During the years 1845–47 this was tested several times but without any encouraging results. Soon afterwards Henson abandoned aeronautical work, but experiments along similar lines were continued for a time by his

mechanically minded friend John Stringfellow (1799–1883), who had earlier designed and built some excellent lightweight steam engines. In 1848 Stringfellow tested another small model bearing a general resemblance to Henson's design, but lacking a fin or rudder. Although the results of these tests have long been a matter of dispute, it seems clear that Stringfellow's model could not remain airborne under its own power.

Imaginative work was carried out in France in the 1850s by a naval officer, Félix du Temple, who in 1857 designed an elegant monoplane based on geometric principles and powered by a single tractor (puller) propeller. A model of this, tested in 1858, had the distinction of being the first heavier-than-air device to sustain itself in free flight. It was not until 1874 that du Temple was ready to try a full-sized machine. Similar in configuration to the powered model, it managed to get airborne after picking up speed down a sloping ramp. Although this 'flight' is sometimes said to give du Temple precedence over the Wright brothers, he achieved no more than a brief hop, assisted by gravity, and unlike the Wright *Flyer*, his machine lacked any effective flight control system. Du Temple's visually pleasing but aerodynamically curious choice of sharply 'swept-forward' wings would in any case have rendered his aeroplane unstable in pitch, giving the pilot a serious problem in maintaining the correct attitude in flight.

An essentially similar experiment, resulting in a powered hop followed by an immediate flop back to earth, was made in Russia in 1884. A large steam-powered monoplane designed by Alexander F. Mojhaiski was launched down a ramp in St Petersburg, with I. N. Golubev at the controls of the elevator (no other control surfaces being fitted.) Three propellers were used, one tractor and two pushers. Like du Temple's hop, this can in no way qualify as a sustained and controlled flight.

During the second half of the 19th century, power plants underwent rapid improvement, tempting more and more experimenters to commit themselves to the air in barely controllable and totally untested machines. Among the most widely publicized of these was the remarkable looking *Eole* of Clément Ader (1841–1925.) Powered by a very fine steam engine, this strange, bat-like aircraft had a semi-enclosed cockpit and a wingspan of about 15m (49ft). On 9 October 1890 Ader gained a minor place in history by becoming the first man to rise from level ground in a powered aircraft. Although the *Eole* managed only a brief flight of about 50m (165 ft), the French military authorities were sufficiently

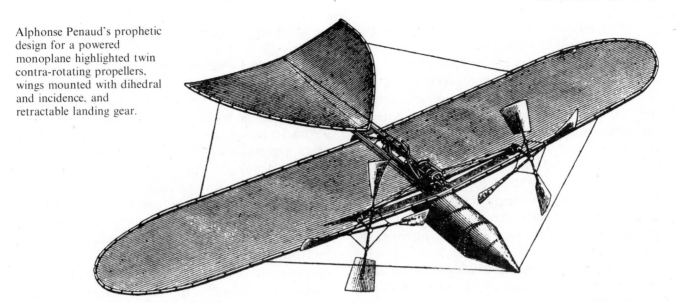

Alphonse Penaud's prophetic design for a powered monoplane highlighted twin contra-rotating propellers, wings mounted with dihedral and incidence, and retractable landing gear.

interested to offer Ader a subsidy for further experiments. In the mid-1890s he began work on a second machine called *Avion II*, which was abandoned before completion. This was followed by *Avion III* which, despite false claims made afterwards by Ader, was quite unsuccessful when tested in 1897.

Meanwhile slow progress was being made in the understanding of the theory of flight. After the work of Cayley, the most important advance in aerodynamic thinking was made by Francis Wenham who, in 1866, delivered to the Aeronautical Society a highly significant paper called 'Aerial Locomotion.' Examining the properties of aerofoils, Wenham continued and further advanced the studies initiated by Cayley. Among other things he investigated the effects of camber, the movement of the centre of pressure, and the relationship of a wing's length to its width (aspect ratio.) Like Cayley, he advocated the use of multiplane structures, with wings superimposed on one another, and in the late 1850s he undertook practical experiments using multiplane gliders. In 1871 he and John Browning made further advances when they tested aerofoils in a wind tunnel, the first ever to be used.

But for his premature death at the age of 24, the Frenchman Alphonse Pénaud (1856–80) might well have made highly significant contributions to the history of aviation. An inventor of talent and imagination, he had experimented as a teenager with rubber-powered models, and in 1876 patented a remarkably prophetic aeroplane which he did not live to build. Large, tapered monoplane wings were attached to a streamlined fuselage fitted with a fully enclosed, glass-domed cockpit. A stick was provided for combined control of the rear-mounted elevators and rudder, and although there was no positive

system of control in roll, steering airbrakes were included on the wings. Twin tractor propellers provided propulsion, and the tricycle landing gear was retractable.

Further useful work was done in England by another original thinker, Horatio Phillips (1845–1924), who undertook a systematic series of experiments with aerofoils of widely differing shape. Having demonstrated conclusively that the greater part of the lifting force of a wing is contributed by the low pressure on top rather than by the high pressure below (a point which had been suspected by Cayley), he patented in 1884 a series of aerofoils of various thicknesses and cambers. His findings were widely publicized and had considerable influence on later workers in the field. After further development of these basic ideas, Phillips built and tested in 1893 a large powered multiplane which travelled around a circular track and demonstrated the validity of some of his theories of lift. This test-bed, sometimes affectionately referred to as 'the flying Venetian blind,' consisted of a great many long, thin wings stacked one on top of the other. It formed the basis of a multiplane aircraft which Phillips tested in 1904. Although he got it to lift from the ground, it was virtually uncontrollable and showed that the multiplane idea could not successfully be pushed to extremes. Only marginally more encouraging results were obtained in 1907, when Phillips flew a still more cumbersome aircraft, consisting of four banks of 'Venetian blinds' arranged in tandem, one behind the other.

Also deserving of mention, if only because of its scale, was the work done by the wealthy American expatriate Hiram Maxim (1840–1916), who lived in England. In the 1890s he built an enormous biplane test-rig at Baldwyn's Park, Kent. Powered by two 180hp compound steam engines, the rig consisted of biplane wings of 31·7m (104ft) span set at a pronounced dihedral angle, twin propellers each 5·43m (17ft 10in) in diameter and provision for a crew of four. The rig was not designed to fly, but was restrained by a twin-rail system intended to allow it to rise only a very short distance from the ground. When tried in 1894 the rig duly lifted from the bottom rail at a speed of round about 65km/h (40mph), and rose so strongly that it broke the upper guard rail. A fore and aft elevator was provided, but no other form of control was used, and, although the test-rig impressed many people, its importance lay more in the encouragement which it gave to others than in any contribution to the advancement of aerodynamic science.

Although the effects of aerodynamic lift had by then been frequently demonstrated and investigated, no one had yet offered a satisfactory physical theory to account for it. In 1894, F. W. Lanchester formulated the first theoretical analysis which he included in a paper called 'The Soaring of Birds and Possibilities of Mechanical Flight,' delivered to the Birmingham Natural History and Philosophical Society. Unfortunately for Lanchester the theory was not given wide circulation until it was incorporated into his book *Aerodynamics*, published in 1907, and even then it was described in a form which did not make it easy to understand. As a result, an independent formulation of lift theory, developed in Germany by Ludwig Prandtl, gained wider currency. To him and to one or two other workers, especially the German Wilhelm Kutta and the Russian Nikolai Joukowski, must go the credit for putting aerodynamic theory on a really sound footing.

Men learn how to handle gliders
During the 19th century public enthusiasm for aeronautics kept growing. Colourful exploits in balloons attracted a great deal of attention, while at the same time a number of intrepid showmen tried to create an impression with flappers and gliders of various kinds. Among the most adventurous of these was the Frenchman Louis Letur, who built a curious, complicated and fragile machine with which he several times descended from a balloon in 1853 and 1854. A colourful combination of parachute and glider, Letur's device did little to advance aeronautical science, but his growing experience of aeronautics might have been helpful to other inventors had he not been killed in 1854, when a mishap

with his release mechanism led to his being dragged over some trees.

A tough French sea captain fared somewhat better than Letur. In 1857 Jean-Marie Le Bris built an elegant glider whose shape was based on the albatrosses he had seen and admired while at sea. With Le Bris on board, the glider was placed on a farm cart which was then driven downhill until take-off speed was reached. A short glide resulted and encouraged Le Bris to continue his experiments. After the glider had again been launched downhill it once more carried Le Bris for a short distance through the air, but as he had no means of control he was unable to avoid a crash-landing in which he broke his leg. Although doubtless somewhat daunted for a time, Le Bris returned to his experiments a few years later. In 1868 he tested a second bird-shaped glider, but on that occasion took the wise precaution of flying it with ballast on board instead of a pilot/passenger. After take-off the second glider also crashed.

These and many other experiments were mostly hit-and-miss affairs, the careful work of Cayley and Wenham being the exception rather than the rule. In the late 1880s, however, gliding was at last put on a proper theoretical and practical footing by the efforts of one of the greatest of all pioneers of aviation, the German Otto Lilienthal (1848–96). Confident that success lay in a really careful study of control systems before applying power, Lilienthal embraced, but in a new and thoughtful way, the old belief that the best procedure was to imitate the birds. Although he envisaged the possibility of flapping, he began with fixed-wing gliders which he flew so as to gain experience of the air. In the 1890s he built a series of 18 different types, all of them hang-gliders from which the pilot was suspended in a harness. Light, and necessarily somewhat fragile, the gliders were nevertheless structurally sound, being built with proper engineering principles in mind. For his earliest work he launched himself from natural heights, but in 1893, after he had had a hangar built to house his gliders, he flew from its roof. Finally he made an artificial mound 15m (50ft) high, from which he could fly in any direction and so launch himself directly into wind.

Lilienthal's gliders were built so that the pilot could support himself on his arms in the centre of the wings. With his hips and legs he could swing his body back and forth or even sideways, so adjusting the centre of gravity and controlling the glider's movement, a technique re-established in recent years by modern hang-glider enthusiasts. The launch was carried out by running forward to gain airspeed and then raising the legs, after which glides of 100–250m (330–820ft) could be made. Despite the inadequacy of the body-swinging method of control, Lilienthal learned to manoeuvre with a remarkable degree of skill. Most of the gliders were monoplanes, with a fixed tailplane and vertical fin at the rear, but in the hope of increasing the range of control he experimented for a while with biplanes. Before his death he was making tentative plans for powered flight and had also begun to consider the use of movable control surfaces.

On 9 August 1896, when Lilienthal was flying one of his monoplanes, he was thrown out of control by a sudden gust of wind. The glider suffered a 'wing-drop stall,' falling sideways out of control from a height of about 15m (50ft), giving Lilienthal insufficient time in which to recover. As a result of his injuries he died on 10 August. Lilienthal's publications, and especially his book *Der Vogelflug als Grundlage der Fliegekunst* (Bird Flight as the Basis

The German Otto Lilienthal, realizing that the means to powered flight lay far in the future, concentrated on the perfection of gliders and of gliding technique. He made literally thousands of flights in hang-gliders such as this.

of Aviation), were based on meticulously kept records and provided valuable data for later experimenters.

Inspired by Lilienthal's example, a young Scot called Percy Pilcher (1866–99) built his first glider in 1895. Pilcher had corresponded with Lilienthal and had twice visited him in Germany. On the second occasion he had been allowed the rare privilege of flying one of Lilienthal's gliders. Retaining the basic hang-glider configuration, Pilcher introduced modifications of his own, including the use of wheeled landing gear and the development of a take-off technique using a tow-line. Although lacking Lilienthal's care and originality, Pilcher was an enormously enthusiastic young man, eager to make fast progress and keen to experiment. Towards the end of his life he was making plans for a powered machine for which he built and successfully bench-tested an excellent light oil engine.

Pilcher's somewhat naive enthusiasm may be gauged from his description of the early difficulties that he encountered. Despite Lilienthal's well founded assurance that a glider is controllable only if an adequate tailplane is fitted, Pilcher's first glider, the *Bat*, completed in the early part of 1895, had a vertical fin and a large degree of dihedral on the wings, but lacked horizontal tail surfaces. Finding in the event that Lilienthal had been quite correct,

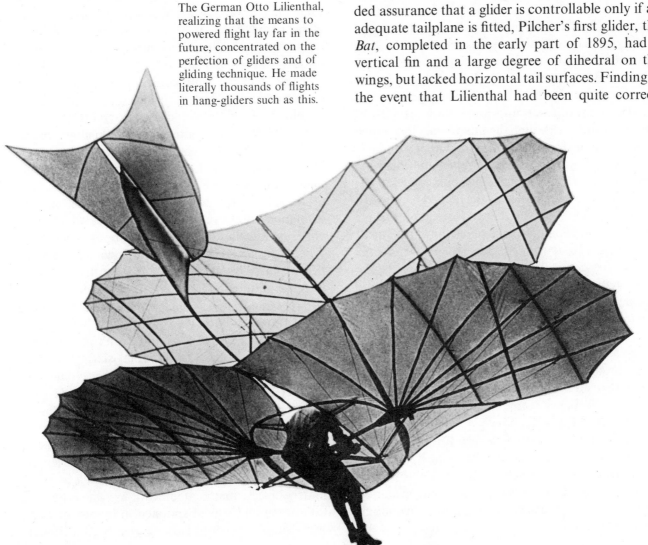

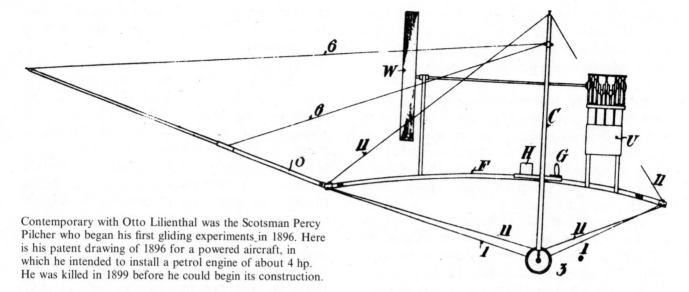

Contemporary with Otto Lilienthal was the Scotsman Percy Pilcher who began his first gliding experiments in 1896. Here is his patent drawing of 1896 for a powered aircraft, in which he intended to install a petrol engine of about 4 hp. He was killed in 1899 before he could begin its construction.

Pilcher modified the *Bat* which then became flyable, though only with difficulty, for the excessive dihedral made the glider rather awkward to control. Two further gliders were tried before Pilcher built the *Hawk*, a famous machine which he flew from 1896 until his untimely death in 1899. Although, like Lilienthal, Pilcher had begun to develop an interest in movable control surfaces, his hang-gliders were all manoeuvred by body movements alone.

Pilcher was nevertheless much more than a slavish imitator of the German master, being prepared to consider many other approaches, including the use of multiple wings. Although the Lilienthal glider which he had tried in Germany in 1896 had been a biplane, he had begun with a prejudice against multiple wing surfaces, tending to rely on the monoplane. In 1897 he was nevertheless impressed by the limited success of a triplane glider being used in America by Octave Chanute, and began to make plans for a triplane or quadruplane (four-winged glider) of his own. At the time of his death he had certainly started on the construction of some kind of multiplane glider, and he may even have finished it.

The speed with which Pilcher worked, the interesting plans which he drew up, and the success of his engine all suggest that, had he lived, he might well have developed a controllable powered aircraft earlier than the Wrights. In September 1899 however, his propensity for taking risks had fatal results. Lord Braye had invited Pilcher to give a demonstration of gliding at Stanford Hall, near Market Harborough. On 30 September Pilcher's glider *Hawk* was launched, using his tow-rope technique: two horses pulled the glider into the air by hauling on a rope attached to a pulley. Although the weather was bad, the glider was sodden, and the

ropes were heavy with water, Pilcher decided to go ahead with his demonstration. On a first attempt to fly the *Hawk*, the tow-line broke, causing Pilcher to glide to earth shortly after take-off. On the second attempt, a bamboo rod in the tail assembly snapped when Pilcher was at a height of about 10m (33ft). The glider plunged out of control and Pilcher died two days later as a result of his injuries, from what might initially have appeared to be an innocuous flight.

While Pilcher's main contribution to the history of aviation lay in his enthusiasm and in his further development of Lilienthal's ideas, Octave Chanute (1832–1910), whose active interest in flight began quite late in his life, made practical advances which materially assisted the Wright brothers to achieve their ultimate success. A French-born American, Chanute first expressed his fascination with the idea of manned flight in a series of articles which were collected together to form a classic book, *Progress in Flying Machines* (1894). By 1896, having like Pilcher been inspired by the example of Lilienthal, Chanute began building gliders. The first of these was a complex multiplane which underwent a strange series of modifications. To begin with it had six pairs of wings, but by July 1896 it had been converted into a five-wing machine, the original sixth wing now serving as a large tailplane. A month later the number of mainplanes had been reduced to four, with one of the original wings still forming a tailplane. Although Chanute records that with this new configuration 'a great many glides were made, with the result of more than doubling the lengths previously attained,' the glider, which eventually underwent six major modifications, was growing a little worn and was abandoned in favour of something simpler. Accordingly he next tried a triplane, which was also soon altered to become a biplane with cruciform tail unit. In its biplane form this was to remain the standard Chanute glider.

In some respects Chanute's gliders incorporated a number of structural refinements, (as did the biplanes used by Otto Lilienthal in 1896) including in particular the light and rigid system of bracing which the Wright brothers were to adopt after the establishment of their friendship with Chanute in 1900. By now Chanute was in his mid-60s and wisely considered himself too old to try glider flying himself. Instead, he employed as his test pilot a vigorous young engineer, A. M. Herring, who was an active collaborator and went on to make his own independent experiments in 1898, using a powered biplane. Flying Chanute's biplane glider, which incorporated one or two ideas of his own, Herring made several successful flights in August and September of 1896. As a result of these flights further modifications were made and, although Chanute remained earthbound, he was a true airman in spirit, always paying careful attention to what Herring told him of his experience of controlling the machine in flight.

Chanute was always keen to develop new ideas, and included, in a new multiplane patented in 1897, a system of control which allowed the wings to be swung back and forth by the use of pedals. Sweeping the wings backwards together would shift the centre of lift backwards in relation to the centre of gravity, resulting in a lowering of the nose and an increase in airspeed. Allowing the wings to come forward would have the opposite effect and slow down the machine. By pushing the wings forward on one side only, a turn would be initiated because of the imbalance which would be created. Although there is no evidence that Chanute ever tried the system in practice, it is an imaginative approach to the matter of control.

Despite his attempts to find ways of manoeuvring his gliders, Chanute shared with most of his contemporaries an undue if understandable concern for stability and safety. Indeed, he always gave absolute priority to stability, seeking, as he put it, 'exclusive equilibrium.' In this he differed fundamentally from the young men to whom he was to offer so much encouragement and support, Wilbur and Orville Wright.

By 1900 men had learned how to make structurally sound gliders which could carry a pilot over distances of several hundred metres at acceptable airspeeds and rates of descent. Although the use of vertical stabilizing surfaces was not universal, the need for a tailplane was generally well understood. Appropriate dimensions for man-carrying wings had been worked out, both in theory and in practice, and several pilots had learned from experience how to keep a hang-glider under control in balanced flight. Straight and gentle descents in good conditions had become relatively commonplace, while a modest degree of success had been achieved in turning. True manoeuvrability was nevertheless still to be achieved, and in order to do this successfully a detailed and systematic study of control methods was essential.

The Wright brothers

From time to time claims are made which attempt to show that the Wright brothers should not be considered the true inventors of the aeroplane. Such arguments often arise from national pride, sensationalism, or sheer malice, and these claims usually offer evidence that someone managed to make a powered machine rise into the air before 17 December 1903. By applying enough brute force it was not very difficult, in the late 19th century, to get a primitive combination of wings and fuselage to rise into the air. What the Wrights achieved, however, was something immeasurably more significant: they designed and built a machine which, after rising from flat ground under its own power, could sustain itself in the air in level flight and could be controlled in all three dimensions of space.

Wilbur and Orville Wright (1867–1912, and 1871–1948 respectively), the sons of a bishop of the United Brethren Church, lived in Dayton, Ohio, USA. Having developed a successful business as bicycle manufacturers, they were familiar with machine design, were skilled with their hands, and had at their disposal a modest amount of money and leisure. In 1896 they learned of the death of Lilienthal, whose brave efforts so caught their imagination that they decided to embark on a methodical study of all the information they could find about experiments in flying. In addition to reading all the available literature, they studied the flight of birds – not in order to build yet another abortive flapping machine, but in the hope of learning the secrets of gliding control. When they judged that their studies had given them sufficient preparation they began, in 1899, a rationally ordered sequence of experiments which led, four years later, to their much-deserved success.

Although the Wrights made a number of design mistakes, and although the fully developed *Flyers* with which they astonished the aviators of Europe in

A glider, built by the Wright brothers, shown in flight over the sand dunes at Kitty Hawk, North Carolina, USA.

1908 contained inherent deficiencies, it cannot be overemphasized that because of their concern for control and manoeuvrability they almost immediately put themselves well in advance of virtually all of their competitors. Instead of trying to build something which would hang stably in the air and would need to be pushed this way and that like an airborne motor car, they wanted to learn to handle a lively machine which would require delicacy and sensitivity on the part of the pilot.

Learning from their observations of soaring birds rather than from flapping flight, they noticed that in gusty conditions buzzards appeared to maintain their equilibrium by, as Wilbur put it, 'a torsion of the tips of the wings.' But how might this idea be applied to a man-made machine? At first they thought of pivoting the aircraft's wings at the fuselage, allowing them to be rotated about the main spar, so that one wing could be made to produce more lift than the other. When this proved to be structurally difficult, they hit on the system of twisting the wings at the tips, an idea which came to them while handling a long cardboard box. To test the effectiveness of this method of control they built a biplane kite, the wings of which could be twisted, or 'warped' at the tips by the use of four control lines. Fore-and-aft stability was ensured by adding a fixed tailplane, and, to carry their investigations further, they also arranged the wings so that they could be moved backwards and forwards in relation to each other thus shifting the relative position of the centre of gravity. This kite, of about 1·5m (5ft) wingspan, had no 'side-curtains' (vertical surfaces between the wings) and therefore had no relationship with the box kite which was to have a strong influence on the design of European aircraft.

Although the Wrights were a little secretive by nature and preferred to do much of their work in comparative isolation, they corresponded with many people, especially with Octave Chanute who recognised their talent and offered to help in a variety of ways. It was Chanute's example which led to their adoption of the biplane configuration and in particular to the light, rigid trussed structure which

Chanute had managed to pioneer so successfully.

Following their initial correspondence with Chanute, and in the hope of applying the experience gained in flying their kite, the Wrights built a full-scale biplane glider which they flew during the autumn months of 1900. Needing to carry out their trials at a place where the winds were favourable and steady, they chose the sand dunes of Kitty Hawk, North Carolina. Their small glider, only 5m (17ft) wingspan, was equipped with a forward elevator which was to remain characteristic of all Wright machines for a decade. The decision to place the elevator in front of the wings rather than behind them to obtain lateral control – a design error which had ultimately to be abandoned – was made in the mistaken hope of avoiding a sudden dive more easily. This time there was no provision for varying the stagger of the wings, and there were no vertical surfaces (fin or rudder.) Although a few glides were made, with the pilot lying prone between the wings so as to reduce the drag, and a few manned kite-flights were tried, the Wrights mainly used this glider as an unmanned kite.

In 1901 they built a second and bigger glider, with a span of nearly 7m (22ft). The pilot was provided with a hip-cradle which he could swing from side to side to operate the warping cables, and, in order to ensure liveliness and sensitivity in the glider, the wings were rigged with 'anhedral' (arranged so that the wingtips drooped a little below the level of the centre section). Glides of well over 100m (328ft) were made when conditions were good, but the Wrights were beginning to experience control problems which led them to be suspicious of the findings of their predecessors and even to doubt the calculations of so careful an experimenter as Lilienthal. In the course of their work with this second glider, they found that although the wing-warping system produced the desired action of banking the wings, there was sometimes an alarming and unexpected tendency for the glider to begin to yaw, or turn horizontally, in the wrong direction. So if the left wing was lowered, with the idea of turning to the left, the glider would, when substantial warping was used, turn to the right instead, with the raised wing in the centre of the circle. This turn against the direction of bank often led to the start of a spin and then to a crash.

Similar problems confronted the Wrights in 1902, when they built and flew their third glider. This had a still larger wingspan of about 10m (33ft), but it differed from its predecessors in a more important respect: for the first time vertical surfaces were used,

in the form of a double fixed fin mounted behind the wings. The Wrights had introduced the fin in the hope of steadying the glider in the turn and so curing the yaw in the unwanted direction. The fixed fins did not, however, provide the necessary degree of correction and the problem was solved only when they took another vitally important step. The brothers had given a great deal of thought to the 'adverse yaw' and had puzzled over its cause. This led to their most important single insight and to what was probably their greatest contribution to the history of manned flight. When the wings were warped so as to vary the lift on the two sides of the glider, therebye making one wing rise while the other was lowered, the increase in lift on the upgoing wing was accompanied by an increase in drag on the same side. As the wing was raised it was therefore also retarded by the drag, swinging the glider into a turn against the bank. Although the fixed fins did not hold the glider against the turn, the Wrights realised that they could cure the problem by converting them into a movable rudder. They therefore pivoted the fins so that they could be turned by cables which were linked to the wing-warping system. This rudder now acted so that whenever bank was initiated to one side or the other, the tendency to yaw in the opposite direction was counteracted. The glider could then both execute balanced turns and roll back from a turn to straight and level flight without loss of control. After several hundred glides had been made, the Wrights justifiably concluded that they had solved their basic control problems and that they were in a position to think about building a powered aircraft.

Having first applied for a patent for their glider, with its revolutionary control system, they set about tackling the problem of finding a suitable engine. Lightweight powerplants being scarce, they decided to design and build their own. Their efforts resulted in a 12hp petrol engine arranged to drive two pusher propellers which were also of their own design. For the *Flyer*, as they called their powered aircraft, they built a new biplane of 13m (40ft) wingspan, with forward biplane elevators and the double movable rudder at the rear, linked as before to the warping system. The light undercarriage consisted of skids which, for the take-off, were laid on a small wheeled truck running along a wooden rail.

In 1903, working with their usual care, the brothers prepared themselves for trials with their powered machine by practising in their 1902 glider to help them brush up their flying technique. Eventually, in December 1903, they were ready to experiment with power. After a minor accident on 14 December, which delayed them for three days, Orville made the first powered, sustained, and controlled flight at 10.35 on Thursday 17 December 1903. Although this lasted for only 12 seconds, it was followed by three others (the brothers alternating as pilot), the last of which covered 260m (852ft) in 59 seconds. The *Flyer* was not an altogether satisfactory machine: among other things the linked rudder did not allow for full harmonization of the controls, the forward elevator was inherently destabilizing, and the engine was capable of delivering only marginally adequate power. The Wrights were nevertheless wise to adhere to the configuration with which they were growing familiar, and the first *Flyer* formed the basis for the aircraft with which they rapidly developed their flying skills over the next few years.

In 1904 the *Flyer II*, using a slightly more powerful engine, completed many successful flights, one of which lasted for more than five minutes. The Wrights, having become the world's first experienced pilots, flew from a 90-acre field eight miles from their home in Dayton, Ohio. About 80 flights were made, during one of which they executed a full circuit of the airfield, the first time such a manoeuvre had been carried out.

Following their pattern of using a new aircraft each year, they built the *Flyer III* for their 1905 season. Similar in size to the previous machines, but with the rudder and elevators placed further from the wings so as to increase their effectiveness, it used the excellent engine that had been installed in the *Flyer II*. Airspeeds of something like 55km/h (35mph) could now be achieved, but these faster aircraft created new problems. For instance, in a tight turn, when the inner wing was travelling more slowly than the outer one, the aircraft had a tendency to overbank with the inner wing suddenly dropping. The cause was the greatly reduced lift on the inner wing, eventually leading to a stall, and the Wright's solution was to fly tight turns at increased airspeed. While they were looking for the cure, they also took the extremely important step of removing the coupling between the wing-warping system and the rudder. Rigged in their new form, all three control surfaces – elevators, wing-tips, and rudder – could now be moved independently, thus making the aircraft fully controllable about all three axes. With the modified *Flyer III* the Wrights could circle, perform figures of eight, and remain safely airborne for as long as half an hour.

After their successes of 1905 the brothers volun-tarily grounded themselves while they tried to arrange the sale of their invention. Only in 1908, when they had secured an agreement to have the *Flyers* built in France, and had arranged an official acceptance test with the US military authorities, did they emerge from obscurity. Using two-seater versions of the *Flyer III*, Orville flew in the USA while Wilbur went to France. From August to December 1908 Wilbur astonished European aviators, who for the first time realised how far ahead of everyone else the brothers were in practical flying skills and in their understanding of methods of control.

Samuel Pierpont Langley

Having achieved some success in 1896 with steam powered model aircraft, one of which flew for over 1km (0.62 miles), S. P. Langley (1834–1906) was commissioned by the US Government in 1898 to build a man-carrying machine. Secretary of the Smithsonian Institution, Washington DC, Langley was a mathematician and astronomer of distinction and had done useful theoretical work on aerodynamics. By June 1901 he had completed a quarter-sized model powered by a small petrol engine. Although this was not a success, being unable to sustain itself in free flight, Langley proceeded with the construction of a full-sized version. These machines which, oddly enough, Langley called *Aerodromes*, were of tandem-wing configuration, with two pairs of wings arranged one behind the other. Elevators and rudder were incorporated, but there was no form of positive control in roll, and the engine was seriously lacking in power.

With the pilot housed in a cockpit slung under the trailing edge of the front wing, the full-sized *Aerodrome* was arranged to be launched, like Langley's earlier models, from a catapult on top of a houseboat on the Potomac River. Not only had the machine never been subjected to any form of flight testing before the first manned launch, but the pilot, Charles Manly, who had assisted Langley in the design and construction of the *Aerodrome*, had no experience of any kind in the air, nor had either of them undertaken any experiments with gliders. Two trials were attempted, on 7 October and 8 December 1903, but on both occasions the *Aerodrome* plunged straight into the river, having apparently fouled the launching mechanism. Manly was very fortunate to escape unhurt, but the failures resulted in the withdrawal of official support, after which Langley, greatly disappointed, decided that the time had come to abandon his aeronautical work.

In 1914, using the hindsight of a decade's experience, the well known pioneer aviator Glenn Curtiss managed to make the *Aerodrome* fly, but only after he had secretly made important modifications to it. As Langley left it, the *Aerodrome* was a thoroughly unsatisfactory and almost uncontrollable machine which contributed nothing to the history of powered flight. Working from the drawing board rather than from experience of the air itself, Langley lacked any intuitive insight into the important matter of control and was unable to give satisfactory practical expression to his very real mathematical ability.

Apart from Langley and the Wrights, very few people were undertaking serious aeronautical work in North America in the first decade of the 20th century. One other enthusiast nevertheless deserves mention, if only because of the grandiose scale of his experiments. Alexander Graham Bell (1847–1912), who invented the telephone, was deeply interested in flight and wrote several perceptive articles about it. Together with Glenn Curtiss and others he founded the Aerial Experiments Association which eventually produced a number of influential biplanes. Prior to that, however, Bell had tried adding power to enormous compound kites of his own design. These fragile structures, containing more than 3,000 tetrahedral (pyramid-shaped) cloth-covered cells, were immensely stable, almost totally unmanoeuvrable, and were characterised by the vast amounts of drag that they produced. Although among the most spectacular kites ever to have been built, they were oddities which had no significant effect on the future development of the aeroplane.

Early powered flight in Europe
Although most of the enthusiasm for aeronautics was centred in Europe, progress with the practical development of powered aircraft was slow in the early years of the century. In 1905, when the Wrights' fully controllable *Flyer III* was remaining airborne for half an hour and more, no European had managed a single sustained and controlled flight, and even by 1908 no one could come anywhere near matching the Wright's flying skill. Although within a few years inventors in Europe were developing aircraft which owed almost nothing to the Wrights, the first encouraging partial successes were undoubtedly the direct result of their influence.

News of the Wright gliders had reached Europe not long after their work began and, despite many expressions of scepticism, a number of imitators were inspired to try similar machines. Among these was Captain Ferdinand Ferber, of the French artillery. In 1901 he was attempting to perfect a hang-glider built on Lilienthal lines and, in search of guidance, began to correspond with Octave Chanute. After reading, on Chanute's advice, the text of a lecture which Wilbur Wright had delivered in Chicago on 18 September 1901, Ferber began experimenting with a glider based generally on the Wright configuration. This initially primitive aircraft, with forward elevator but no vertical surfaces, was substantially modified and improved in the course of the next two or three years. By 1904 Ferber had, among other things, made one highly significant change which was to have an important influence on the history of aircraft design. While retaining the Wright-style forward elevator, he added a fixed horizontal tailplane at the rear. Departing, as so many others did, from the Wright's idea of inherent instability and liveliness of response, he also introduced a degree of lateral stability by adding dihedral to the wings.

In May 1905 Ferber added a 12hp motor to a glider built along these lines and attempted to fly it at Chalais-Meudon. Although it could not sustain itself, Ferber made a successful powered glide from an overhead cable, thus achieving a place in history as the first man in Europe to make a real free flight in a powered aircraft.

In 1903 the European aeronautical scene again experienced direct American influence when Octave Chanute paid a visit to France. In April he gave a lecture on flying machines to an audience which, as well as Ferber, included a wealthy Parisian lawyer, motor car enthusiast, and devotee of flying, Ernest Archdeacon (1863–1957). Inspired by Chanute's accounts of the Wright brothers achievements, Archdeacon built a modified form of their *No. 3* glider and tested it in 1904, the pilots being Ferber and a young colleague who was later to exert a powerful influence on European aviation, Gabriel Voisin (1886–1973). Unfortunately, Archdeacon appears to have been unable to understand the nature and importance of the Wrights' control system and incorporated no wing warping or other lateral control into his glider.

Abandoning this unsatisfactory machine, Archdeacon collaborated with Voisin to try something different, a large float-glider designed to include a high degree of inherent stability. The principles which they followed were those included in the design of the box kite, invented in Australia by Lawrence Hargrave (1850–1915). In the 1880s and 1890s, Hargrave, who kept in close touch with

Alberto Santos-Dumont, flying his odd-looking 14-*bis* aircraft, recorded the first sustained flight by a powered aeroplane in Europe on 23 October 1906.

aeronautical progress in Europe, experimented with many gliders, flapping models, and kites. After trying a great variety of kite shapes, he perfected the simple, rectangular box kite in the early months of 1893. Confident that the rigidity and stability of the box kite could make it the basis of a practicable aeroplane, Hargrave used a train of four of them to lift himself into the air in 1894, and proceeded to draw up a number of tentative designs for aircraft using box kite wings. Although Hargrave himself lacked the resources to try out these ideas, the news of his kites soon became widespread in Europe and led to the creation of several practical flying machines.

The Archdeacon–Voisin float-glider, built in 1905, was the first significant aircraft to apply the Hargrave principle. Large box kite wings with four 'side-curtains' (vertical surfaces) were attached to an open framework fuselage carrying a small double box kite tailplane. The only control surface was a forward elevator of the Wright type, though a rear rudder was later tried. In June 1905 Voisin twice piloted this glider when it was towed off the Seine by motorboat and flown 'captive' as a kite. Two long hops of 150m and 300m (about 500ft and 1,000ft) were made without mishap.

A second float-glider was tried in the same year by Gabriel Voisin, this time in collaboration with Louis Blériot (1872–1936). Although built along essentially similar lines, it was distinguished from its predecessor by the sharp angling of the outer side-curtains on the wings, a modification which was intended to provide additional stability. Like the Archdeacon–Voisin float-glider, this also was flown captive in June 1905, but it crashed. After rebuilding, it was modified several times by Blériot, who added a

powerplant. Although he learned much from these trials, no real success was achieved with the glider in any of its forms.

Neither Archdeacon's modified copy of the Wright *No. 3* glider nor the two float-gliders incorporated any form of control in roll. The Wright brothers' system of control, explained by Chanute at his lecture and fully publicized in 1905 and 1906, was something of a mystery to European experimenters, who were in any case reluctant to accept claims made on the Wrights' behalf. Before 1908, when Wilbur's brilliant performance in France in a *Flyer III* put an end to doubts, a good deal of independent progress was nevertheless being made. In 1904, Robert Esnault-Pelterie experimented with ailerons in place of wing warping, which at that time he considered to be structurally dangerous. Light, powerful engines were becoming available, especially the 50hp Antoinettes of Léon Levavasseur (1863–1922), and the introduction of wheeled landing gear made European aircraft easier to handle on the ground than the *Flyers*, which for some years continued to use skids.

Success of a qualified kind finally crowned European efforts when, in 1906, Alberto Santos-Dumont, a Brazilian living in Paris and a highly enterprising designer of airships, tried his hand at heavier-than-air flight. Using the Hargrave box kite idea, he built a grotesque 'canard' machine, with forward box kite elevator, multiple box kite wings set at a sharp dihedral angle, and an enclosed rectangular fuselage. After trying the aircraft suspended beneath his airship *No. 14*, he undertook flights from the ground with the 14-*bis*, as he called it. When the need for lateral control surfaces became apparent, Santos-Dumont added large octagonal ailerons between the biplane wings. With this unwieldy, semi-controllable machine he managed in October and November 1906 to make a series of flights, the longest of which, lasting 21·2 seconds, covered a distance of 220m (720ft).

At about the same time, another expatriate also living in Paris, the Romanian Trajan Vuia, experimented with two little aeroplanes which, although themselves unsuccessful, had a profound influence on the history of aviation. They were bat-like monoplanes with tractor propellers, the first driven by a carbonic acid motor and the second by a 24hp Antoinette. Primitive wing warping was fitted, together with forward fin and rear rudder, but only after a heavy landing in his first machine did Vuia add a long rear-mounted elevator. In addition to this important modification, he decided to abandon a device which he had originally incorporated to make it possible to vary the angle of the wing in flight. The pilot sat well below the wing on a framework to which a four-wheeled undercarriage was attached. Between March 1906 and March 1907 Vuia made several short hops from level ground using this first monoplane, and, although it could barely fly, it was destined to be the immediate ancestor of the many monoplanes which flourished in Europe until shortly before World War I, when biplanes eclipsed them for a time. Vuia's second monoplane, tested in June and July 1907, managed only two short hops.

Among the earliest of those to be influenced by Vuia was Santos-Dumont, who soon abandoned biplanes in favour of an elegant and diminutive monoplane design known as the *Demoiselle* (dragonfly). The first of these (1907) had a wingspan of only 5m (17ft), weighed only 110kg (243lb), and used a 20hp 2-cylinder engine. A combined elevator and rudder, like that on Cayley's kite-glider of 1804, was fitted at the rear, but lateral control was achieved by the pilot leaning his body from side to side as he sat on the undercarriage frame.

Also in 1907, Robert Esnault-Pelterie, who had previously done pioneering work with ailerons, changed his mind about wing warping and incorporated it in a strange monoplane known as the *REP No. 1*. This had wings of 9·6m (31·5ft) span set at an anhedral (drooping) angle, landing gear consisting of two main wheels set one behind the other, with additional wheels on the wingtips, a wide tailplane and elevator, a fuselage which provided a large ventral keel area, but no vertical fin or rudder. Despite inadequate stability and control, the *REP No. 1* managed five short flights in November and December 1907. In 1908 Esnault-Pelterie built a

second monoplane on similar lines but this time added a large dorsal fin and a ventral rudder. After further modifications the aircraft made several quite creditable flights.

The man who most firmly established the monoplane configuration was nevertheless Louis Blériot, who tried three machines of his own design in 1907. The last of these, his *No. VII.*, influenced the design of monoplanes for many years to come. It used a 50hp Antoinette engine to drive a tractor propeller, had a long, enclosed fuselage with landing gear consisting of two main wheels and a rear wheel, and was controlled by rear-mounted rudder and elevons. (Elevons are tailplane surfaces which combine the function of elevators and ailerons controlling both lateral movement and roll). Half a dozen flights were made in November and December 1907, the machine being abandoned after the last flight ended in a crash-landing.

For a time, however, the development of the biplane was to prove more important. Having started up a factory at Billancourt, Gabriel Voisin and his brother Charles standardised a configuration which was to dominate European aviation until 1910. Using their experience with the float-gliders of 1905, they built biplanes with box kite tails, pusher propellers, and forward elevators. It was in such an aircraft that Henry Farman, an Anglo-Frenchman, made the first fully controlled European flights. Tentative trials in September and October 1907 were followed by more adventurous flights, including a full circle on 9 November and a prize-winning kilometre circuit on 13 January 1908. A skilled and sensitive pilot, Farman used his experience to make many modifications to his aeroplane. For about a year he nevertheless flew without any form of control in roll, having to execute turns with rudder alone, making wide, yawing circles. In October 1908, having recognized the need for proper lateral control, he added four large ailerons which made his aircraft a great deal more responsive and manoeuvrable. On 30 October 1908 he was able to make the world's first real cross-country flight in a heavier-than-air machine, covering a distance of 27km (about 17 miles) in 20 minutes, and the next day he established an official altitude record when he reached a height of 25m (about 80ft).

Britain, so often a leader in 19th century technology, lagged uncharacteristically behind in aircraft design. Despite the enthusiasm of many members of the Aeronautical Society, very little was achieved between the death of Pilcher in 1899 and the flight of the first British designed aeroplanes in 1908. Slow progress began to be made in 1905 when Samuel F. Cody (1861–1913) built and flew a glider. A flamboyant, engaging, and adventurous expatriate American, Cody had for some years been concerned with man-lifting kites. His well designed and efficient system, primarily intended to increase the observational range of military observers, was officially adopted by the British War Office. Profiting from experience with the large, winged box kites which he had perfected, Cody designed a kite-glider incorporating ailerons attached to outriggers.

Following his gliding trials in 1905, Cody embarked on the design and construction of a powered aircraft. Eventually called the *British Army Aeroplane No. 1*, it was basically a Wright biplane, with forward elevator, rear rudder, a supplementary rudder on top of the wings, and ailerons between the wings. Powered by a 50hp Antoinette engine driving two pusher propellers, it was first flown at Farnborough on 16 October 1908.

Cody was an excellent mechanic and soon became a first-rate pilot. His gusto and drive did much to promote aviation in England and his death in an accident near Aldershot in 1913 was a great loss. He was not, however, a particularly original thinker and he made no really significant contributions to the development of aircraft design.

Sporadic aeronautical experiments were carried out in other European countries, often by enthusiasts working more or less unaided and alone. In Germany a civil servant called Karl Jatho made running jumps of up to 60m (200ft) with a curious little 'semi-biplane' powered by a 9hp motor and a pusher propeller. He used a rather rudimentary rudder and elevator for his control surfaces, he tested his machine near Hanover and became the first German to be raised from the ground in a powered aircraft. Nothing further seems to have resulted from these trials.

In Denmark J. C. H. Ellehammer caused some excitement with another semi-biplane consisting of a monoplane with a loose sail stretched over the top to form an additional lifting surface. A tractor propeller was driven by a three-cylinder 18hp motor. On 12 September 1906 Ellehammer tested this machine by tethering it to the centre of a circular track around which it was allowed to run while he sat in a pilot's seat slung low on the undercarriage. Although airborne for about 42m (138ft), Ellehammer's biplane lacked any true control system and could not undertake free flight. During the next three years he went on to build a mono-

plane, another biplane, a triplane, and a seaplane, none of which was successful.

The aeroplane comes of age

If any year may be said to mark the final emergence of the aeroplane as a fully practical vehicle, it must be 1909. This year has a special place in aeronautical history because of the great encouragement provided by the first and most famous aviation meeting, held at Reims, France, from 22 to 29 August. The many contests were financially supported by the champagne industry, which offered handsome prizes.

In all, 38 aircraft were entered, though in fact only 23 took off. Over a period of 8 days, more than 120 take-offs were made and a number of records established. Of the 23 aircraft to fly, 15 were biplanes and 8 monoplanes. Top speeds of about 75km/h (47mph) were reached, the altitude record was 155m (508·5ft), and the greatest non-stop distance flown was 180km (111·8 miles), covered by Henry Farman in 3hr 4min 56·4sec. Although many of the spectators were disappointed by the failure of the Wrights to attend, three of their aircraft were flown by French pilots.

The maturity of aeroplane design as represented by the entrants at Reims may be judged from the close similarity between their basic characteristics and those of the aircraft which were to dominate for the next three or four decades. First, there were no multiplanes of the kind that had been tried by Wenham, Phillips, Chanute, and others, and although triplanes were to have a brief period of success a few years later, they soon faded from the aviation scene. Second, all the aircraft except the rather cumbersome Voisin biplanes had full control about all three axes: that is, they had rudder, elevator, and either ailerons or wing warping. Third, they were all powered by lightweight petrol engines, for alternative powerplants such as steam and carbonic acid engines had by now disappeared.

The Reims meeting clearly established the future importance and viability of the aeroplane. Competitions, meetings, and races became popular and aircraft were produced in increasingly large numbers. Both monoplanes and biplanes continued to flourish, with many manufacturers trying to outdo each other in improvements and refinements. Once the basic problems of control and reliability had been solved, attention could be given to other important matters, such as streamlining to reduce drag and increase efficiency. Although the first monoplane made in 1909 by Edouard Nieuport

(1875–1911) owed much to the *Blériot XI*, it had the advantage of a fully enclosed cockpit streamlined into the fuselage. The best of the monoplanes, most of which still used the original Wright wing warping system of lateral control rather than ailerons, were beginning to show the benefits inherent in their simpler lines.

Léon Levavasseur's beautiful monoplane, *Antoinette* which had gained the altitude record at the Reims meeting, remained a particular favourite. The very large tapering wings were 12·8m (42ft) long and 3m (9·8ft) wide at the base, and there was a long triangular tailplane with a triangular fin and rudder. The slim braced fuselage of hollow V section had a clear veneer casing at the front. An 8-cylinder 50hp Antoinette engine directly connected to the propeller gave it a speed of about 70km/h (44mph). A present-day light aircraft pilot would find the control system decidedly tricky to operate. In addition to the familiar rudder bar, the pilot had to manipulate two hand wheels placed on either side of the cockpit. The one on the right controlled the elevators – rotating the wheel forward lowered the nose, as with a modern trim-wheel. The wheel on the left operated the warping mechanism whereby a forward rotation lowered the left wing.

Biplanes, sometimes thought to have greater structural rigidity, also continued to be popular, in particular those aircraft made by Henry Farman. The standard Farman *No. 3* had a span of 10m (33ft), a biplane tail with twin rudders, a forward elevator and a 50hp Gnome engine directly coupled to a pusher propeller. The pilot's controls consisted of a stick and rudder bar, as in many modern aircraft. An ungainly looking machine, the *No. 3* was a good all-round performer capable of about 60km/h (37mph).

The Wrights continued to produce aeroplanes both in Europe and in America. The *Wright B*, which first appeared in 1910, abandoned the forward elevator in favour of a rear elevator, while ground handling was made easier by the adoption of a wheeled undercarriage. A racing aeroplane, the Wright *Model R*, could achieve speeds of over 110km/h (over 70mph). By now, however, the Wrights had been overtaken by many later designers who were not only making considerable advances in respect of structural simplicity and streamlining, but were also developing a better balance between stability and manoeuvrability.

While the original success of the Wrights was the result of their concentration on control, their choice of an inherently unstable design meant that their aircraft had to be 'flown' at all times, the pilot needing to make continuous small adjustments to maintain straight and level flight. They could not be left to fly themselves for a moment, and could certainly not be flown 'hands off.' Although it was to be some years before a fully satisfactory compromise emerged between the instability of the Wright

The most graceful of the early French monoplanes were the *Antoinettes*, designed and built by Léon Levavasseur, pioneer aviation engineer and designer (1883–1912).

design philosophy and the almost immovable stability of some of the early European machines, progress was rapidly being made in this direction.

Aeroplanes began to be used in an ever greater variety of roles. The birth of the aircraft carrier may be said to have occurred on 14 November 1910, when an American pilot, Eugene Ely, flew a Curtiss biplane from the cruiser *Birmingham* off Virginia, USA. The foredeck had been covered by a platform to give Ely a suitable runway, and the short take-off performance of the aeroplane may be judged from the platform's length, which was only 25m (82ft). The first float-plane was flown in 1910, and in the same year bombing trials were made over targets marked-out like ships. The first mid-air collision also occurred in 1910, when an *Antoinette* dived into a *Farman* near Milan. Both pilots survived, though one was seriously injured.

Less than seven years after the invention of the aeroplane, speeds of well over 100km/h (62mph) were common, and the altitude record stood at 3,193m (10,476ft). Not only were more and more people learning to fly, with more or less formal instruction being offered, but pilots' certificates had been established.

Among the most exciting developments in aviation in the years before World War I was the creation by the Frenchman Adolphe Pégoud (1889–1915) of the art of aerobatics. Although the first loop was flown by Petr Nesterov, a Russian lieutenant, in a Nieuport monoplane on 20 August 1913, it was Pégoud who first seriously explored the possibilities of aerobatic manoeuvre. In September 1913, flying his *Blériot*, he gave the first demonstration of sustained inverted flight, and flew loops, bunts (half outside loops), rolls, and a tail-slide. The bunt and the tail-slide can be especially dangerous, since they subject the aircraft to unusual stresses, and a number of pilots subsequently lost their lives when flying these manoeuvres. Pégoud's courage is all the more remarkable in that very little was known at that time about the best way to recover from the most common results of mishandled aerobatics: the stall and the spin. In 1912 Lieutenant Wilfred Parke had successfully recovered from a spin when he was flying the newly developed Avro cabin plane. Although he later recorded and analysed the sequence of his handling of the controls, it was some time before the process of spin recovery was properly understood and the technique standardized.

From about 1912, when designers and governments alike were giving increased attention to the military potential of the aeroplane, the biplane configuration with tractor propellor began to predominate. In view of the technology at this time this is not surprising, since it was a good deal easier to build strong and compact biplane wings, especially if they had to support the weight of twin engines. This trend was nevertheless unfortunate in slowing down the development of the monoplane which, as early as 1911, had appeared in beautifully streamlined form when Levavasseur introduced his Latham design. This had fully cantilevered wings (without external struts), a completely enclosed fuselage with seating for three, and 'spatted' landing gear to reduce drag. Although it was not in itself successful, it influenced other designs, such as the shapely *Monocoque Deperdussin* of 1912. It was not until the 1920s, however, that the growing interest in speed led to a renewed development of streamlined monoplanes.

The first powered passenger flights in heavier-than-air machines had been made by the Wrights on 14 May 1905, and after the Reims meeting of 1909 passenger flying became increasingly common. A scheduled airline service using a flying-boat was started in Florida, USA, in 1914, though not until after World War I did regular services begin to proliferate. Converted de Havilland 4s and 4As (converted light biplane bombers) and Handley Page 0/400 twin-engined bombers were used on the first daily service between London and Paris, which started on 10 January 1919.

World War I led to a rapid growth in aircraft technology. By 1920, speeds of about 300km/h (185mph) were being achieved, and by the mid-1920s these had been increased to about 450km/h (280mph). Altitudes of more than 10,000m (32,800ft) had been reached. Aircraft size was growing, with planes capable of carrying 15-20 passengers and a crew of two or three. As well as being improved in size and performance, aircraft were diversifying. Flying-boats, first flown in 1912, grew popular with many airlines, and there was a renewed upsurge of interest in gliding. With the achievement of controlled, powered flight there had naturally been a period of decline in glider development. After World War I, however, when the Treaty of Versailles prevented Germany from making powered aircraft, this trend was reversed. German aerodynamic engineers put a great deal of intensive effort into the improvement of gliders, making rapid advances both in the theory of drag and in wing design. These developments were of immediate benefit to gliding as a sport, and a little later they were of immense value to the designers of Germany's new air force.

The early history of the parachute

As the parachute is a comparatively simple device, it is likely that in the remote past many unrecorded attempts were made to descend gently from heights with the aid of cloth canopies. The earliest known drawing of what is undoubtedly a parachute dates, however, from the late 15th century. The British Library now holds a notebook, compiled by a Siennese engineer, which contains a fully detailed sketch of a parachute consisting of a conical cloth canopy attached to a circular wooden base across which two cross-members are fixed at right angles. A man with a harness around his waist is holding one of the cross-members to steady himself. This picture, which has no accompanying text, predates by a few years the much better known drawing by Leonardo showing a four-sided cloth pyramid, also with a rigid wooden base. Leonardo added a few notes to his sketch, indicating that the base would be well over 7m (23ft) square, giving the canopy sufficient size to be effective.

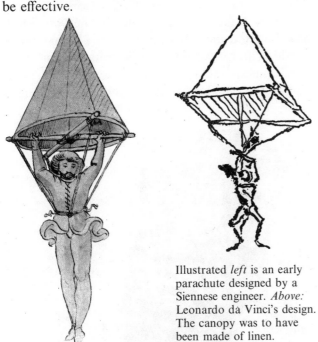

Illustrated *left* is an early parachute designed by a Siennese engineer. *Above:* Leonardo da Vinci's design. The canopy was to have been made of linen.

A book of machines published about 100 years later by a Venetian engineer, Faustio Veranzio, shows another square-rigged parachute being used by a man dropping from a tall tower. It was not until the first hot-air balloons had flown, however, that anyone is known to have attempted a jump. On 22 October 1797 Jacques Garnerin released himself from a balloon over Paris at an altitude of about 1,000m (3,300ft). His successful descent, using a ribbed circular canopy from which a wicker basket was suspended, caught the imagination of balloonists and started a minor craze among stunt artists of

the 19th century. Not all parachute drops ended so happily. In 1837, Robert Cocking, then 60 years old, jumped over Kent, England, using an untested parachute of new design. Cayley had suggested that the violent swinging from side to side, characteristic of some of the earlier parachutes, could be avoided by using a canopy shaped like a shallow cone with the apex pointing downwards. Cocking's parachute, made in this shape, was insufficiently strong and broke up in flight, which resulted in his plunging to his death.

The first time a parachute was used in a real emergency was on 24 July 1808, when a hot-air balloon being flown by the Polish balloonist Jordaki Kuparento caught fire over Warsaw. The technique of jumping from powered aircraft which was to save so many lives during the two world wars, was pioneered in the United States by Captain Albert Berry, who jumped from a Benoist aeroplane at 460m (1,500ft) over St Louis on 1 March 1912. The first man to make a jump in Europe was Adolphe Pégoud, on 19 August 1913.

Three great flights across water

During the first 25 years of the aeroplane's existence as a practical vehicle, many exciting pioneer flights were made. Among the most courageous were those in which pilots used fragile and unreliable aircraft to cover considerable distances over which they flew with minimal navigational aids, very poor weather information, and no established search-and-rescue system. Flights over water were particularly dangerous, since over reasonable terrain a forced landing was a much safer proposition than a ditching.

The crossing of the English Channel

Before 1909 the Channel had been crossed several times, but only in balloons. The first crossing had in fact been made less than two years after the invention of the hot-air balloon. On 7 January 1785 Jean-Pierre Blanchard and John Jeffries flew from Dover, Kent, to the Forêt de Felmores in two and a half hours. Among the many prizes put up by the London newspaper the *Daily Mail* to promote the development of heavier-than-air flight was the sum of £1,000 for the first crossing of the English Channel in an aeroplane. During the summer of 1909 a number of well known pilots, including Wilbur Wright, contemplated trying to win it, and but for an unfortunate engine failure the first man to succeed would have been the Anglo-Frenchman Hubert Latham. Associated with Levavasseur, Latham was a brilliant performer on the highly successful Antoinettes, two of which he was to fly with great success at the Reims meeting in August that year.

On 19 July 1909 Latham set off from Sangatte, near Calais, flying an *Antoinette IV*. This aircraft had wings rather smaller than those of the more familiar later models and used ailerons for control of roll. When he was less than half way across, however, the 50hp engine failed and Latham had to ditch. Both he and the aircraft were rescued by a French destroyer which was escorting him, and after it had been rebuilt Latham flew the *Antoinette* again at the Reims meeting.

The Channel crossing was finally achieved six days later by Louis Blériot, who had recently collaborated with Raymond Saulnier in the design of a new and highly successful monoplane. Known as the *Blériot XI*, it had a wingspan of about 8m (26ft), could make a top speed of 60km/h (37mph), and included a number of interesting features giving it a comparatively modern appearance. The pilot sat over the braced monoplane wings in a partially enclosed fuselage of rectangular section. A 25hp, 3-cylinder Anzani engine drove a wooden two-bladed tractor propeller fixed directly to the crankshaft. The now familiar stick and rudder bar were used to operate the control surfaces, consisting of warped wingtips, hinged tailplane tips acting as elevators, and a large all-moving rudder (the whole surface turning, with no fixed fin in front of it).

Although confident that his *No. XI* could manage the crossing, having used it to make a 41km (25 miles) cross-country flight on 13 July 1909, in the course of which he flew for 44 minutes with one stop,

Blériot needed first class conditions and good luck if he were to succeed. Despite some discomfort which he suffered as a result of a crash a little while before, he set off from Les Baraques, near Calais, at 04.41 on 25 July 1909. The morning was calm but misty, and Blériot, who had no compass or other navigational instruments, had to navigate by visual reference and fly 'by the seat of his pants.' Official timekeepers recorded the duration of the flight as 37 minutes 12 seconds at the end of which Blériot made a heavy landing on a grassy hillock in Northfall Meadow near Dover Castle. The distance covered over the surface was about 38km (23·5 miles). This remarkable feat of courage astonished and delighted those who were keen to promote the future of aviation, but it also frightened many people who rightly foresaw the way aeroplanes might in future be used for aerial warfare and invasion.

The first non-stop crossing of the Atlantic

Like the Channel crossing, the first non-stop flight across the Atlantic was promoted by the *Daily Mail*, which offered the generous prize of £10,000. After World War 1, which delayed serious thought of so hazardous an undertaking, many teams began planning flights, and several hastily organized attempts

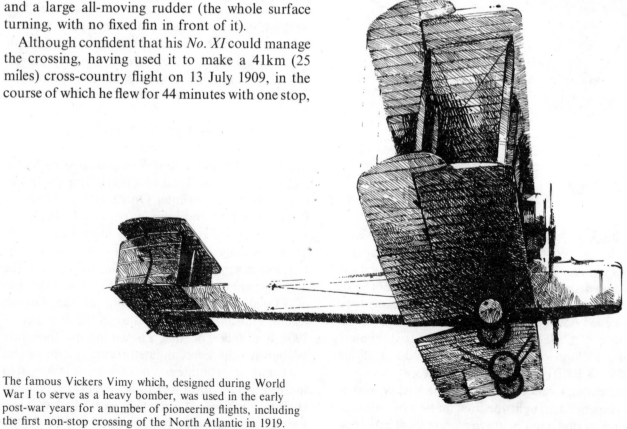

The famous Vickers Vimy which, designed during World War I to serve as a heavy bomber, was used in the early post-war years for a number of pioneering flights, including the first non-stop crossing of the North Atlantic in 1919.

ended in crashes. Success was eventually achieved in June 1919 by two British pilots, John Alcock and Arthur Whitten Brown.

The aircraft chosen by Alcock and Brown was the twin-engined Vickers *Vimy*, a large biplane bomber first flown in 1918. The *Vimy*, which weighed 6,035kg (13,290lb), was fitted with extra fuel tanks for the long journey. Essentially a wooden aeroplane, it was entirely fabric covered. The four-bladed propellers were powered by Rolls-Royce V-12 engines. It had a biplane tail with twin fins and rudders, and powerful aerodynamically balanced ailerons on both upper and lower wings.

As the prevailing wind across the Atlantic blows from west to east, Alcock and Brown, in common with other crews, began their flight from Newfoundland. On 14 June 1919, in very poor weather, they took off. The aircraft was so overloaded with fuel that the aircraft could only just climb away, and, once airborne, the pilots had to contend during most of the flight with reduced visibility because of fog and drizzle. These conditions are ideal for the formation of ice, and for a period of about four hours the whole of the *Vimy* was coated in frozen sleet. The airspeed indicator could not work because of the icing-up of the pitot tube which conveys the outside air pressure to the instrument, and, perhaps also because of icing, the armature of the wind-driven wireless generator sheared off. Most dangerous of all was the icing-up of the engine air intakes which would have led to engine failure unless they had been cleared. In order to do so, Brown climbed out along the wings to clear each engine in turn. On one occasion the poor visibility caused the pilots to lose control of the aircraft, which entered a spin from which they recovered only just above the surface of the water. Despite the difficulties of navigation, they maintained a good course to Ireland, flying at about 1,200m (about 4,000ft) and covering the distance of 3,041km (1,890 miles) in 16 hours 27 minutes. Their landing in Ireland was rather anticlimatic for the field which they had chosen proved to be so soft that the four-wheeled undercarriage dug in and the aircraft nosed over. Not only did Alcock and Brown win the *Daily Mail* prize, but in recognition of their courageous flight they were both knighted by King George V.

The first solo crossing of the Atlantic

One of the most astonishing flights in history, requiring not only an outstanding pilot but also an aircraft of unusual endurance and reliability, was the famous first solo crossing of the Atlantic by Charles Lindbergh on 20–21 May 1927. Before Lindbergh achieved the crossing which won him a prize of $25,000, several pilots had failed in the attempt, and six had died. So as to prepare himself for the many hours that he would spend alone in his single-engined aircraft, Lindbergh first undertook a number of long flights to develop his own endurance.

Lindbergh's famous *Spirit of St Louis* was a highwing Ryan monoplane equipped with the then recently-developed Wright Whirlwind radial engine, which for its day was highly efficient, had a good power-to-weight ratio, and was very reliable. The aircraft had nevertheless to undergo substantial modification before the flight was possible. Extra tanks had to be installed to accommodate the fuel and these not only made the aircraft extremely heavy on take-off but also completely obscured the forward visibility. In order to see ahead Lindbergh had either to look obliquely forward through the side windows or use a specially-installed periscope.

The take-off, early on the morning of 20 May 1927, was a delicate matter. Rain had softened the field, further slowing down the acceleration of the already sluggish aeroplane so that it only just managed to get airborne. The long, demanding trip across the Atlantic was flown with minimal navigational equipment which obliged Lindbergh to descend, on one occasion, almost to surface level to watch the waves and spray so that he could check his estimates of wind speed and direction. One of his greatest problems was to try and keep awake, and once, when he became very tired, he lost control of the aircraft which then went into a spin. However he finally reached Paris after two nights in the air, but large crowds at Le Bourget airfield added to his problems by impeding his touch-down.

Lindbergh's aircraft, which was never flown by anyone but him, now hangs in the National Air and Space Museum in Washington DC. Compared with the airliners of today it is tiny, with a wingspan of only 14m (46ft) and an empty weight of less than 1,000kg (2,150lb). The average speed for the journey of 5,810km (3,610 miles), which took $33\frac{1}{2}$ hours, was 173km/h (107·5mph). Even with modern navigational aids and the generally more reliable engines of today, many pilots are reluctant to fly single-engined aircraft over a stretch of water as wide as the English Channel. It is thus not surprising that Lindbergh's flight should have aroused enormous interest in 1927. His achievement may be considered a fitting conclusion to the pioneer days of aviation.

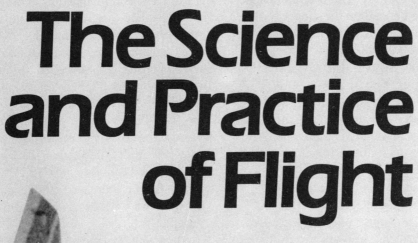

The Science and Practice of Flight

Aerodynamics

Aerodynamics is the science of air in motion, and is relevant to aviation because of its effect on aircraft moving through the atmosphere. The forces exerted by air on moving bodies depend on the square of the relative speed; very roughly, the drag (resistance) of a projection on an aircraft is quadrupled if the speed is doubled. Aerodynamic forces also vary directly with the density of the air. The drag of a body at a height of about 6,700m (22,000ft) is half what it would be at the same speed at sea level. Anyone with a taste for arithmetic can thus calculate the way aerodynamic forces vary with aircraft speed and the height at which the aircraft is flying.

These variations are at the root of aircraft design. Aircraft that fly extremely fast tend to have small aerodynamic surfaces – wings, tails and control surfaces – though as they may also tend to fly high up, the extremely low air density has the opposite effect. A Concorde flying at 2,250km/h (1,400mph) at nearly 18,300m (60,000ft) is controlled by very large elevons on the trailing edges of the wings, as described in the section on systems. If the pilot

ably, and so has the aerofoil section, the 'profile' of the wing seen from the end. Early aeroplanes had thin wings with acute camber (curvature), nearly all the strength residing in external bracing wires. In the 1920s designers began to build unbraced cantilever monoplanes, as explained in the next chapter, and to make the wing strong enough it had to be very deep. Such an aerofoil gave excellent lift at low speeds, the air being sharply accelerated over the long curving upper surface. Nearly all the lift of a wing comes from the reduction in pressure over the upper surface caused by this speeding-up of the airflow, and within limits the thicker the wing, the more the air is speeded-up. But very thick wings have high drag as speed is increased, and few of the aeroplanes with very thick wings could exceed about 200km/h (125mph).

In the 1930s designers therefore searched for ways of making wings thinner, without losing lift. While structural designers solved the problems of preserving strength without so much depth, aerodynamicists not only found improved profiles that gave better ratio of lift to drag (called L/D ratio, one

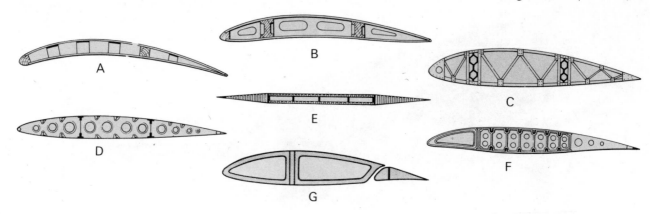

moved these elevons through the same angles at sea level they would tear the wings off; but the Concorde would in any case rip itself apart if it tried to fly so fast in the dense air at sea level. One can see at once that a ground-attack aircraft needs very small wings, which provide ample lift at high speed at sea level, whereas an air-combat fighter needs much larger wings in order to outmanoeuvre other aircraft at great heights. It is partly to meet these opposing demands that some aircraft have variable-sweep wings, held on by pivots. The wings can be spread out for take-off, landing and low-speed flight, and for most operations at high altitude, but progressively folded back as either height is reduced or speed is increased until, in supersonic flight at tree-top height, they are folded alongside the body.

The basic plan shape of wings has varied consider-

of the fundamentals of wing design) but invented ways to change the profile to suit the wing to different flight conditions. The most important new idea was the flap. For take-off the rear of the wing could be hinged down, to increase effective upper-surface camber and deflect the air more sharply downwards, giving greatly increased lift at all speeds. A little depression of the flaps gave extra lift and not much extra drag, and was useful for takeoff. Lowering the flaps to a greater angle gave not much more lift but greatly increased drag, and was best for landing. By the late 1930s there were many kinds of flap. The split flap has only the underside of the wing hinged. The slotted flap leaves an important small gap between the flap and the wing. The double-slotted flap has, in effect, two flaps and two slots, and gives powerful increase in lift. The Fowler flap at first moves out on tracks behind the wing, increasing the area, and only slowly hinges downwards until near the end of the flight.

Another device which increased both lift and safety was the slat. The lift of a wing varies not only with the area, with air density and the square of the airspeed, but also with the angle at which the wing meets the air, called angle of attack. As a pilot flies slower and slower, he has to keep pulling back on his control column or yoke to keep raising the nose higher and higher, to increase angle of attack and thus keep the lift equal to the weight of the aircraft. Eventually, when the wing has an angle of attack of about 16 , it cannot continue to work; suddenly the airflow ceases to flow smoothly, writhing back in giant eddies and whirlpools of air. Immediately the lift drops right away, and the aircraft falls like a stone. Properly designed aircraft automatically rotate nose-down, to pick up speed, reduce angle of attack and continue flying at a lower level. This sudden loss of lift is called the stall, and pilots who do it too near the ground have no way of avoiding a crash. In the early days of flying many pilots stalled in turns near the ground (the apparent weight of an aircraft is multiplied in a turn, just as in pulling out from a dive) and were killed. Handley Page in Britain invented the slat to postpone the breakaway of the airflow. A small auxiliary wing along the top of the leading edge is sucked open near the stalling angle of attack, or driven open by internal power. It leaves a narrow slot through which the air shoots at increased speed, keeping the flow 'attached' to the upper surface longer, to an even greater angle of attack. This allows aircraft to land more slowly and safely.

In the more recent jet age engineers have gone much further in making aircraft fly more slowly and efficiently, largely because the gas turbine can readily make available copious supplies of compressed air. The air immediately surrounding any moving body is called the boundary layer, and the closer one approaches the body's surface the more one finds the air being pulled along with it, until the air molecules actually touching the surfaces are moving along with the body, at the same speed as the body through the rest of the air. Controlling this layer has occupied the attention of many workers. BLC (boundary-layer control) can increase the mass of air flowing into the engines, by diverting the sluggish layer from the inlet ducts and using only high-speed air from further away from the aircraft. By blowing compressed air from thin rearward-facing slits along the wings, flaps or tail, the airflow can be prevented from breaking away, and the desired lift or control effect multiplied. Many warplanes, such as the F-104, MiG-21, Phantom and Buccaneer, have powerful BLC across the flaps to allow them to be lowered to a sharp angle without stalling, thus giving much greater lift and drag. The Buccaneer has also been provided with BLC slits over the leading edge, ailerons and tailplane.

For the longest ranges it is important to reduce drag. In theory this can be done by preventing the boundary layer from breaking down into turbulent flow, and keeping it laminar (smooth). One way to do this is to keep sucking the layer away through millions of fine holes, but this type of BLC – though marvellous in special test aircraft – has never been practical. In everyday use aircraft get covered in drops of oil, dust, flies and other small fragments,

The wing sections which are illustrated *left* include (A) Wright *Flyer*, all wood, (B) Sopwith Camel, all wood, wire trailing-edge, (C) Armstrong Whitworth interwar type, light alloy fabric-covered, (D) Hawker Hunter, pressed ribs. (E) B-70, panels and spars, (F) A300B, machined ribs and skins and (G) modern glass-fibre sailplane.

Right: Airflow (A) at high angle of attack, (B) improved by addition of leading-edge slat, (C) in normal flight, (D) deflected flap increases lift and drag, (E) plain flap, (F) split flap, (G) Fowler flap and (H) double-slotted flap.

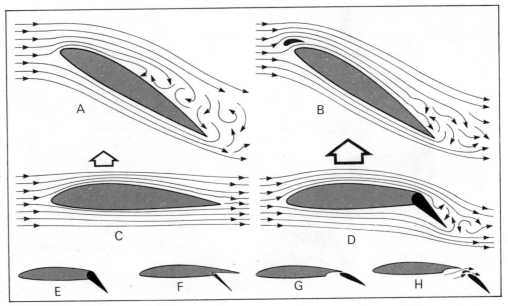

and also are covered in small manufacturing imperfections such as badly seated rivets or the edges of skin plates. All these destroy the ideal laminar boundary layer. Anyone who could make ordinary aircraft fly with a laminar boundary layer could open the way to non-stop services to Australia or other distant places, without having to carry any extra fuel.

Today we can guess what aeroplanes have to do from their wing plan shape. A slender delta, such as the Concorde or Mirage 2000, is built for speed, and despite clever design features needs a long runway for high-speed take-off and landing. Its thickness/chord (t/c) ratio, the ratio of the thickness of the wing to the breadth from front to back, may be three per cent. Variable-sweep aircraft, such as the F-14 and Tornado, may also reach three per cent at maximum sweep, but at minimum sweep the wings are spread out and sprout advanced slats and flaps, giving much greater lift than the plain delta and thus enabling the aircraft to land on aircraft-carriers or on short airstrips, and to go on many types of combat mission. A traditional highly swept wing, such as on the 707, is today seen to be inefficient; modern jetliners, such as the A300B and 757, have 'super-critical' wings with greater depth and much less sweep. These wings have higher aspect ratio, the ratio of span to area. High aspect ratio, a long slender wing, is most efficient at subsonic speeds. So the high-performance sailplane and man-powered aircraft have wings of aspect ratio up to 30 or more, as against 10 for typical aeroplanes and two for a slender delta, so that they can lift with extreme efficiency at low speeds. Trying to make a supersonic aircraft with slender wings would be structurally

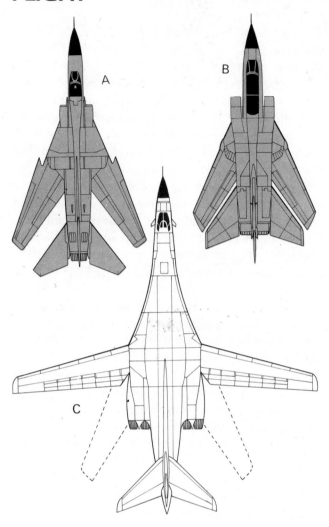

Below left: (A) High aspect ratio, wide span and narrow chord, (B) low aspect ratio, short span and wide chord and (C) wing-tip vortices with high aspect ratio.

Above: contemporary variable-geometry aircraft include (A) the Tornado, (B) Mikoyan MiG-23 Flogger and (C) the Rockwell International B-1.

impossible, like building an elephant with the legs of a gnat. The supercritical wing achieves its efficiency by escaping from the traditional aerofoil profile which generates nearly all its lift along a narrow strip, just behind the leading edge where the suction is extreme. By redesigning the wing to have a flatter top the speeding-up of the air is spread out, thus reducing the sharp peak and replacing it with gentler lift right across the upper surface. This results in a wing that not only has higher aspect ratio for greater efficiency, with reduced sweepback because of the lower airspeed over the top, but also has greater depth and thus can be made lighter which therefore enables it to hold more fuel.

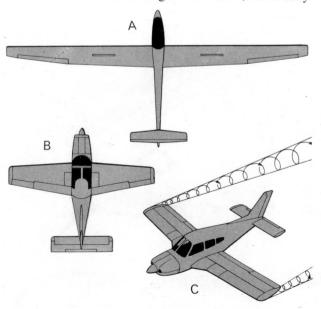

Right: During 1908 Wilbur Wright demonstrated a new Model A *Flyer* in France: its performance acted as an enormous spur to European designers.

Le Petit Journal

Le Petit Journal
CHAQUE JOUR — **6 PAGES** — **5 CENTIMES**

Administration : 61, rue Lafayette

Les manuscrits ne sont pas rendus

5 CENTIMES SUPPLÉMENT ILLUSTRÉ **5** CENTIMES

Le Petit Journal agricole, 5 cent: ~~ **La Mode du Petit Journal, 10 cent.**
Le Petit Journal illustré de la Jeunesse, 10 cent.

On s'abonne sans frais dans tous les bureaux de poste

ABONNEMENTS

	SIX MOIS	UN AN
SEINE et SEINE-ET-OISE..	2 fr.	3 fr. 50
DÉPARTEMENTS..	2 fr.	4 fr. »
ÉTRANGER	2 50	5 fr. »

Dix-neuvième Année DIMANCHE 30 AOUT 1908 Numéro 928

L'AEROPLANE DE WILBUR WRIGHT EN PLEIN VOL

Overleaf: A typical scene in European skies during World War I: a group of D.H.4 day-bombers under attack from German biplane and triplane fighters.

Below: A replica of the Bleriot XI monoplane, which was the first heavier-than-air craft to cross the English Channel.

Swept wings were originally introduced in the early days of jet propulsion in World War II, because like the triangular delta they offered a way of making the wing appear to have a reduced t/c ratio, compared with ordinary 'straight' wings, and thus give less drag at speeds near that of sound. Today we find some of the fastest aircraft have returned to the 'straight' wing, though made with low t/c ratio for high speed and fitted with powerful slats and flaps. Often the leading edge does not have a slat but instead is hinged down to form a leading-edge flap. There are many other leading-edge devices, such as the Krüger flap that swings down from the under-surface. The first supersonic fighter with unswept wings was the F-104 of 1954, which had fast-landing wings almost like the sharp-edged wings of missiles (which do not have to land slowly). Today the F-15 Eagle, F-16 and F-18 show the way the seemingly ordinary unswept wing has returned as the best answer for aircraft able to fly at over $1\frac{1}{2}$ times the speed of sound.

In the 1950s designers were especially busy trying new ideas, which were often merely urgent 'fixes' to cure problems caused by the increased speeds made possible by jet engines. Turbulators, or vortex generators, were groups or rows of small blades about the size of bus tickets or playing cards, mounted at alternate left and right diagonal angles to the airflow wherever the flow round the aircraft was bad; they caused spiral wakes that stirred up the boundary layer and improved the airflow. Dog-teeth were sudden kinks in the leading edge, usually at the start of an outer wing section of increased chord, the extra leading-edge portion being invariably sloped downwards. The sloping leading portion improved airflow at high altitude or in tight manoeuvres when angle of attack is high, and the dog-tooth acted as a turbulator to improve airflow. Fences were thin walls running back across the wing to prevent the air from flowing sharply out towards the tips on the upper surface (which was a problem with early swept wings); some Russian warplanes had as many as four or six large fences. Leading-edge sawcuts did the same job with rather less drag. The area rule was a discovery for minimum transonic drag, at speeds close to that of sound; if the overall cross-section area is increased, for example by adding a wing,

Previous page: The superb Sopwith *Pup* fighter, which could fly at a top speed of 110mph (177km/h).

Top left: A replica of the Vickers Vimy which, in 1919, was the first aircraft to fly non-stop across the North Atlantic.

Left: The odd-looking configuration of the Fokker Dr. 1 Triplane had high manoeuvrability in combat.

engine nacelle or drop tank, the area should be reduced at that point in some other place, for example by making the fuselage slimmer.

It is strange that, while today's designers have managed to avoid the need for obvious 'fixes' to cure problems, they have stuck to traditional shapes and even returned to unswept wings. Throughout the history of aeroplanes it has been known that in theory it ought to be possible to build more efficient long-range machines that had everything inside the wing, thus doing away with the non-lifting but drag-producing fuselage and tail. All-wing aircraft have failed to beat the traditional kind, and today's superbly efficient vehicles such as the 747 and A300B have gigantic fuselages and tails.

Propulsion

Fundamentally, all aircraft propulsion – other than flapping wings, which have never worked – relies on jet propulsion. The propeller, the first successful method, generates a very large jet moving at relatively low speed. Attempts to fly fast with propellers eventually run into difficulties at speeds around 800km/h (500mph) where the tips of all propeller blades reach the speed of sound, causing intense noise and greatly reducing efficiency. Aeroplanes escaped from this absolute limitation on their speed by the invention of the turbojet, the first power plant to be popularly called by the vague term 'jet engine', which opened the way to speeds faster than sound. This ejects a relatively small jet, but moving much faster than that behind a propeller. For higher speeds still the turbojet can have an afterburner, which accelerates the jet even more. For the highest speeds of all the rocket must be used with the smallest jet, so establishing the greatest velocity. Between the propeller (which can be driven by a piston engine or turboprop) and the turbojet we now have a spread of intermediate engines called turbo-fans which fit any desired task. Those with single fans of very large diameter are almost like turbo-props, while the low by-pass ratio turbofans are closely akin to the turbojet.

Throughout the first 40 years of the aeroplane there was no alternative to the piston engine. These began as lightened versions of car engines, but soon acquired distinctive features and often had totally different design. By 1920 there were three basic families. The most numerous were air-cooled radials, with seven or nine cylinders arranged like spokes of a wheel with all the connecting rods acting on one crankpin. This gave a cheap, efficient and compact layout, making best use of the material, unaffected by extremely hot or cold climates. For some aircraft, especially those for racing, the in-line water-cooled engine was favoured, because though these tended to be long and heavy they could offer less drag, especially in racers which did not need large radiators projecting into the airstream; but they were difficult to use in the Arctic or tropics. For lightplanes the small air-cooled in-line or radial engine was almost universal.

After World War II the liquid-cooled engines almost vanished and the only remaining species were large air-cooled radials for a few slow transports and helicopters, and air-cooled horizontally opposed engines for lightplanes. Today the latter families are still in full production, while the rapidly growing worldwide fleets of agricultural aircraft are almost exclusively powered by air-cooled radials. The United States dominates the modern piston-engine scene, though it has competition from Poland and to a lesser degree Czechoslovakia, especially in the important agricultural market. Since 1970 the water-cooled car engine of American type has emerged as a strong agricultural contender, with excellent efficiency in the 400hp class.

The mass lightplane market has been almost entirely captured by flat-four or flat-six engines of two American makers, Lycoming and Teledyne Continental. These are used mainly in power classes from 150 to 400hp, and at the upper end of the scale are often fitted with a turbocharger. This fitment comprises a gas turbine driven by the hot exhaust gas, spinning a supercharger to pump more air into the cylinders and thus reduce the loss in power as the aircraft climbs to high altitudes. Turbocharged engines are relatively costly and would never be seen on a club or training machine, but aircraft flying long distances are today often fitted with pressure cabins and can cruise at well over 6,100m (20,000ft). At such heights the turbo can almost double the engine power and increase efficiency by flying faster and thus further on the available fuel. Another feature that has become predominant in the past 20 years is direct fuel injection. Instead of having a float-type carburettor to release a fine spray of fuel into the airflow entering the inlet manifold (the air pipe feeding the cylinders), the direct-injection engine supplies the manifold with plain air and squirts individual measured doses of fuel into each cylinder at the appropriate times. Though still uncommon in cars, direct injection increases power and efficiency, eliminates the spectre of carburettor icing and allows the engine to run equally well, whether it be inverted or in any other attitude.

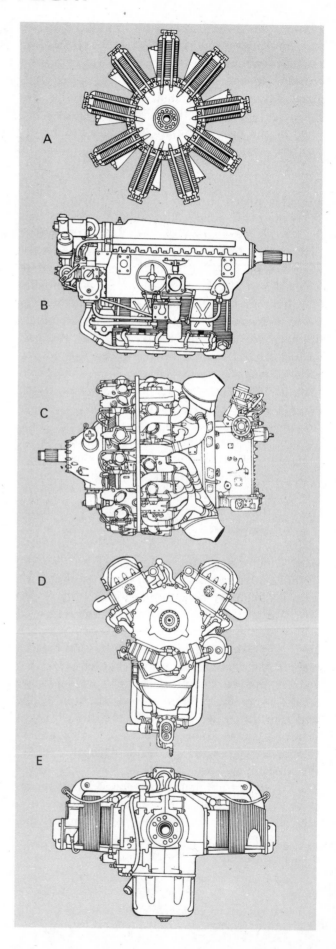

A

B

C

D

E

Piston-engine layouts *left* include (A) Clerget nine-cylinder rotary, (B) DH Gipsy Major inverted in-line four-cylinder, (C) turbo-compound 18-cylinder twin-row with turbochargers, (D) Rolls Royce Merlin 12-cylinder vee and (E) Continental 10-360 'flat-six'.

Right: Turbo-jet engine showing (A) front bearing, (B) inlet guide vanes, (C) compressor blades, (D) stator blade, (E) diffuser section, (F) centre bearing, (G) fuel spray manifold, (H) combustion chamber, (I) rear bearing, (J) turbine guide vane, (K) turbine blade, (L) exhaust duct.

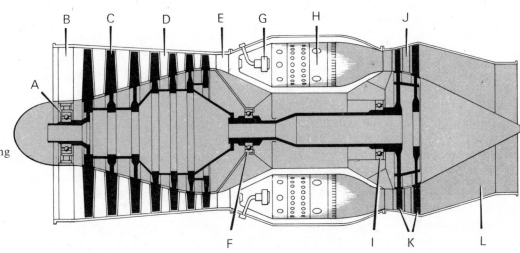

The once great family of high-power piston engines has all but disappeared. In Montreal Canadair builds the CL-215, a unique amphibian water-bomber for putting out fires. This flies low and relatively slowly, and uses two 2,500hp Pratt & Whitney engines of a type that made a great contribution to World War II. It appears to be the last survivor.

Unlike the piston engine, where combustion takes place in an array of separate cylinders which each pass through successive cycles – induction of air, compression of the air, ignition and expansion of hot gas to do useful work, and exhaust of the waste gas – the gas turbine passes the same steady airflow through different parts in turn, each of which remains in a steady condition except when the throttle is opened or closed. The air is sucked in through an inlet into a compressor, where its pressure is increased. The compressed air is then passed to a combustion chamber where fuel is sprayed in and burned. The resulting hot gas is then allowed to escape past the blades of a turbine and along a jet pipe. The turbine is fixed to the same shaft as the compressor. Though some kind of external power must be used to spin the engine to start it, once it has reached a certain self-sustaining speed it will run smoothly as long as fuel is supplied. The gas turbine is the heart of the turbojet, turbofan, turboshaft and turboprop, which power almost all modern aircraft of over 600hp.

Some of the earliest gas turbines had centrifugal compressors, very much like piston-engine superchargers, because these were well known and relatively easy to make. The centrifugal compressor is still important in the smallest sizes of gas turbine, because it is cheap, robust and adequate for the task, but almost all other aircraft engines use the more complex axial compressor. This is basically a drum, or series of discs, on which project hundreds of small radial blades each behaving like a small wing. The surrounding casing is provided with matching rows of fixed 'stator blades' projecting radially inwards. When rotated at very high speed the compressor draws in air at the front, where the rotor and stator blades are large, and compresses the air stage by stage until at the downstream end, where the blades are very small, the pressure may be up to 30 times atmospheric. In recent years very high compression has been achieved without difficult engine-handling by making 'two-shaft' engines. The compressor is divided into an upstream LP (low-pressure) portion and a downstream HP (high-pressure) portion. Each is then driven by its own turbine, the HP turbine being coupled to a tubular HP shaft surrounding the LP shaft. The LP and HP assemblies are each allowed to run at their own speed, the smaller HP usually running the faster. This gives excellent results, and in Britain Rolls-Royce has even produced quite remarkable world-beating three-shaft engines.

An alternative to the multi-shaft layout is to use variable stators. Instead of being fixed to the inside of the compressor casing, some of the stator blades are mounted in pivoted bearings, connected to a surrounding drive ring and positioned by a ram driven by fuel pressure. The angle of each row of blades can thus be adjusted to suit the operating conditions, so that despite having a long compressor with a very high pressure-ratio the engine always works properly instead of suffering airflow stalls and turbulence. Almost all modern aircraft engines have at least one row of variable stators, often upstream of either the LP or HP compressor to feed the airflow in at the most advantageous angle.

Combustion chambers or combustors have developed amazingly since the monster chambers necessary in the early turbojets of World War II. Those old chambers showed the size of the combustion problem: parts were cool while others glowed white-hot and melted, and few lasted more than a few hours. Today ten times as much heat can be released in one-tenth of the volume, and with the whole chamber at uniform temperatures. Some engines still have two or more separate tubular chambers, but most have an annular combustor in the form of a ring round the middle of the engine, with hot highly compressed air entering at one end and much hotter gas being delivered from the other. The kerosene or other fuel is sprayed in from a ring of burners and the designer attempts to achieve a uniform high gas temperature all round the ring where the gas enters the turbine.

The turbine is the key component in the engine, because while it is almost as hot as the combustion chamber, it has to bear intense stresses caused by spinning at full speed. The hottest portion is the turbine stator, or nozzle guide vanes, which do not rotate but direct the white-hot gas on to the rotor. The latter is very like a stage from the compressor, but the materials are designed to withstand white heat whilst spinning at anything up to 50,000rpm (833 revolutions per second!) whilst driving the compressor and, in some engines, additional items. A simple low-pressure turbojet can manage with a

single turbine stage, while good two-shaft turbojets have a single-stage HP turbine and single-stage LP turbine. But some large turbofans have as many as four or five stages of blades on the LP turbine in order to extract as much power as possible.

Chronologically the earliest aircraft gas turbine was the turbojet. This is also the simplest form, and after passing through the turbine the hot gasses simply accelerate down the jet pipe and out of a nozzle to the atmosphere. In doing so the gas generates thrust to propel the aircraft. The turbojet was ideal for early fighters, but it was gradually realised that its noise and inefficiency could be improved upon for other classes of aircraft. Airliners, in particular, caused effort to be made to make the gas turbine drive a propeller, to produce a turboprop. This is essentially a turbojet with several additional turbine stages which extract as much energy as possible from the hot gas before it escapes down the jet pipe. The surplus power, over and above that needed to drive the compressor, is taken through a speed-reducing gearbox to drive a propeller. The pioneer Rolls-Royce Dart turboprop was a single-shaft engine, with turbine, compressors and propeller all rotating together. Amazingly, it is still in full production after well over 30 years. Many small turboprops today are simple single-shaft engines, but some have two shafts, with the propeller gearbox driven off the front of the LP compressor. Another engine, the Proteus (also used for ACV

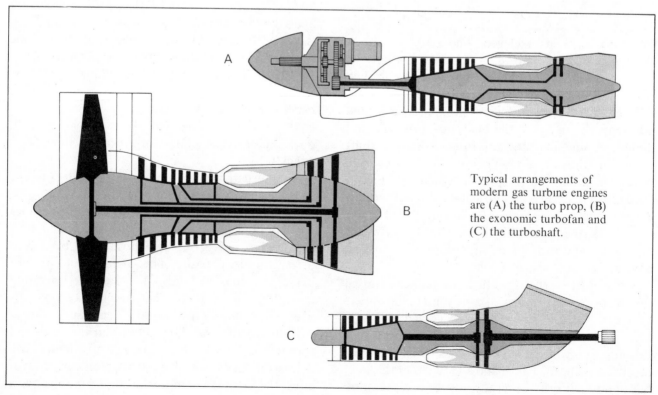

Typical arrangements of modern gas turbine engines are (A) the turbo prop, (B) the exonomic turbofan and (C) the turboshaft.

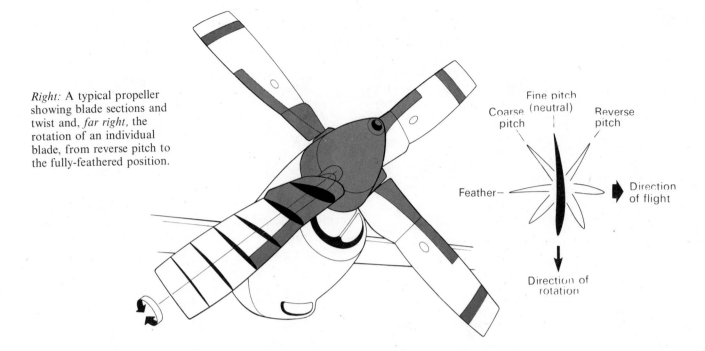

Right: A typical propeller showing blade sections and twist and, *far right,* the rotation of an individual blade, from reverse pitch to the fully-feathered position.

Fine pitch
Coarse (neutral)
pitch   Reverse
                pitch

Feather—   Direction
                  of flight

Direction of
rotation

hovercraft and warships) has the compressor driven by one turbine and the propeller by another.

Modern propellers have wide blades made of GRP (glass-reinforced plastics), densified wood laminations, solid light alloy or hollow steel. Invariably, except in the smallest aircraft, the blades can be set to fine pitch for take-off, coarse pitch for high-speed cruising flight, and, usually, to a feathered (no rotation) position if the engine should cease operating. Some propellers also have a reverse-pitch position at which opening the throttle gives a powerful braking effect after landing. A very recent development is the ducted propulsor, pioneered by Dowty, in which a multi-blade propeller of unusually small diameter is run inside a large surrounding shroud or duct. This dramatically reduces noise and also improves aircraft performance, and there is little doubt it will gradually sweep away traditional propellers as aircraft are designed to take it.

The turbofan is in many ways like a turbojet driving a ducted propulsor, but it came about by a different kind of engine, the by-pass turbojet. Pioneered by Rolls-Royce, this was a two-shaft turbojet with the LP compressor made larger than necessary. The surplus air was by-passed round the combustion chamber and turbine to mix with the hot jet and give performance like a turbojet but with reduced noise and higher propulsive efficiency. Gradually the oversize LP compressor became a fan, a very large single stage like a ducted multi-blade propeller. Today these great turbofans give thrust in the 25-ton class and propel the world's largest

aircraft at close to the speed of sound with unrivalled fuel economy and less than one-hundredth as much noise as the small early turbojets.

Supersonic aircraft are a special class which need an afterburning turbojet or ramjet, which are engines designed only for very high speeds. They need complex inlet and nozzle systems able to alter their area and shape to match the engine to different speeds and altitudes. Unlike the large turbofan they have to generate a supersonic jet and so are inevitably noisy.

Structures

It is self-evident that all aircraft pose severe structural problems because of the compelling need to save weight. In general lighter-than-air craft are able to spread the loads over large areas and thus use very light construction. The pioneer Montgolfier balloons were, in fact, made of paper, despite their immense size. In contrast, aeroplanes inevitably have to be strongly made, and they have always taxed the stressman (the structural designer) to the uttermost. In many cases the problems need all the resources of technology and improved materials to find a solution, while with general aviation the need is to make the airframe simple and cheap.

Early designers favoured the biplane because this could easily be braced by wires into a light yet rigid structure based closely on that of the boxkite. Upper and lower wings sometimes had no strong structural members except along the leading and trailing edges and the ribs that joined them and preserved the

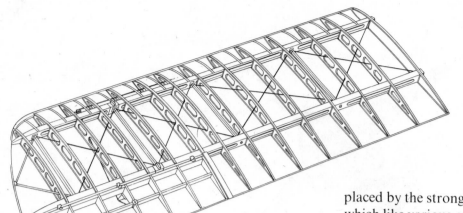

Below left: A typical fighter of World War I was the Sopwith Camel. It was compact, highly manoeuvrable, and had a lightweight but robust structure. Above it is its all-wooden wing, wire-braced internally so as to provide adequate fore and aft rigidity.

desired aerofoil profile of each wing. The upright interplane struts that joined such wings thus had to link the leading and trailing edges, and all that was needed to complete the rigid box was a mass of diagonal wires linking the wing/strut joints and pulled tight until the 'twang' sounded right. By 1909 designers had begun to add extra spanwise members called spars, and gradually these became more important than the leading and trailing edges. In fact many aeroplanes had all the strength in the spars, and the trailing edge had a scalloped effect where the tight fabric covering had made the trailing-edge wire sag between each pair of ribs.

Fuselages were, in comparison, almost simple. Almost everyone used four strong 'longerons', joined by upright and transverse members, all made from high-quality spruce or other wood free from knot-holes, joined by various methods of glueing and pinning or metal connections, and again braced throughout with tight wires. The worst problem was the monoplane, which had to be braced by a mass of diagonal wires fixed to the upper and lower ends of vertical king-posts projecting above and below the fuselage. Sometimes the lower king-post was re-

placed by the strong landing gear, or undercarriage, which like various other parts was made of bamboo, steel tubes or aluminium, with bungee (stretched rubber) springing. The wing and tail were covered in thin but strong fabric, such as silk varnished to make it taut and weatherproof. Increasingly the fuselage was skinned as well, and gradually a few designers introduced fuselages of a new type in which the strength was in an outer shell made of layers of thin wood veneer. Such a structure was called a monocoque, and it increased the room inside, with no interfering bracing wires, and also made possible streamlined shapes; but it was expensive, and much more difficult to repair after a crash.

By the end of World War I many engineers had begun to make aircraft mainly out of metal, while Fokker (with wood) and Junkers (with corrugated light alloy) introduced deep monoplane wings that were strong enough to be cantilevered, with no external bracing. Most designers introduced metal gradually, while sticking to the old style of structure with fabric stretched over a braced skeleton. The cantilever monoplane remained uncommon, and because of the depth of such wings was seldom faster than other types. But during the 1920s designers in Germany and the United States gradually perfected a new kind of monocoque construction made in metal. By skinning the wings, as well as the fuselage, in strong but light alloy such as Duralumin (mostly aluminium, but with a little copper and magnesium), it was possible to transfer most of the strength from the underlying skeleton to the skin, which had previously done nothing to add strength. Though the basis of the wing remained the spar(s), with chordwise ribs to give the aerofoil shape, the stressed skin made it possible to build wings much stronger despite being thinner. These new thin cantilever monoplane wings made possible a revolution in aircraft design. Airliners and bombers doubled in speed in the early 1930s from 161 to 322km/h (100 to 200mph), and as engines became increasingly powerful speed kept rising beyond 483km/h (300mph) without the stressed-skin structures running out of strength.

To this day the light-alloy stressed-skin structure has remained dominant except for sailplanes and lighter-than-air craft. But the modern airframe is greatly refined, and in any case there have been many important alternatives. Junkers stuck to corrugated skin until 1934 when it was realised that the drag was much worse than had been thought (despite this, the Ju 52/3m transport remained the standard Luftwaffe utility vehicle until 1945, and it was used by the RAF and many other air forces and airlines after the war). The Vickers company persevered with a metal basketwork derived from airships, called geodetic construction. The airframe was assembled from thousands of small curved strips pinned together at their ends, with just a single wing spar to make the wing stiff in bending. Covered in fabric, geodetic construction was tough and easy to repair, but it proved unable to match the needs of faster aircraft and faded after World War II. The de Havilland company, which had built a metal stressed-skin airliner, surprised everyone in 1940 by creating the Mosquito, a 644km/h (400mph) bomber faster than the Spitfire yet made entirely of wood. It was constructed of thin sheets of plywood formed to the right shape and then made into a double-layer 'sandwich' with a core of spruce strips or light balsa wood. In the United States there were other attempts at moulding wood or plastic aircraft, but all they managed to do was to demonstrate that they were the 'odd men out'.

With the B-29 Superfortress bomber the Boeing company took a large leap ahead into a realm where structural loads were several times greater than before, due to much-increased aircraft weight and higher speed demanding a thinner wing. The stressed-skin panels on the wing were doubled in thickness, doubled again, and then doubled again – an eightfold increase in a decade. Immediately after 1945 the same company built the B-47 six-jet bomber which doubled the skin thickness yet again, in creating a giant swept wing much smaller in area

than that of the B-17 Fortress heavy bomber, and less than half as thick, yet supporting a 966km/h (600mph) bomber weighing four times as much. One dramatic new feature of the B-47 was that all the engines were hung outside the airframe in pods under the wing, and though jets could all be grouped close beside the fuselage (in a way impossible with propellers), the B-47 engines were distributed right across the wing to act as weights that would reduce the bending stress on the spars and prevent the dangerous vibration called flutter. Ever since, designers of large aircraft have often hung the engines well outboard under the wings to improve structural and aerodynamic efficiency. A few hang them on the sides of the rear fuselage, because though this tends to make the wing heavier and also adds weight in the fuselage it reduces noise in the cabin and can have other benefits (for example, by allowing large flaps right across the wing without causing gaps behind the engines).

Designers have always experienced problems with landing gear. Until World War II airfields were constricted grass areas unsuitable for large or fast aircraft, and the fastest racers and biggest transports tended to be seaplanes and flying boats. Seaplanes, with strut-mounted floats, tended to be fairly small, but the flying boat could be very large and the strange 12-engined Do X of 1929 was the largest aeroplane of its day. The flying boat is inevitably

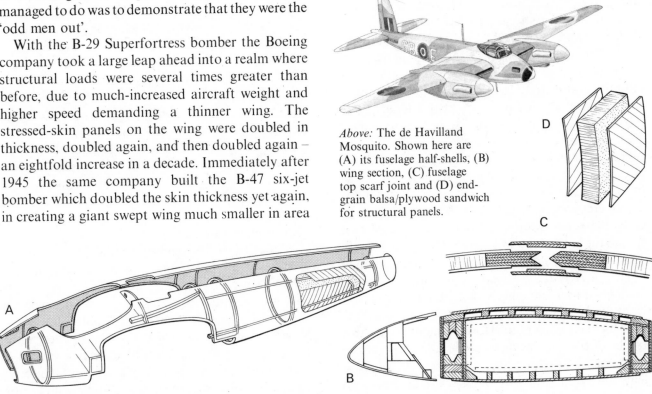

Above: The de Havilland Mosquito. Shown here are (A) its fuselage half-shells, (B) wing section, (C) fuselage top scarf joint and (D) end-grain balsa/plywood sandwich for structural panels.

aerodynamically inferior to the aeroplane in that its fuselage has to serve as a hull able to take off from, and alight on, quite rough water. The underside, called a planing bottom, has a vee shape in cross section, with one or more steps like racing motor boats to stop the water sticking to the surface right back to the tail as the boat takes off. Made of non-corrosive Alclad (aluminium-clad light alloy) or waterproofed plywood, marine aircraft were always tricky to operate and, despite a nostalgic appeal, faded from the scene quickly after 1945 as aircraft were built to suit the excellent runways that had sprung up everywhere during the war.

Wheeled landing gear inevitably involve concentrated loads, which structural designers try to avoid. Since 1950 catastrophic accidents have focussed attention on fatigue and the weakening of metal structure as a result of repeated loading by endless bending back and forth, such as is experienced by wings vibrating in turbulence or a pressure cabin repeatedly pumped up and deflated. To combat fatigue designers try to use 'redundancy' or 'multiple load paths' and similar techniques which all mean that, should any part break, there is always another part left to carry the load. When intense concentrated loads are met, this cannot be done. The 'fail safe' philosophy cannot be adhered to, and the landing gear and its attachments have to be made so strong that they never break.

A further influence on landing gears is the strength of the airfields. In 1935 grass fields demanded monster wheels with fat low-pressure tyres on the largest aircraft. By 1945 heavy aircraft had twin-wheel gears on each leg, and were increasingly of the tricycle or nosewheel type which, unlike the old tailwheel layout, gives perfect pilot view and is easy to steer on a runway, with no tendency to pirouette or 'swing' in the way that damaged thousands of tailwheel aircraft in World War II. Increasing weight and reduced stowage space for the retracted gears meant higher tyre pressures, so tarmac runways gave way to concrete. But weights kept increasing, and the Comet, B-36 and 707 of the early 1950s had four-wheel bogie main gears that spread the load over a wider area. Today the DC-10-30 has an extra centreline gear, the 747 has four main bogies, and the C-5A Galaxy has four six-wheel main gears and four wheels on the nose leg. In many cases modern landing gears are unconventional, the tandem main gears of the Harrier and four steerable trucks of the B-52 being examples, but these are to meet particular requirements. More universal problems have been posed to military-aircraft designers by the increasing need to operate from rough fields and frontline strips, and this calls for large tyres, tall and strong landing gears and a generally very robust structure. Similar problems are met in designing carrier aircraft, which also have to be stressed for catapult take-offs and arrested landings; but at least they can use small high-pressure tyres because flight decks are strong.

In the past 20 years many new structural methods have been introduced, either to save weight, save cost or meet the difficult demands of highly supersonic flight in which the whole airframe is heated by kinetic effects, which can loosely be thought of as air friction. One of the first major answers was the introduction of a new metal, titanium, which is slightly heavier than aluminium but much stronger and well able to retain its strength at raised temperatures. At high altitudes, where the supersonic aircraft tend to fly, the atmosphere is intensely cold, at about minus 60 C (minus 76 F), but despite this a Mach 2 aircraft with enough fuel to hold this speed for a long period will settle down at about 150 C (302 F), which is hotter than ordinary ovens. This is right on the limit for light alloys, but titanium can go on to about 250 C (482 F), which easily takes care of all ordinary aircraft. Yet a few very fast aircraft needed something even better. The XB-70 Valkyrie bomber, designed to cruise at Mach 3 3,200km/h (2,000mph), was made of extremely costly stainless steel fabricated into complex corrugated or honeycomb-filled sandwiches by advanced welding. The YF-12 and SR-71 'Blackbird' aircraft, which hold the current world speed record at 3,530km/h (2,193mph), considerably in excess of Mach 3, are made almost entirely of special titanium alloy which necessitated the solution of thousands of new problems before it could be fashioned by new tools and joined by new methods. The X-15, the fastest known aeroplane in history, which at 7,297km/h (4,534mph) became almost red hot in seconds, was constructed of high-nickel alloy similar to that used in turbojet turbine blades. When completed, the whole airframe was cooked twice at almost white heat in giant furnaces.

While GRP (glass-reinforced plastics) are among many plastics increasingly used for all kinds of subsonic aircraft, with the largest such part the vast wing/body fairing of the 747, the development in Britain of graphite and carbon fibres has revolutionised structures of many kinds. In the United States a rival series of fibres have been produced in which boron is deposited on a white-hot filament of tungsten. All these fibres have to be bonded by

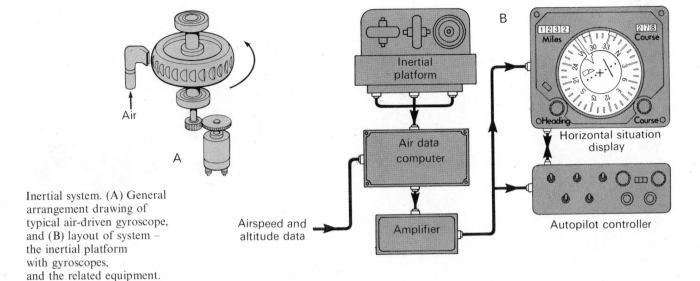

Inertial system. (A) General arrangement drawing of typical air-driven gyroscope, and (B) layout of system – the inertial platform with gyroscopes, and the related equipment.

adhesives to make useful structures, either by 'filament winding' or by use of intermediate stages with adhesive-impregnated sheet or strip. The floors and furnishings of airliners have long been made of carbon fibre plastics, and now CF materials – visually obvious by being jet-black – are used for complete control surfaces, tail sections and wing panels of the latest American fighter and attack aircraft. They are both stronger and lighter than equivalent parts in metal. Later, when the present high cost has been reduced by mass-consumption, they will be increasingly used in other aircraft.

Systems

People who do not actually come into contact with aircraft may well overlook their 'systems', but without them most aircraft would be helpless. Typically the systems cost from 50 to 100 per cent as much as the airframe itself, and in a few specialised military aeroplanes they account for three-quarters of the total cost. Like ourselves, with our blood system, nervous system, digestive system and respiratory system, the systems of aircraft ease the load on the pilot or structure, permit safe flight through bad weather and perform countless other important duties.

The most universal of all is the flight-control system (those for rotary-wing aircraft are discussed in the Rotorcraft and V/STOL chapter). As explained earlier, lack of such a system made heavier-than-air flight fatal prior to the Wright brothers, whose patient experiments eventually hit on a method of wing-warping to keep the wings level, or to bank into a turn, with a foreplane for control in pitch and a rear rudder. Today a few aircraft have a foreplane, but most have a tailplane, either moved

as a single 'slab' or fitted with hinged elevators. A few supersonic aircraft have 'slab' vertical surfaces, but most have a fixed fin and hinged rudder. And instead of wing-warping modern aeroplanes have special control surfaces for roll. The most common arrangement is to fit each wing with an aileron, a hinged surface at the outer trailing edge. To roll to the left, the left aileron is moved up and the right aileron down, thus reducing the lift on the left and increasing it on the right. Modern jets sometimes lock the outer ailerons in cruising (high-speed) flight and use only small inboard ailerons, and nearly all high-speed aircraft have spoilers. Spoilers are hinged plates above the wings which can be driven open upwards by hydraulic jacks. When all are opened together they destroy lift and increase drag, allowing the aircraft to lose height quickly without reaching excessive speeds. When the pilot wishes to roll the aircraft the spoilers open on one side only, to supplement the effect of the ailerons. Yet another roll control in supersonic aircraft is the tailplane, the left and right halves of which can be driven in opposite directions; such a tailplane is now known as a taileron.

In simple aircraft all control surfaces are positioned by wires from the pilot's stick or wheel and rudder pedals. In the 1940s very large aircraft became extremely tiring to fly, and powered controls had to be introduced. At first these were often electric motors which boosted the pilot's forces, but today the flight-control system is so complex it thinks for itself. Invariably it is linked with an autopilot, which is in principle based on a giroscope, a spinning mass like a top whose axis tends to remain fixed in space. The autopilot can keep the aircraft in any desired attitude, quickly restoring it

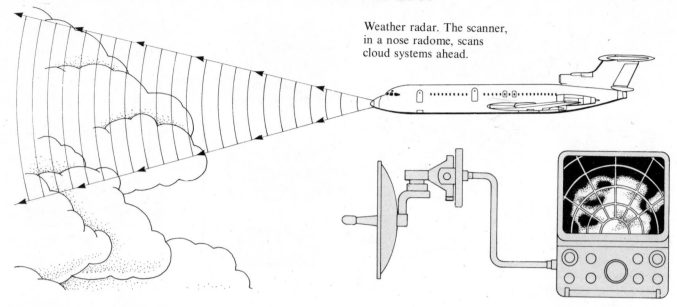

Weather radar. The scanner, in a nose radome, scans cloud systems ahead.

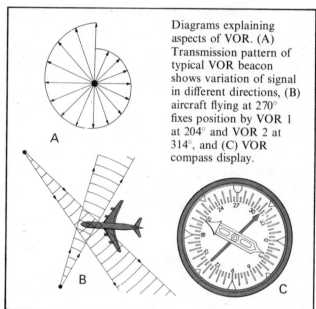

Diagrams explaining aspects of VOR. (A) Transmission pattern of typical VOR beacon shows variation of signal in different directions, (B) aircraft flying at 270° fixes position by VOR 1 at 204° and VOR 2 at 314°, and (C) VOR compass display.

after any external disturbance, and modern ones can make the aircraft climb to any desired height and remain there, fly a preset course (heading) or approach to land along an ILS radio beam as described later. The pilot's demands are fed from the cockpit into the automatic flight-control system (AFCS) which after digesting them, comparing them with any safety limits or other factors, and possibly receiving inputs from other systems such as a stall-warning system or GPWS (ground-proximity warning system), sends signals mechanically to hydraulic power units driving the control surfaces. Sometimes the signals are sent electrically, in what is called FBW (fly by wire). The surface power units again are clever enough to think for themselves and keep the control surfaces moving in the exact manner needed. Often large aircraft have the surfaces split into two to six segments each driven independently to avoid any danger from incorrect operation by one 'autopilot channel' or surface power unit. An extra feature in combat aircraft is an automatic ability to chase hostile aircraft or follow the undulations of the ground at the lowest safe height.

The electric system has already been mentioned, and chronologically this was one of the first to be needed. In 1920 it comprised a lead/acid battery, like that in a car, serving a cumbersome radio and three navigation lights, red on the left wing-tip, green on the right and white at the tail. The battery was charged by a d.c. (direct-current) generator driven by a small windmill. By 1930 the engine(s) were being arranged to drive generators, and today the generating capacity of large aircraft can be similar to that needed by a substantial town. Most of the power is a.c. (alternating current), massive quanti-

ties of which may be consumed in heating the leading edges or windscreens in icing conditions, or in heating food in the galley or water for washing. A proportion of the a.c. is rectified into d.c. for a multitude of services and for charging batteries which are of advanced types usually based on nickel and cadmium. All except the simplest aircraft need large amounts of specially precise a.c. at exactly 400Hz (400 cycles per second) to serve the avionics (aviation electronics). Usually the engine drives an IDG (integrated-drive generator) which keeps spinning at exactly the correct speed for 400Hz current no matter what speed the engines may run at.

Avionics account for from 10 to 40 per cent of the cost of most aircraft, and are so incredibly diverse that only a few examples can be mentioned here. The fundamental need has been radio communications,

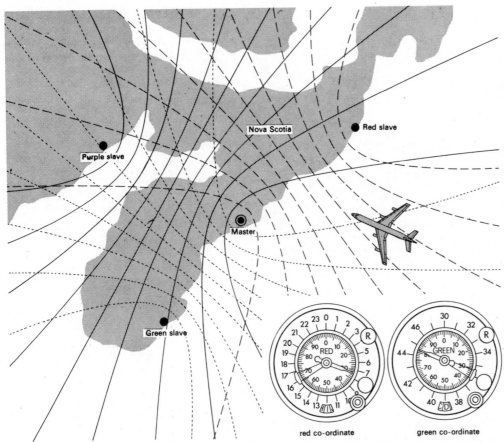

Purple slave

Nova Scotia

Red slave

Master

Green slave

red co-ordinate

green co-ordinate

The principle of the original Decca Navigator system is illustrated by a map of the Nova Scotia chain. Continuous radio signals emitted in synchronization by the master and slave stations are picked up and analysed by the aircraft, which compares the time-delays from the various stations. Matching the times from any two stations fixes the aircraft on a hyperbolic position-line, some of which are plotted. In the 1945 system the navigator had to read the position-lines off dial instruments called Decometers, two of which are shown. Soon the system was made automatic, one method being to make the signals drive a pen across a roller-blind moving chart to give a continuous readout of the aircraft position. Today Decca is combined with other systems to give foolproof navigation.

and today these neat boxes also show the direction to radio beacons on the ground, provide DME (distance-measuring equipment) readings to the destination or a 'waypoint' ahead, provide communication with any other compatible aircraft, automatically send information on the flight when requested by ground ATC (air traffic control) and do many other things including storing all R/T (radio telephone) traffic on crashproof tape. The ILS system installed at airports needs special receivers in the aircraft, one for the G/S (glide-slope) to control aircraft height on the glide-path and the second for the LOC (localizer) to control its direction along the runway centreline. Special receivers are also needed for the numerous kinds of ground-based radio navaid (navigation aid). Gee, Loran and Decca were the pioneer R-Nav (area navigation) systems which allow aircraft to fly with precision anywhere, without having to follow congested fixed routes. Omega is one of the later systems, notable for being able to cover the entire world with great accuracy, even the previously very difficult regions near the poles. Doppler radar is carried in the aircraft, sending out four beams diagonally to the ground and, by measuring the speed of the four reflections, giving the exact ground speed (speed in relation to the ground) and thus the wind at the height at which the

aircraft is flying. Weather radar points the beam straight ahead to pick out turbulent clouds in bright light, with especially dangerous storm 'cores' shown black (a few modern radars use brilliant colours), usually with further facilities such as the ability to avoid mountains or watch the ground below – no matter how much cloud or blizzards may obscure it – or project bright all-weather pictures on the windscreen in a head-up display (HUD). Most HUDs present the pilot with bright lines, numbers, words and symbols which help him to fly accurately in any way he wishes whilst still looking ahead with his eyes focussed on infinity. Perhaps the most amazing navaid is the inertial navigation system (INS) which is totally self-contained yet can guide aircraft, missiles, submarines or spacecraft with great accuracy just by sensing their accelerations. INS giros and accelerometers are among the most precise of all human engineering products.

Yet one could not scorn the precision in the non-electronic systems. Hydraulics, the muscles of the modern aircraft, frequently call for surfaces machined with tolerances measured in millionths of an inch or centimetre, notwithstanding the fact they may be a component exerting forces measured in tons. There were few hydraulic systems before 1930, but the largest modern aircraft may have more than

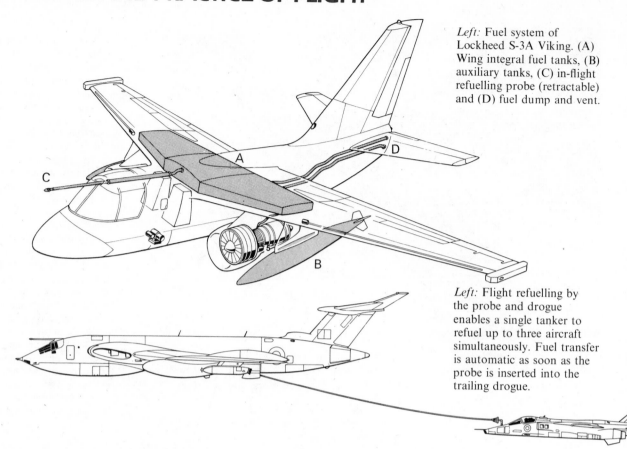

Left: Fuel system of Lockheed S-3A Viking. (A) Wing integral fuel tanks, (B) auxiliary tanks, (C) in-flight refuelling probe (retractable) and (D) fuel dump and vent.

Left: Flight refuelling by the probe and drogue enables a single tanker to refuel up to three aircraft simultaneously. Fuel transfer is automatic as soon as the probe is inserted into the trailing drogue.

300 separate hydraulic items, forming components of four or more separate systems. The earliest systems were just closed circuits of piping filled with oil, which were controlled by valves and hand-pumps worked by the pilot. Such systems retracted the landing gear, pushed down the flaps against the airstream, opened bomb doors and moved any other items in inaccessible places. But such new features as gun turrets or powered flight controls might have to take hydraulic power for long periods, and the obvious next step was to add a hydraulic pump driven by the engine(s), with control valves worked either electronically or automatically. The modern hydraulic system may be able to generate or absorb more than 1,000hp, as much as the most powerful aircraft engines a few years ago. The fluid is no longer mineral oil but non-inflammable liquids such as 'phosphate-esters', circulated at a pressure of 3,000 pounds per square inch. The systems in modern aircraft may be remarkably complicated, and predictably clever. Just to take landing gear alone, one of the simplest items, this begins with at least two hydraulic systems to raise and lower it, the first for normal operation and the second for emergency stand-by. An emergency system is always provided energised by large storage bottles, and usually with pumps driven electrically and/or by ram-air turbines which can be extended into the

airstream if all engines should fail. Further hydraulics drive the nosewheel steering when the aircraft is taxiing, and some main gears have a 'kneeling' facility which can lower the main floor level for loading heavy freight. Yet other hydraulics operate the brakes, with automatic anti-skid systems to ensure maximum retardation on slippery runways without allowing the wheels to lock.

Another system concerned with liquid is that handling fuel. In simple aircraft this is just a single tank, filled by gravity like that of a car and with gravity feed to a pump driven by the engine. Most modern aircraft need much more. Even the tankage is far from simple. The largest aircraft carry virtually all their vast capacity in integral tanks, which are not separate tanks at all but parts of the airframe that are carefully sealed. The main tankage is usually in the long box forming the structural basis of the wing, bounded by front and rear spars and upper and lower skins. This is coated on the inside with sealant and divided into multiple cells linked by pipe systems with numerous electrically driven on/off valves, one-way valves, jettison valves, fuelling valves, flow proportioners to feed exactly right amounts from each tank to each engine, and in the tanks themselves high-power booster pumps and amazing contents-gauging systems which do not just measure the height of the fuel but tell the pilot the exact mass of fuel in each tank, after allowing for the type of fuel being used and the fact the aircraft may be upside-down! Many military aircraft have in-flight refuelling capability to allow them to fly very long distances without having to land at inter-mediate points to refuel. Some use a forward-pointing probe, which the pilot thrusts into a drogue on the end of a special hose trailed by the tanker, while others have a receptacle on top into which a skilled crew-member in the tanker thrusts a tele-scopic rigid boom.

Special measures are needed to minimise the danger of fuel explosion or fire. Whenever tradit-ional fuel cells are used, not an integral part of the structure, they are as far as possible made crashproof. Each cell can be smashed violently but it will not leak. Combat aircraft have self-sealing tanks; when they are pierced by cannon shells the punctures quickly seal themselves off with layers of material which expand on contact with fuel. The space above the fuel in the tank is never air but nitrogen or other dry inert gas which will not support combustion. A new material called re-ticulated foam, which can fill complete volumes with very little weight, can be filled in between tanks or inside them to isolate the fuel and prevent any fire from spreading after the most severe crash or combat damage. Of course there are also fire-suppression systems which automatically or on pilot command discharge various kinds of extinguishant into the tanks, the engines or other hazardous parts of the aircraft. This is just one of many emergency systems, which also include escape slides, automatic dinghy release and inflation, auto-broadcast bea-cons and the ejection of special recording devices packed with information to help rescuers find the site and discover the cause.

Many one-shot systems operate on compressed air from bottles, or high-power gas supplies from cartridges or devices like enclosed rockets. Such sudden one-shot power may be needed to fire off a cockpit canopy, shoot out an ejection seat or eject any of a host of items from combat aircraft. Continuous-operation pneumatic systems are also used, some involving vast airflows at quite low pressure to drive turbines or air-conditioning sys-tems and others supplying very small flows at ex-tremely high pressure. In contrast, some cockpit instruments are driven by a vacuum, while extremely hot air from the engines is used for de-icing.

Almost all jet aircraft, and the upper strata of private and executive machines, are pressurized. As they climb into thinner air, to go faster and fly further on each gallon of fuel, air is gradually pumped into the passenger cabin partly to warm it and partly to keep the pressure as nearly as possible at the original level at take-off. Piston and turbo-prop aircraft usually need special compressors, while jets usually have abundant air supplies (though these are not often fed to the cabin but are used to drive complex plant which supplies air at the correct temperature, pressure and humidity). Even the task of sealing the cabin is quite large, and requires such extra items as special seals where the dozens of pipes, wires and cables pass through the cabin walls, and inflatable seals round all the doors – which, of course, must be proof against any attempt to open them in flight, whilst opening outwards instantly on the ground. The pressure differential (dP) across the cabin wall of most jets is 4 to 8 pounds per square inch, but Concorde has a dP of 10.7 to keep the interior comfortable in the almost airless stratosphere.

At high altitude an oxygen system is also neces-sary, though in pressurized and air-conditioned civil aircraft only as an emergency stand-by. Military aircrew, and one pilot in high-flying jetliners, wear their oxygen masks all the time, while the other masks in civil aircraft are stored above each occu-pant and drop down automatically if the cabin pressure should suddenly fall. The gas is increasingly stored not in very high pressure bottles but as a liquid, at minus 183°C, subsequently being warmed in evaporator coils and piped to the users. Modern oxygen masks are of the 'demand' type, which shut off the vital gas except when the wearer inhales. Some incorporate R/T microphones, but nearly all civil aircrew wear the boom type in which the sensitive cartridge can be swung down on a curved arm hinged beside the head and then swung back out of the way to facilitate normal speech on the flight deck.

De-icing systems have already been mentioned, but are important enough for a brief description. Strictly, items on which no ice must form are 'anti-iced', while de-icing systems keep shedding ice already present. Some aircraft, such as Concorde, pass up or down through icing levels so quickly that only certain items such as engine inlets need protec-tion, and this often is provided by hot air from the engines, or electric heaters recessed into the outer metal skin. Most large aircraft have hot-air de-icing of leading edges, though the Transall freighter uses electric mats. Slow aircraft use pneumatic rubber strips which are continually inflated and defladed in order to break up and shatter the ice.

Warplanes

Reconnaissance Aircraft

Prior to World War I few national or military leaders gave much thought to the possible use of aeroplanes in war. This was to some degree understandable because war in the air was a completely novel idea that seemed best fitted to the pages of science fiction (where, in fact, it had already appeared), and the weak and flimsy aeroplanes of the day had enough difficulty getting into the air, without having to operate in a war. The only task for which a few experts thought the flying machine might be suited was reconnaissance. This was because for over 100 years observers in balloons had proved that they could usefully watch the enemy from their lofty position and report his movements in a way impossible from ground level.

In fact prior to the outbreak of World War I in August 1914 there had already been numerous combat missions by aeroplanes, most of them concerned with reconnaissance. The first took place on 23 October 1911 when Capt Piazza took off in his Blériot XI of the Italian Army and spent an hour writing details of the Turkish forces between Tripoli and Aziziyah. On 10 January 1912 the first 'psychological warfare' mission showered Arabs with leaflets urging them to support the Italians, and from 24 February aerial cameras were used for the first time, several missions even being flown by an airship taking ciné film. Further photographic missions were flown in the Balkan War later the same year.

By August 1914 most major powers had experimented with airstrips and aeroplanes used in the reconnaissance role, discovering that the best reconnaissance machine appeared to be a two-seater with an 'observer' who did nothing but attend to cameras or write notes. In the summer of 1912 Britain had held trials to find the best military aeroplane, the only duty considered being reconnaissance. The only really good competitor was the B.E.2 tandem-seat biplane, but as this had been designed at the government's own Royal Aircraft Factory at Farnborough it was barred from taking part. Despite this, the B.E. was adopted as virtually the standard aeroplane of the Royal Flying Corps, and was produced in quantities that grew month by month through the first three years of warfare. Though outwardly a serviceable and efficient machine, the B.E. in fact was destined to cause the deaths of more British airmen than any other World War I aircraft. One of the basic requirements when it was

designed by Geoffrey de Havilland back in 1912 had been that it should possess powerful natural stability, so that it could safely be left to fly by itself while both crew-members carried out reconnaissance duties. This stability made it very hard to manoeuvre, and, as recounted later, once enemy aircraft began to come after the B.E.s with machine-guns there was little the luckless Britons could do about it. The observer in the B.E. was foolishly put in the front cockpit, and when in self-defence he was given a Lewis machine-gun there was no way he could fire this without the struts and wires around it getting in the way. Most crews did not bother to fit the gun, judging the small extra speed more useful; and some even tried dangling lead weights on thin cables to try to wreck the enemy propellers! The sky did not become deadly until the late summer of 1915, and by that time all the ground army commanders on the Western Front had learned by experience the immense value of aerial reconnaissance. In fact, the quick evolution of the fighter was not to shoot down other fighters but to shoot down the enemy reconnaissance aircraft, which had shown they could make the difference between winning and losing a battle. In fact many of the armed combat aircraft were themselves intended for reconnaissance and were called fighting scouts, or just scouts. While the sky increasingly resounded not only to the drone of engines but also to the stuttering of machine-guns, air crews of many nationalities learned how to spot the things that mattered and not to make silly mistakes – in August 1914 an 'enemy bivouac encampment' turned out to be headstones in a graveyard in a friendly village! Of course, armies learned to do their best to camouflage their activity and if possible their presence, but from above it was often ineffectual. Digging, fresh tracks, and especially activity on roads and railways, all betrayed the enemy's plans and intentions before he could carry them out.

By 1915 most air reconnaissance was being done with specially designed cameras, which were either mounted vertically in the floor of the aircraft or held by the observer over the side – no easy job with frozen fingers, 'archie' (anti-aircraft) shell-bursts and the drag of the massive camera in the slipstream. Gradually the cameras were made more reliable and automatic in operation, so that frozen fingers in heavy gloves did not have to do 10 or 11 distinct tasks between exposing each glass plate. Likewise, the aircraft got better, and France, Britain and Germany all produced excellent specially designed reconnaissance machines able to hold their own

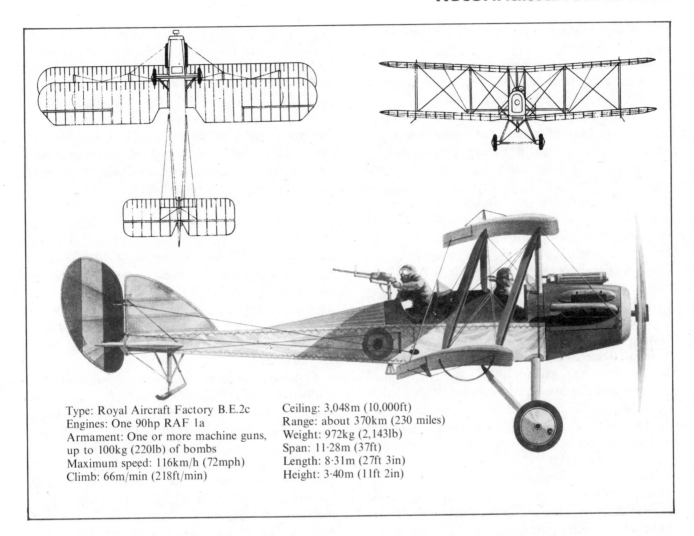

Type: Royal Aircraft Factory B.E.2c
Engines: One 90hp RAF 1a
Armament: One or more machine guns,
up to 100kg (220lb) of bombs
Maximum speed: 116km/h (72mph)
Climb: 66m/min (218ft/min)

Ceiling: 3,048m (10,000ft)
Range: about 370km (230 miles)
Weight: 972kg (2,143lb)
Span: 11·28m (37ft)
Length: 8·31m (27ft 3in)
Height: 3·40m (11ft 2in)

against individual enemy aircraft. By the time of the final German offensive in March 1918 the Idflieg (Imperial German air force) was taking about 4,000 photographs a day on the Western Front alone, while the newly formed RAF had no fewer than 9,000 cameras in action. Some were in such aircraft as the D.H.4, D.H.9 and O/400 which could penetrate deep into Germany bombing and taking pictures.

A related duty was artillery observation. Gunners on the ground could seldom see accurately where their shells were falling, and the aerial observer soon had to fulfil this extra task. The standard method was the 'clock code': the air crew were told the position of the target and placed a transparent sheet over a map, centred on this position and arranged like a clock but with added circles at different radii from the target. Each shell-burst was then quickly related to the position on the 'clock' so that the result could be sent to the gunners (often by Aldis lamp using Morse code) as a simple letter/number code. This 'spotting' duty was to remain an important one for co-operating with both armies and

navies until after World War II.

In the 20 years between the world wars there was gradual progress in the design of aerial cameras, in detailed mapping and aerial survey and in the aircraft themselves. From its birth in March 1935 the Luftwaffe concentrated strongly on reconnaissance, and about one-fifth of its strength comprised specially equipped reconnaissance machines which often were not conversions but specially designed types such as the Hs 126 and Fw 189. German photographic coverage of British military installations was complete by the end of the 'phoney war' on 10 May 1940, and thousands of missions had also been flown over Poland, Norway and France. But once battle was joined in earnest the German crews had little time for reconnaissance and for the remainder of the war sorties over Britain were rare, except for lone missions by extremely high-flying aircraft such as the Ju 86R. Tremendous efforts were made to build better strategic reconnaissance aircraft, but Allied command of the air denied them any opportunity until in the closing months before VE Day the twin-jet Ar 234B and Me 262A appeared in reconnaissance versions that no Allied fighter could catch.

In contrast the Allies had thousands of reconnaissance aircraft airborne daily. The RAF pioneered the PR (photo-reconnaissance) fighter, using the Allison-Mustang for low-level work and such superb camera platforms as the Spitfire PR.XI for high-level sorties. The PR.XI had several times the fuel capacity of most fighter Spitfires and could cover targets as far as Berlin or Prague in daylight. Another valuable PR aircraft was the Mosquito, and RAF PR aircraft took many millions of photographs all over Europe on which appeared the first V-1 flying bomb, the first V-2 rocket and countless other Nazi secrets. Meanwhile the US Army and Navy used camera-equipped versions of many famous fighter and bomber aircraft, while the Japanese consistently accorded reconnaissance the honour of calling for purpose-designed aircraft rather than mere versions of combat types. The best-known were the Mitsubishi Ki-46 twin-engined machine of the army and the Nakajima C6N carrier-based aircraft of the navy. Both were outstanding in performance, and generally superior to Allied counterparts – which was emphatically not the case with combat aircraft.

An aspect of reconnaissance that might be overlooked is Elint (electronic intelligence). From the earliest days of radar it had been appreciated that radar and military electronics would soon become an important branch of warfare, with nations doing their best both to find out about enemy equipment and interfere with its proper operation. A simple instance occurred on 22 February 1941 when an RAF Spitfire raced at tree-top height past a German radar station to bring back the first pictures. This was traditional reconnaissance, but true Elint began in a heroic way on 3 December 1942 when a specially equipped Wellington succeeded on its eighteenth attempt to lure up Luftwaffe radar-equipped night fighters. Before being shot to ribbons it radioed back full details of the German fighter radar, including the vital wavelength and pulse-repetition frequency. Ever since, Elint has been one of the most vital reconnaissance tasks of all service arms, and during the 1950s the USAF sent out many large jets packed with Elint equipment to bring back details of the radars, communications and other installations round the Soviet Union. Several of these aircraft were shot down, though, so far as is known, none crossed the Soviet frontier.

But on 1 May 1960 an American aircraft did penetrate over the Soviet frontier. A civilian pilot of the US Central Intelligence Agency, F. G. Powers, took off from Pakistan to Norway, nearly 4,000 miles (6,440km) away, taking photographs and Elint recordings all the way (the May Day date was specially chosen). Over the Siberian city of Sverdlovsk he was shot down by a missile. Thus did the true purpose of his aircraft, the Lockheed U-2, become revealed. Instead of upper-atmosphere research, the American cover-story, it was for strategic reconnaissance at ultra-high altitudes where it was hoped it would be immune from interception. Subsequently the USAF more openly used a descendant of the U-2, the Lockheed SR-71, for even more remarkable strategic reconnaissance missions at speeds exceeding 2,000mph (3,218kph) at heights in the order of 80,000ft (24,624m). But so far as is known nobody today would try to cross an unfriendly frontier except with a satellite, as will be explained in the chapter on Missiles and Space Exploration.

Today nearly all reconnaissance is performed by multi-role tactical combat aircraft fitted with quickly attached pods, and visual-light cameras are no longer enough. Other cameras are needed with film sensitive to IR (infra-red, or heat) radiation. Others use films sensitive to other parts of the EM (electro-magnetic) spectrum, so that when all the photographs of each target are compared the results defeat all enemy attempts at camouflage. Further pictures are taken by IR linescan, which records the 'picture' even further into the infra-red to indicate such things as the place where a car was parked a few minutes before. SLAR (side-looking aircraft radar) is another valuable sensing tool, carried by most advanced combat aircraft to search for scores of miles on either side of the flight path and bring back pin-sharp photographs showing a vast amount of detail invisible in pictures taken by optical or IR cameras. Elint accounts for a further burden of 'black boxes' that must today be carried by the specialised reconnaissance aircraft.

The list of tasks for reconnaissance platforms is seemingly endless. In 1962 photographs taken over Cuba showed Russian ballistic missiles and triggered a near-nuclear crisis. For the next ten years thousands of reconnaissance aircraft operated in south-east Asia, many of them of a new type called the RPV (remotely piloted vehicle) which contains no human pilot but, as its name suggests, is flown by a pilot in another aircraft far away, or on the ground. The RPV can zoom on to a difficult target making repeated photographic runs, and is difficult to shoot down because of its very small size.

Other even stranger aeroplanes were lightplanes made so quiet that even in the still of the night they could pass low overhead with no more noise than the rustling of leaves in the wind. These piloted 'quiet penetrators' carried high-definition cameras and other equipment for night use without being detected. A totally different kind of reconnaissance activity is the monitoring of unfriendly missile testing. The Soviet Union has for many years flown remarkable full-range missile tests from various launch sites to target areas in the mid-Pacific. The USAF and USN have done their best to monitor these tests with radar, Elint, and various cameras which are often fantastic. They have photographed the brilliant trail from missile re-entry vehicles from 100 or more miles away and discovered many vital facts about the missile and its terminal guidance or penetration aids.

Today yet another important class of reconnaissance aircraft is the AEW (airborne early warning) or Awacs (airborne warning and control system). This is always a large aircraft, with a crew of 18 or more, able to lift many tons of radars and electronics to a height of seven miles or more, remaining there for up to 12 hours. Such machines as the USAF Boeing E-3A, USN Grumman E-2C, Russian Tu-126 'Moss' and RAF Nimrod AEW-3 not only intently study friendly and hostile airspace but can direct all military air effort, both defensive and offensive, in a way never before possible.

Bombers

The idea that flying machines might drop bombs has been written about for centuries in what was then mere science-fiction. Once man began to travel above ground in powered airships and aeroplanes the bomber was only a matter of time. In fact, because early aeroplanes were so puny and limited in lifting capability, the first effective bombers were airships, and the Zeppelin, Schutte-Lanz and Parseval ships of the Imperial German Navy and, to a slightly lesser degree, Army, were engaged in practice bombing before the start of World War I, as related in a later chapter. The first heavier-than-air bomber was nothing more than a prank by high-spirited young officers. At an air meeting near San Francisco in 1910 US Army Lts Paul Beck and Myron T. Crissy concocted two live bombs, one made from an artillery shell and the other from 'two-and-a-half-inch pipe', and dropped them from a Wright Flyer. They detonated successfully, to the

consternation of thousands.

On 1 November the following year Sub-Lt Gavotti of the Italian Army reached into the cockpit of his Rumpler Taube, picked up a Cipelli heavy grenade, withdrew the safety pin and dropped it over the side on Turkish troops at Ain Zara, near Tripoli. He then flew to Taguira oasis and dropped three more. Nearly a year later, in the Balkan war, specially designed aerial bombs were dropped in fair numbers. But the methods were still crude. Most were thrown out by hand, while one Russian pilot used to have bombs looped by string over his shoes, releasing them by sharp kicks.

Gradually methods were made less rough and ready. Bombs began to be hung from specially designed racks, at first along the side of the fuselage outside the cockpit and later from the underside of the fuselage or wing, with the bombs carried horizontally to reduce air drag. Release was effected by pulling a wire or string. Having the bombs already fused to explode on impact was dangerous; they might be jolted free on take-off. Some had safety pins which could be withdrawn by the pilot pulling another string, while from 1915 properly designed bombs came into use with good aerodynamic and ballistic properties, so that they fell accurately point-first, with a small fan driven by the slipstream to unscrew a safety device and make the fuse live.

More important even than developing bombs was learning how to aim them. Despite dropping bombs from low altitude at speeds of under 110km/h (70mph), early attacks were usually ineffective except in causing harrassment. Anyone who attempts dropping objects accurately from a fast car or other vehicle will soon realise that it is difficult to judge how far ahead of the target the missile should be released. But any mathematician could quickly work out the basic rules for designing a bomb sight. Clearly the aim depends on both the height and ground speed of the aircraft (speed actually made good relative to the target, in other words air speed corrected for wind). The first person to make a bomb sight was another US Army Lieutenant, Riley E. Scott, who constructed a device involving a pivoted telescope and tables of figures. Finding no interest in the United States he took it to France and in 1912 won the first-ever bombing contest with the Michelin Prize of 75,000 francs. His crude sight was the starting point for those of almost all other air forces.

As already related, most officials and generals in the period before World War I were unanimous in doubting that aeroplanes could serve any useful purpose in war except possibly as reconnaissance platforms. Yet a few far-sighted manufacturers proposed 'combat aircraft', and some actually constructed machines capable of carrying a significant load of bombs. Unquestionably the pioneers were in Czarist Russia and Italy.

Pride of place must be accorded to Igor Sikorsky, born in 1889. As early as 1912 he was designing a giant biplane, the world's first four-engined aeroplane, which was completed in May 1913. Named *Russkii Vitiaz* (Russian Knight), it flew with complete satisfaction, and needed very little development other than progressive increase in engine power to permit operation at greater weights. The second example, flown in December 1913, was named after a 10th-century folk hero, *Ilya Mourometz*. This became the prototype of the famed IM class of giant warplanes which not only equipped the world's first heavy-bomber unit – the EVK, Squadron of Flying Ships, formed in December 1914 – but pioneered the whole concept of strategic bombing. The RBVZ, the Russo-Baltic Wagon Works, built at least 80 of these great bombers in seven main versions. Though apparently flimsy, they proved tough and serviceable, operating by day and night with multiple wheel or ski landing gear and making long bombing missions lasting up to six hours. Bomb load varied up to well over a tonne, and defensive armament was progressively increased until the later (1917) models carried eight machine-guns and sometimes a 50mm cannon. Guns were carried in nose and tail, above and below the fuselage and sometimes in cockpits on the wings. Self-sealing fuel tanks were introduced, for the first time on any aircraft, and the lower part of the pilot's cabin and the backs of their seats were armoured with thick slabs made from compressed metal shavings. It is said that only one IM was shot down in air combat, and then only after it had shot down three of the attacking fighters.

Italy's bomber pioneer was the wealthy industrialist Count Gianni Caproni. He built several aircraft from 1908, and in 1913 completed one that was intended to surpass all others. A neat biplane, it had three Gnome rotary engines in tandem in the rear of a central nacelle. The first two engines drove tractor propellers on the front of the two long booms carrying the tail, and the third drove a pusher screw at the rear of the nacelle. Soon he built an improved bomber with three 100hp Fiat water-cooled engines, one in the rear of the nacelle and the others mounted at the fronts of the tail booms. This eliminated the unsatisfactory indirect transmission to the tractor

propellers. The improved aircraft, the Ca 32, was put into immediate production and 164 were built for the Corpo Aeronautica Militare. Entering service from July 1915, they were courageously flown on long bombing missions across the Alps and other hazardous obstacles against Austro-Hungarian strategic targets. From the Ca 32 stemmed many later Capronis for bombing, torpedo dropping and other duties. All the first families of Capronis had twin tail booms. This left a short central nacelle, almost always containing a pusher engine. Inches from the propeller of this engine, and standing almost on the cylinders, was a gunner who had nowhere to sit, was surrounded by slipstream at often sub-zero temperature and sometimes, after several hours, was suddenly needed to take accurate aim with triple Revelli machine-guns aimed by hand. Life for the first bomber crews was no picnic.

The brothers Voisin, pioneers of aviation after the Wrights, were among the chief suppliers to the French Aviation Militaire. In 1914 four squadrons were equipped with the Type L pusher biplane, which though puny (80hp, carrying up to 60kg [132lb] of small bombs) was strongly made in welded and bolted steel, and proved itself able to take punishment. Voisins flew the first bombing mission

Igor Sikorsky's *Ilya Mourometz* giant four-engined biplane which flew in early 1914 with a total of 14 people on board.

of World War I on 14 August 1914 when a small force raided Zeppelin hangars at Metz-Frascaty, but it was May 1915 before proper missions were organized. By this time large numbers of Voisins were in use, some of them armed with Flechettes, simple steel darts dropped 250 at a time on enemy troops. This form of bombing was no less random in 1915 than the dropping of most other bombs, and dart-like weapons have been considered many times since for use against 'soft' (unarmoured) forces. Subsequently Voisin bombers grew in power and capability, but gradually gave way to the equally slow but more powerful Breguet-Michelins and then the excellent Breguet XIV tractor machine which in various versions delivered 2,000 tonnes of bombs in the final year of the war.

In Britain there were no bombers at the start of the war, and the Royal Flying Corps had no thought of such machines; but the Royal Naval Air Service was more awake. Cdre Murray Sueter, Director of the Admiralty Air Department, issued a specification in December 1914 for a bomber with two engines and able to carry six 50.8kg (112lb) bombs at 116km/h (72mph). An immediate response came from Frederick Handley Page, with a biplane bomber with two 120hp Beardmore engines. But when he showed the drawings to Sueter in January 1915 the naval officer knew that Rolls-Royce were developing engines that would be much more powerful. Sueter said as much to the young planemaker, and added an oft-quoted command 'Go and build a bloody paralyser of an aeroplane'. The result was the O/100, first flown on 18 December 1915. Powered by two 250hp Eagle II engines, it carried double the requested bomb load at 129km/h (80mph), with defensive armament of up to five machine-guns. HP built 40 with the Eagle II, followed by a batch with the 320hp Sunbeam Cossack. Many companies then built over 400 of an improved version, the O/400 with more powerful engines and speed nudging 161km/h (100mph). These great bombers carried bombs of up to 748kg (1,650lb), the maximum load being 907kg (2,000lb). At the end of the war the much larger V/1500 was entering service with the RAF's Independent Air Force, a specially created strategic arm to bomb Berlin and other centres deep in Germany. The V/1500 had four of the latest 375hp Eagle VIII engines and could carry a bomb load of 3,402kg

(7,500lb), including special bombs weighing 1,497kg (3,300lb) each.

In some respects even the V/1500 was beaten by the largest of the Staaken 'Giants' built for the German Idflieg (air force) by Zeppelin Werke. Known as the R series, from Reisenflugzeug (giant aeroplane) these were the greatest bombers of the first world war. The most important model was the R.VI, with two tandem pairs of 260hp Maybach or Mercedes engines, 18 wheels, a crew of seven, and a load of 18 bombs of 100kg (220.5lb). The Giants were extremely well equipped with radio navigation direction-finders, oxygen, map tables, comprehensive flight instruments and even a simple autopilot. The wing span of most Giants was 42.2m (138ft 6in), largest of any service bomber until just beaten by the B-29 Superfortress 30 years later.

Between the world wars bombers developed in tune with improved engines, new kinds of structure and such advances as flaps, retractable landing gear, variable-pitch propellers and the power-driven gun turret. Throughout the 1920s most nations progressed slowly, with metal structures instead of wood but still braced with wires and covered with fabric. Only in the Soviet Union was there dramatic change, and this was because of the setting up by the German Junkers company (which could not build in its own country) of an aircraft plant at Fili, near Moscow. This was an example of what is today called 'technology transfer'; the new Soviet government, which later called the factory State Aircraft Factory No. 1, found itself possessed of the ability to build large cantilever monoplanes with all-metal construction, using Kolchug alloy (named from the Russian town where it was developed) which was claimed to be superior even to the German Duralumin. The obvious leader of the team to put the Junkers technology to use was Andrei N. Tupolev. His bureau produced an excellent twin-engined bomber in November 1925, the ANT-4, put into service by the VVS (Soviet air fleet) as the TB-1. Powered by 500 hp M-17 water-cooled engines, it was noteworthy for its completely unbraced all-metal wing, designed by Vladimir Petlyakov (who later became a design bureau leader in his own right). Large numbers of TB-1s made many notable missions in 1925–35, and there were also torpedo-bomber and seaplane versions.

Only a month after the TB-1 first flew, the government Ostechbyuro issued a specification for

Ground crew working on this German bomber of World War I, the Zeppelin-Staaken R.VI, give an idea of its enormous size.

a bomber with even greater capability, with engines totalling 2,000hp. Tupolev did the obvious, though no other constructor followed suit until the mid-1930s: he doubled the number of engines to produce the first four-engined cantilever monoplane, the type that dominated heavy-bomber design in the Second World War. The ANT-6, designated TB-3 by the air fleet, was first flown on 22 December 1930. Temporarily powered by four imported 600hp Curtiss Conqueror engines, it was basically a scaled-up TB-1, with a monster all-metal wing of very deep section, tandem-wheel landing gears and no fewer than five machine-gun stations each with two DA-2 rifle-calibre weapons. Two of the gun positions were beneath the wing. Normal bomb load was 2,200kg (4,850lb), but for overload missions some versions of the great load-carrier could be burdened with 5,800kg (12,800lb), a load not even approached by any other bomber of the period. The TB-3 was built in several quite different versions generally called 'Type 1932, 1934 and 1936' from the year of service issue. The original production engine was the 730hp

M-17, but this gave way to the 900hp M-34R and then later M-34 versions of up to 1,280hp, and a few even had the AN-1 diesel. In the mid-1930s some of the 800-plus fleet of these great bombers pioneered mass use of paratroopers, carried the first airborne tanks, and landed on skis at the North Pole.

In the United States the most significant advance of the inter-war period was the Martin Bomber, a popular name for the Martin 123. By 1930 German development of stressed-skin structures, which unlike the Junkers/Tupolev technology did not suffer the drag penalty of a corrugated skin, had sparked construction of truly modern aircraft in the United States. Boeing's B-9 bomber unfortunately appeared a few weeks ahead of Martin's 'dark horse' which, unlike the B-9, had not been designed to any official specification. Martin put into the Model 123 every new feature, including the 600hp Cyclone engine in a long-chord cowling and driving the new Hamilton propeller. Even in its original form the Martin reached a speed of 317km/h (197mph), which exceeded that of any single-seat pursuit (fighter) in the Army Air Corps! When re-engined with 675hp Cyclones and a streamlined front turret it reached 333km/h (207mph). The Army promptly

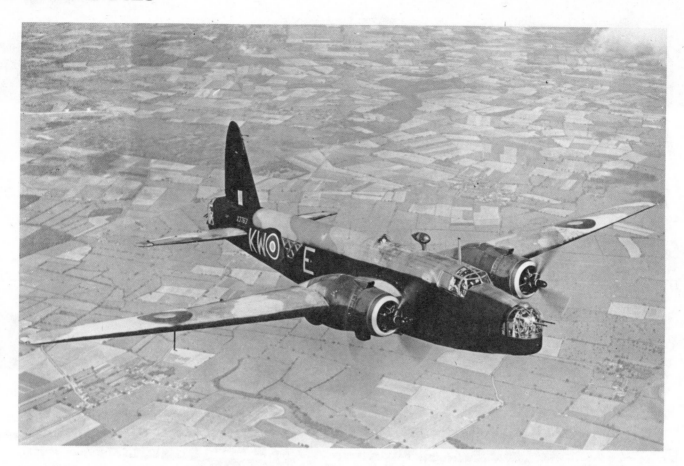

bought a version of the B-10, and ultimately received over 150, a useful number for the depression era; and 189 were sold to foreign customers despite the fact that the United States Army would not permit exports until 1936.

By this time Boeing had flown the Model 299, a startlingly large and advanced aircraft with four engines and so many gun positions a newspaperman called it the Flying Fortress, a name which stuck. Because of its geographical position the United States judged that its only bomber targets might be a hostile fleet. Range was thus not very important, and Boeing chose four engines purely for greater over-target height and speed. By this time there was a strong belief in the Army Air Corps in unescorted mass attacks in daylight at the greatest height possible. The Norden bombsight promised unprecedented accuracy, and the General Electric turbocharger promised greatly increased power at heights as high as 9,000m (29,530ft). Gradually the impressive Boeing, which was ordered with both the new devices in 1937 as the B-17, became the focal point of an entirely new kind of air power in which attacks were made from the stratosphere.

Germany's reborn Luftwaffe had the misfortune to lose its first Chief of Staff, General Wever, in a crash in June 1936. With him died plans for four-

The Vickers Wellington bomber (*above*) can be regarded as typical of long-range bombers at the start of the war. Its unusual geodetic form of construction enabled it to take a great deal of punishment. The Bristol Blenheim (*right*) derived from the Bristol Type 142 named *Britain First*, built for *Daily Mail* proprietor Lord Rothermere.

engined strategic bombers. Wever had called them Die Ural-Bomber because they could hit the Urals, but the north of Scotland was also written into their specification. His successor, Kesselring, believed in smaller twin-engined bombers for tactical support in a Blitzkrieg campaign. So the Luftwaffe equipped itself exclusively with medium bombers, the biggest being the broad-winged He 111, the lightest the twin-finned Do 17 and the newest the Ju 88. The first two saw action in Spain and, because they encountered no determined modern fighters, left the Germans satisfied that their bombers were entirely adequate for any future war despite their light defensive armament of three hand-aimed machine guns. The Luftwaffe also put into service the little Hs 123 biplane and then the Ju 87 monoplane as purpose-designed dive bombers, able to put down heavy bombs with considerable precision in dives as steep as 80°.

The Soviet Union, which also had aircraft in Spain, likewise decided to concentrate on smaller tactical bombers, and in the Tupolev SB-2 had one of the best, with modern stressed-skin structure, and both heavier bomb-load and higher speed than the Blenheim, the RAF's first modern bomber, which entered service 18 months after the SB-2. Even less known by the world outside, Japan had by the mid-1930s moved totally beyond the stage where she produced modest aircraft either copied from foreign designs or led by imported design engineers, and had begun to create aircraft of outstanding quality. Where bombers were concerned the Japanese were confronted by a war theatre of vast extent, so that bomber range tended to be twice as great as in other countries. This inevitably meant something else had to give. As bomb-load and speed were vital the penalty tended to be the acceptance of light construction and lack of armour or fuel-tank protection. This affected both the Army Ki-21 and the otherwise remarkable Navy G3M, both of which saw extensive action in China in the late 1930s. Both had range problems, but pushed the available technology to the limit.

Though courageously operated throughout World War II the Ki-21, like the He 111, had to soldier on long after it was obsolescent and highly vulnerable. The G3M was progressively replaced by the G4M, but not before it had flown many historic missions. On 14 August 1937 a strong G3M force based on Formosa (Taipei) made the first trans-oceanic bombing raid in history, flying 2,010km (1,250 miles) each way to Hangkow and Kwangteh. On 10 December 1941 G3Ms sank the British capital ships *Prince of Wales* and *Repulse* at a range the Royal Navy thought impossible. On this occasion they were joined by G4Ms, and for this bomber, the most important of the Imperial Navy, the problem of extreme range was so acute that it ended up being extremely vulnerable (Allied pilots called the G4M 'the one-shot lighter' because of its habit of bursting into flame when hit). Japan did build much larger four-engined bombers, which it so obviously needed, but never put any into service. And in the closing stages of the war its problems were so impossible that it not only began mass use of all kinds of aircraft (not only bombers) in bomb-laden suicide attacks but also produced special aircraft for this role. The MXY-7 Ohka (Cherry Blossom) was a true manned missile, released from a G4M and steered at an Allied ship, the pilot igniting three rocket motors to boost the speed in the terminal dive to 927km/h (576mph). The warhead was a formidable 1,200kg (2,646lb) charge. The Nakajima Ki-115 Tsurugi (Sabre) was a cheap-piston-engined machine able to carry an 800kg (1,764lb) bomb recessed under the belly. Most observers judge such 'bombers' to be aberrations resulting from unique circumstances; other nations were concerned with improving the *safety* of aircrews.

While the US Army painfully developed the B-17

and its later partner the B-24 Liberator, Britain built a comprehensive range of two- and four-engined bombers (far too many, in fact) without a clear idea of how they should be used. The best of the early twins was the Vickers Wellington, designed for 980hp Pegasus engines but offering far better capability with the 1,600hp sleeve-valve Hercules and remaining in production with the latter engine until after the end of the war when the total of 11,461 exceeded that of any other bomber in Europe (nearly 15,000 Ju 88s were built, but 40 per cent were not bombers). Though Wellingtons, despite their 'geodetic' basketwork structure with fabric covering, flew over 180,000 missions in World War II and served in Coastal and Transport commands, the RAF had by 1940 decided to concentrate on four-engined heavy bombers for its main strategic attacks. There were three types, the Short Stirling, Handley Page Halifax and Avro Lancaster. The Stirling was foolishly restricted to a span of under 30m (100ft) and this made it a very poor performer at altitude. Like the Halifax it had bomb bays divided into numerous small cells which could not accommodate the large light-case bombs of 1,814kg (4,000lb) and multiples thereof which had not been considered when the aircraft were designed. But the Lancaster, a four-engined derivative of the unsuccessful twin-engined Manchester, could carry all the new bombs and, in addition, had excellent performance at all altitudes and was simpler to make and maintain.

Bomber Command's early missions in 1939 were by unescorted small groups in daylight, and suffered such casualties that daylight raids became very rare. But night raids failed to find targets until in 1941–43 electronics came to the rescue with the Gee and Oboe precision navaids, and H$_2$S radar to paint a picture of the ground beneath. The power-aimed turret of RAF bombers carried multiple rifle-calibre guns, but could be aimed only ahead, above and to the sides and rear. Luftwaffe night fighters learned to home on the various radio emissions from the bombers and attack from below, and losses became increasingly heavy. Likewise the USAAF heavy bombers suffered severe losses on their daylight attacks, despite being much more heavily armed than the RAF aircraft with 12 or 13 guns of 12·7mm (0·5in) calibre. This heavy armament, with three tonnes of ammunition for prolonged running battles, inevitably reduced bomb loads to about two tonnes on typical long-range attacks, compared with more than five for most RAF heavies. At the same time, the RAF bombers could not have flown the

day missions at all without crippling losses, and in any case seldom attempted to put their bombs on particular factories.

By far the most advanced bomber of World War II was the B-29 Superfortress, begun in 1938 with exceptionally farsighted vision at a time when the US Congress would not even keep funding the B-17. In 1939–40 the Army Air Corps boldly planned a gigantic nationwide manufacturing programme, though the first XB-29 did not even fly until September 1942. In every respect the B-29 thrust ahead into new areas. It was defended by electrically driven turrets armed with heavy machine guns and cannon, aimed from manned sighting stations via a complex control system which enabled gunners to control whichever turrets they most needed; the turrets themselves were exceptionally small and offered little drag. All crew compartments were fully pressurized, for comfort and unimpaired efficiency at cruising heights of not less than 10,000m (32,000ft). The wing was more advanced than any flown previously, with aspect ratio of 11·5 (a very slender shape for long-range efficiency) and the almost unprecedented loading of 398kg/sq m (81·1lb/sq ft), at a time when half as much was considered dangerously high. Inevitably the B-29 needed long and strong runways, and it did more than most aircraft to educate aircrew and airport planners to the modern long-runway era. Thickness of wing-skin was more than double anything previously attempted. Fuel capacity was by any previous standard enormous. To fly the extremely weighty aircraft four of the most powerful engines available, the 2,200hp Wright R-3350 Duplex Cyclone, were equipped with two turbochargers each and fitted into nacelles the size of a fighter.

By combining resources of many companies B-29 production got into its stride in 1944 and by VJ-day deliveries totalled 3,667. Subsequently the B-29 was developed into the even more powerful and efficient B-50, while in 1946 Convair flew the world's biggest-ever bomber, the gigantic B-36, designed in 1941 to bomb Germany after the possible defeat of Britain. Originally powered by six 3,000hp Pratt & Whitney R-4360 Wasp Major piston engines, the B-36 was in 1950 boosted by the addition of pods under the outer wings each containing two J47 turbojets, which added speed and height. Vulnerability was still a problem, and protracted experiments were made with parasite fighters carried inside the bomb bay. McDonnell made the grotesque little XF-85 Goblin fighter to defend the B-36, but difficulties were such that the idea never became operational. The F-84F

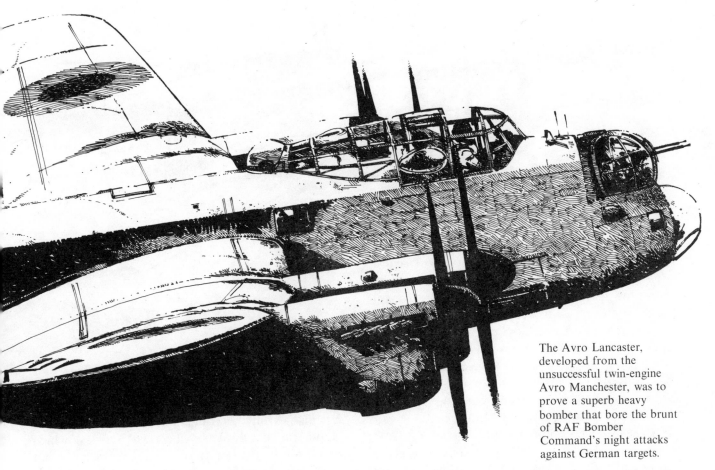

The Avro Lancaster,
developed from the
unsuccessful twin-engine
Avro Manchester, was to
prove a superb heavy
bomber that bore the brunt
of RAF Bomber
Command's night attacks
against German targets.

fighter-reconnaissance aircraft was also hung under
the B-36, and towed from the wingtips, in further
attempts to reduce the monster's vulnerability by
allowing it to stand-off at a safer distance from its
targets.

By the end of World War II it was obvious the
future lay with the gas-turbine engine. Germany had
actually built jet bombers and got them into action
in the shape of the purpose-designed Ar 234B and
the Hitler-inspired Me 262A-2a. Both of these were
relatively small single-seaters, but they posed problems to the Allies because of their high flight
performance. By 1945 a whole range of much larger
jet bombers were on the drawing boards in the
United States. The conventional North American B-
45 Tornado entered service with the newly created
USAF in 1948, and was followed by the dramatically unconventional Boeing B-47 Stratojet. This
introduced swept wings and tail, podded engines
and many other changes among which were bicycle
landing gears on the centreline and a radar-directed
tail turret. Over 2,000 of these impressive bomber,
reconnaissance and electronic-warfare aircraft were
delivered, in what in monetary terms was probably
the largest manufacturing programme in history to

that time. The B-47 also briefly introduced another
concept, the strategic stand-off missile. In 1943 the
Luftwaffe had used two types of radio-guided
missile, the Hs 293 and FX 1400, in attacks on ships
and other important targets. Other nations did not
develop such weapons until much later, and by the
1950s the need was not only to hit the target
accurately but also to allow the bomber to stand off
outside the heavy target defences.

This new requirement caused cancellation of
Britain's first precision-guided bomb, the Blue Boar,
and its replacement in 1953 by the long-range
winged Blue Steel. Until a few years previously it
had been believed that a bomber flying at 16,000m
(52,500ft) at Mach 0·9 (0·9 times the local speed of
sound) would be safe from AA guns and almost
immune to interception, and three types of 'V-
bomber', the Valiant, Vulcan and Victor, were built
to drop free-fall nuclear and conventional bombs.
By 1953 it was clear that a stand-off missile such as
Blue Steel would be needed to help deliver nuclear
warheads to heavily defended targets, though the
quest for greater speed and height persisted. New
versions of the Vulcan and Victor were produced
with greater power and larger wings. In the United

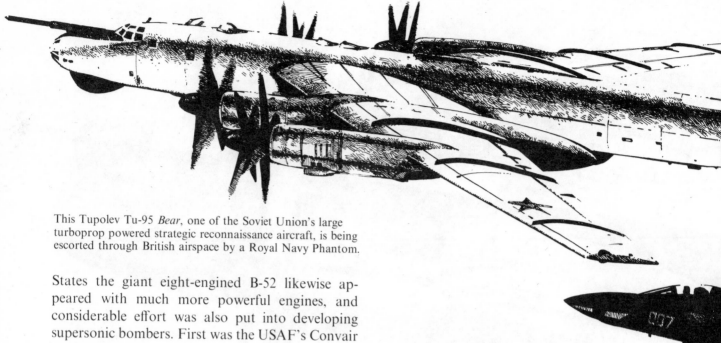

This Tupolev Tu-95 *Bear*, one of the Soviet Union's large turboprop powered strategic reconnaissance aircraft, is being escorted through British airspace by a Royal Navy Phantom.

States the giant eight-engined B-52 likewise appeared with much more powerful engines, and considerable effort was also put into developing supersonic bombers. First was the USAF's Convair B-58 Hustler, an outstanding technical achievement, able to fly over 8,000km (5,000 miles) without air refuelling and to exceed Mach 2. France built the much smaller, shorter-ranged Mirage IVA.

In the Soviet Union one of a series of large strategic bombers from the Tupolev bureau was the Tu-95, called Tu-20 by the DA (long-range aviation force), in which the apparent impossibility of securing adequate range with turbojets was answered by using turboprops. A very large aircraft with swept wings and tail, it has four turboprops of 15,000hp each, and after more than 20 years is still in service for strategic reconnaissance and electronic duties. Like other Soviet bombers it once carried a large stand-off missile, of aircraft-type configuration. Gradually improvements in aircraft and engines led to strategic jet bombers, and in the 1960s to long-range aircraft not much larger than fighters. The gigantic American Mach 3 XB-70 Valkyrie was cancelled, the last of the high-speed high-altitude bombers. To penetrate defended airspace the bomber had to hug the ground, to try to evade detection and accurate plotting by enemy radar. Though the XB-70, like the B-58, was good on low-level missions it was inferior to aircraft designed specifically for them.

After more than a decade of study the Rockwell (North American) B-1 was finally flown in 1974 as the next bomber for the USAF. The gap of more than 20 years since the B-52 was the longest in history between successive generations of aircraft, and the B-1 was designed for specific purposes. Its main need was to carry a nuclear deterrent that could unfailingly strike back after a Russian attack by missiles on the United States. Such an attack could knock out America's own missiles, which dared not be fired on the mere appearance of what looked like Russian warheads on American radar screens. But the B-1, designed to start engines and take off within two or three minutes of an enemy attack being detected, could get away from its airfields into the safety of the sky. From there it could hit back, should the Russian attack prove to be real. The B-1 was designed solely for the low-level mission, flying intercontinental ranges at tree-top height with its variable-sweep wings at their most acute angle. Its final attack could be made with SRAM, ALCM or Tomahawk missiles fired at great distances from the targets.

In the summer of 1977 the US Government terminated the B-1 as a bomber, giving as its reason the development of cruise missiles. No explanation was given of why the cruise missile had suddenly become of such interest; such weapons had been flown in World War I and in World War II the so-called 'V-1' had been a major weapon. The modern cruise missile is more sophisticated and can take evasive action, shelter behind ECM (electronic countermeasures) and try to protect itself with various decoys against missiles sent to shoot it down. But in theory it should be much easier to shoot down than was the V-1 in 1944, and in any case no deterrent philosophy could be based on

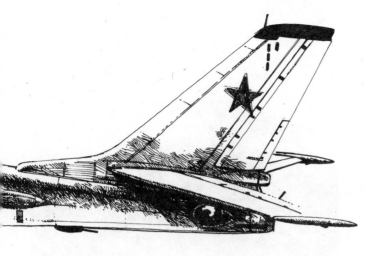

cruise missiles without aircraft to carry them and, despite enormous expenditure, the 16- to 20-year-old B-52 force will be hard-pressed to soldier on into the late 1980s.

Increasingly the missile is being left to handle the strategic delivery of nuclear warheads, launched from either submarines or from land launchers which must be either so 'hardened' (protected against attack) as to be invulnerable, which is impossible, or capable of moving over great distances and thus very difficult to pin down to a particular time and place. The only remaining bombers are used chiefly for strategic reconnaissance, mostly of an electronic nature and often at very low altitudes. Modern air-defence systems linking radars, missiles and computers, are so quick-reacting and deadly that the traditional type of bomber mission is almost suicidal unless the bomber has quite exceptional 'defensive electronics' able to confuse and mislead enemy defence systems.

All modern bombers are packed with such defensive electronics, which were designed into the B-1 and are an integral part of the latest Soviet strategic platform, the Tupolev Tu-26, called 'Backfire' by

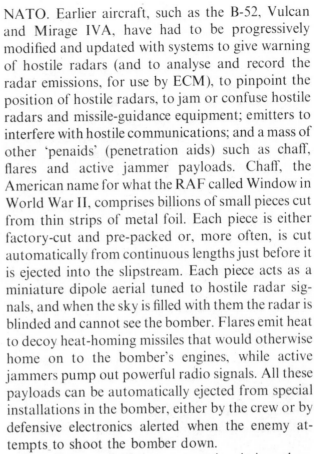

NATO. Earlier aircraft, such as the B-52, Vulcan and Mirage IVA, have had to be progressively modified and updated with systems to give warning of hostile radars (and to analyse and record the radar emissions, for use by ECM), to pinpoint the position of hostile radars, to jam or confuse hostile radars and missile-guidance equipment; emitters to interfere with hostile communications; and a mass of other 'penaids' (penetration aids) such as chaff, flares and active jammer payloads. Chaff, the American name for what the RAF called Window in World War II, comprises billions of small pieces cut from thin strips of metal foil. Each piece is either factory-cut and pre-packed or, more often, is cut automatically from continuous lengths just before it is ejected into the slipstream. Each piece acts as a miniature dipole aerial tuned to hostile radar signals, and when the sky is filled with them the radar is blinded and cannot see the bomber. Flares emit heat to decoy heat-homing missiles that would otherwise home on to the bomber's engines, while active jammers pump out powerful radio signals. All these payloads can be automatically ejected from special installations in the bomber, either by the crew or by defensive electronics alerted when the enemy attempts to shoot the bomber down.

For all except the longest strategic missions there is no need for purpose-designed bombers. Modern fighters can carry such heavy weapon loads that they can fly all the required attack missions, backed up by certain specialised close-support or anti-tank aircraft which can be as diverse in character as the Harrier, Thunderbolt II and anti-tank helicopters.

Fighter and Attack Aircraft

This section deals with the third type of warplane to be developed. First came the reconnaissance aircraft, then the bomber and third the fighter. Some of the earliest of this class were called fighting scouts, or just scouts, because they also often had to fly scouting (reconnaissance) missions in between dogfighting with the enemy. Another duty was bomber or reconnaissance-aircraft escort, and by the 1930s some fighters had been specially designed for long-range escort duties. Another name is the pursuit, but this was just the common American term for a fighting aeroplane until after World War II. Night fighters usually differed from other fighters only in detail until World War II when the invention of airborne radar led to a class of large multi-seat radar-equipped night fighters which would have been of little use in a daytime dogfight. The interceptor was at first a fast-climbing fighter to protect home targets, and today this term describes all fighters used in a defensive role. Today's interceptors all carry radar and are both day dogfighters and night fighters rolled into one. To add even further complexity, almost all modern fighters can carry many tonnes of bombs or missiles and may frequently be called upon to strike at surface targets.

A few far-sighted people thought through the possibility of air warfare several years before World War I and came up with proposals for fighters, to shoot down the aircraft belonging to enemy nations. Seven or eight inventors proposed 'synchronization gear' with which a machine gun could be interlinked with an aeroplane engine in such a way that it could safely fire straight ahead, the bullets passing between the revolving blades of the propeller. Others proposed pusher aircraft, in which the problem of mounting a machine gun disappeared. The early Wright Flyers were pushers, and it was one of these which in June 1912 may have been the first aircraft to complete official firing trials with a machine gun. Two months later the British FE.2 had a Maxim gun in the nose of its pusher nacelle, and in France two aircraft carried guns before 1912 was out, one of the guns being a shell-firing cannon. There arose a growing number of novel schemes for fitting aircraft with guns, not necessarily for use against other aircraft, but when World War I broke out there was no such thing as a fighter, a fighter pilot or any accepted doctrine for air combat.

Two-seaters soon began to carry such weapons as rifles, carbines and revolvers, fired by the observer, and on 5 October 1914, a French Voisin III shot down a German aircraft, the first aerial victory. The Voisin was not a fighter but an observation and bombing aircraft which happened to carry a machine-gun. This event spurred the already widespread attempts to produce an aeroplane specially suited to shooting down other aircraft. The ideal appeared to be a small single-seater able to fire a machine-gun straight ahead. The pusher machines had a decided advantage in this respect, because there was no propeller in the way. But it was still a new field in which most things had to be learned by painful experience. Capt de Havilland designed the D.H.2 as a true single-seat fighter but still provided the pilot with left and right mountings on which he could mount a Lewis machine-gun. His problems in a deadly combat can be imagined, for he had to work

Aircraft of 'pusher' configuration such as the FE.2b (*above*) were evolved to provide a clear field of fire for a machine-gun mounted in the bow position. *Top right:* The French Morane-Saulnier Type-L saw early use as a bomber.

the rudder pedals, control column, throttle, aim the gun, work the trigger and change magazines, whilst looking through the gunsights.

Inventors, aircraft manufacturers and squadron pilots all sought the most effective armament. One aircraft had two rotary engines side by side with guns in the centre. A second had one engine with guns on each side, firing beyond the tips of the propeller blades. A third had guns inclined outwards to miss the propeller, but this was difficult to aim. A fourth had two rotary engines mounted fore-and-aft on each side of the nose, driving a central propeller through the centre of which fired a 37-mm cannon. A fifth had five Lewis guns fixed to the landing gear. A sixth had a 37-mm cannon, almost 3m (10ft) long, mounted obliquely, firing up through the centre of the upper wing, loaded by the observer in the rear cockpit. A BE.2c carried Lee Prieur rockets, another anti-aircraft darts and a third the frightening Fiery Grapnel device, a kind of explosive anchor dropped on enemy aircraft. Meanwhile a few crack shots,

such as Lanoe G. Hawker of the Royal Flying Corps, kept on shooting down enemies though armed with nothing but a single-shot Martini rifle.

Before 1914 was out several pushers had appeared with guns fixed to fire straight ahead, thus leaving the pilot free to fly the aircraft, and aiming the gun by aiming the whole aircraft. Later, in 1915, this seemingly obvious answer partly solved the problem of the British D.H.2. But the thread of development tha. led to the main stream of fighters began in France. Before the war began three French aircraft constructors, Deperdussin, Esnault-Pelterie and Saulnier, had all independently invented methods of safely firing a machine-gun through a revolving propeller disc. Saulnier's famed demonstration pilot, Roland Garros, joined the Aviation Militaire, but could get no official to show interest in Saulnier's invention. As a second-best solution he personally spent months developing and refining steel channel-shaped deflectors on his propeller blades, and from November 1914 until early 1915 overcame one difficulty after another. Eventually he received permission to try his deflectors in action, and after prolonged bad weather at last saw a German aircraft on 1 April 1915. Garros shot it down in flames, and he shot down more Germans on 15 and 18 April. Later on the 18th he went out to bomb Courtrai railway station but was shot down and captured.

Like the tank, an idea of the very greatest military importance had been allowed to fall into the enemy's hands prematurely, without any advantage having been taken of it except on an unimportant and purely local scale. The Germans quickly assessed Garros's deflectors and asked their Dutch aircraft constructor, Anthony Fokker, to copy it. Fokker did better: his German engineers designed an interrupter gear which linked the engine and a fixed Parabellum machine-gun so that the latter could fire only when there was no propeller blade in front of the gun barrel. In three days Fokker had developed, tested and perfected an installation for a forward-firing gun on one of his M.5 monoplanes, generally known as E-types from *eindecker*, German for monoplane. By chance the Fokker monoplane was a direct copy of a pre-war Morane-Saulnier type, the Model H, which could in fact have been in service, with Saulnier's own gun-firing gear, before the outbreak of war.

What happened next was the first phase of organised air warfare, and it was a particularly bloody one. The little Fokker was just sufficiently faster and more manoeuvrable than the early aeroplanes of the Allies to have the latter completely at its mercy. Moreover, some of the pilots of the German Idflieg (Imperial Air Force) were outstandingly gifted and far-sighted officers, who devoted their working hours not only to seeking out the enemy but also to working out how best the new breed of fighting aeroplanes should be used. Oswald Boelcke was the leader, and his chief assistant was Max Immelmann (though the first actually to shoot down an Allied aircraft was Kurt Wintgens). Using their little 100hp monoplanes they worked out much of the basis of air tactics, deflection shooting, how to work in pairs with a wingman, how to avoid being 'jumped' from behind and many other totally new ideas. Usually the Fokker E-types had a single Spandau (Parabellum 08) on the centreline, but a few had two guns and Immelmann once tried three. Some pilots fitted a headrest to assist in accurate aiming, Garros having fitted a forehead rest in front (which in a crash might have killed him).

So heavy were Allied casualties that newspapers wrote of 'the Fokker Scourge', calling Allied pilots 'Fokker fodder'. Gradually it was realised the little Fokker was really quite primitive. Even the D.H.2 and FE.2b could beat it, and when such purpose-designed Allied fighters as the Nieuports, the Sopwith Pup, Bristol Scout, Hanriot and Spad designs came into use the Fokker swiftly faded. In its place Germany uniformly relied upon biplanes with six-cylinder water-cooled engines, and fighters of this layout dominated the German and Austro-Hungarian air forces for the remainder of the war. But at the end of 1916 the Sopwith company produced a triplane (three superimposed wings) which did well in combat. It caused such a stir on the German side that copies were immediately started, and the Fokker Triplane, the Dr.I, was actually made in greater quantity than the British machine. Triplanes were at least equal to contemporary biplanes with regards to climb and manoeuvrability, and the greatest ace of the war, Baron Manfred von Richthofen, was flying one when he was killed in April 1918. But they were only a temporary phenomenon.

The most successful fighter of World War I was the British Sopwith Camel, which shot down over 3,000 enemy aircraft (until 1978 the figure was thought to be 1,294, and even that was more than the score achieved by any other fighter). The Camel was a tricky biplane, rather short and hunchbacked and powered by a 130hp rotary engine over which fired two Vickers guns in most versions. But the old-fashioned rotary engines were having their final fling, and were incapable of being developed to give the powers of 300 or 400hp that were being de-manded. One of the best Allied engines was the Hispano-Suiza of 250–300hp. and this was used in a superb British fighter, the SE.5a, but its unreli-ability, especially when fitted with a reduction gear to the propeller, caused severe delays and casualties. Greater power meant higher speed and more rapid climb to greater altitudes, and instead of staggering off the ground with one gun, and being sluggish with a big pilot, fighters were by late 1917 climbing purposefully to over 6,000m (20,000ft) with two

guns and a heavy load of ammunition. Many of the best fighters were two-seaters. The first was the versatile but curiously named Sopwith 1½-strutter, the first aircraft to have a properly thought-out scheme providing for synchronized fixed guns for the pilot, rear guns aimed by the observer and bombs. Upgraded to twice the horsepower, the Bristol Fighter of late 1916 was an outstanding type which could behave as a single-seater at the front while the observer shot down additional aircraft from the rear.

In the immediate post-war years there were few changes in fighter design, apart from elimination of rotary engines and general consolidation on more reliable engines of 400hp and above. Two fixed machine guns remained common armament, and though most of the guns were reasonably reliable 'stoppages' of various kinds were still common. It was therefore almost universal to mount the guns close enough to the cockpit for the pilot to reach them and, using a mallet if necessary, clear the offending round, re-cock the gun and rejoin combat. Having to clear a stoppage during combat was often fatal, but it was still important to be able to attempt it.

From 1916 fighters had also increasingly to take on various missions against ground targets, and these often became accepted fighter duties. One was close support of friendly troops, using the fixed guns against hostile land forces. Several important types of ground-attack fighter were built, notable for their heavy frontal firepower and considerable armour protection, which of course had to be gained at the expense of reduced performance. In Germany the Junkers company built both cantilever monoplane and cantilever biplane attack aircraft of all-metal construction, known by such names as 'Tin Monkey' and 'The Furniture Van' but nevertheless strong enough to stand up to severe front-line missions. Between the two world wars many fighters had to fly ground-attack missions, both in major wars and in policing colonial empires. In the latter duty flight performance became relatively unim-portant, the dominant requirements being supreme reliability and versatility when operating in remote areas in possibly very harsh climates.

Two new factors in the 1920s were, first, the emergence of naval fighters and, second, the emer-gence of new air forces. The mighty United States had been unprepared for war in 1917 and filled its air squadrons with types bought, borrowed or given by France and Britain, and later made under licence. But in the 1920s its growing home industry began to flex its design muscles and produced increasingly competitive fighters, at first with water-cooled in-line engines and then almost exclusively with air-cooled radials. Almost all had all-metal structures, and after 1930 the American designers led the trend towards stressed-skin metal covering. The American Browning machine gun also existed in two sharply differing sizes, a 7·62-mm (0·30in) and 12·7-mm (0·5in), and the bigger gun increasingly began to supplement or replace the rifle-calibre weapon. A common American fighter armament was one gun of each calibre.

Japan at first relied totally upon imported designs and designers, but after 1930 was sufficiently confident to produce totally original fighters which were at first water-cooled biplanes and then, from 1935, air-cooled radial monoplanes. Like Italy, Japan willingly sacrificed almost everything to the attainment of the highest possible manoeuvrability. The two leading fighters of the late 1930s, made in very large numbers, were the Army Ki-27 and Navy A5M. Very alike in design, they had just two rifle-

calibre guns, and were lightly built, but could make a tighter turn inside every foreign fighter they met – even the Soviet 1-15 biplane.

Russian designers tried more contrasting designs of fighters than those of any other nation. Water and air cooling, single and twin engines, biplanes and monoplanes and a fantastic variety of armament schemes were all to be found on Russian fighters of the 1930s. The Russian gun designers produced the 7·62-mm ShKAS and later the 12·7-mm Beresin and 20-mm VYa, all of them guns of outstanding quality and exceptional performance. Despite this extensive experiments were made with large-calibre recoilless weapons, usually of 75mm (2·95in) calibre, which probably owed much to the recoilless Davis and the Vickers (Crayford) rocket guns used to a limited extent in the larger fighters of World War I. Russian designers also produced reliable and accurate rocket projectiles which are very effective against armoured vehicles and similar 'hard' targets, and though these were used in large numbers at public manoeuvres from 1935 onwards they still managed to surprise

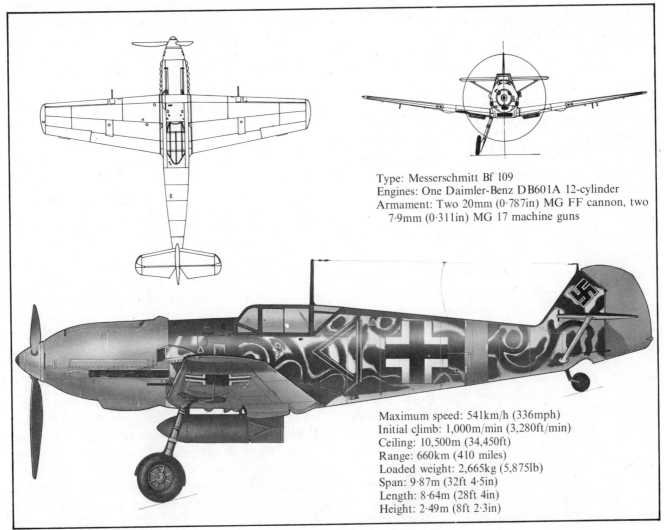

Type: Messerschmitt Bf 109
Engines: One Daimler-Benz DB601A 12-cylinder
Armament: Two 20mm (0·787in) MG FF cannon, two 7·9mm (0·311in) MG 17 machine guns

Maximum speed: 541km/h (336mph)
Initial climb: 1,000m/min (3,280ft/min)
Ceiling: 10,500m (34,450ft)
Range: 660km (410 miles)
Loaded weight: 2,665kg (5,875lb)
Span: 9·87m (32ft 4·5in)
Length: 8·64m (28ft 4in)
Height: 2·49m (8ft 2·3in)

both Allies and enemy in World War II.

Many fighter pilots continued to insist long after 1935 that a fighter had to be a biplane with an open cockpit, in order to offer the best possible all-round view and the greatest possible manoeuvrability. Many countries, especially the Soviet Union, conducted prolonged comparative trials, and the Russians deliberately kept the I-15, I-15bis and I-153 in production long after the I-16 monoplane was in service in numbers. The I-16 was the first cantilever monoplane fighter with retractable landing gear to go into service with any air force, and though tricky and only marginally stable longitudinally, its outstanding speed and firepower made it a formidable opponent (though in the Spanish Civil War it was badly flown because of rigid Russian political control). In fighting against Japan, however, the agile Japanese monoplanes beat the I-16 and put fresh impetus behind advanced I-15 biplanes such as the 1,100hp I-153 with retractable landing gear. The Russians even built a biplane fighter which could retract the lower wing into the upper! (During World War II Britain built a version of the monoplane Hurricane with a biplane upper wing which could be jettisoned if necessary.)

By 1934 the British Air Ministry had satisfied itself that in a future war the time available for shooting down enemy bombers might be as little as two seconds of accurate shooting, and the belief that an average of 250 rifle-calibre bullets would be needed naturally led to a search for armament capable of scoring 250 hits in two seconds. (Curiously, no calculations appear to have been done on the effect of different larger calibres, such as the 20mm cannon that was finally adopted.) The answer, translated into about 8,000 shots per minute, appeared to be eight machine-guns. The Vickers that had served the RFC and RAF so faithfully was basically a 19th-century gun, and after looking for a more modern weapon the British chose the American Browning (which was only slightly later in origin but far more reliable). New fighters with eight Brownings of 7·7mm (0·303in) calibre were going to need at least 1,000hp, and several suitable engines were available; British designers mainly chose the liquid-cooled Rolls-Royce Merlin, which later proved eminently suitable for development to give greater power at all altitudes. Around this engine Hawker Aircraft produced the Hurricane, with its eight guns in compact groups. This fighter, flown in 1935, was a traditional fabric-skinned machine of rather large dimensions, but it was tough, manoeuvrable and available in large

numbers in the nation's dire hour of need in 1939–40. Supermarine produced the Spitfire, considerably smaller and of stressed-skin construction. Few of these were available in 1939, but it was to become by far the most important British fighter of World War II.

Germany's leading World War II fighter had a most inauspicious beginning, its designer, Willi Messerschmitt, being unpopular with the Nazis and totally discounted as a fighter designer. His Bf 109, flown in 1935, was even smaller than a Spitfire, and likewise a modern stressed-skin machine. Fitted with powerful slats and flaps it managed to use a wing far smaller than those of the British fighters, and it was continuously developed until 1945 with more powerful engines and devastating armament. But, unlike the Spitfire, later Bf 109 versions of the G series (popularly called Gustav) were even more full of shortcomings than the early models had been. At low speeds excellent, the controls tightened up with increasing speed until at really high speeds the manoeuvrability was very poor indeed, further handicapped by the attitude of the pilot and difficulty of applying large forces to the control column to roll the aircraft. The best feature of the 109 was its weaponry, which included outstanding guns of 7·92, 13, 15, 20 and 30mm calibre. In many versions a shell-firing cannon was mounted on the engine to fire through the hub of the propeller, an arrangement pioneered by the Hispano-Suiza company from World War I onwards. From 1942 most versions of 109 could carry *Rüstsätze* (field modification kits) for a wide variety of different extra guns and other devices, such as rocket launchers or extra fuel tanks which could be jettisoned when empty (so-called drop tanks). This fighter was also one of the first to be used as a bomber.

In fact the chief pioneers of the fighter-bomber were the US Navy, as explained in the section on carrier-based aircraft. On land airfields the pressure to make one aircraft do two or more jobs may have been less intense, and until well into World War II there was a general rather cosy belief that first-line land combat aircraft could be divided into specialised bombers and specialised fighters. A few of the new breed of monoplane bombers were at first thought so fast and manoeuverable as to be capable of fighting other aircraft, but actual experience soon swept away this notion. There remained, however, the notion of the twin-engined long-range escort fighter, while the reborn Luftwaffe became polarised around the *Zerstörer* (destroyer), which was a similar machine but used not so much to escort bombers as to penetrate enemy airspace and destroy aerial opposition. It was recognised that large twin-engined machines could not manoeuvre as tightly or quickly as small single-engined types, but they were expected to compensate for this by their heavy armament and good all-round performance.

Many nations built these big fighters, which all had two engines and a crew of at least two. Two of the earliest were the Bristol Bagshot and Westland Westbury, built in Britain in 1927, mainly for shooting down bombers. These big multi-seat machines each carried two of the enormous Coventry Ordnance Works 37mm cannon, aimed by hand or fixed at an oblique angle and aimed by the pilot from below and behind the target. (This method of shooting was extensively explored in Britain in the late 1920s, but was never used operationally by British aircraft in World War II – though it was by her enemies.) These lumbering machines were aberrations, as were the succession of 'multiplace de combat' (multi-seat warplanes) built in France which were supposed to be large bombers that were also fighters. By 1935 many countries were building small, powerful twin-engined aircraft as fast as the other new fighters, but Britain showed no interest until near the start of World War II when, suddenly recognizing the omission of a long-range fighter, urgent conversions were made of the Blenheim bomber which proved grossly deficient in speed, manoeuvrability and firepower. The purpose-designed twin-engined fighters included the Bf 110, Fw 187, Potez 63, SE 100, Breguet 690, Fokker G.I, PZL Wilk, Lockheed P-38 and Bell XFM-1. The last-named was the strangest of all, for it had the new nosewheel landing gear, two pusher engines in large over-wing nacelles and a gunner with a 37mm cannon in the nose of each nacelle. The P-38 was also unusual in having the tail carried on twin booms, mainly in order to find somewhere to put the rear wheels of the nosewheel landing gear, the engine cooling radiators and the turbochargers to give power at altitude. The armament of one 37mm cannon, two 12·7mm guns and two rifle-calibre guns was grouped in the nose ahead of the single seat. Britain's Westland Whirlwind also had two engines and guns in the nose – the very heavy armament of four 20mm cannon – but this machine had low-powered and unreliable engines and accomplished little.

As related later, in the Battle of Britain in 1940 the Bf 110 came up against the RAF's Hurricanes and Spitfires, and quickly suffered heavy casualties. Within two weeks it was recognised by the Luftwaffe that their big twin-engined fighter, the pride and élite of the nation, could not even survive in combat with modern single-engined fighters flown with skill and determination, and for the rest of World War II the Bf 110 was never again used for dogfighting in daylight except in the few theatres where opposition was weak. Another shock, this time to the RAF, was the realization that a fighter with its guns in a power-driven turret was if anything even easier meat. The RAF had devoted immense effort to turrets in the

De Havilland's 'wooden wonder', the superb Mosquito, was designed as a bomber of such high performance that it needed no defensive armament. This level of performance enabled it to be used in a wide variety of roles.

late 1930s, and by 1940 not only had large numbers of Defiant turret-armed fighters coming into service but also planned a series of bigger fighters with monster turrets armed with four 20mm cannon. When it met the Luftwaffe the Defiant was apparently often mistaken for a Hurricane and attacked from above and behind, with disastrous results for the enemy. When it was recognised however, it was attacked from below or ahead, and as it could not rival the Bf 109's performance and manoeuvrability it was shot down. None of the projected turret fighters entered service, though later the US Army P-61 night fighter often carried a turret – however, the results of this change did not come up to the original expectations.

In 1940 the fighting in Europe caused American designers to tear up their fighter plans and instead double the horsepower. Vought had already built the Corsair XF4U prototype for the US Navy, with the 2,000hp Double Wasp engine swinging a four-

Lockheed's P-38 Lightning was first evolved to meet a US Army requirement for an interceptor with an unprecedented rate of climb, namely 6,100m (20,000ft) in six minutes. It was later developed as an outstanding escort fighter.

blade propeller larger than any previously fitted to a fighter. Despite its great size this prototype in 1940 became the first American aircraft – even including racers – to reach 644km/h (400mph). Republic built the P-47 Thunderbolt and Grumman followed in 1942 with the F6F Hellcat, all using the Double Wasp engine which grew to 2,300, 2,500 and finally 2,800hp in war emergency condition with water injection. With such power these large fighters could outfly their opponents despite having heavy fuel loads, heavy armament and considerable armour protection. They were produced in great numbers; there were over 12,000 Hellcats, over 12,000 Corsairs and over 15,000 Thunderbolts.

At first their enemies were much smaller and lighter. The Japanese Navy mainstay was the

Mitsubishi A6M, popularly called the Zero, and despite having barely 1,000hp it could outfly all its rivals in 1941–42 and destroy them with cannon and machine guns. A special quality of the A6M was its very long range, which enabled it to conquer vast areas of the Pacific and south-east Asia even when operating from very distant airstrips. Only gradually was it realised that, far from being invincible, it was actually deficient in flight performance (compared with the less-obsolete fighters the Allies eventually supplied to the Pacific theatre) and ill-protected, so that it was easy to shoot down. The corresponding Army fighter, the Nakajima Ki-43, was even more manoeuvrable but totally lacking in everything else and usually carried only two machine guns. By 1942 most Allied fighters either had six 12·7mm or four 20mm guns, which could blow the Japanese fighters apart. Gradually the Japanese saw the need for a complete rethink. A generation of fighters totally unlike their predecessors, with small wings and everything sacrificed to performance and firepower (examples were the Ki-44 and J2M) were unimpressive, and only near the end of the war were such excellent machines as the Ki-84 and Ki-100 produced – and then it was too late.

In Europe the Luftwaffe produced ever-greater quantities of successive versions of the Bf 109, despite the fact it had in 1941 introduced a vastly superior machine of completely new design, the Fw 190. This was one of the most outstanding fighters of the war, with unsurpassed technology and engineering design, cleverly packaged with a 1,700hp radial engine into an extremely compact airframe. It had no deficiencies, and its fantastic capability is shown by the ability of later versions to carry torpedoes and bombs weighing up to 1,800kg (3,968lb). By 1943 the Fw 190 was by far the chief tactical attack machine of the Luftwaffe, replacing the Ju 87 'Stuka' dive bomber and serving on all European fronts in large numbers. Like many of today's attack aircraft

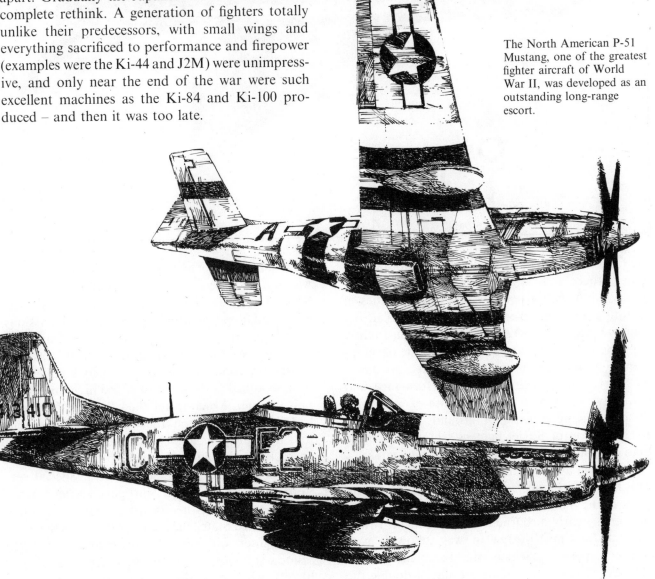

The North American P-51 Mustang, one of the greatest fighter aircraft of World War II, was developed as an outstanding long-range escort.

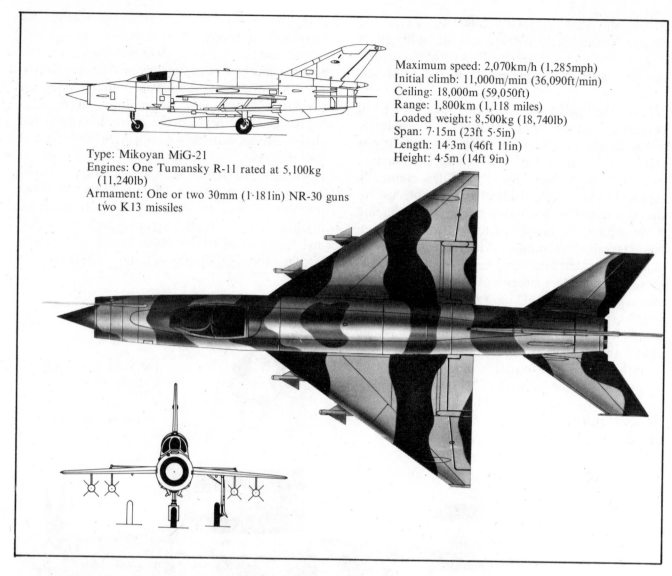

Type: Mikoyan MiG-21
Engines: One Tumansky R-11 rated at 5,100kg
  (11,240lb)
Armament: One or two 30mm (1·181in) NR-30 guns
  two K13 missiles

Maximum speed: 2,070km/h (1,285mph)
Initial climb: 11,000m/min (36,090ft/min)
Ceiling: 18,000m (59,050ft)
Range: 1,800km (1,118 miles)
Loaded weight: 8,500kg (18,740lb)
Span: 7·15m (23ft 5·5in)
Length: 14·3m (46ft 11in)
Height: 4·5m (14ft 9in)

the Fw 190 served as an offensive carrier of bombs, mines, rockets, heavy cannon and even guided missiles, but with absolutely first-class qualities in the quite different role of dogfighter. Russian designers rated the Fw 190 highly. They consistently kept their own fighters small, despite the development of more powerful engines, and put every ounce of effort into achieving greater performance and manoeuvrability even at the expense of firepower. Despite having engines of around 1,600hp, the 1943–44 crop of Soviet fighters seldom had more than one cannon and two heavy machine guns.

One thing all Russian fighters lacked was radar. This had been pioneered in Britain, and the saving grace of the Blenheim fighter was that it could carry the clumsy early airborne radar equipment. In the late fall (autumn) of 1940 it was followed by a greatly superior machine, the Beaufighter, with Hercules sleeve-valve engines of almost double the power and the devastating armament of four 20mm

cannon and six machine guns. This became the most widely used Allied night fighter, with the RAF and USAAF, and though a little on the slow side its range and firepower fitted it for long attack missions with guns, rockets and torpedo in all European theatres. Another outstanding aircraft was the Mosquito, conceived as an unarmed bomber but soon modified as the greatest Allied night and long-range fighter. Much faster than the Beaufighter, it had four 20mm cannon (the first fighter version had four machine guns as well) and also carried bombs or rockets with enough range to hit the furthest areas in the Balkans or Scandinavia. Special versions carried a 57mm (2·24in) gun, various radars and extended wing-tips for use at extreme altitude. In the United States the complex P-61 was the main night fighter, with powerful radar and four 20mm cannon (often plus four heavy machine guns in a turret), while even the small carrier-based single-seaters often carried new radars which operated

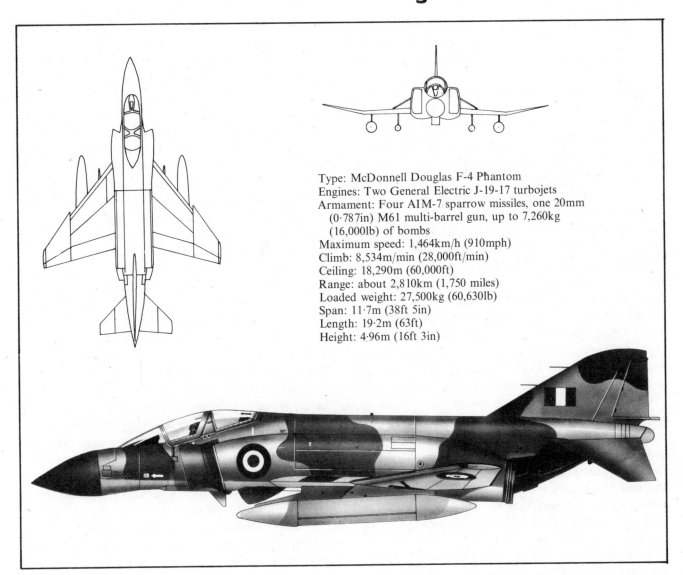

Type: McDonnell Douglas F-4 Phantom
Engines: Two General Electric J-19-17 turbojets
Armament: Four AIM-7 sparrow missiles, one 20mm
  (0·787in) M61 multi-barrel gun, up to 7,260kg
  (16,000lb) of bombs
Maximum speed: 1,464km/h (910mph)
Climb: 8,534m/min (28,000ft/min)
Ceiling: 18,290m (60,000ft)
Range: about 2,810km (1,750 miles)
Loaded weight: 27,500kg (60,630lb)
Span: 11·7m (38ft 5in)
Length: 19·2m (63ft)
Height: 4·96m (16ft 3in)

on very short wavelength.

Luftwaffe night fighters were at first modified Bf 110, Do 17 and Ju 88 aircraft, later joined by the big Do 217. Despite being burdened by cumbersome aerial arrays the later night-fighter Bf 110 and Ju 88 versions were able to inflict extremely heavy losses on the RAF bombers, largely because the latter did all they could to help; they broadcast radar signals continuously, so that night fighters could home on the bombers automatically, and were totally devoid not only of defensive armament underneath but lacked even a window that could look down and to the rear. Increasingly the night fighters were armed with oblique upward-firing guns, basically like those tried by the RAF 15 years previously, which offered a perfect no-deflection shot. Purpose-designed night fighters generally failed to see service, the exception being the outstanding He 219 Uhu (owl).

By mid-1944 jet propulsion was fast opening up new horizons in aircraft performance, and especially relevant to fighters. First to reach the squadrons was the radical German Me 163 Komet, which was a tailless rocket interceptor with very short endurance. Though superb to fly (much better than a Bf 109G) it was designed to take off from a jettisoned trolley and land on a skid, and this, coupled with the danger of its highly reactive fuels, made it exceedingly tricky and dangerous. The bigger Me 262 was an even more formidable aircraft, a conventional twin-jet with 900kg (1,984lb) thrust Jumo 004 engines and armament of four 30mm (1·18in) cannon, the heaviest then flown. Roughly 161km/h (100mph) faster than Allied fighters, the 262 was impossible to catch unless it could be 'bounced' (taken by surprise) while taking off or landing. Near the end of the war in Europe the little He 162 was put into production as a 'volksjäger' (people's fighter) to be built at the rate of 4,000 a month and flown by the hastily trained Hitler Youth. Though potentially formidable, and far in advance of other nations in its concept, with multiple rocket launchers and X-4 wire-guided missiles backing up the heavy cannon, this desperate last-ditch measure merely showed Germany had lost the war.

In fact the first jet fighter actually to become operational with regular squadrons was the RAF's Gloster Meteor (616 sqn, July 1944), but few of these saw action. After the war the Meteor was given much more powerful engines, gaining a world speed record in full fighting trim, and also became a stop-gap night fighter. Far more significant was the American XP-86 (later F-86) Sabre. This was designed by North American Aviation, which in 1940 had been asked by the British to build the old Curtiss P-40 fighter and instead built a better fighter of their own, the Mustang. When fitted with the British Merlin engine the Mustang became the leading Allied fighter, combining outstanding performance at all heights with enough range to escort bombers from England to Berlin or Czechoslovakia. There were nearly 15,000 P-51 Mustangs, and they did more than any other aircraft to defeat the Luftwaffe even over the heart of Germany. It was natural to use it as the basis of a jet, but in fact the result (the Navy FJ-1 Fury) saw only limited production. The key to the XP-86 was swept wings. German documents captured in 1945 showed how speed could be increased, typically from 900 to 1,100km/h (560 to 680mph), by sweeping back the wings and tail like an arrowhead. The XP-86, flown in October 1947, was the first all-swept fighter.

In June 1950 war broke out when North Korea invaded South Korea, and soon various Allied fighters were in use helping the South Koreans. The jets were handicapped by their inability to use short rough airstrips or carry heavy bomb loads and fly long distances at low levels, because early turbojets burned fuel very rapidly. Such machines as the Mustang and carrier-based Skyraider and Sea Fury came back into prominence as attack aircraft, with bombs and rockets, and jet combats lagged until suddenly the North Koreans appeared with advanced swept-wing jets. These were the MiG-15s, quickly produced in the Soviet Union in 1947 as a result of a British Government free shipment of the latest jet engines. At once the tables were turned, and though the F-86 Sabre in improved versions achieved ascendency over the MiG-15 this was mainly because of better Allied pilot skill and aggressive tactics, allied with superior shooting. Subsequently the MiG bureau, which had taken a back seat in World War II, produced an outstanding succession of advanced fighters. The MiG-19 twin-jet was one of the first supersonic fighters, and its outstanding manoeuvrability makes it formidable in close combat to this day. The delta-winged MiG-21 has been built in enormous numbers as a small multi-role attack fighter. The much later MiG-23S is a night and all-weather interceptor with swing-wings, and the MiG-27 is the corresponding tactical attack version bristling with advanced sensors and weapons. The MiG-25 is the fastest fighter in service, with a speed of about 3,200km/h (2,000mph) and radar and long-range missiles for destroying aircraft at a distance; it is in no sense a dogfighter.

Of the profusion of new jet fighters built in the ten years after World War II the most innovatory were the interceptors of the US Air Force. These carried totally new radar fire-control systems by the Hughes Aircraft Company, in which powerful radars, auto-pilots and computers were linked to steer the aircraft automatically to shoot down intruders at night or in bad weather, without the pilot of the fighter even having to see the enemy. Previously fighters had invariably tried to get on the tail of their opponent, but these new interceptors used the collision-course technique from the side. The enemy aircraft could be seen better by the radar from the side, and the computer could steer the aircraft automatically so that it could shoot the enemy down, at first with guns, then with a battery of rockets and finally, from 1956, with air-to-air guided missiles.

These developments revolutionised fighter design and operation. By the mid-1950s it could be seen that there was little point in regarding the radar-equipped night fighter as a special class; all fighters were going to need radar. The last examples of the old-style night fighter were the gun-armed British Javelin and Russian Yak-25, both quite large twin-jets of the mid-1950s. By this time the US Air Force was receiving the Convair F-102 Delta Dagger, a supersonic delta with automatic fire-control and guided-missile armament. Since then the developments have merely been ones of degree, in particular (through advances in electronics) easing the work-load of the crew. Modern fighters tend to fall into one of two groups, the first being relatively large multi-engined machines such as the McDonnell Douglas F-15 Eagle and F-4 Phantom, and the second being the smallest single-engined aircraft that can carry the required radar and weapons, such as the F-104 of the 1950s, the Mirage of the 1960s and the F-16 of the 1970s. Thanks to sustained development of jet engines, notably the high-pressure afterburning turbofan, today's fighters no longer suffer from the limited endurance and weapon load of the Korean

War period. Even the smallest can use short airstrips, carry five tonnes or more of bombs or missiles and fly a four-hour mission. Equipment always includes an ejection seat, comprehensive ECM and other defensive decoys, and a flight-refuelling probe giving almost unlimited endurance.

Most of the missions flown by 'fighters' in the past 20 years have been against surface targets. In World War II a few special attack aircraft were built, such as the Il-2 Stormovik of the Russians and the Hs 129 of the Luftwaffe with poor flight performance but thick armour and a heavy gun for knocking out tanks. Such machines tended to be forgotten in the 1950s (though a few significant prototypes were built, such as the French Potez 75 with its pioneer wire-guided anti-tank missiles) but the simmering succession of so-called brushfire wars at last spurred the development of completely new kinds of tactical aircraft for use against battlefield surface targets. There could hardly be a more complete contrast than between the Lockheed AC-130, the Fairchild A-10 Thunderbolt II and the Rockwell OV-10 Bronco. Yet all are in a way 'tactical fighters' in that they are designed to fight in a land battle.

The AC-130 was a specially equipped version of the most widely used large turboprop freight transport, packed with guns, ammunition, radar, infrared, television and other sensors for finding hostile troops, trucks or armour at night and destroying them with withering fire. In a situation that completely turned the tables, it was not uncommon in the war in Vietnam for supersonic fighters, such as the F-100 Super Sabre, F-105 Thunderchief and F-4 Phantom, to spend their time finding and marking ground targets that would then be attacked by the lumbering AC-130s! The Thunderbolt II is a carefully designed battlefield weapon platform with the most powerful tank-killing gun ever fitted into an aircraft. Its speed is no faster than that of World War II fighters, and it relies largely on thick armour and the ability to accept severe damage. In return it offers more first-shot lethality than has ever before been built into an aeroplane, with its deadly gun backed up by heavy loads of precision-guided missiles. The OV-10 is an even slower turboprop machine able to dodge about over the battlefield and fire guns and rockets, guide 'smart' (laser-homing) bombs and even land and pick up casualties. The only tactical aircraft that are even more versatile, able to operate without a purpose-built airstrip, are the various versions of the Harrier, described in the section on carrier-based aircraft and in the Rotorcraft and V/STOL chapter.

Anti-Submarine Warfare and Maritime Aircraft

In World War I the existence of naval air arms as separate services in almost all the warring powers prompted the development of air war at sea. A further factor was the suitability of stretches of sheltered water for the take-off and alighting of marine aircraft much larger than could safely use the short and often bumpy landing grounds. Quite apart from this there was the additional problem of how to operate aircraft from ships, but this is discussed in the next section.

To some degree military requirements over land and naval requirements at sea ran parallel; aircraft could fly reconnaissance missions, spot for the guns, drop bombs and endeavour to shoot down enemy aircraft. Anti-submarine warfare (ASW) was something quite separate which in the course of time was to lead to purpose-designed aircraft. In general, larger machines could fly further than small ones, and so shore-based aircraft tended to be either landplanes or float seaplanes suited to coastal and short-range work only, and large flying boats able to make extended flights over the open sea. In World War I the non-rigid and rigid airship were also of great importance in over-sea missions.

Before 1914 the pre-eminent builder of seaplanes and flying boats was the American Glenn Curtiss. A British naval officer, John C. Porte, intended to fly a Curtiss, called the America, across the Atlantic in 1914 to win a prize, but war came and he had to return to his duties. As a squadron commander in the RNAS (Royal Naval Air Service) he managed the construction at Felixstowe of a series of improved flying boats with Curtiss-type wings and tails but his own designs of hull. From early 1917 the excellent F.2 series were produced as the standard RNAS boat for open-sea reconnaissance, and they proved to be tough and seaworthy. Usually powered by two 345hp Rolls-Royce Eagle engines they could reach 153km/h (95mph) and fly six-hour patrols when carrying four Lewis guns and two 104kg (230lb) flat-nosed anti-submarine bombs under the lower wings. The guns were aimed from bow and amidships cockpits and from left and right waist positions, giving good all-round arcs of fire. To show how 'prickly' were these sedate boats, on 4 June 1918 three of them fought off 14 German seaplanes, shooting down six. One F.2A had non-standard 'fighting tops', also called elephant howdahs, on its upper wing, each being a streamlined nacelle for a Lewis gunner. Many F.2-series were painted in individual schemes of bright dazzle stripes and checkerboards, partly for identification at a distance and partly to show up on the water in the event of being forced down. The larger F.3 was a development able to carry four AS bombs and, being less manoeuverable than its predecessor, was used exclusively on ASW missions out of range of hostile aircraft. The final production Felixstowe boats were the F.5 series, some of which were built in the USA with 400hp Liberty engines and carried as many as 11 machine guns.

Most of the German naval seaplanes stationed around the North Sea in World War I were Hansa Brandenburgs, designed by an engineer later to go into business on his own account – Ernst Heinkel. The KDW (*Kampf-doppeldecker wasser*, war biplane seaplane), W.12 and W.19 were trim biplanes which had excellent performance and up to three or even four machine guns or, exceptionally, the new 20mm Becker cannon. By 1918 the extremely modern W.29 monoplane seaplane was in production, and from this Heinkel developed a long succession of outstanding reconnaissance/fighter seaplanes from his own company operating in Scandinavian 'cover' countries after the war. A few of the seaplanes were still operating in Norway when the Luftwaffe arrived in 1940, one batch having been made there under licence. The Friedrichshafen company, noted for its large bombers, also made almost 500 coastal seaplanes.

In Austro-Hungary the Löhner and M-series of flying boats were used throughout the Adriatic Sea and northern Italy, and spurred the development of Italy's own Macchi flying boats which were mostly used as single-seat fighters of quite high performance, often being able to outstrip hostile landplanes with their speed of up to 210km/h (130mph). Many other types of flying boat and seaplane were produced in France, Britain, the United States, Russia and Japan. Some carried torpedoes and most could fly ASW patrols with AS bombs or depth charges, but aircraft still lacked performance, (notably speed and endurance to cover large sea areas) and the absence of any way of finding submarines from the air, other than actually sighting them in shallow water, made kills few and far between.

Between the World Wars most countries had maritime patrol and ASW squadrons, but technical progress was modest other than the basic development of aeroplanes with all-metal stressed-skin structure and other refinements. In the mid-1930s

these combined with sudden jumps in available engine power, from 500hp to 1,000 and then almost to 2,000, to revolutionize the capability of maritime aircraft. The US Navy, for example, used the Naval Aircraft Factory's PH series flying boats from June 1932, with two 550hp engines, weight of 7,430kg (16,379lb) and range of 2,400km (1,490 miles). By 1936 the Consolidated PBY Catalina was being delivered – the most-produced flying boat in history, with over 4,000 of all versions built by 1945 – with two 1,200hp engines, weight up to 15,420kg (34,000lb) and range over 3,860 km (2,400 miles). In 1940 Martin began deliveries of the PBM-1 Mariner, with 1,600hp engines, weight of 18,660kg (41,139lb) and range of 5,552km (3,450 miles). In 1942 Boeing flew the XPBB-1 Sea Ranger, with two 2,300hp engines, and though this was not put into service it was a truly remarkable performer, able to carry 20 bombs of 454kg (1,000lb) and fly a 72-hour mission after taking off at a weight of 45,873kg (101,130lb) covering a still-air distance of 10,140km (6,300 miles). Seldom has aviation made such fantastic progress in one ten-year period.

Nearly all the patrol flying boats of the 1930s had been biplanes, but these disappeared swiftly after the start of World War II, with a few notable exceptions such as the RAF's faithful Walrus amphibian, with a 775hp pusher engine and a design that could be traced straight back to 1919. The Walrus, popularly called the 'Shagbat', was the chief ASR (air/sea rescue) aircraft of the RAF and also served with the Fleet Air Arm, and, though seldom armed with more than two hand-aimed machine guns and possibly some light bombs, was used as a fighter, dive bomber, minelayer, target tug and for almost everything else in all theatres of war. Another ancient-looking biplane that kept on proving its value was the German Heinkel He 59, a big twin-float seaplane, which though obsolete as a combat machine long before 1939 was kept in service as an ambulance and ASR aircraft, then began to fly minelaying missions, then served as an electronic-warfare platform, and on 10 May 1940 suddenly arrived in force in the heart of Rotterdam to disgorge airborne troops who captured the main bridge over the Waal river on which the old biplanes landed!

Probably the nation that suffered most from submarines (German U-boats) was Britain, and it was Britain that led in the development of ASV (air/surface vessel) radar. Radar capable of detecting surface ships was easy, but the main objective was radar that could detect the extremely small tip of

The Consolidated Catalina was originally used as a patrol boat, but was later used as a patrol bomber. RAF Catalinas were used to locate the Bismark after the loss of surface contact. It was the best-known flying boat of World War II.

a U-boat periscope and, later, a schnorkel breathing pipe. The early ASV radars were very bulky, operated on wavelengths as long as 10m (32ft), and needed large aerial arrays with 40 dipoles on the rear fuselage, looking like prominent washing lines, and horizontal dipole arrays slung under the outer wings. These were carried by the famous Sunderland flying boats in the 1941–43 period (Sunderland III) and the Coastal Command Liberator III and V which were the first VLR (very long range) aircraft able to fly patrols long enough to close the previous gap in the mid-west Atlantic where U-boats had been able to surface unmolested. By 1944 the ASV Mk V was in use, on a centimetric wavelength with a rotating scanner in a streamlined radome. This was carried under the nose of Wellingtons, under both outer wings of Sunderlands and above the hull of the PBM-3D and -3S Mariner. All these, and many other aircraft, could detect a U-boat at periscope depth in extremely stormy seas.

At first a U-boat was helpless if caught by a radar-equipped aircraft, though at night or in bad weather there remained the problem of how to aim AS bombs or depth charges at it. An officer in Coastal Command invented a brilliant searchlight, the Leigh Light, and for a while this provided the answer. The aircraft would detect the U-boat with radar, dive at it at full throttle, switch on the light at the last moment and let go the weapons on a visual aim. Then kills began to fall off; the U-boats had been equipped with a passive receiver that warned of the presence of radar signals. The answer was the new ASV operating on totally different wavelengths, which the U-boat could not detect. The only remaining problem was that by spring 1943 the U-boats had begun to bristle with flak (anti-aircraft cannon), and when they fought it out on the surface the struggle was grim indeed. Many aircraft were shot down, but the others were given heavy batteries of forward-firing guns which with luck could knock out the U-boat flak gunners, often as they were running from the conning tower to their guns. The standard RAF 27·2kg (60lb) rocket and US Navy 127mm (5in) rocket were often used to pierce U-boats, as was the 57mm gun of the Mosquito XVIII.

In the Pacific the Imperial Japanese Navy used a

diversity of seaplanes and flying boats, the most popular seaplane being the trusty Aichi E13A1 and the best flying boat the exceptionally powerful and well-armed Kawanishi H8K. Many of the Japanese seaplanes were single-seat fighters, intended to sweep enemies from the sky in areas where the Japanese had no airstrips available. Although remarkable technical achievements, they were inevitably outclassed by the best Allied land- or carrier-based fighters, and the same has been true of water-based fighters ever since. Britain built the Saro SR.A1 twin-jet fighter flying boat just after World War II, and in 1954–57 Convair and the US Navy worked hard at twin-jet delta fighters with hydro-skis which failed to see service.

The Luftwaffe was orientated towards land battles, but nevertheless produced many maritime aircraft. Two formidable seaplanes were the cannon-armed Ar 196B and the extremely tough twin-engined He 115, extensively used for minelaying and other oceanic duties. The curious diesel-engined Bv 138 flying boat had a short hull and high tail carried on twin booms. Like Britain's Coastal Command versions of the Wellington it was equipped with large electrically conductive rings through which very large pulses of current could be passed to detonate magnetic mines – often a hazardous duty because it meant flying very low across the minefield. The chief long-range oceanic aircraft of the Luftwaffe was the four-engined Fw 200C Condor landplane, which despite structural weakness earned the title 'Scourge of the Atlantic' from Churchill and in its final sub-types carried the Hs 293 radio-guided missile for use against Allied ships. An even larger patrol aircraft was the Ju 290, and two examples were built of the six-engined Ju 390, one of the largest and heaviest landplanes of the war. The six-engined Bv 222 Wiking was used mainly as a transport flying boat, and the Bv 238, largest aircraft flown during the war, never saw service.

Numerically the majority of maritime aircraft in World War II were relatively small twin-engined land-based aircraft used for ASW patrols, torpedo carrying, rocket attacks on surface ships and all kinds of photographic and electronic reconnaissance. Types included the British Beaufort, Beaufighter, Mosquito, Whitley, Wellington and Warwick, and the American Mitchell, Ventura, Hudson, B-18 and Harpoon. Italy was proud of its S.M.79 Sparviero three-engined torpedo-bombers, which were often preferred to the proposed replacement, the S.M.84, and extensive use was also made of the Cant Z.506B three-engined seaplane and the

Z.501 single-engined flying boat.

In the Soviet Union float seaplanes were few, but the Beriev MBR-2 and MDR-6 flying boats were used in large numbers for coastal reconnaissance and many other duties though, so far as is known, seldom on combat missions. Both had high-mounted stressed-skin wings, but the MBR-2 had a wooden hull and was already ten years old when war came to the Soviet Union in 1941. Since 1945 the Beriev bureau has been the only one in the world to continue to produce military water-based aircraft in large numbers. The Be-6, with two 2,000hp piston engines and the first complete set of oceanic and ASW equipment to fly in a Russian aircraft, served with the AV-MF (naval air force) from 1951 until the mid-1960s. The record-breaking Be-10 twin-jet was an exceptional design, with large hull and small highly swept wings and swept tail, but is not thought to have served in large numbers. The less advanced Be-12, with two 4,000hp turboprops, has in contrast been built in very large numbers as the M-12 for all kinds of ocean patrol and ASW missions. An amphibian, with retractable tailwheel-type landing gear, the Be-12 carries radars, MAD (magnetic-anomaly direction gear for finding submarines by sensing their disturbance of the terrestrial magnetic field) and sonar detection equipment, as well as extensive electronic systems for communications, navigation and intelligence gathering.

In contrast, almost all Western maritime aircraft since the 1950s have been landplanes. Numerically the most important has been the large twin-engined Lockheed P2V (later P-2) Neptune, which in 1946 set a world record by flying 18,083km (11,236 miles) non-stop from Australia to Ohio. Eight major versions of Neptune were used by almost all NATO and many other air forces in 1950–79, the later variants usually having two jet pods under the wings to boost the two basic 3,700hp Wright Turbo-Compound engines. A contemporary of the Neptune was the fractionally larger Shackleton developed in Britain from the Lancaster bomber and powered by four 2,455hp Griffon piston engines (in the final version boosted by two small Viper jets). Canada used the Britannia turboprop airliner as the basis for its extremely large and capable CL-28 Argus, powered by four Neptune-type Turbo-Compounds and in service from 1958. Several European NATO partners collaborated to build the Br.1150 Atlantic, with a large pressurized fuselage with cavernous weapons bay, two 6,100hp Tyne turboprops and with an overall efficiency which exceeded that of any rival.

Lockheed won a US Navy competition for a new off-the-shelf patrol aircraft in 1958 with the P-3 Orion, based on the Electra airliner and powered by four 4,910hp Allison T56 turboprops. By a constant updating process this aircraft has been kept competitive right up to the present time, and its large sustained production run has helped it fight inflation and remain at prices below that of rival aircraft. Though propeller-driven, an Orion has set a speed record at over 805km/h (500mph), and the latest versions have exceptional equipment, including a large tactical compartment manned by a crew of at least eight who manage the whole local war situation. Digital computers interlink the numerous sensors, communications, navigation and recording systems, while a diverse array of ordnance can be carried in an internal bay and ten external pylons. The corresponding Soviet aircraft is the Il-38, likewise derived from a civil airliner, in this case the Il-18. Powered by four 4,250hp Ivchenko turboprops, the Il-38 has a large pressurised fuselage like the Orion but has a smaller internal weapon bay and, in versions so far seen, fewer external pylons. Britain uses an outstanding and unique jet for ocean reconnaissance, the British Aerospace Nimrod. Originally based on the de Havilland Comet airliner, the Nimrod has a much larger fuselage of 'double-bubble' cross-section, much of the lower (under-floor) area being enormous weapon bays. These bays are so large the Nimrod has seldom had to carry any weapons or other stores externally, though it does have a searchlight in the front of a wing tank-pod and various special fairings for its numerous electronics aerials. In 1979 the RAF's Nimrod force was being recycled back through the manufacturer's factory in a major rebuild and updating programme, while the AEW.3 (mentioned earlier) is a grossly rebuilt version with powerful surveillance radars including giant aerial domes on the nose and tail. These radars are expected to be the best in the world for surveillance over ocean areas, with the capability of seeing even a sea-skimming missile above rough waves which earlier radars would have been blind to.

Since the final year of World War II the helicopter has been increasingly useful for maritime missions. Since the advent of the deadly nuclear submarine, with very high underwater speed and almost unlimited endurance, there has been demand for the development of specially equipped ASW helicopters. The helicopter has the great advantage over the aeroplane that it can fly slowly enough to stay in close contact with its quarry, yet compared with a submarine it has considerably greater speed and

agility. A further advantage is that, whereas aeroplanes have to drop expensive sonobuoys into the sea, each thereafter to 'listen' for a submarine and send back what it hears by radio, the helicopter can 'dunk' a single sonobuoy into the sea, listen and then dunk it in a different spot. The sonobuoy dropped by aeroplanes cannot be retrieved, and both cuts into the disposable load of the aircraft and represents a large financial waste as well as ocean pollution. Moreover, as the helicopter can go on using a single sonobuoy, this can be made much larger and more powerful than the small buoys strewn by aeroplanes, and so more likely to find its prey.

In the early 1960s the US Navy devoted many years to developing the DASH (drone anti-submarine helicopter) system. Small ships were to carry miniature pilotless helicopters, carrying two AS torpedoes, and direct them by radio to the location of an enemy submarine on the instructions of the ship's sonar submarine-detection system. This was used in numbers, but caused hair-raising incidents and eventually the US Navy decided an ASW helicopter had to have a man aboard. Today there are many excellent ASW helicopters, all tailored to operations from small ship platforms as well as shore bases, and equipped with radar, sonar and other sensors as well as AS weapons. Another important role of the overwater helicopter is surface-ship attack, using sea-skimming missiles. This is one of the few ways of dealing with today's crop of extremely small missile boats which, though only the size of small fast patrol boats, can fire large missiles against large warships or cities from a considerable distance, which in the case of some Russian missiles can exceed 160km (100 miles).

Carrier-based Aircraft

The first man to take off in an aeroplane from a ship was Eugene Ely, of the United States, who on 14 November 1910 successfully piloted his Curtiss off a makeshift 25m (83ft) platform over the foredeck of the US cruiser *Birmingham*. Two months later, on 18 January 1911, he flew out from San Francisco and successfully landed on a platform built at the stern of another cruiser, *Pennsylvania*. Ropes had been stretched across this platform with sandbags at each end, and the Curtiss caught these ropes with hooks built on the landing gear. In principle this was very like the deck arrester system of a modern carrier.

One might have thought that the next step would be the construction of flat-topped ships, fitted with

Fairey's classic biplane, the Swordfish torpedo-bomber, renowned for the attack on Taranto harbour in November 1940 which disabled the majority of Italy's capital ships.

this simple arrester system, so that aeroplanes might go to sea in ships. But this did not happen. For the early part of World War I the warring powers did little with aeroplanes at sea apart from carrying small groups of seaplanes on a tender (small craft such as converted ferries). To launch a seaplane the tender had to stop and hoist it by crane into the water. The recovery was the same procedure in reverse. Apart from making the stationary tender vulnerable there was always a likelihood of smashing the flimsy seaplane against the ship. In November 1915 a British Royal Naval Air Service pilot took off from a ship, and in August 1917 another RNAS pilot 'landed-on', in each case with the ship at speed in the open sea; but it was not until 1918 that the first aircraft carrier was commissioned. Even then the 1911 scheme of arrester wires was not followed-up. At first sailors ran out and tried to cling on to the wing tips to hold the aircraft down and bring it to a halt. Then longitudinal wires were fixed along the deck, the pilot trying to rub along them on landing. Some aircraft, such as the Fairey Campania (named for the first long-deck seaplane tender), were seaplanes which took off on deck trollies and landed in the water alongside the ship. Landplanes often dispensed with their wheels and operated on plain skids, which were unsprung and jarred the

pilot. By 1918 many manufacturers had produced special ship-planes, such as the Sopwith 2F.1 Camel, the naval version of one of the most famous fighters of World War I. The 2F.1 had several changes including steel centre-section struts instead of wood and a quickly detachable rear fuselage. The latter feature enabled more aircraft to be stored in constricted spaces below the flight deck, a problem that has persisted with aircraft carriers ever since.

Deck take-offs using seaplanes on trollies endured into the 1920s, and some seaplanes were constructed as true amphibians with wheels built into the floats. Increasingly the carrier-based fighter diverged from its land counterpart, nearly always because the life aboard a ship was tougher or made additional demands. By the mid-1920s it was almost universal for wings to be made with hinges and locking pins so that they could fold back to occupy less space. The structure had to be strong enough to stand up to constant manhandling in small areas, manoeuvring on pitching decks and the stresses of storms at sea. As wooden construction was unsuitable the structure was almost always non-corrodible or Alclad, a special light alloy coated with pure aluminium. Many navies continued to use fabric-covered aircraft until almost the end of World War II (the Royal Navy went on long after) but most welcomed stressed-skin as a good way of avoiding most superficial tears and dents. Landing gears had to be stronger than for land aeroplanes to stand up to landing on a pitching or rolling deck.

By 1930 several carriers in Britain, France, the United States and Japan were in operation with two basic new installations, both initially of a hydraulic nature. In the bows was at least one catapult, to shoot aircraft into the air in a short distance. At the stern was a series of arrester wires to bring landing aircraft rapidly to a halt. Both made severe new demands on aircraft. The addition of catapult hooks or 'spools' not only demanded massive local strengthening but also called for a rethink throughout the aircraft; every single part or item of equipment had to be fixed more firmly than before, and it also gave the pilot a new experience, forced back into his seat almost like being shot from a gun. The catapult usually had a large hydraulic ram, whose travel was often multiplied by a cable and pulley system to drive a sliding shuttle or trolley on the deck. The aeroplane was connected to the shuttle by a coil of rope or cable, called a strop, which in most navies was allowed to fall into the sea. Catapults of a somewhat different type, often arranged on a large pivot to face the wind, were added to most of the world's newer or larger warships, such as battleships and heavy cruisers. These carried seaplanes which in a naval engagement could be shot off to spot for the heavy guns. Afterwards they landed alongside and were picked up by crane. A few shipboard aircraft even carried torpedoes to do their own bit to hit the enemy, but in general torpedo aircraft operated only from carriers.

The landing system took a long time to become efficient and reasonably safe. By 1933 the universal system was to stretch about eight heavy cables transversely across the rear of the flight deck, each fastened at the left and right ends to an energy-absorbing system. At first each end was fixed to a drum, which was spun round by the landing aircraft against a brake, the aircraft lowering a hinged hook before making its landing. Pilots sometimes landed well to one side, making the nearer drum run out of cable before the other, and eventually the best method was found to be to make the arrester wire a closed loop, half over the deck and half inside the ship carried on pulleys, and fix the centre of the lower portion to a single arrester gear. Since the mid-1930s this has invariably been a device which forces hydraulic oil through holes to compress air in a large container. An arrested landing would pull the tail off ordinary aircraft. To take care of landings that fail

to pick up a wire and do not 'bolt' (go round again for another try) a barrier is raised across the deck. In 1917 the Royal Navy tried barriers with a flexible fence of vertical ropes, and this kind of scheme is still used. Yet again the aircraft has to be extremely strong to minimise damage from a barrier crash.

In the 1920s it became accepted that all these severe requirements inevitably resulted in carrier-based aircraft being slower, clumsier and less-manoeuvrable than land-based counterparts. Some machines were additionally burdened by flotation gear, usually in the form of large bags of rubberized fabric which in an emergency could be inflated from air or gas bottles, and topped up by a hand-pump in the cockpit. Naval aircrew were also the first to have buoyant life-preservers, of the type later called a Mae West (after the film actress), and inflatable rubber dinghies – which again had to be installed somewhere on the overburdened aircraft. The Blackburn Blackburn was a typical fleet aircraft of the 1920s, ugly and sluggish yet tough and capable of doing its job. A few fighters, such as the American Boeing F4B and British Hawker Nimrod, were naval versions of land aircraft and did their best to minimise their penalties. Of course, carrying a torpedo inevitably meant a large and heavy aeroplane which until World War II could only be fitted into carriers with difficulty.

By 1935 the strong trend towards the monoplane extended to naval aircraft, and though inferior performers the new breed of carrier-based machines had almost double the speed and greater load-carrying capability. One exceptional fabric-covered biplane, the British Swordfish, proved so tough and useful that it outlasted various intended replacements and continued throughout World War II, despite having an engine of only 775hp. In the US and Japanese navies great emphasis was placed on the dive bomber, and large numbers of such excellent machines as the Douglas SBD Dauntless and Aichi D3A wrought havoc in 1942 because of the consistent ability to put heavy bombs accurately on to warships. But the early torpedo monoplanes, such as the Douglas TBD Devastator and Nakajima B5N, were extremely vulnerable to fighters and intense anti-aircraft fire. The best Allied torpedo bomber of the war was the Grumman TBF Avenger, which had an internal bay for a 0·56mm (22in) torpedo and a heavy machine-gun in a power-driven dorsal turret. The new torpedo could be dropped at a speed double the 161km/h (100mph) limit of the old Mk 13.

Britain failed to build modern naval fighters, but during World War II produced carrier-based versions of two proven land fighters, known as the Sea Hurricane and Seafire. American design teams fortunately doubled engine power to over 2,000hp in 1940, resulting in such formidable machines as the F4U Corsair and F6F Hellcat, which completely mastered the Japanese A6M 'Zero'. By 1944 the Grumman F7F Tigercat was being cleared for carrier use with two 2,000hp engines and not only devastating armament but also air-interception radar. British thinking preferred two-seat fighters, despite their poorer performance, because at sea it was judged that a professional navigator would be less likely to fail to find the way back to the carrier, especially at night or in bad weather in war conditions when no helpful radio signals could be broadcast. The first two-seat monoplane for the Royal Navy was the Fulmar, which though slow could defeat most aircraft in the Mediterranean in 1940. Its successor was the Firefly, which again could hold its own at the end of the war and continued in later versions throughout the 1950s in ASW and other missions. Last and fastest of the piston-engined fighters was the 752km/h (467mph) Sea Hornet, which at last had adequate range and in later form carried radar and an observer. But most useful machines were the American TBF and the F4U and F6F fighters whose range and ordnance load made dive bombers no longer necessary.

Early naval jet fighters were short on range and endurance, and also suffered from long take-off runs. By 1950 Britain was revolutionizing carrier operations with three great advances. The steam catapult was a long tube recessed into the deck with travel long enough to need no pulleys yet with dramatically greater power to match the needs of any kind of aircraft. The American aircraft designers mated it with a nose-tow system which pulls the aircraft by the bottom of the nose landing gear, which demands vastly greater structural strength but greatly improved the rate at which aircraft can be launched and replaces the old strop by a short metal link. The angled deck allows aircraft to land diagonally while the foredeck is full of aircraft either parked or being catapulted. The mirror sight helps landing pilots stay on the correct glide-path, with further assistance by the LSO (landing signals officer) or 'batsman' who instructs the pilot to make necessary corrections. Good forward view on the approach is vital, and in the first supersonic naval fighter, the F-8 Crusader, resulted in the wing being pivoted in a variable-incidence arrangement so that the wing stays nose-high while the fuselage rotates into a nose-down attitude.

Today's US Navy carriers are the largest warships in history, with displacements up to more than 90,000 tonnes and a complement exceeding 5,000.

Barracudas and Corsairs being prepared for operations aboard HMS *Illustrious*. Note methods of wing-folding.

They can operate aircraft such as the F-14A Tomcat multi-role fighter, weighing over 31 tonnes and capable of Mach 2·3. Another large carrier aircraft is the E-2C Hawkeye, like the F-14A a Grumman product, with two 4,910hp turboprops and carrying the largest load of surveillance radars, ECM, computers and communications ever to go to sea in the Awacs role. A transport version is the C-2A Greyhound, called a COD (carrier on-board delivery) aircraft, used to maintain liaison between ship and shore. Standard ASW aircraft is the S-3A Viking, with two 4,200kg (9,275lb) thrust T34 turbofan engines and a crew of four, advanced radar, IR (infra-red) seeker, MAD gear, large batteries of sonobuoys, comprehensive ECM and a wide range of ASW weapons in an amazingly ingeniously packed fuselage much shorter than the F-14. Another compact machine is the A-7 Corsair II attack aircraft, with a weapon load greater than that of World War II heavy bombers.

Britain has abandoned air power at sea apart from helicopters (the naval Lynx HAS.2) and the unique Sea Harrier, flown from the *Invincible* class of 'through-deck cruisers'. The latter are neat vessels of 19,810 tonnes which will operate ten Sea King ASW helicopters and five Sea Harriers. The vectored-thrust Harrier is discussed in the V/STOL chapter, and in its Sea form has much more comprehensive equipment for fighter, reconnaissance and strike roles. The foredeck of the ship has a curved 'ski jump' so that when a rolling take-off is made the aircraft is projected high into an upward trajectory. This improves safety in the event of take-off engine failure and allows a much heavier weapon load to be carried without risking the aircraft dipping close to the sea surface. The Soviet Union also uses vectored-thrust (in this case pure VTOL) aircraft at sea, but the operational roles of the Yak-36 developed in the new *Kuril*-class ASW cruisers appear to be limited. The standard Soviet ASW helicopter is the Ka-25, and the next-generation multi-role helicopter of the US Navy will be the Sikorsky SH-60B.

Trainers and Transports

Trainer and transport aircraft also have important civilian roles, but in this section only military examples are considered. To some extent all the earliest aeroplanes were trainers, and often the designer was also the builder who taught himself to manage the machine. Many aspects of flight, especially the stall and spin, remained worrying enigmas, and until well into World War I there was no formal system of training at all. The pupil simply flew with a qualified pilot and watched the latter's actions. The so-called instructor was often a front-line pilot who had been wounded or for some other reason was not on operations, and his mental attitude was frequently one of bitter disinterest. To a considerable degree the training was decided by whether or not the pupil was killed. If he survived, he was sent into action. At many schools the casualty rate exceeded 30 per cent, and usually nothing was learned from the crashes.

It was one of the wounded RFC pilots who, appalled at the way instruction was done, told the RFC leader General Hugh Trenchard. The latter, in his characteristic way, replied 'Go and do something about it'. The result was a completely systematic syllabus of instruction, called after its originator the Smith-Barry scheme, together with properly run schools and specially designed trainer aircraft. From 1917 until well into the 1930s the most numerous RAF trainer was the Avro 504, a trim and safe biplane made in prodigious quantities (exceeding 10,000), seating the instructor behind the pupil with a speaking tube intercom (intercommunication system) consisting of a simple pipe system with mouthpieces and earphones. Trainers were almost all biplanes of 100–150hp until World War II. Even then, biplanes dominated the primary or *ab initio* phase, but the mass-produced (10,346) Boeing Kaydet was exceptionally large and powerful with various engines of well over 200hp. The Soviet Union's Po-2 biplane, dating from 1926, was built in enormous numbers and, despite having an engine of only 100hp, was used for front-line tactical harassment and attack missions. The Luftwaffe soon did the same, using over 6,000 primary trainer biplanes in action on the Eastern front. These aircraft were armed with various machine guns, light bombs and rockets, and operated on wheels, skis and floats.

There had, since World War I, been various advanced trainers which were often dual-control versions of operational types. In the mid-1930s specially designed monoplanes came into the picture, and by 1945 more than 20,000 examples had been produced of the North American AT-6 (later T-6) Texan, called Harvard by the British. Powered by a 550hp engine, this all-metal stressed-skin monoplane was ideal for all forms of advanced instruction including aerobatics, gunnery, simple bombing and navigation, and many survived into the 1970s. After World War II Britain toyed with three-seat trainers, and for a time used Harvards as ab initio machines, subsequently using an all-jet sequence such as the Jet Provost followed by the speedy Gnat. The USAF continues to use lightplanes, such as the Cessna T-41 Mescalero, to weed out at an early stage pupils lacking natural pilot aptitude. Then instruction proceeds on the Cessna T-37 twin-jet, with side-by-side seating like the Jet Provost, before going on to the tandem-seat Northrop T-38 Talon, the only purpose-designed trainer with supersonic performance. Almost all modern trainers have the instructor seated behind and above the pupil. Such machines as the Hawk, Alpha Jet, Macchi 339, CASA 101, Fouga 90 and Albatros have sufficient power to fulfil a secondary role as light attack or reconnaissance aircraft.

Most modern combat aircraft also exist in dual-control versions. Sometimes, as in the case of many versions of the F-4 Phantom, this is so that an operational crew of two rated pilots can share the flying, or return safely from a mission after either has been incapacitated. More often the dual version is for converting a qualified pilot to fly the new warplane, and learn to use its complex equipment efficiently. Examples of combat aircraft of which special dual-control versions exist include the F-15 Eagle, F-16 and F-18, A-4 Skyhawk, A-7 Corsair, Jaguar, Harrier, Su-7, MiG-23, MiG-25 and all single-engined versions of Mirage.

Until well into the 1920s there were hardly any military transport aircraft, and those that did exist were converted bombers used for opening up pioneer air routes or conveying statesmen to peace conferences. One of the earliest purpose-designed transports was the RAF's Vernon, of 1922, which led to the Victoria and Valentia, each seating about 17 equipped troops and not only flying every kind of transport mission in distant outposts of empire but on occasion even dropping bombs. As early as 1923 American Air Service pilots had managed to refuel each other in flight, but this way of extending range or endurance did not become an operational method until after World War II (except for civil flights by Imperial Airways in 1939). Few flight refuelling experiments took place outside Britain and the United States.

In the 1930s the Soviet Union pioneered many aspects of aerial assault, included the use of mass parachute troops, air-landed troops and air-portable armoured vehicles including T-26 tanks. The only country that appeared to notice was Hitler's Germany, which in all its Blitzkrieg campaigns made very extensive use of airborne assault, using such cargo aircraft as the Ju 52/3m 'Tante Ju', He 59 seaplane and DFS 230 glider. Assault gliders were a new idea, and though small at first they put teams of crack troops in exactly the right spots. Britain promptly built a series of gliders, the most important being the wooden Airspeed Horsa seating 25 troops, while the United States mass-produced the steel-tube, wood and fabric Waco CG-4A, seating up to 15. But if Germany decided to invade Britain (after the defeat of the Soviet Union) gigantic gliders would be needed. The answer was the Messerschmitt Me 321 Gigant, which despite lack of any suitable tugs (so that many men died using the tricky Troika-Schlepp tow requiring three Bf 110s on each Gigant) was put into wide service with the ability to carry well over 100 troops or an 88mm gun and ammunition. By 1943 production had switched to the powered Me 323, with six 1,140hp engines, which represented a complete break with traditional transport design. Features included full-section nose doors, an enormous unobstructed interior with a level floor at truck-bed height, and multiple soft-field landing gears.

In fact the most numerous transport of World War II was a version of the DC-3 airliner, but in the post-war years the military transport made great strides. Via the C-74 Douglas created the C-124 Globemaster seating up to 200 troops and with great clamshell nose doors. Douglas later built a much more powerful aircraft, the C-133 Cargomaster with four 6,000hp turboprops, a pressurized interior and full-section rear doors and ramp, which in the early 1960s was kept busy carrying the new breed of intercontinental missiles. But before this aircraft appeared Lockheed had produced the C-130 Hercules, and for 25 years this has been the leading military transport, sold to almost all major non-Communist nations and still in production. Powered by four Allison T56 engines which were at first of 3,450hp but now are 4,910hp, the C-130 has been continually improved and updated. Its design introduced all the desirable features which had never before been united in one aircraft: pressurization, high-speed turbine propulsion, unobstructed cargo hold with low level floor, soft-field landing gears, full-section rear doors and ramp for ground or air unloading, flight refuelling, all-weather equipment and size matched to 92 troops or 23,133kg (51,000lb) cargo. By 1979 over 1,600 Hercules had been delivered, with fresh orders coming in. During the programme the same factory also delivered 285 considerably larger C-141 StarLifters, with turbofan engines, and 81 of the gigantic C-5A Galaxies, in many respects the largest aircraft ever produced in quantity. Powered by the first 'wide-body' turbofans, the TF39s, the Galaxy carries up to 120 tonnes (265,000lb) of cargo including almost every kind of US Army vehicle.

In the late 1940s the need of USAF Strategic Air Command to fly unprecedented global missions led to the development of a production system of flight refuelling, and the construction of a mighty fleet of tankers. The selected system was the Boeing Flying Boom, in which the receiver aircraft flies below and behind the tanker, at very close range, and a 'boomer' crewman in the tanker 'flies' an extensible boom (a pipe with flight controls for aiming its direction) into a receptacle on the receiver. The first tanker was the Boeing KC-97, a version of the civil Stratocruiser, of which 888 were delivered to SAC, some later being fitted with booster jet pods. In 1957 production deliveries began of the much larger and more capable KC-135, a smaller version of the civil 707, of which 792 were delivered at a very high rate. Later versions of the C-135 family had the original J57 turbojets replaced by more powerful and more fuel-efficient TF33 turbofans, though not as powerful as the TF33 version fitted to the C-141. Boeing also delivered a diverse array of other military versions of the 707, some of them configured for flight refuelling by the British probe and drogue method in which the tanker merely trails a long hose with a conical drogue on the end into which the receiver thrusts a probe connected to its fuel system. Since 1960 large numbers of the C-135 family have been rebuilt into diverse special-purpose aircraft concerned with electronic intelligence, spaceflight, strategic command and control, research and many other tasks, often with grotesque protuberances and aerial arrays.

Today the most powerful aircraft in any air force are the E-4 airborne command posts of the USAF, based on the 747B airframe, in which a crew of about 60 would form the seat of government of the United States in time of crisis. Their high cost is a problem, and inflation has also halted plans to replace the evergreen 'Herky-bird' (C-130) with an AMST (advanced medium STOL transport). The YC-14 and YC-15 both completed extensive test flying, the former with USB (upper-surface blowing, an advanced way of using the engines to add lift) and the latter with four engines blowing straight back into very powerful heat-resistant flaps. Though their wings are much smaller than that of the C-130, and the aircraft are very much heavier and carry double the load, these remarkable transports can take off in less distance and climb with amazing speed and steepness. In the course of the 1980s their technology will reach across into advanced civil transports.

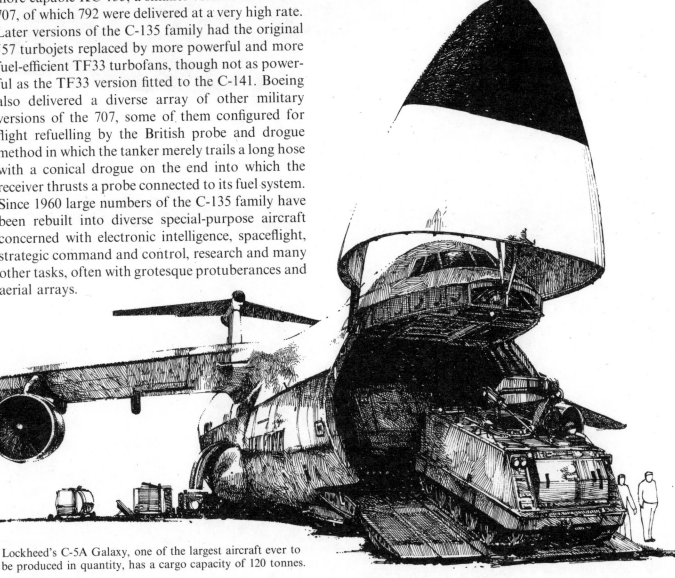

Lockheed's C-5A Galaxy, one of the largest aircraft ever to be produced in quantity, has a cargo capacity of 120 tonnes.

Aerial Warfare

World War 1

At the beginning of World War I, flight in heavier-than-air craft was still in its early stages of development. The Wright brothers produced the *Flyer III* in 1905, and early aviation only really came of age in 1909 with *La Grande Semaine d'Aviation*, sponsored by the French champagne industry, held outside Reims. The realization that flight had become a practical proposition received widespread acceptance in 1909 with Louis Blériot's pioneering flight across the English Channel in his modified *Type XI* monoplane. The Wright's *Flyer III* was more advanced than any aircraft participating at Reims, but as the two American brothers were working in almost total isolation, the true age of aviation can be properly dated from the Reims meeting in 1909. This was a direct result of Wilbur Wright's magnificent demonstration tour in 1908, which led the French aviator Leon Delagrange to exclaim '*Eh bien, nous sommes battus! Nous n'existons pas!*'

For all practical purposes, therefore, heavier-than-air flight can certainly be considered only five years old by the time World War 1 began in August 1914. Great strides had been made in the design and construction of aircraft since the Reims meeting, but the primary consideration of most designers and aviators was whether or not the aircraft would fly satisfactorily. However, public pressure had led to the introduction of rudimentary air forces in most European nations, and to the limited development of aircraft as military vehicles. The first specification for a military aircraft was made in 1907 by Brigadier General James Allen of the US Army Signal Corps, a specification met in 1909 by the Wrights' Modified Type A biplane. Other countries, notably France, Germany, Great Britain and Italy, had followed with military trials for aircraft, which entered service in increasing numbers during the three years before the start of World War 1.

In general there was a lack of constructive thought and development about military aircraft, largely the result of the fact that it was deemed sufficient that aircraft could take-off and remain airborne with a crew of one or two, and land again in comparative safety. In such circumstances, it was inevitable that the only real role envisaged for aircraft was that of reconnaissance for the army and navy. Yet the way forward had been shown by the activities of a handful of pioneers – many of them Americans – despite the fact that their country had lost the lead in aeronautical development established by the Wright

brothers: in June 1910 Glenn Curtiss had 'bombed' the buoyed-out outline of a battleship with weights; two months later Jacob Fickel had fired a rifle from an aircraft for the first time; in 1911 Myron Crissy and Phillip Parmalee first dropped HE bombs from an aircraft; in the same year *Capitano* Guidoni of the Italian air service air-dropped the first torpedo; in 1912 Captain Charles de Forest Chandler fired a Lewis light machine-gun from a Wright biplane, and in 1913 the Coventry Ordnance Works 37mm cannon was fired from a Vickers biplane. While these practical experiments were being performed, more advanced work, was being undertaken, notably in France and Germany, especially on the development of a mechanism to interrupt the fire of a machine-gun at precise moments, and so allow such a gun to be located on the fuselage of an aircraft along the pilot's line of sight, without fear of shooting off the propeller blades.

The main experimenters were Raymond Saulnier in France and Franz Schneider in Germany, but the work of both men foundered on the problem of 'hang-fire' rounds, which detonated a little late and might hit a blade, and so practical development work came to a halt before World War I. Also of great importance was the initial work on the development of air-to-ground wireless telegraphy – the first airborne wireless was fitted to the Zeppelin LZ6 in 1909, and the first air-to-ground transmission from an aircraft was made in 1910 in the United States. By 1914 air-to-ground transmissions over a range of 160km (100 miles), were possible, as were air-to-air transmissions over a range of 16km (10 miles). With the aid of such wireless gear, it was realized, practical spotting for the artillery would be possible from the air. Yet by 1914 such control had to be effected by air-dropped messages, for the wireless was not at that point an entirely reliable instrument. There had, however, been advances in aerial photography, so making possible the advent of photographic reconnaissance by aircraft, a factor that was to be of paramount importance after the establishment of static lines during the winter of 1914–1915.

All these aspects of 'air warfare' were tested in the Italo–Turkish War of 1911–1912 fought in Libya, but the limited nature of the aircraft available made it impossible to gauge with any accuracy whether or not armament and other aids would be of use in future wars. Nevertheless, the foundations had been laid, and although reconnaissance was the only real role expected of aircraft by the high commands, future growth was possible, perhaps even inevitable.

The need to prevent enemy
aircraft from photographing
lines and troop positions
often led to the classic
dogfight.

At the outbreak of war in 1914, the aerial line-up of
the opposing nations was as follows: Germany 246
aircraft and 11 airships; Austria–Hungary 35 aircraft and one airship; France 160 aircraft and four
airships, Great Britain 110 aircraft and six airships,
Russia 300 aircraft and 11 airships, and Belgium 25
aircraft. There was virtually no standardization of
types, spares and replacement aircraft were virtually
nonexistent, pilots and groundcrew were wholly
incapable of engaging in prolonged operations, and
no clear tactical precepts had been worked out for
the effective use of aircraft in war.

Nevertheless, despite these obvious shortcomings,
it was not long before the air services became
effective. On 22 August 1914 the German airship
SL2 penetrated 480km (300 miles) into Russian
territory on the world's first long-range strategic
reconnaissance mission. Early in September 1914 it
was aircraft of the British Royal Flying Corps that
brought back the first indications that the German
1st Army, against expectations, was in fact edging to
the east of Paris. This information led directly to
the decisive first Battle of the Marne, in which the

Germans were finally halted and then driven back by the Allies. It was this victory that led, during the next three months, to the 'Race to the Sea' as each side strove to outflank the other to the north.

So, reconnaissance by air was established as a vital adjunct to army operations very early in the war. Inevitably, therefore, it became clear that it would be of use to prevent the enemy securing similar advantages, and so air fighting gradually began to appear. Few aircraft were capable of carrying machine-guns and so early air combat was often a matter of using a bewildering number of practical and impractical weapons ranging from rifles and shotguns to bricks, steel darts and grappling hooks on the end of rope. The first air-to-air combat appears to have taken place as early as 22 September 1914, when a British aircraft in attempting to bomb a German artillery observation balloon was attacked by a German aircraft, and the British pilot was wounded in the leg. A more momentous event occured on 5 October 1914 when a German Aviatik reconnaissance biplane was shot down by a Hotchkiss machine-gun in a French Voisin biplane. It is from this date that real air combat can be dated.

Meanwhile, both sides had been moving towards the beginning of sustained bombing operations, the Germans using airships as well as aircraft. But so great were the losses suffered by the airships that the Germans soon decided that greater benefit would accrue from the use of these expensive lighter-than-air craft in a maritime reconnaissance role. Nevertheless, the threat of the airships led to a number of daring raids by aircraft of the French air service and the British Royal Naval Service on German airship sheds. The French made their first attack, on the base at Metz, on 14 August 1914; the RNAS made its first strike, from Belgium against the base at Düsseldorf, on 22 September. Neither of the raids was successful, but an RNAS attack on 8 October succeeded in destroying the Zeppelin ZIX, which had been filled with hydrogen in readiness for a sortie. The RNAS raided Friedrichshafen, where the Zeppelin company's works were located, from Belfort on 21 November, and the base near Cuxhaven from seaplane carriers in the Heligoland Bight on 25 December. Thereafter, with the threat of the German airships receding temporarily, and the base defences greatly strengthened, Allied raids ceased.

At the end of 1914 the importance of 'strategic' bombing loomed large in the minds of the French and Germans, and in November each established a strategic bombing force: the *1 Groupe de Bombardement*, under Commander de Göys, and the *Fliegerkorps des Obersten Heeresleitung*, led by Major Wilhelm Siegert. Neither unit had any real stretegic, or even tactical importance at first, but serve to illustrate that the high commands on each side were becoming more appreciative of the value of aircraft as offensive weapons. At the same time the Germans realised that their airship fleet, with good load-carrying ability and considerable range, was ideal for a strategic offensive against London, and this began in January 1915. The scale of these raids, which had a great nuisance value, rose during the next 18 months, and it was only by the diversion of significant forces from the British air effort over the Western Front that defences effective enough to deter the Germans were built up. By the end of 1916 the threat of the airship was ended, and thereafter the Germans sought to bomb London mainly with aircraft.

During the early part of 1915 air operations over the Western Front grew more frequent for factories began to turn out aircraft in increasing numbers, and the air arms of the combatants began to receive increased numbers of aircrew and ground staff. Artillery spotting and reconnaissance were still of the greatest importance, and during the battle of Neuve Chapelle in March 1915 the British brought into action a sophisticated scheme of air support, with artillery observation aircraft using the new 'clock' method of spotting 'the fall of shot'. Photographic reconnaissance aircraft provided more accurate information of the German trench system, reconnaissance aircraft patrolled the German side of the lines to warn the British high command of the arrival of German reinforcements, and British planes bombed the German defences and lines of communication. At the same time, the first 'contact patrols' were flown: they were designed to locate the positions of the forward British ground forces, so that commanders could assess the success of the operation in terms of ground captured in a given time, and arrange for reinforcements, supplies and artillery support for the leading wave of infantry. The 'contact patrols' proved of limited value in the Battle of Neuve Chapelle, but the problems were eased during the Battle of Aubers Ridge in May 1915 by providing the infantry with cloth markers to lay out along their forward position.

The increasing scale of French bombing attacks at this time led the Germans to form the world's first point interception force of fighters, the *Briefabteilungen Metz*. The efficiency of the units in this force was minimal, however, and none of the aircraft had weapons capable of inflicting great damage. Yet in April 1915 the world's first moderately effective fighter, a modified Morane–Saulnier Type L with Saulnier deflector plates on the propeller, saw action in the hands of the celebrated prewar pilot, Roland Garros. With the wedge-shaped deflector plates warding off any bullets that might hit the propeller, Garros was able to aim his aircraft, and hence the gun situated along the pilot's line of sight, at the enemy aircraft and fire. Garros had everything his own way, and immediately started to shoot down German aircraft. Inevitably, though, the vibration of the occasional bullet striking the deflector plates finally caused the engine to fail, and Garros was forced down behind the German lines, where he was captured before he could destroy his aircraft. Realising the significance of their capture, the Germans at once instructed a Dutchman, Anthony Fokker, to copy the device. Fokker went one better: the designers at his aircraft company developed the world's first effective gun synchronization mechanism. This was fitted to the M-5 monoplane, which thus became the Fokker E-1, the world's first true fighter aircraft. The type and its successors started to enter service in July 1915, and by the autumn the 'Fokker Scourge' was firmly under way. At the same time the German reconnaissance and general-purpose aircraft were becoming more dangerous, for the unarmed B types in service at the beginning of the war were being superseded from the start of 1915 by armed C types. Together, the C and E types posed a formidable threat to the Allies' two-seater plane. The E types were inevitably the greater threat in the hands of ace pilots such as Oswald Boelcke, Max Immelmann and Kurt Wintgens, but the C types were soon fitted with a fixed forward-firing armament.

Throughout the winter of 1915–1916 and the spring of 1916 the Fokkers reigned supreme: as aircraft the E types lacked any real distinction, but by virtue of their fixed armament they prevailed over everything the Allies could put into the air. The only aircraft to have forward-firing machine-gun armament were two-seater pushers, whose performance was inferior to that of the Fokker E types, while high-performance single-seaters had to rely on expedients such as a machine-gun angled to fire outwards from the fuselage in order to clear the propeller blades. Types fitted with deflector plates were so unreliable as to be ineffective as a long-term solution to the problem.

Thus the E types gave the Germans total air superiority until at last the Allies replied with two new aircraft: the French Nieuport 11 *Bébé*, with a machine-gun on top of the wing and firing over the propeller, and the British DH2, a trim pusher single-seater with a forward-firing machine-gun. Both these aircraft could outfly the Fokker E types, and had an armament of equal efficiency. Squadrons of these aircraft, along a tactical pattern suggested by the master of air fighting, Oswald Boelcke, finally turned the tide in favour of the Allies in the summer of 1916. With the arrival of the Sopwith $1\frac{1}{2}$-Strutter two-seater, the Sopwith Pup single-seater, and the SPAD S-7 single-seater in the summer and autumn of 1916, the Allies had a numerical and technical superiority over the Germans. All three of the Allied aircraft featured a synchronized forward-firing machine-gun, and the $1\frac{1}{2}$-Strutter also had a flexible machine-gun for the observer.

Earlier in the year, the swing of air superiority towards the Allies had been hindered by the inability of aircraft manufacturers to produce anything like the volume of new aircraft required. Front-line units therefore had to make do with what they had, and the few replacement aircraft they could get. Meanwhile, manufacturers expanded their production capacity so that by the summer of 1916 production could keep up with the losses suffered during the Battle of Verdun, which began in February 1916. Here the Germans sought to prevent the French from gaining any aerial reconnaissance by standing patrols of fighters over the front-line; but the arrival of the Nieuport 11 allowed the local commander, du Peuty, to commit his air forces to the offensive and batter the Germans severely, preventing them in turn from enjoying the benefits of air reconnaissance and support. The Germans tried to develop their own contact patrols during this bitter struggle, but without success. The French offensive tactics found favour with the new commander of the RFC in the field, Brigadier General Hugh Trenchard, who had taken over in France in August 1915. Firmly convinced that his forces should always remain on the offensive, Trenchard was prepared for the RFC to suffer heavy losses, which it did, for whereas the Germans generally operated over their own territory from 1916 onwards, the RFC flew over enemy territory to engage in combat. This meant that crippled aircraft had to run the gauntlet of the German AA defences to get home, generally against the prevailing westerly winds. Though costly, Trenchard's tactics were feasible, given that the flying schools and aircraft manufacturers could make good the losses, in numbers if not in calibre. Trenchard's methods have

always been a source of controversy, but they certainly seemed to pay handsome dividends during the great first Battle of the Somme, which began so disastrously for the ground forces on 1 July 1916. The Germans could muster only 129 aircraft (19 of them fighters) against the Allies' 386 aircraft (138 of them fighters). Allied ground support tactics had reached a peak, and while the fighters roamed behind the German lines to prevent the interference of German fighters, other fighters operated over the battlefield at low-level on contact patrol work, recording the progress of the infantry and attacking German strongpoints with machine-gun fire and light bombs; artillery observation aircraft controlled the work of the guns, reconnaissance aircraft probed deep into the German line of communications zone to warn of the arrival of reinforcements, and light bombers attacked marshalling yards, railways, and bridges to hinder the arrival of German reinforcements and supplies.

The widespread use of these Allied tactics in the first Battle of the Somme may be judged a limited success, but in a way proved counterproductive as it spurred the Germans to reorganise their air force under the command of General Ernst von Hoeppner, with Hermann Thomsen, formerly field commander of the air service, as his chief-of-staff. This was a clever move, and ensured both continuity of command and the accession of new high-ranking officers to the highest position in the air service. Hoeppner and Thomsen had already appreciated that the key to Allied superiority lay in adequate

numbers of the right fighter aircraft, and the new command structure made it easier for the Germans to respond with the introduction of a new fighter, the Albatros D.I/II, during October 1916, and a new type of unit, the *Jagdstaffel* or fighter squadron of 14 aircraft, which would receive the best pilots, and be used as mobile 'fire-brigade' units where air superiority was needed. The success of the German tactics and aircraft may be gauged from the fact that between September and November 1916, Oswald Boelcke's *Jagdstaffel No. 2* shot down 76 Allied aircraft for the loss of only seven of its own aircraft. While the importance of the new type of fighter unit must be stressed, so too must that of the new D type fighters, armed with two synchronized machine-guns and possessing performance generally superior to that of Allied types. Thus the Germans attained total air superiority during the winter of 1916–1917, culminating in the RFC's direst days during 'Bloody April' 1917, when Trenchard's continued insistence on the offensive, combined with the generally low standards of replacement pilots and the excellence of German aircrew and aircraft, meant that pilots' lives over the Western Front could be measured in terms of hours rather than days. During this period, the Germans were altering fundamentally the nature of the air war by increasing the importance attached to artillery spotting aircraft, by rapidly developing their 'contact patrol' tactics into an effective system, by building up a powerful force of ground-attack aircraft and their relevant tactics, and by scaling

Germany's Albatros D.VA (*left*), one of the remarkable D-series, dominated air fighting over the Western Front until the introduction of new-generation Allied fighters, such as the Sopwith Camel (*above*), ended German air supremacy.

down the airship offensive against Great Britain as bombers capable of carrying a worthwhile bomb-load over the necessary range became available. Combined with an increase in the production capacity of the German aircraft industry, this gave the Germans total air superiority during the spring of 1917, and the attacks of *Bombengeschwader Nr. 3* against England proved a severe set-back to the British war effort. Indeed, so severe was the threat of German bombing, especially to civilian morale, that a committee to investigate the RFC was established. This finally recommended that an air force independent of both the army and the navy should be established: the Royal Air Force, an amalgamation of the former RFC and RNAS, was formed on 1 April 1918.

Throughout the winter of 1916–1917, the Allies prepared new combat aircraft for service, but the weather prevented any large-scale air activity, although the Allies tried to get air reconnaissance of the Germans' new defence line, the so-called 'Hindenburg Line.' However, both the weather and the activities of the formidable *Jagdstaffel Nr. 11*, led by Manfred, *Freiherr* von Richthofen, made this an extremely risky business. The Allied problem was made more difficult by the fact that the German elite fighter units took it upon themselves to roam over to the Allied side of the lines, a tactic untried since the

winter of 1915. During this period the Allies introduced aircraft such as the Bristol F2A, the Royal Aircraft Factory SE 5a, the Sopwith Triplane and others, while the Germans brought in only the Albatros D.III. Most of the new Allied aircraft, and two others that entered service in the middle of the year, the Sopwith Camel and SPAD S.13, were more than a match for the German fighters, but failed to win air superiority from the Germans. During 'Bloody April' 1917, the Allies lost some 300 aircraft to the Germans' 140, despite the fact that in overall terms the Germans were outnumbered by about 3·4 to 1 in terms of fighters. Committing only their most experienced pilots, the Germans concentrated on the Allied two-seaters, which British and French tactics kept over the front line in large numbers. But despite the Germans' numerical victory, it should be noted that the Allies achieved their objectives of supporting their ground forces during the spring offensives, albeit only at a high cost. In the long term, moreover, the Allies learned much of the German tactical methods, and so devised means of countering them, and this was to stand them in good stead in the months to come.

From May 1917 onwards, the Allies gradually gained overall air superiority: more of the latest aircraft, which the Germans could not match in calibre or in number, arrived at the front, and their crews worked hard on effective tactics. The growing success of the Allied fighters is reflected in the German reorganisation of their fighters on 26 June 1917. The high command realised that individual

Jagdstaffeln could not be expected to cope with the large formations of Allied fighters now beginning to roam the German skies, and so it was decided to form *Jagdgeschwader* or fighter wings. The first was *Jagdgeschwader Nr. 1*, commanded by the redoubtable von Richthofen, consisting of *Jagdstaffeln Nr. 4, 6, 10 and 11*.

Despite the introduction of the supremely agile Fokker Dr.I triplane fighter, the Allies continued to have things very much their own way, the Germans being able to regain only temporary air superiority with the limited numbers of *Jagdgeschwader* available. The Allied superiority overall was well displayed in the great third Battle of Ypres, which culminated in the Battle of Passchendaele. Here Allied reconnaissance two-seaters were able to go about their work virtually unmolested, so effective was the screen of Allied fighters.

The Germans meanwhile, had been working to improve their chances of winning a decisive land victory before the arrival of large-scale American forces made this impossible. All was subordinated to the land battle, therefore, and in the Battle of Cambrai (November 1917) there occurred the first mass use of the German *Schlachtstaffeln* or ground attack-squadrons. The British had achieved a tactical advantage in the opening stage of the battle by the first mass use of tanks, but this was more than balanced during the second stage by the vehemence and speed of the German riposte, aided quite considerably by the deployment of specialist *Schlachtstaffeln* with special aircraft, covered by the fighters of *Jagdgeschwader Nr. 1*.

This established the pattern for the Germans' last major efforts in the spring of 1918, when five major offensives were launched with a view to defeating the British and French before US ground forces could reach the battle area. All available air forces were committed, including the first production examples of Germany's best fighter of the war, the first-class Fokker D.VII. In general the Germans were able to secure short-term local air superiority, but the Allies could deploy significantly greater numbers of aircraft, and could thus regain air superiority, if only with heavy losses. Ground attack by both sides predominated, and most losses were attributable to such operations. Losses themselves were heavy, Germany's totalling 180 aircraft during May, with Allied losses exceeding 400. Gradually, the enemy's advances were halted, and the Allies then set about their own great offensives designed to end the war by crushing the exhausted Germans.

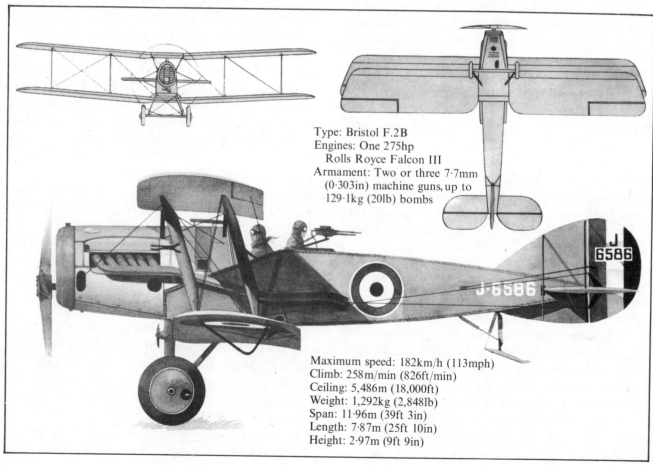

Type: Bristol F.2B
Engines: One 275hp
  Rolls Royce Falcon III
Armament: Two or three 7·7mm
  (0·303in) machine guns, up to
  129·1kg (20lb) bombs

Maximum speed: 182km/h (113mph)
Climb: 258m/min (826ft/min)
Ceiling: 5,486m (18,000ft)
Weight: 1,292kg (2,848lb)
Span: 11·96m (39ft 3in)
Length: 7·87m (25ft 10in)
Height: 2·97m (9ft 9in)

By August 1918, when the offensives started, the Allies were all but invincible in the air, having large numbers of qualitatively superior aircraft, while the Germans had few of their latest fighters and ground attack machines. The German effort was hampered, moreover, by lack of fuel and inadequate aircrew. The Allies therefore dominated the skies, although the very numbers of Allied aircraft allowed the better German pilots to score heavily. For the rest of the war, the Allies swept all before them, their fighters generally keeping the German fighters at bay while the light bombers, artillery spotters and reconnaissance two-seaters went about their tasks virtually unhindered.

Throughout the period from spring 1917, the Germans had been raiding London with Gotha twin-engined bombers, later supplemented by a few of the giant Zeppelin Staaken R.VI four-engined bombers. So great was the public anxiety about this new type of warfare that during the summer of 1917 a combined force of RFC and RNAS bombers was established to undertake similar raids against Germany. This culminated, in June 1918, in the formation of the RAF's Independent Air Force, under the command of Major General Sir Hugh Trenchard, who had recently resigned as Chief of the Air Staff at the new Air Ministry. For the rest of the war, the Independent Air Force launched a series of raids on German munitions factories and other targets designed to lower Germany's war-making potential and affect civilian morale. Overall results were small, and losses at times heavy, but the Independent Air Forces' effort is of great importance as the first properly conceived strategic bombing effort.

Also of great importance was the French *Division Aérienne*, formed on 18 April 1918 as a mobile air force of two *groupements*, one having 12 *escadrilles* (squadrons) of SPAD XIII fighters and nine of Breguet Br.14 light bombers, and the other with 12 SPAD and six Breguet *escadrilles*, for a total of nearly 600 aircraft. This major air formation played a decisive part in the checking of the last stages of the German spring offensive, and then in supporting the French and American offensives during the final months of the war.

On other fronts, such as the Eastern, Italian and Macedonian, the pattern of air warfare followed that of the Western Front for the most part, and only in the Palestinian and Mesopotamian campaigns were Allied aircraft able to play a more decisive part, largely as a result of the lack of Turkish air opposition.

The Spanish Civil War

The Spanish Civil War, which began on 18 July 1936 and ended on 28 March 1939, was significant in the history of military aviation as the testing ground for the aircraft and tactics of the first part of World War II for it was in Spain that German, Italian and Russian aircraft of the latest designs were first used in action. The Germans and Italians in Spain also adopted the tactics characteristic of their air efforts in World War II. These tactics contributed to the series of dazzling successes achieved by German arms in the period up to the end of 1941 (though the writing appeared on the wall, so far as the Luftwaffe was concerned, after Germany's defeat in the Battle of Britain). Having been a relatively strong advocate of strategic bombing, the Luftwaffe during the Spanish Civil War became the world's main advocate of tactical air power, to the exclusion of strategic bombing after the death of the Luftwaffe chief-of-staff, General Walther Wever, in an air accident in 1936.

In 1936 the Spanish Air Force was both obsolescent and small. There were some 40 Nieuport ND 52 fighters (about 13 of them in Nationalist hands), 60 Breguet Br. XIX light bomber/reconnaissance aircraft (some in Nationalist hands), 20 Vickers Vildebeest torpedo-bombers (all in Republican hands), and miscellaneous other types. Most aircraft were in the hands of the Republicans (the government), but the majority of the air force's flying crew swiftly joined the Nationalists (fascists).

The Nationalists, better prepared for the war of their own making, were soon able to make important but limited land advances in continental Spain. However, the Nationalists were outnumbered by the Republicans, so they recruited reinforcements from the pro-Nationalist Spanish North African territories. But with the Spanish Navy loyal to the Republican cause, a large-scale airlift, never before attempted, seemed the only way to reinforce the Nationalists in southern Spain. The Nationalists had a number of miscellaneous transport types, including DC-2s, Fokker F-VIIs and others, and these were soon ferrying troops across the Straits of Gibraltar. Though this first airlift of major proportions provided useful additions to the Nationalist strength in southern Spain, the lift capability of the Nationalist aircraft on their own was too limited to bring in sufficient troops to alter significantly the balance of power. The Nationalist leaders therefore asked Germany and Italy, both pro-fascist states, for assistance. By 27 July the Germans had sent to Spain 20 Junkers Ju 52/3m bomber/transports, joined on 3 August by nine Savoia-Marchetti SM-81 bomber-transports. These were used principally to patrol the Straits of Gibraltar against Republican naval interference. The effort by the Ju 52/3m transports was to have a decisive effect: some 13,500 men and 260,000kg (570,000lb) of supplies were airlifted to Spain.

Meanwhile, the first air combats had taken place over Spain, the Nationalists generally prevailing: on 23 July 1936 Lieutenant Narcisco Bermudez de Castro, flying a ND-52, shot down a Republican aircraft of the same type. Between 25 and 31 July Lieutenant Miguel Guerrero Garcia, also flying a ND-52, shot down a Dornier Wal, a Vildebeest and a BR.XIX. By the end of the month the Nationalists in southern Spain, commanded by General Franco, had gained local air superiority, largely because most of Spain's experienced aircrew had joined the Nationalist cause. The Nationalist cause was strengthened, moreover, by the arrival of the first German fighters, six Heinkel He-51 biplanes, by sea on 6 August. Unloaded and assembled at Barcelona, the He-51s were soon in action, largely as escorts for the Ju 52/3m force commanded by Lieutenant Rudolf von Moreau. The SM-81 force had set off from Sardinia for the base at Tetuan in North Africa, and were joined by 12 Fiat CR-32 biplane fighters on 14 August. Nominally part of the Spanish Foreign Legion, this Italian air component became known as the *Aviazione Legionario*, or *Aviacion del Tercio* to the Spanish. The German air component, made up of 'volunteer' aircrew, eventually became the *Legion Condor* and soon expanded from the initial 20 Ju 52/3m and six He-51 aircraft. The Nationalist air arm also grew considerably during the war, largely as a result of a major influx of German and Italian aircraft.

Up to the middle of August 1936, the Germans had confined their efforts to the airlift and to advising the Spaniards. However, the Spaniards were not particularly successful in battle when flying the German aircraft and they requested the Germans to fly their own aircraft in combat, which they agreed to do. The Italians had already paved the way on 5 August when some SM-81s bombed a Republican cruiser shelling Nationalist positions in Morocco, and on 14 August the Germans followed with a Ju 52/3m raid on the Republican battleship *Jaime I*, which was crippled in Malaga harbour by two bombs dropped from 450m (1,475 ft).

With the situation turning decisively against them, the Republicans appealed for foreign aid to balance that supplied to the Nationalists by Germany and Italy. The French government, headed by the socialist Léon Blum, responded quickly by allowing the dispatch of seven Potez 540 multi-seat combat aircraft to Spain. At the same time a volunteer *Esquadra Espana* (Spanish Squadron) began to form in Toulouse under the inspiration of the celebrated writer André Malreaux, with 17 Dewoitine D-372 parasol-wing fighters. Fourteen of the aircraft finally reached Madrid, where the squadron's personnel had been flying any aircraft they could find since their arrival on 5 August. The Republican cause also had other volunteer personnel in the form of the International Squadron, formed at Getafe under Captain Martin Luma with a miscellany of aircraft, including Dewoitine D-372s, D-371s, D-501 monoplanes, D-510 cannon fighters, five Loire-Nieuport LN-46 monoplanes and some 25 SPAD 510 biplanes.

The Nationalists had also realized the importance of a formal air organization, and the first squadron of He51 fighters was formed under the command of the celebrated aerobatic pilot Captain Joaquin Garcia Morato, at Seville on 15 August. For the most part, though, Spanish pilots performed only indifferently, and it was only when the German 'advisers' were allowed to fly operationally from the middle of the month that Nationalist successes began to increase rapidly. So far the Nationalist air effort had been in the south, there being only one type of aircraft, the ND-52, in the northern sector, which was also a Nationalist stronghold. During August, therefore, reinforcements were sent north, the first being a bomber squadron of de Havilland Dragons and Fokker F-VIIs.

The Nationalists were the first to go into action with their extemporized bomber force, but the Republicans also formed a bomber unit, manned by international crews, with the 49 Potez 540s to reach Spain during 1936, plus some 30 Bloch MB-200 and MB-210s, and four Potez 543s. With this force the Republicans were able to undertake some significant raids against the rather haphazardly organized Nationalists, the squadron commanded by Rambaud and Morato proving inadequate. In German hands, however, the He-51s began to show their effectiveness as did the Ju 52/3m bombers of the Pedros y Pablos Escuadrilla. During September a second squadron of CR-32 fighters became operational, with Italian and Nationalist pilots. During October 1936 the Nationalists reorganized their air arm to meet the needs of the primary front around

Madrid, held by Republican popularist forces against strong Nationalist pressure. On this front the bomber forces and ground attack units had a considerable part to play. Five new Breguet units were formed, as was a flying-boat unit with Dornier Wal aircraft, and the Italian IMAM Ro-37*bis* gradually replaced the indifferent Heinkel He46 as the Nationalists' prime army co-operation aircraft. Another Fiat squadron arrived from Italy, bringing the Nationalists' inventory of the type to 30.

During October the Republicans' SPAD 510 fighters had regained air superiority over the Madrid sector, but as a result of strenuous Italian efforts the Nationalists were once more in command at the end of the month. Thereafter, the Nationalists decimated the Republican air force and flew over the battlefields and Republican rear areas at will against negligible fighter opposition. At the same time, however, the first Russian aid for the Republicans was beginning to arrive at Cartagena, where 'volunteer' aircrew and groundcrew of the Red Air Force had been preparing since 10 September. First to arrive was a shipment of Tupolev SB-2 monoplane bombers of the latest design, far superior to anything deployed by the Nationalists. The *Katiuskas*, as the SB-2s were dubbed in Spain, made their first raid on 29 October, and immediately proved their overall superiority. Commanded by Lieutenant-General Yakov Smushkevich, the Russian aircrew (all of the pilots being of the rank of captain or above), proved

capable in combat as did the fighter pilots, whose aircraft began to arrive in mid-October: 30 Polikarpov I-15 biplanes arrived by sea at Bilbao, and then 25 at Cartagena. The Russian aircraft formed the equipment of four *escuadrillas*, two in the northern sector with Russian aircrews, and two in the southern sector.

By the end of October 1936, all the protagonists of the Spanish Civil War (Nationalist Spanish, German, Italian, Republican Spanish, Russian and

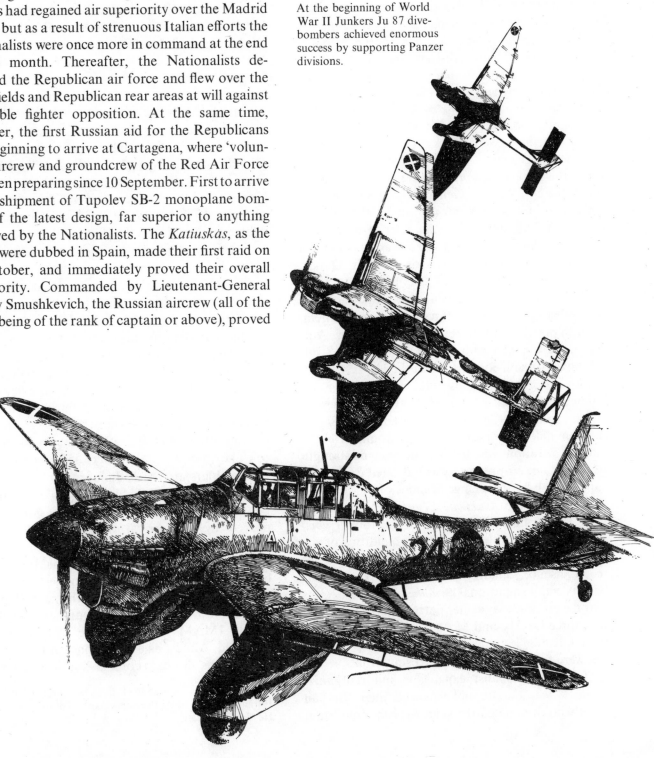

At the beginning of World War II Junkers Ju 87 dive-bombers achieved enormous success by supporting Panzer divisions.

a number of other supporters of the Republican cause) had arrived and the scene was set for the introduction of larger numbers and more modern types of aircraft. This was indicated at the end of the month by the arrival of the very latest Russian fighter, the Polikarpov I-16, dubbed *Rata* by the Nationalists and *Mosca* by the Republicans. Some of the Type 5 mark were the first to arrive, and these were immediately issued to three all-Russian squadrons specially formed to use them. The Republicans now had five modern fighter units in the Madrid sector, and two in the Bay of Biscay sector, where they had by this time launched an offensive, and the Russian aircraft were the main aerial support against the small Nationalist air force in this region. Thus in the northern and central sectors, the Republicans reigned, the I-15s and I-16s proving more than a match for the obsolescent He51s and the slightly more modern CR-32s.

The Germans were quick to respond however, and the main elements of the German expeditionary force, the *Legion Condor*, arrived in mid-November under the command of General Hugo Sperrle, with Lieutenant Colonel Wolfram von Richthofen (a cousin of Manfred) as his chief-of-staff. The reinforcements consisted of various army personnel and tanks, 4,500 Luftwaffe personnel, 20 Ju 52/3ms, 14 He-51s, six He-45s and two floatplanes. The German strength in Spain thus amounted to Kampfgruppe/88 with three bomber Staffeln, Jagdgruppe/88 with three fighter Staffeln, Aufklärungsgruppe/88 with two reconnaissance Straffeln, one of them equipped with the Heinkel He-70 monoplane, Aufklärungsstaffel See/88 with reconnaissance float-planes, plus two batteries of 20mm light Flak guns, four batteries of 88mm medium flak guns (Flak/88), a signals unit (LN/88) and an operations staff (S/88). The growing strength of the Italian air presence was indicated during the same month by the change of name of the *Aviacion del Tercio* to the *Aviazione Legionaria*, which had three CR-32 fighter squadrons of the 16th *Gruppo 'Cucaracha'*, two SM-81 bomber squadrons of the 24th *Gruppo 'Pipistrelli'* and one IMAM Ro-37*bis* reconnaisance squadron. The Spanish Nationalists had six small squadrons (*escuadrillas*) of Ju 52/3ms in three groups, two fighter patrols (*patrullas*), each with three He-51s, and Morato's *Patrulla Azul* with three CR-32s. In January 1937 two groups of He-46s were formed, each with two squadrons.

Despite Nationalist expansion and reorganization, it was the Republicans and their Russian allies who dominated the skies over the important

sectors, and who took the air war to the fascists' airfields. Nationalist morale suffered a severe blow, particularly in the Italian camp, where Colonel Bonomi ordered his men not to cross the lines. The swing to the Republicans was emphasized during the December 1936 offensive in Aragon. The two Nationalist He-51 *patrullas* were rushed to the scene of the Republican advance, but were soon destroyed, one in the air, and four of the surviving five on the ground. On each occasion Russian aircraft were responsible. Also significant towards the end of 1936 was the gradual disappearance of many of the 'international' aircrew, who perhaps found the savagery of the Spanish Civil War too much for them. This left a number of dedicated men, most of them Yugoslavs, who were then gradually replaced by more volunteers, many of them British and American. In December Malreaux's *Esquadra España* was disbanded. Both sides had by now had time to assess the combat performance of their aircraft, and it was clear that the Republicans' Russian aircraft were superior to the Nationalists' German and Italian aircraft, even if the calibre of the nationalist pilots was higher. The I-16 monoplane fighter was better than the He-51 and CR-32, and the I-15 biplane fighter was superior to the Italian fighter in some respects, and to the German fighter in all respects. So far as bombers were concerned, there was no doubt at all that the SB-2 was markedly better than the Ju 52/3m and SM-81. Fortunately for the Nationalists, however, better German aircraft were ready for dispatch to Spain: the He-51, which had proved quite successful as a light ground attack aircraft once its inadequacies as a fighter had been pointed out, was supplemented and then replaced in 1937 by the Henschel Hs-123, used as a ground attack aircraft rather than in its designed role of dive-bomber; the first of 45 Messerschmitt Bf 109B fighters arrived in January, and soon proved themselves the best fighters of the war; and the Nationalist bomber force was considerably improved by the arrival of Heinkel He-111 and Dornier Do-17 monoplane bombers, in most respects the equals of the SB-2. The Italians also brought in newer aircraft, though they were content with the CR-32 fighter. Among the newer types was the Savoia-Marchetti SM-79 bomber, and the adequate IMAM Ro-37*bis* force was considerably bolstered. Almost as a by-product of the Germans' and Italians' desire to combat-test their new aircraft

Two of the most famous RAF fighters from the Battle of Britain era. *Top right:* the Supermarine Spitfire and *right* the Hawker Hurricane.

Lockheed P-38 Lightnings were designed to fill the absence of a high-performance interceptor. No single power plant then available could have provided the rate of climb demanded by the USAAC; note its rather unusual configuration. It has proved to be a highly efficient aeroplane.

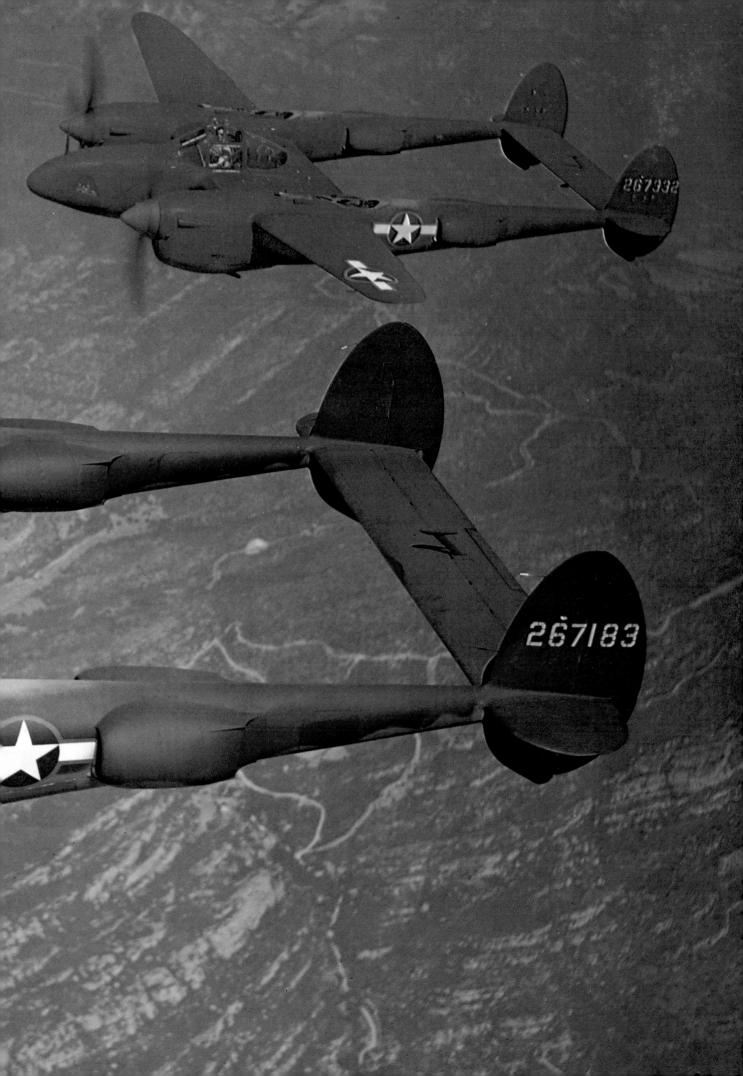

All of the major air arms have procured aircraft such as the four-engined Lockheed Orion and Two-engine Neptune (*left*), which are equipped with advanced electronics to locate deep-diving nuclear-powered submarines. The Gloster Meteor (*top*) was the first turbine-powered fighter to enter service with the RAF in 1944. The Hawker Sea Hawk (*below*), here in use with the German Navy, is typical of the next generation of turbojet-powered combat aircraft which became operational in the 1950s.

The small world of an aircraft carrier can be home to more than one type of aircraft. Seen here on board HMS *Ark Royal* are two Fairey Gannets, equipped to provide early warning against enemy intrusions, four Hawker Siddeley Buccaneer strike aircraft near the bow, and three McDonnell Douglas Phantom II multi-role fighters over on the starboard side.

and give their aircrew battle experience in Spain, the Spanish Nationalist air arm, the *Agrupacion Española*, received the aircraft lately used by the Germans and Italians, which greatly strengthened it, especially in the ground attack role. The Republicans had also received a considerable number of Russian aircraft, paid for in gold, but the presence of Russian aircrews sometimes caused friction, for the Russians wished to run things their own way, and sometimes bombed targets of their own choice, or refused to give the Republicans adequate fighter cover unless certain conditions were met.

In 1937 there was a basic balance in the air war, though the Republicans enjoyed a marked superiority during the Nationalists' February offensive to cut the Valencia–Madrid road in the region of the Jarama river, and also during the Italian ground forces offensive that culminated in the battle of Guadalajara in March. In the latter battle the *Aviazione Legionaria* was forced to operate from fields so poorly sited, relative to the land battle, that the Republican aircraft, operating in all weathers, could play a decisive part in halting the Italian offensive.

At this stage General Franco, now commander-in-chief of the Nationalist forces, decided to switch the main weight of the fascist attacks to the north, to secure the industrial regions of Asturias and Vizcaya during April and May 1937. Both the *Legion Condor* and *Aviazione Legionaria* deployed their main strength in support of this offensive, which met little Republican air opposition apart from a few I-15s and some 25 Koolhoven FK-51 biplane fighters, joined later by 30 Avia A-101 biplane reconnaissance bombers. The Nationalist offensive opened on 31 March 1937, and was notable for some swift advances with excellent air support. The object of the offensive was Bilbao, a key city in Republican defence plans, and provided with a series of fortified outer defence towns, one of which was to go down in history as one of the key names in military history: Guernica, a market town with an important river bridge. For reasons that are not clear to this day, the *Legion Condor* decided to destroy the town, possibly to assess the effect of area bombing on a town and its civilians with a view to future operations. On 26 April the centre of Guernica was destroyed by the bombs of a wave of He-111s, followed by a wave of Ju 52/3ms. The psychological impact of the raid was

enormous, both in Spain and elsewhere, and it helped to foster the myth that the bomber was invincible, and that attacks on civilian targets would dominate any future war. Be that as it may, Guernica fell to the advancing Nationalists two days later, and Franco's forces were able to drive on towards Bilbao. The Republicans could not halt the drive in the area, but could only hope to persuade the Nationalists to divert their attentions elsewhere. This they achieved by a breakthrough from Guadalajara by the International Brigades, beginning at the end of May and directed towards Segovia. Nationalist air strength was diverted from the Basque region to stem the Republican drive, with the result that the offensive towards Bilbao faltered. Near Guadalajara the International Brigades were finally driven back with the aid of air power.

Again, the Republicans tried to divert the Nationalists, this time with an offensive at Brunete, north of Madrid, starting on 6 July. The Republican air force could muster some 200 aircraft, and the Nationalists diverted about 150 from the northern front. Both sides were now well equipped and fully trained, and fierce air battles raged over the Brunete front until the Republicans were defeated by the end of the month. The relative combat efficiency of the two sides may be gauged from the fact that the Republicans lost about 50 of their 200 aircraft, and the Nationalists about 25 of their 150. With the defeat of the Brunete drive, the *Legion Condor* returned north to cover the final stages of the offensive against Santander, which fell on 22 August, putting the key areas of the northern coast in Nationalist hands and freeing considerable forces for operations on the Madrid front.

Yet again the Republicans sought to divert the nationalists, this time with the Belchite offensive from Barcelona towards Saragosa, supported by 150 aircraft. At first the Republicans advanced well during the closing days of August, but although Franco released some air units, which caused the Republicans heavy losses, he concentrated his main strength against the last northern stronghold in Asturias, which collapsed in early October, finally freeing all Franco's forces for operations against Madrid and in Aragon and Catalonia.

By this stage of the war, most of the foreign volunteers in the Republican air force had gone, either killed or having left the country, though their places were taken by Spanish aircrews trained in Russian flying schools. In September 1937 the Republicans still had about 470 aircraft to the Nationalists' 350, but the strategic initiative was now

A Hawker Harrier V/STOL (vertical/short take-off and landing) strike aircraft is being prepared for a sortie.

firmly in the latters' hands as they controlled the key economic areas of Spain. During the last months of 1937 and the whole of 1938, the Nationalists gained total air superiority, allowing them to use their air forces as a powerful adjunct to their land forces. Here the Germans experimented very widely under the eagle eye of von Richthofen, and evolved the fighter and ground attack tactics that so characterized the German air operations at the beginning of World War II.

The Republicans held out with enormous tenacity for the rest of the war, while the Russians realised that defeat was inevitable and so pulled out before the loss of Barcelona on 26 January 1939.

The Battle of Britain

With the fall of France in June 1940, just six weeks after the beginning of the western offensive, Hitler had realized most of his goals. It remained only to arrange a compromise with the British, or failing that, to force them into submission. France's defeat would have opened the way for an immediate invasion of England had it not been for two facts. First, adequate preparations for an aerial and seaborne invasion of the British Isles had not been made prior to the launching of the western offensive. Second, Hitler hestitated to order the preparation of such an offensive, hoping that a peaceful settlement of Anglo-German differences might be reached despite the events of September 1939–June 1940.

The mounting of an assault on England involved an effort for which the German army had few if any troops trained for seaborne landing operations and the navy had few if any transports to carry the men across the Channel. Neither the General Staff nor the Admiralty had seriously considered plans for such an operation before the fall of France, nor had they been prodded to do so by Hitler; more important still, there was little enthusiasm for such a venture. The military leaders of the Reich had prepared only for continental war. In their minds, a crossing of the Channel could only be undertaken as a last resort and even then they were not willing to guarantee the result. Only the Luftwaffe seemed anxious to initiate the conflict, but Goering could not persuade his peers or the Fuehrer to take immediate action.

Hitler continued to cling to the notion that he could reach a compromise with Churchill. Even after the British Prime Minister refused to accept his proposal that England recognize German hegemony on the continent in return for German recognition of the sanctity of the Empire, Hitler appears to have been reluctant to force an invasion of England. His military advisers were only too happy to follow his lead in the matter, hoping that Churchill's stubbornness would give way and rapprochement could be reached; they dallied for almost a month after the fall of France before ordering the preparation of a comprehensive scheme to invade England, and even then discussed the operation with little enthusiasm.

The German plan, 'Operation Sealion,' was finally commissioned on 16 July 1940, but was not ready for consideration until the end of the month. As presented to the chiefs of staff, it called for a hurried accumulation of shipping from Belgian, French, and Dutch ports to transport Rundstedt's Army Group A across the Channel for landings near the Thames estuary along the southeast coast. Following the initial landing of some 10 divisions plus one airborne unit in this area, the second and main force would be ferried across the Channel. This would be the major part of the operation, necessitating the requisition of thousands of ships both large and small to transport several hundred tanks and other armoured vehicles plus an additional 17 infantry divisions. Once landed, these forces would join up with the first wave, isolate London, and cripple British resistance. If carried out, the operation would be by far the most complex and difficult the Germans had undertaken.

The original target date for launching Operation Sealion was set for mid-August 1940 at the insistence of the Fuehrer, who was growing increasingly angry at Churchill's intransigence. However, since actual preparations for the manoeuvre were not begun until the last week in July, the sheer magnitude of the task of assembling hundreds of thousands of men and thousands of transports for the venture defied Hitler's timetable and the military and naval commanders involved were forced to ask for a postponement until September. Reluctantly, Hitler acceded to their appeals and the invasion was pushed back some two weeks; in the meantime the Luftwaffe was told to go ahead and soften the British for the eventual kill.

Goering's anxiety to annihilate the Royal Air Force as soon as possible delighted his peers in the other services; they continued to be sceptical about the whole affair and were only too happy to have the Luftwaffe prepare their way. Admiral Raeder

Five Spitfires cruise in formation. One of the two types of defending fighters during the Battle of Britain, it derived from R. J. Mitchell's Schneider Trophy winning designs.

was frightened enough at the possibility of having to engage the British fleet while ferrying thousands across the Channel, and if Goering could neutralize the RAF in advance of the effort, his task would be much easier. If, on the other hand, the Luftwaffe failed to carry out Goering's mission, the Fuehrer might think twice and cancel or drastically alter the operation. In any event, by encouraging the Luftwaffe to 'carry the ball', the resources of the army and navy remained intact, at least for the moment.

When Hitler issued his order on 5 August authorizing massive air strikes against England, Goering was more than ready to respond; by the beginning of August, he had already gathered a force of nearly 2,700 aircraft for this purpose. Like his colleagues in the Admiralty and on the General Staff, Goering placed little faith in Operation Sealion but he was anxious to employ his aircraft not simply in support of the objectives of the operation, but in a total effort to prove the supremacy of air power in modern warfare. During the months of the Luftwaffe attack on England this was to cause considerable friction between the Air Marshal and the heads of the other services who feared that Hitler might continue to insist on the execution of Operation Sealion whether Goering had correctly followed his orders and prepared the ground for them or not. The bickering and confusion continued until the operation was superseded by the invasion of the Soviet Union and proved to be of indirect but immense aid to the British, who were spared the worst consequences of the attacks by Goering's frequent changes of tactics and targets.

At the start of the Battle of Britain, the Luftwaffe possessed a numerical superiority over the Royal Air Force. Numbers, however, proved to be somewhat misleading; although the Germans had more bombers and dive bombers, the two sides were nearly equal in numbers of fighter aircraft. Germany's slight edge at the outset was more than compensated for by the superior output of British aircraft plants which were eventually able to build planes faster than they were destroyed. It was the fighter and not the bomber which held the key to victory or defeat in this encounter, and here the British had one great advantage: their nimble Spitfires and Hurricanes did not have so far to fly. The German Bf 109, like its British counterparts, had a radius of action of only about 201km (125 miles), and thus, after leaving a fuel margin for combat, could barely cover the distance from Calais to London and back. Even to do this much required a fairly inflexible routing plan,

which greatly aided the British defenders. What the RAF did lack as the battle progressed was not aircraft so much as trained pilots. Had RAF training schools been producing more pilots, or had there been a larger reservoir of trained men available at the beginning of the battle, the situation would have been far less critical for the British.

The first phase
Although German aircraft had appeared over England in July, the Battle of Britain was not begun in earnest until 13 August 1940, when the Luftwaffe launched the first in a series of concentrated attacks on British fighter bases and radar installations. Such attacks were to last until the middle of September when tactics and targets were changed and a second phase of the battle began.

The raid on 13 August was the largest aerial attack to that date; over 1,400 aircraft were sent aloft over England and German pilots were instructed to destroy the forward bases of the Royal Air Force southeast of London. This was to be the first in a series of four raids, to be launched on consecutive days, with which Goering hoped to destroy the aerial defence in the greater London area, thus opening the way for uncontested German bombardment of British naval and merchant marine targets. The success of Operation Sealion would depend upon the accuracy of the Air Marshal's predictions.

Adlertag, or 'Eagle Day' proved to be somewhat disappointing for the Luftwaffe and its commanding officer. The weather cooperated with the British defenders and this plus the early warning radar system allowed them to 'scramble' into the air and meet the attacking German fleet before the planes could successfully unload their cargo of bombs and explosives. By the end of the afternoon the Luftwaffe had managed to seriously damage only two of the forward bases at a cost of 45 of their aircraft. The RAF lost only 13 planes, managed to prevent the destruction of all but two fighter bases and, most important, kept the Germans from destroying the radar installations that dotted the coastline. That night, however, the Luftwaffe returned and succeeded in attacking several aircraft plants, thereby exposing the RAF's inability to combat night attacks as efficiently as those flown during daylight hours. Such night flights, therefore, were to become increasingly common and devastating.

The Luftwaffe launched another massive attack on 15 August, the largest raid to be mounted by Goering during the battle for Britain. Over 1,800

Dornier Do 17 'Flying Pencil' bombers, a nickname derived from their slim lines, were used extensively in Luftwaffe attacks on British cities.

aircraft, including Scandinavian based units, took part in this effort to destroy RAF fighter bases in southeast England, but these bases were not the Luftwaffe's only target. Goering and his subordinates also hoped to destroy installations in the Midlands and Yorkshire while RAF units from these bases were sent south to aid Fighter Groups 10 and 11, who were protecting the southern approach to London and the Channel coast. If the plan was successful, England's fighter defence would be incapacitated in one stroke and the Luftwaffe would be able to bomb military and naval targets with impunity.

The sheer number of aircraft in the German armada dwarfed the resources of the British defenders and Germany might well have scored a dramatic victory were it not for the early warning provided by British radar installations and aircraft spotters, and the superb co-ordination of RAF units by Air Vice-Marshal Keith Park, the commander of Fighter Group 11 and his counterpart in the north, Air Vice-Marshal Leigh-Mallory, commander of Fighter Group 12. By the time German task forces arrived over the Channel coast, British units were already in the air and what they lacked in numbers was more than compensated for by their spirit, and the information they had received from flight headquarters.

In the north the Luftwaffe was badly mauled. There a fleet of over 100 bombers accompanied by

35 fighters attempted to attack airfields near Newcastle and in Yorkshire, but met with little success. Savage fighter resistance and effective anti-aircraft fire cost the invaders over 15 aircraft while little damage was done to British bases. A second attack by some 50 bombers met with better luck later in the day and managed to circumvent RAF defences in the north, destroying at least one base in the process; on the run back to Denmark, however, RAF pilots downed at least 10 of the 50 with no loss to themselves. Such losses proved the vulnerability of bombers without adequate fighter cover, a fact which the Luftwaffe was slow in comprehending. In any case, 15 August marked the first and last major attempt to destroy British bases north of London.

In the south, the RAF was less successful in turning back the German attackers. Faced with far greater numbers of enemy aircraft than their colleagues in the north and with more bases to defend, about the best that Fighter Groups 10 and 11 could do was to soften the German blow by engaging the enemy wherever and whenever possible; even this might not have sufficed were it not for the fact that the two German airforces sent over southern England, Luftflotten 2 and 3, had not adequately co-ordinated their efforts. Thus, although confusion at the radar stations caused by the sheer number of German planes threatened to deprive the RAF of the time they needed to prepare their defence, the Luftwaffe's inability to deliver a single well co-ordinated attack insured that the RAF could at least hold its own. By the end of the day the Luftwaffe had not succeeded in its major goal of eliminating England's southern aerial defence. Thirty-nine British planes had been lost and several bases were badly damaged, but the RAF still flew. Churchill might well say of this encounter that: 'Never in the field of human conflict was so much owed by so many to so few.' Unfortunately, however, there was still more to come.

The loss of 75 planes on 15 August did not stop Goering's campaign against British bases. On 16 August yet another attack was launched and almost 1,700 planes were sent over England. Based upon intelligence reports which estimated current British strength at no more than 300 planes, Goering remained confident that the end of the RAF was near. He was so confident, in fact, that he suspended attack on British radar sites on the grounds that the few aircraft left to the RAF could hardly benefit by such installations, and that by using all of his aircraft against the bases instead of diverting some to the radar sites, he could hasten the collapse of England's aerial defence, guaranteeing absolute German air supremacy within a matter of weeks. Neither of these assumptions proved to be correct.

Although British losses had been heavy during the first days of the Battle of Britain, they were not nearly as high as German leaders believed; thus, the German raiders suffered heavier losses than were necessary in their raids of 16 August and later because they were not prepared for the stiff resistance they encountered. During the first three weeks of August alone, Germany lost over 450 planes, an average of about 150 per week. During the weeks to come, this number was to increase considerably.

Despite the heavy losses suffered by the Luftwaffe, Goering still pressed the attacks on British fighter installations, and by the beginning of September they were beginning to bear fruit. Although British morale was buoyed up by exaggerated reports of RAF successes, the almost daily pounding of their facilities by Goering's air force was not without grave consequences. During the month of August the RAF had lost 359 planes while many more were damaged on the ground, and the loss per week had increased rather than decreased by the beginning of September. Since the aircraft production schemes introduced by Lord Beaverbrook had not yet reached the point where replacements exceeded losses, the situation looked bleak as summer drew to a close. Had Goering not been diverted to another course of action early in September, the RAF might

well have been destroyed.

The campaign against the forward bases of the RAF was brought to an abrupt end by the British bombing attack on Berlin on 25 August. This attack, itself precipitated by the accidental bombing of London the night before, had a stunning effect on the people of Berlin and their leader, as did renewed bombings of the German capital during the last days of that month. An infuriated Hitler, speaking before the German people on 4 September, vowed to repay the British in kind. 'The British,' he said, 'drop their bombs indiscriminately on civilian residential communities and farms and villages. . . . If they attack our cities, we will rub out their cities from the map. . . . The hour will come when one of us breaks and it will not be Nazi Germany.' Three days later, on 7 September 1940, Goering was ordered to begin the blitz on London. Although countless thousands would suffer in England's largest city, the RAF would be spared.

The attack on London

Hitler's order initiating the Blitz on London was welcomed by Goering's lieutenants in the Luftwaffe who did not share the Air Marshal's enthusiasm for the campaign against RAF installations. Ironically, their attitude was even beginning to affect Goering who appeared to be wavering just as the effort was on the brink of success. When he and his commanders met at the Hague on 3 September, it did not take them long to persuade him to abandon the first phase of the offensive in favour of a new course of action. The main advocate was Field Marshal Albert Kesselring whose Luftflotte 2 had borne the brunt of the original attack plan; having sustained extremely heavy losses, his advocacy of the bombing of London was not difficult to understand, as it afforded the Luftwaffe an easier target and promised the collapse of British resistance at an early date. With his own commanders clamouring for a change of tactics and Hitler demanding revenge for the British raids over Berlin, Goering had little choice but to accept their suggestions.

Once the new plan was agreed, the Germans lost little time in beginning raids on London; on 7 September, the same day that Hitler had ordered Goering to seek reprisals, the first was launched; while Goering and Kesselring watched from the cliffs at Blanc Nez, a thousand planes (almost 400 bombers and slightly more than 600 fighters) were sent against England. Arriving over the Channel coast in the late afternoon, the German fleet met little opposition and was able to unload its cargo of

Hawker Hurricanes intercept
a formation of He 111
bombers. During the Battle
of Britain these fighters
destroyed more German
aircraft than all other British
defences combined.

bombs over the docks, central London, and the East End. By the end of the afternoon, much of the city was in smoke and over 1,600 Londoners were either killed or injured.

As if the daylight attack on London was not enough, the Luftwaffe returned that evening to continue its devastating work. Over 250 bombers maintained 'a slow and agonizing procession over the capital' which lasted from early in the evening until 4 a.m. the following morning. With inadequate numbers of anti-aircraft guns to protect the capital and few if any aircraft equipped for night fighting, there was little that could be done to resist the attack; only one plane was shot down and much of the city was aglow with flames.

From 7 September to 3 November, the Luftwaffe attacked London nightly. At first there was little that the British could do to protect themselves, but as they grew accustomed to the raids, more adequate precautions were taken. People took to the underground railway stations, evacuation was stepped up, anti-aircraft installations were increased, and the RAF began to master the technique of night fighting. Soon after the start of the blitz the Luftwaffe was forced to discontinue daylight bombing runs, and even the nightly raids over the capital became increasingly costly as anti-aircraft gunners became more and more accurate with their weapons. Most important, the abandonment of daylight raids prevented the Luftwaffe attacking military and naval targets whose destruction was absolutely necessary if Operation Sealion was to be launched before winter. That possibility, however, seemed increasingly remote as the campaign over London was obviously not succeeding.

Operation Sealion's target date was 14 September and over on the French side of the Channel, everything was now ready. British reconnaissance aircraft confirmed the presence of thousands of barges and other seagoing vessels. It remained only for Goering to destroy London and the RAF and the invasion of England could begin, but since neither of these tasks had been accomplished the operation had to be postponed yet again. Although Hitler remained confident that the attacks on London would succeed, Goering and his associates were under pressure to produce signs of progress. If more time was lost, bad weather would indefinitely postpone the venture; if the RAF was not quickly neutralized, they would inevitably strike back. There was little time to lose.

Goering and Kesselring decided to make an overwhelming effort to destroy London, and that effort, launched on 15 September 1940, proved to be a major turning point in the Battle of Britain. On that day 1,000 aircraft were sent over the British capital. The weather was good, the time was right, the Luftwaffe was ready. But so too, fortunately, was the RAF.

The fact that the raid took place in daylight nullified the Luftwaffe's numerical advantage over the Royal Air Force, and German bombers, even though accompanied by five fighter aircraft per bomber, did not succeed in breaking through the British defences. Luftwaffe formations were broken up, scattered, and doggedly pursued by RAF pilots. The battle raged all day and took every plane the British could muster; the sky above London was a 'bedlam of machines', but by the end of the day the attack had been repulsed. The Germans lost 60 planes, most of them bombers, at a cost of 26 aircraft to the RAF, once again proving the value of the early warning system. Had it not been for this and the skilful co-ordination of the defence by Air Chief Marshal Sir Hugh Dowding, the Luftwaffe raid might indeed have been successful. As it was, not only was the RAF still intact, but it was soon able to take the offensive, successfully hitting the landing craft that the German navy had assembled for Operation Sealion.

Raids spread to the provinces

The Luftwaffe's failure on 15 September sealed the fate of Hitler's plans for invasion. The weather had already begun to turn cold, turbulent conditions in the Channel were predicted for the rest of the month, and most important of all, the Luftwaffe had failed to destroy the RAF or significantly devastate other military and naval targets. On 18 September, the craft assembled in French coastal posts were dispersed lest additional losses be suffered from RAF attacks and on 12 October the campaign was officially postponed until the spring of 1941.

The death of Operation Sealion effectively eliminated the threat of invasion but the aerial harassment of England was to continue for another three months. If Goering could not succeed in defeating the RAF, Hitler would still attempt to destroy the morale of the British and humble his hated rival, Winston Churchill.

Night raids had become almost commonplace and by the beginning of October the residents of London were dug in and resigned to suffer more of the same. The Germans however, had a new surprise in store for them; on 15 October and every night thereafter, incendiary bombs were added to the

A typical post-raid scene in Britain during the winter of 1940-41, when the Luftwaffe was sent across the Channel to attack and destroy British cities.

Luftwaffe's manifesto. Londoners had become accustomed to seeking shelter when the Luftwaffe approached, but now they were forced to leave their basements and take to the shelters. Incendiary bombs were far more destructive than the more conventional high explosive bombs, as they set fire to the structures they hit instead of just pulverizing them. Fires were difficult to control in the midst of an attack and even the organization of civilian fire brigades and spotters could not contain the destruction wrought by these weapons. By the end of October, fire bombings were extracting a greater and greater toll. Each night the Luftwaffe destroyed the homes of thousands, hundreds were maimed, and business and government were disrupted.

On 3 November, for the first time in nearly two months, no air raid alarms sounded in the British capital. After the nightly ritual of Luftwaffe attacks the silence must have seemed strange indeed, giving rise to conjecture as to what the Germans might be up to next. Did this mark the end of the Battle of Britain or was there more to come? Would the Germans return to London or were other cities to be subjected to the blitz? The answers to these questions came soon.

A new phase in the air war over England was initiated on 14 November when German bombers attacked Coventry. Failing to obliterate London, Goering and his lieutenants had decided to change policy and refocus the German attack on provincial cities and industrial installations. The capital had been too large and 'vague' a target to destroy, but the provincial cities might be more efficiently devastated. Also, attacks on munitions facilities, especially aircraft plants, would decrease or destroy their capacity to provide replacements and new equipment for the RAF which was daily becoming bolder in its defence.

The attack on Coventry destroyed much of the centre of the city and took the lives of over 400 of its residents. Other cities were also to be attacked during this phase of the battle. Birmingham was struck from 19–22 November, and Bristol, Liverpool, and Southampton were attacked the following week. Glasgow, Leeds, Manchester, Plymouth, and Sheffield were added to the list during the first weeks of December and, of course, London was not forgotten. Thousands lost their lives as a result of these attacks; tens of thousands were made homeless; commerce and industry were disrupted but British resistance was not destroyed.

The end of the Blitz

At the start of the new year (1941), there was no sign that England had been humbled by the Luftwaffe. The programme of raids over provincial cities and munitions works had been no more successful than previous attacks on London and RAF installations. The German offensives had, perhaps, caused more damage than was publicly admitted at the time but the Luftwaffe had failed in achieving its main objectives, the destruction of the RAF and the breaking of British morale. Although attacks over England would continue for several months, the major blitz was over. Bad weather and faltering enthusiasm would soon abort the entire venture and, having failed to defeat England, the Fuehrer would look to the east for satisfaction.

Bombing in World War 2

At the beginning of World War II, the idea of being bombed was one to strike terror into the heart, thanks to the propaganda surrounding events such as the bombing of Guernica and Shanghai, and films such as *The Shape of Things to Come*. And so it proved, for in the later stages of the war indiscriminate bombing was a decisive weapon against civilians; with the dropping of the atomic bombs on Hiroshima and Nagasaki, bombing became the terror weapon infamous for its ability to end civilisation.

Although various air forces in 1939 prided themselves on the possession of strategic bomber forces such as those advocated by the Italian General Guilio Douhet and the British Lord Trenchard, in reality these forces were merely tactical and supportive. They were incapable of carrying a sufficient weight of bombs over a great enough range with adequate navigational and bomb-dropping accuracy to break through determined fighter defences. Most of these nations were developing new generations of bombers with more powerful engines, better performance and load-carrying capability, and improved defensive armaments.

The Germans, on the other hand, had for a variety of reasons opted out of the strategic bomber race: the type's main protagonist General Walther Wever had been killed in an air accident and the lessons of the Spanish Civil War seemed to indicate that mechanised ground forces, together with tactical air power acting as flying artillery, was the best solution to the age-old military search for a combination of both speed and local superiority of forces. In addition the Luftwaffe high command, under the urging of the Nazis, came to the conclusion that a given number of engines, all that the German factories could produce, would be really much better employed in double the number of twin-engined bombers than in half the number of four-engined bombers.

By 1 September 1939, therefore, when the German invasion of Poland started World War II, the Luftwaffe was fairly well equipped with such bombers: Dornier Do 17, Heinkel He 111 and Junkers Ju 88 level bombers (the last being capable of dive-bombing operations as well), and Henschel Hs 123 and Junkers Ju 87 dive-bombers (the former being available only in limited numbers, and used almost exclusively as ground attack aircraft). Notable, though, is the fact that only the Ju 88 was of the very latest design, and that replacements for these types were being developed only as a very low priority.

Nevertheless, for the Polish campaign, the Luftwaffe deployed two *Luftflotten*, mustering between them 648 level bombers, 219 dive-bombers and 30 ground attack aircraft, against a Polish air force deploying only 154 bombers, most of them PZL P.23 *Karás* single-engined light and PZL P.37 *Los* twin-engined medium bombers. Strategically and tactically the Germans prevailed, though only as a result of fairly heavy losses. While their fighters engaged any Polish aircraft in the air, the German bombers concentrated on the elimination of airfields and aircraft on the ground before turning their attentions to repair and production facilities for aircraft, fuel supplies and the like. With the Polish air force thus effectively destroyed or grounded, the Luftwaffe bombers could turn their attentions to aiding the army without hindrance: lines of communication were cut, dumps and reinforcements attacked, headquarters bombed, and front line centres of resistance crushed at the request of the army. Comprehensive means of communication between the ground forces and Luftwaffe had been worked out before the war, but not yet implemented due to lack of adequate vehicles and ancillary equipment. But despite this lack, the Luftwaffe was able to speed the advance of the army in general, and of the armoured columns in particular, throughout the Polish campaign. On 13 September 1939 there began a determined effort to destroy Warsaw, the Polish capital. On 17 September, as a result of a prior agreement with the Germans, the Russians invaded eastern Poland, and the remnants of the Polish air force pulled back into Romania, where

Ground crews of the RAF prepare an Avro Lancaster heavy bomber for a night attack on Germany, or on German occupied territory.

they were interned. Without air opposition, therefore, the Germans set about the systematic destruction of Warsaw, which reached its culmination on 25 September: 240 Ju 87 dive-bombers were followed by wave after wave of level bombers and 30 Ju 52s. Out of the latter aircraft tons of incendiaries completed the work of the high explosive dropped by the level and dive-bombers. Poland surrendered on 27 September, by which time her air force had lost 327 of the aircraft with which she had started the war; Germany, on the other hand, had lost 285 aircraft, a remarkably high proportion in view of the lack of effective aerial opposition in the second half of the campaign. Although many of Germany's losses were attributable to accidents, the scale of the losses was still a huge problem to senior Luftwaffe commanders.

As a result of Germany's invasion of Poland, France and Great Britain had declared war on Germany. Afraid that an attack on land targets, where civilians might be killed, would bring retaliation from bombers each of the combatants generally refrained from all bombing except that which was directed against naval installations and ships. In this way the Royal Air Force on 4 September sent a force of Bristol Blenheims and Vickers Wellingtons to attack Wilhelmshaven: pounced on by German fighters, the British bomber force lost five obsolescent Blenheims and two newer Wellingtons, indicating that in daylight raids bombers were incapable of warding off the attentions of fighters armed with cannon and machine-guns. This lesson was proved, so far as the RAF was concerned, on 18 December 1939 when 24 Wellingtons were intercepted over the Heligoland Bight by Bf 109 and Bf 110 fighters: 12 bombers were shot down, and another three were written off in crash-landings on their return to Britain. The RAF high command decided that in future losses of this magnitude would have to be expected in daylight raids against Germany, and so night bombing, less accurate but also less costly, was decided upon as the RAF's operational norm. Such 'attacks' had already been pioneered by both British and French heavy bombers, though the offensive load carried consisted only of leaflets. A British bombing raid against targets on the island of Sylt, in retaliation for a German raid on the Orkneys, immediately brought protests from the French lest it lead to retaliation by the Luftwaffe against all the Allies.

The liaison between the Luftwaffe and the army for the tactical use of air power, especially bombing, continued during the German invasion of Denmark and Norway, which began on 9 April 1940, and in the German onslaught against the West, which began on 10 May 1940. As in the Polish campaign, Luftwaffe attacks against poorly defended airfields decimated the opposing air forces, allowing the German aircraft to turn their whole attention to the

Boeing B-17 Flying Fortresses, together with Consolidated B-24 Liberators, became responsible for the daylight bombing of strategic targets on German-held territory in Europe.

tactical support of the army except when British fighters arrived to operate from Norway, and the RAF made a few daylight raids on Scandinavian targets. Here again the vulnerability of unescorted bombers against efficient air defences was made crystal clear: on 13 August 1940, for example, German fighters downed 11 of 23 Blenheim light bombers attacking Aarlborg.

Operations during the early stages of the battle for France and Belgium made it apparent that day bombing, with its far greater accuracy could be a decisive weapon only when total air superiority was assured. On 11 May, for example, the Belgian air force committed almost all its Fairey Battle light bomber force, with fighter escort, in an attempt to destroy the vital bridges over the Albert Canal in the vicinity of Maastricht: German fighters and AA guns shot down nearly the whole force. A similar fate befell a British and French attack during the afternoon of the same day. An attack by RAF Battles on the next day suffered 100 per cent losses. The trouble lay in the fact that the Allies lacked sufficient fighters to tackle the German fighters, and so allow the bombers to get through to the target, and also to suppress the highly efficient German flak arm, whose light and medium AA guns were always to be found in large numbers protecting important targets. Thus bombers which managed to get by the defending fighters were usually downed by the heavy flak concentration round the target area.

What could be achieved in a situation of air superiority was demonstrated on 13 May, when the Luftwaffe's II and VIII *Fliegerkorps* unleashed 310 level bombers and 200 dive-bombers in a number of waves against the French artillery defending the

west bank of the Meuse river in the Sedan sector, chosen as the spot where the main German armoured thrust was to divide the Allied defences and push on to the English Channel coast. There were few defending fighters and AA defences were almost non existent, so the German bombers were able to knock out a few guns, and so concuss and demoralise the French with a downpour of high explosives that the assaulting ground forces met with little opposition.

The lesson was driven home the next day, as the Allies committed major air forces in an effort to destroy the bridges thrown across the Meuse by the Germans immediately after their success. Some 115 British aircraft, both fighters and bombers, attacked the Meuse bridges without success, losing 70 of their number in the process. Of this total, 36 were obsolescent Battle single-engined bombers. In the same attack the French also lost 40 aircraft. At about the same time, the old quarter of Rotterdam was razed in an attack by 86 Heinkel He 111 medium bombers just as the Dutch defence capitulated. The threat of such bombing was intended to scare the Dutch into surrender, but failure in the German communication system caused the threat to be implemented, confirming the world's opinion that the Germans were staunch advocates of the concept of 'terror bombing' of civilian targets. While it cannot be denied that the Luftwaffe was later the instrument of such attacks, it must be emphasised that at this stage of the war attacks on civilian targets, or rather on targets with civilians in the area, were very much linked to military objectives.

With the German arrival on the Channel coast splitting the Allied defence in two, the British command decided that the retention of the British

and the Germans lost 190 aircraft in about 4,000 sorties, compared with the RAF's 100 in 2,740 sorties.

In general the Luftwaffe had air superiority for the rest of the French campaign, which ended on 25 June. The French at times managed to inflict severe losses on the Luftwaffe: on 5 June, for example, the French shot down 66 German aircraft for the loss of only 24 of their own. But, in general, the Germans were able to provide their ground forces with all the air support, especially from the bombers, that they could desire. By the end of the campaign, though, it cannot be denied that the Luftwaffe had suffered heavily: admittedly the Allies had lost some 2,000 aircraft, but the Germans had lost 1,200 in combat and another 300 due to operational reasons, with a high percentage of the survivors unserviceable as a result of damage. This last factor applied particularly to the bomber force. Anticipating swift and relatively bloodless victories, however, the Germans had tailed off aircraft production, and replacements were not available as quickly or in the numbers required.

Notwithstanding the losses of the Western campaign, the Luftwaffe was now committed to the Battle of Britain. As previously described the result was Germany's first defeat of the war, and a clear proof that the Luftwaffe, only designed for tactical operations, was wholly inadequate for a strategic task such as eliminating Britain from the war.

The Germans were forced to learn the lesson they had themselves inflicted on the Poles, Norwegians, Dutch, Belgians, French and British: that without air superiority over the target, daylight bombing is a prohibitively expensive business. After the last day-

Expeditionary Force in northern Europe was merely courting certain disaster, and so the evacuation of these and other Allied forces was started at Dunkirk and other ports close by. Hermann Göring, the commander-in-chief of the Luftwaffe, was so certain of his force's invincibility that he persuaded Hitler to slow the land attacks on the Dunkirk perimeter and so give the Luftwaffe the opportunity to prove itself capable of winning the battle single-handed. The result was a disaster for the Germans, for most of the Allied forces in the area escaped back to England, though most heavy equipment had to be abandoned,

raid against London on 27 September, the Luftwaffe concentrated its efforts on night bombing of London and other important industrial cities, the so-called 'blitz'. A new type of bomber battle began that was to continue throughout the war: on the one side, bombers tried to find their targets in the dark and inflict serious damage; on the other side, ground defences attempted to conceal targets by decoys and used ever more sophisticated methods to jam or distort the attackers' electronic aids, so that the enemy bombers could be destroyed by AA guns and increasingly potent night-fighters fitted with radar. Throughout the winter of 1940 and 1941 the German bombers attacked targets in southern and central England for the most part, the most significant being that of 14 November 1940 when the pathfinder force (*Kampfgruppe* 100) led 437 other bombers to Coventry, whose centre was devastated. In comparison, RAF Bomber Command's efforts seemed puny, for the number of aircraft available rarely exceeded 100, and the electronic navigation devices pioneered by the Germans were as yet not available to the British. And lacking the extensive pre-war emphasis on night navigation featured in German flying schools, the British system had not prepared its aircrews for long-range night flights

with any real hope of arriving over the target area accurately.

Nevertheless as there was no other real way in which Great Britain could carry the war to Germany, it was decided that a great bomber fleet should be built up to strike at Germany's war-making potential and civilian areas, in the hope of forcing Germany to come to terms, or of so weakening her that an eventual invasion would be met with a weakened opposition.

Meanwhile the ball was still very much in the Germans' court, and the blitz reached its climax between 19 February and 13 May 1941, when tremendous damage was caused. During the second half of May the mass night raids ceased in favour of small nuisance raids, and the blitz was believed defeated: losses suffered by the Germans were still only 3·5 per cent, however, and the real reason for halting the attacks became clear with the German attack on the Balkans, starting on 6 April 1941, and on Russia, starting on 22 June 1941. As in the earlier 'blitzkrieg' operations of the war, the Luftwaffe once again proved itself the master of tactical air operations in support of the ground forces, and the German army swept all before it up to the end of 1941, the Balkan countries (Yugoslavia and Greece) being captured with little trouble. Great inroads into Russia were made before strategic indecision and the onset of winter halted the Germans just outside Moscow and Leningrad.

The first real evidence of the potential scale of Bomber Command's effort was the arrival in the early part of 1941 of the force's initial four-engined bomber, the Short Stirling, which made its maiden raid on Rotterdam during the night of 10–11 February 1941. The Stirling, which was not particularly successful as a heavy bomber with short range and low ceiling, was soon joined by the RAF's next heavy bomber, the four-engined Handley Page Halifax, which made the service's first four-engined bomber raid on Germany on 12–13 March 1941. The Halifax had been preceded into service by the Avro Manchester, but the two Rolls-Royce Vulture engines of this type proved unreliable, and it was soon replaced by a four-engined version, the excellent Avro Lancaster. The Halifax and Lancaster were to be the mainstays of the RAF's strategic bombing effort right up to the end of the war, supplemented by the ubiquitous Wellington, and the magnificent de Havilland Mosquito which entered service later in 1941.

After a delay occasioned by the use of Bomber

Left: Photographed from the weapons bay as they fall, a couple of bombs en route to their target. *Below:* The Bristol Beaufighter, intended to provide an answer to the RAF's need for a long-range fighter. It was to prove a most versatile aircraft, used for a wide variety of roles, including night fighting, reconnaissance and bombing of shipping.

Command in raids against the German battle-cruisers *Scharnhorst* and *Gneisenau*, which arrived in Brest after a sortie into the Atlantic at the end of March 1941, the aircraft of Bomber Command started an offensive against the German industrial areas of the Ruhr and Rhineland on 11 June. Twenty raids were launched, but all proved totally useless thanks to the inaccuracy of their navigation. At the same time the Germans were beginning to prove themselves well able to defend themselves in the air, the night-fighters of the Kammhuber Line using the *Himmelbett* interception system, taking an increasingly heavy toll of the RAF's bombers: between May 1940 and July 1941 Bomber Command lost 543 aircraft, and the German night-fighters claimed about half of these.

The strength of Bomber Command continued to grow as more of the four-engined types entered service, but raids by more than 150 aircraft were rare, and success very limited. During the autumn the British position began to improve with the introduction of a new radio navigation aid, 'Gee,' the withdrawal of German night intruder missions over England and the British decision to make cities the primary target, whereas up to 8 July oil refineries had been given this priority. However, losses also increased, to the extent that improved bombing accuracy, itself minimal, could not balance the scales. Between August and November 1941 Bomber Command lost slightly over 5 per cent of aircraft dispatched. During November, Bomber Command's efforts were temporarily suspended to

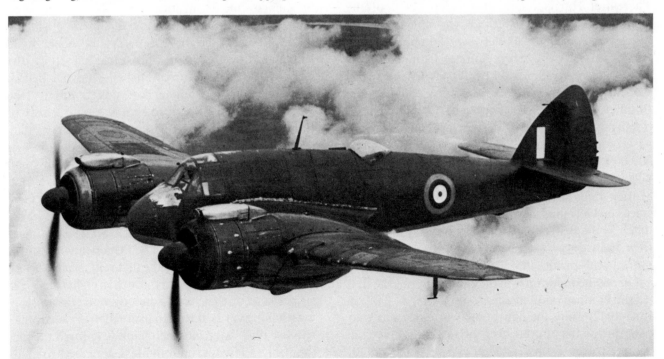

allow the expansion of the bomber force with new aircraft, and to allow extra training.

Raids were started again in February 1942 with the appointment of Air Marshal Arthur T. Harris as the chief of Bomber Command. A blunt and forceful man, Harris expounded the concept of area bombing to the virtual exclusion of all other kinds of bombing, and managed to get the Air Staff and War Cabinet to endorse his opinions. Thereafter Harris was almost independent of the rest of the British armed forces, and able to pursue his chosen course at will. Between 13 February 1942 and 6 March 1944, when Bomber Command was subordinated to General Dwight D. Eisenhower's tactical command for the invasion of Europe, Bomber Command dropped some 195,000 tons of bombs on German cities, destroying 23,000 acres of 42 German cities and causing the Germans about 500,000 civilian casualties. Bomber Command lost 4,285 aircraft during the campaign – the effectiveness of their raids was not all that had been hoped for, since the Germans learned to live with the bombing much as the British had done in 1940 and 1941, and their industries were removed to less accessible or more bomb-proof sites.

One of the milestones in the campaign was the first '1,000-Bomber raid,' against Cologne on 31 May–1 June 1942, using the new bomber stream tactics to concentrate the aircraft over the target and give the defence forces as small a target as possible to aim at;

earlier tactics had allowed pilots to approach the target over a course of their own choosing, but the individual crews had provided the line of German night-fighter defences with ideal targets. Also notable were the devastating raids on Essen and Bremen during June. Thereafter Bomber Command entered into a see-saw competition with the German night-fighter arm as first one and then the other gained a measure of superiority, to be counterbalanced only by improvements in the other side's tactics or electronics. Another milestone at this time was the first raid launched by US heavy bombers from England, the precursor of a great effort by the USAAF's 8th Air Force. Disbelieving the British assessment of the dangers of day bombing, the Americans were convinced that the tight boxes of Boeing B-17 and Consolidated B-24 bombers they proposed could protect each other from determined fighter attacks, and so get through to bomb small industrial targets of vital importance to the German war machine. So the double offensive against Germany was begun, the RAF bombing cities and whole industrial areas by night, and the USAAF striking at pinpoint targets by day. The American optimism seemed to be confirmed by the first raid, when 12 B-17 Flying Fortresses dropped more than eight tons, over 50 per cent of their total, directly onto the railway marshalling yards at Rouen in northern France on 17 August 1942. All concerned happily ignored the fact that only a few German fighters appeared to chal-

Boeing B-29 Superfortresses, undoubtedly the most important of the USAAF's strategic weapons in the war against Japan's home islands. They were also used to deploy the world's first operational atomic bombs.

lenge this wholly new form of attack.

The main objectives for the US effort at the beginning of 1943 were U-boat construction yards and pens, and aircraft factories. For this purpose it had some 155 aircraft, against which the Germans could pit 200 fighters. The RAF's main objectives were still German cities, for which it had some 740 four-engined bombers and 25 Mosquitoes, against which the Germans could match 390 night-fighters. However, the efficiency of the RAF's effort was greatly improved by the introduction of a new navigation aid, 'Oboe,' new centimetric airborne radar, and special bombs to mark the target area. Between March and July 1943 Bomber Command launched 43 major raids in the Ruhr, but failed to disrupt the output of the area significantly. Some 18,000 sorties were dispatched, but losses amounted to 872 lost over Germany and 2,126 damaged, Bomber Command's heaviest losses of the war. However, Bomber Command came closest to success in late July and early August 1943, when a large proportion of Hamburg was razed in four raids, the second of which caused the first 'firestorm', in which some 40,000 people died.

The 8th Air Force, meanwhile, had suffered its first major reverse in an attack on Bremen in April 1943, when 15 of 107 bombers were lost, and another 48 damaged. However, the threat of the US effort, which was still in its infancy, caused the Luftwaffe to call back to the Reich fighter units from Russia and the Mediterranean fronts, easing the task of the Allies in those areas.

Another major US raid was the double attack on Schweinfurt and Regensburg on 17 August 1943. Considerable damage was caused but 59 of the 363 bombers dispatched were lost, and very many others were severely damaged. The USAAF had already begun to appreciate the need for fighter escorts to accompany the bombers, but as yet these were limited to the Lockheed P-38 Lightning and Republic P-47 Thunderbolt, neither of which had the range to roam deep into Europe. The need for escort fighters was further emphasised on 14 October 1943, when 291 B-17s were sent out against the ball-bearing factories at Schweinfurt. The target was badly damaged, but only at the cost of 60 bombers shot down and another 138 severely damaged. After this, deep-penetration raids by the USAAF were called off until adequate numbers of the superlative North American P-51 Mustang escort fighter became available at the beginning of 1944.

Harris now tried to seal his success in the Battle of Hamburg with the Battle of Berlin, which saw 16 main raids between 18 November 1943 and 24 March 1944 against the capital. Harris expected that for the loss of only some 500 aircraft, the German capital would be crushed and Germany forced to sue for peace. However, civilian casualties were only 10,000, material damage relatively slight, and the losses to Bomber Command and the USAAF were 581 four-engined bombers. At the same time, 19 other major raids were sent out against other German targets, resulting in serious, but not crippling damage to the cities involved, although only at the cost of 565 bombers. 'Window,' tinfoil strips to confuse the German's radar, had proved excellent at the time of its introduction over Hamburg, but by the time of the Battle of Berlin the Germans had devised means of evading its worst effects, and so their night-fighters scored heavily. The general failure of Harris's area attack tactics was emphasised right at the end of the campaign, when 794 four-engined bombers attacked Nuremberg: some 300 German night-fighters intercepted, and 94 bombers were lost, this being Bomber Command's most serious defeat of the war.

At the beginning of 1944, the USAAF embarked on an offensive to destroy the Luftwaffe as an effective weapon: while the heavy bombers eliminated aircraft production facilities, their escort fighters would destroy the Luftwaffe's fighter arm as it tried to intercept the heavy bombers. The campaign got off to an inauspicious start on 11 January 1944, when the bomber and escort forces failed to link up.

The Gloster Meteor was the only Allied jet fighter to see service during the Second World War. It went into service with the RAF in July 1944.

resulting in the loss of 60 out of 667 bombers attacking targets in the Brunswick area. However, the 'Big Week' between 20 and 26 February was far more successful, 26 factories being attacked, for the loss of only 228 of the 3,800 sorties dispatched. US fighter losses were only 28, but the Luftwaffe lost 355 of its 1,100 fighters.

At the end of March 1944 the strategic forces of Britain and the US came under the control of Eisenhower for the final run-up to the invasion of France, which took place on 6 June 1944. Only after this had clearly succeeded did the strategic forces revert to their previous 'owners,' though it must be emphasised that the tactical use of the heavy bombers had proved devastatingly successful, along the lines pioneered by the Germans with medium bombers between 1939 and 1941. After the strategic bombers had returned to their original tasks, the Luftwaffe's decimation at the hands of the USAAF continued until Germany was all but defenceless in the air, despite the introduction of new fighters such as the jet-propelled Messerschmitt Me 262 and rocket-propelled Me 163. Germany's power supplies and oil production installations became the primary targets, and so effective were the heavy bombers, operating by day and night, that Germany's armies were starved of fuel by the beginning of 1945. Bomber Command continued with its area bombing policy, which was completed with ghastly and indeed quite useless results, on the night of 13–14 February 1945, when 804 bombers

attacked Dresden, causing an enormous firestorm that resulted in the destruction of this militarily unimportant city with the deaths of at least 100,000 people. Only three bombers were lost. During the war Bomber Command had lost 7,449 bombers during night operations, and the USAAF lost 8,067 four-engined bombers.

The strategic bombing war in the Pacific was exclusively an American affair. Both sides made extensive use of medium bombers for tactical support, as had all parties in the European war, but only the Americans had the technological expertise and production capabilities, coupled with the determination, to use strategic bombing in the war against Japan. During 1942–3 it was only from China that a beginning could be made against Japan's cities and industries, with the main part being played by Major-General Claire Chennault's 14th Army Air Force. However, with the advance across the central Pacific by the forces under the command of Admiral Chester W. Nimitz, islands from which the new Boeing B-29 Superfortress bombers could operate against Japan fell into American hands during 1944. The first B-29 arrived on Saipan in the Marianas on 12 October 1944, and it is from this date that the true air offensive against Japan may be dated. Based in the Marianas was the 21st Bomber Command, part of the 20th Army Air Force, whose other formation, the 20th Bomber Command, was in India for operations in South-East Asia. The scale of operations was soon built up, and the B-29s made their first decisive raid on 19 January 1945, when the Kawasaki factory near Tokyo was hit, reducing Japan's output of aircraft engines by 12·5 per cent and that of aircraft by almost 20 per cent. The performance of the 21st Bomber Command was generally inadequate until Major-General Curtis LeMay assumed command on 20 January 1945. LeMay diagnosed that night raids with incendiaries, flown at low altitudes, would produce better results than high-explosive raids by day at high altitude. The diagnosis and remedy were borne out by the results of the raid on Tokyo on 9–10 March 1945: 334 bombers dropped napalm and incendiaries, burning out 16 square miles of the city, killing 124,000 people and making another one million homeless. Thereafter the B-29s, in ever increasing numbers devastated Japan's cities, food production and distribution centres, power system, transportation network, and industrial areas, making it all but impossible for Japan to sustain her war effort.

However, despite the success of these conventional attacks, it was to take a new weapon to put an end to the war. On 6 August 1945 the port and industrial city of Hiroshima, hitherto only lightly touched, fell victim to the first atomic weapon used in anger: some 4·7 square miles of the city were destroyed, 71,400 people killed, and another 68,000 injured. A similar fate befell Nagasaki on 9 August, although the geography of the city, built on hilly ground, prevented the damage from being so devastating: about 30,000 were killed.

Yalu River, Korean War

At dawn on 25 June 1950, the armed forces of the Democratic People's Republic of Korea (North Korea) swept south over the 38th parallel and invaded the Republic of Korea (South Korea) in an effort to overrun the country and annex it to the North before the South could receive aid from the Western world. The invasion was on a wide front, and at first seemed to have an excellent chance of success.

Origins of the Korean War

The origins of the Korean War (1950 to 1953), as the conflict came to be called, were both complex and yet basically simple. After her success in the Russo-Japanese War (1904 to 1905), Japan had annexed the Korean peninsula. During World War II, however, the Allies had promised Korea freedom once Japan had been defeated. First made at the Cairo

Conference (December 1943), the promise was re-affirmed at the Potsdam Conference (July 1945). But Korea was not liberated by force of arms during World War II, and when Japan surrendered in August, a hurried Allied agreement allowed for the temporary 'partition' of Korea: the 38th parallel was chosen as an arbitrary dividing line, and the Japanese forces to the north of this line would surrender to the Russians, those south of it to the Americans. The idea was then that democratic elections should be held, and that the country should be unified under an elected government. After two years of fruitless negotiation on the subject with the Soviet Union, the United States took the problem to the United Nations Organization, which adopted the American point of view and called for free elections. The USSR refused this and on 15 August, 1947, the second anniversary of the Russian and American partition agreement, the Republic of Korea was declared in Seoul. Shortly afterwards the Democratic People's Republic of North Korea, a Russian puppet, was declared in Pyongyang. The Russians set about developing North Korea's army and air force on Russian lines with Russian equipment, but purportedly left North Korea in December 1948; a similar role was played by the Americans, albeit on a far smaller scale, for South Korea. They had pulled out their forces (with the exception of an advisory group) by June 1949. As soon as the Americans had gone, the North Koreans started a low-key campaign of sabotage, terrorism and guerilla raids against South Korea.

Seeing that there was no chance of the southern half of the Korean peninsula being joined with the northern one under a communist victory in free elections, the Russians, with the agreement of the Communist Chinese (who had gained final victory over the Nationalist forces of Chiang Kai-shek in 1949), set about masterminding a military campaign that would give the North Koreans victory in one fell swoop. The Russians themselves provided thousands of technical personnel for both combat and logistical purposes.

The course of the war

North Korea's armed forces consisted of some 130,000 men in 10 divisions and a brigade of Russian T-34 medium tanks. Air power consisted of some 150 Russian aircraft, mostly Yakovlev Yak-7s, Yak-9s and Yak-11s, with a fair number of Ilyushin Il-10s. The South Koreans, on the other hand, were barely trained for para-military operations, and a further setback was the fact that they only had

100,000 men in eight divisions with negligible artillery, armour and air support.

It is not surprising, therefore, that the North Koreans made rapid progress in their drive south after 25 June. Although the Americans quickly decided to aid the South Koreans, there were few troops and equipment available in the Far East, so little of material use to the South could be achieved. The United Nations, meanwhile, decided to take a united stand against the North Koreans and authorized its member states to send troops and support elements to Korea. General of the Army Douglas MacArthur was appointed to lead the United Nations' forces, predominantly American in composition. The effect of these political moves would in the long run produce results, but in the short term the South Koreans, with minimal US support, were forced back towards the south-east tip of Korea, where a final defensive perimeter was established around the port of Pusan at the beginning of August. With the stabilization of this perimeter the North Koreans' only realistic chance of finishing the war in one bold stroke was gone, and the allies of each side could now be expected to take a major part in the war.

By the end of July the North Korean air force had been reduced to about 18 aircraft, and there was little that the North Koreans could do when the United Nations launched a joint offensive on 15 September, the US 8th Army breaking out from the Pusan perimeter and the US X Corps pulling off one of the great military coups of all time by landing in hazardous conditions at Inchon, half way down the North Koreans' lines of communication and harassing their rear. The North Koreans fell back in disarray, hotly pursued by the United Nations forces, and by the end of November the UN forces were lodged deep inside North Korea. Some of them even managed to get as far as the Yalu river, which denotes the boundary between China and North Korea.

China had threatened to enter the war on overt terms should the UN forces cross the 38th parallel, and on 25 November 18 Chinese divisions, some 180,000 men strong, tore into the UN forces and sent them reeling back towards the 38th parallel. The North Korean and Chinese offensive was resumed on 1 January 1951, but after a series of offensives and counter-offensives, the front gradually stabilized along the 38th parallel by 22 April. Thereafter fighting continued almost unabated in the same general area, both sides suffering very heavy casualties.

Russian and American air involvement

The success of the United Nations offensive in the autumn of 1950 had the effect not only of bringing Chinese ground forces into the war, but also of persuading the Russians that if they were not to see the crushing of their client, considerable *matériel* would have to be supplied. Among other items, this aid took the form of powerful air reinforcements, together with the necessary groundcrew and aircrew. As a major part of the UN offensive scheme was based on the interdiction of the communist supply routes, by the destruction of all means of communication and convoys using them, the Russians decided that the best application of their air power would be in a defensive role. Thus there began to appear in North Korea the first examples of the new generation of Russian fighters, the Mikoyan-Gurevich MiG-15. With the arrival of the new fighter the air war over Korea intensified considerably, and from November 1950 to April 1952 the MiGs flew a monthly average of over 2,000 daytime sorties. The highest average for a month came in December 1951, when over 4,000 sorties were flown. During 1953 the MiGs also flew night sorties against raiding B-29 and B-26 bombers, but their lack of onboard equipment of a sophisticated nature limited them to co-operation with ground-control radar and searchlights, which severely hampered their effectiveness.

The Americans had already deployed one jet fighter, the Lockheed F-80 Shooting Star in Korea, but this was a design of strictly limited performance, dating from the closing stages of World War II. Nevertheless, the Americans were confident of their overall air superiority in the closing months of 1950.

The first jet combats

Late on in the afternoon of 1 November, six North American P-51 Mustangs patrolling along the southern shore of the Yalu river, the crossing of which was denied to the Americans for political reasons, were attacked by six swept-wing jet fighters that emerged from Manchuria across the river, attacked and then returned without shooting down any of the Mustangs. It was clear that the Americans, for the first time in Korea, had to face a superior combat aircraft.

A second generation turbojet-powered interceptor, based on Russian and captured German research into the amelioration of compression problems at high subsonic speeds by the sweeping back of the flying surfaces, the MiG-15 had entered design in March 1946, and had first been seen by westerners at the 1948 Soviet Aviation Day display just outside Moscow. It was realized only in Korea, however, how impressive the type's performance was: this was due in part to the advanced and carefully thought-out design, and in part to the use of a turbojet engine derived from the Rolls-Royce Nene, 45 examples of which had been supplied to the Russians in 1946 by the British government through an Anglo-Soviet Trade Agreement.

The MiGs increased in numbers rapidly in Korea, and on 8 November there occurred the world's first combat between jet fighters, when MiG-15s attacked the F-80 escort of a group of B-29s on a bombing mission. In every respect the MiGs were technically superior, but the Americans escaped without loss, one of the F-80s, flown by Lieutenant Russell J. Brown of the 51st Fighter Interceptor Wing, actually managing to shoot down a MiG. Here at least there was some consolation for the Americans, for it was clear even from these early encounters with the MiGs that the American pilots were in every respect superior to their communist counterparts. There thus quickly developed the odd situation in which one side possessed considerably superior *matériel*, but could not take real tactical advantage of this because the other side possessed far superior pilots, enabling the American aircraft to keep out of harm's way.

For the most part, American pilots enjoyed the legacy of aerial expertize and success in World War II. There were many veterans of the air war against Germany and Japan, and newer pilots had benefited from training under combat-experienced pilots and instructors, and from close association with World War II pilots in their squadrons. Unlike the communist pilots, moreover, the Americans were well suited and trained to the full use of sophisticated electronic equipment such as radar gunsights, this latter in particular giving them a decided combat 'edge'. It is possible to see, therefore, that although they enjoyed a tactical advantage because of the restrictions on UN pilots, and possessed slightly better aircraft, this was more than balanced, when combat was joined, by the overall superiority and experience of the American pilots. This was the key to the communists' failure to win overall air superiority and the Americans' excellent kill:loss ratio.

The tactics used by the communists were simple and should have proved effective: having deduced that the Americans were not allowed to cross the Yalu, the communists climbed to altitude on their own side of the river, and then drove across the river in a dive at maximum speed to make a single firing pass before turning to recross the river, still in a dive. The fighter could then climb in safety again, and repeat the attack. The tactics were used by both piston- and turbojet-powered fighters. but the losses suffered by the Yakovlev and Lavochkin fighters soon meant that only the MiGs posed a real threat to the Americans.

By December 1950, the strength of the MiG-15 deployed by the Communist Chinese Air Force had risen to about 150, evidence of the large-scale production of the type in the Soviet Union, and of the importance the Russians placed on both the need to stave off US bombing and reconnaissance attacks, and the desire to acquire combat experience with this new venture in fighter design. To a great extent the Chinese air force was an extension of the Red Air Force, with its aircraft almost exclusively of Russian origins. At the same time the instructors, senior tactical leaders and many technicians were Russians. Thus the aircraft could be tested in combat under 'Russian' conditions, but without the expenditure of too many Russian lives. Early in 1951, in addition, the Russians started setting up a comprehensive radar network on the north side of the Yalu. This enabled the MiGs to operate more efficiently, and also allowed the Russians to develop the equipment and techniques of radar-controlled interception without any fear of reprisals by the US forces.

It was clear, though, right from the first encounters with the MiGs over North Korea, that the F-80 could not meet the new Russian jet on anything like equal terms, and so the latest American fighter, the elegant North American F-86A Sabre, would have to be deployed in the theatre. Like the MiG-15 the Sabre was a second-generation jet fighter, and featured swept flying surfaces. Thus in December 1950 the 4th Interceptor Wing was rushed over to Korea via Japan.

The Sabres made their first operational flight, a sweep along the Yalu, on 17 December. The events of the day were to prove typical of the air fighting along the Yalu in the months to come. Led by the commander of the 336th Squadron, Lieutenant-Colonel B. H. Hinton, a detachment of P-86A-5s from Kimpo airfield set off on an armed reconnaissance of the Yalu. Flying at 9,750 metres (32,000 feet) at a speed of 760 km/h (472 mph), the patrol spotted four MiG-15s patrolling some 2,135 metres (7,000 feet) under them. Hinton led his aircraft down, and in a brief combat one of the MiGs was sent down, its rear fuselage swathed in flames; the

remaining three escaped in high-speed dives towards the Manchurian border. The one untypical factor in this brief clash was that it was the Sabres that 'bounced' the MiGs.

More typical, perhaps, was another sortie by four aircraft under Hinton just two days later. The Americans clashed with six MiGs which, however, raced through the American formation in head-on passes. Not a shot was fired, though, and by the time the slightly less agile American fighters had turned in pursuit of the MiGs, the latter had escaped across the Yalu.

The communists did at times stay to fight, as the Americans found out three days later. This time eight Sabres met a larger formation of MiGs at 12,190 metres (40,000 feet). The MiGs swept in with determination, the muzzle flashes of their cannon in striking contrast with the rather subdued flashes at the mouths of the Americans' machine-guns. Before the MiGs dived away one of the Sabres had been knocked down in flames. Later in the day, though, the Sabres returned to the scene, where this time they met some 15 MiGs. A short but savage battle followed, and in this encounter the Americans more than evened the balance, disposing of six MiGs without loss to themselves.

Escalation of the air war

Combat in the following weeks, though, showed that the Sabres would not have things entirely their own way, for a variety of tactical considerations gave a certain advantage to the communist pilots. For a start, it soon became apparent that in many respects the MiG-15 was a better combat aircraft: it possessed a higher rate of climb, speed, ceiling, acceleration and turning radius than the Sabre. To balance this slightly, the F-86 had greater stability (thus making it a considerably superior gun platform), a better weapons system (in its radar gunsight), longer range, quicker rate of fire (though its armament was only six ·5-inch machine-guns, compared with the MiG-15's two 23-mm cannon), superior medium altitude performance and better diving characteristics. Technically, therefore, the MiG-15 had slightly superior attributes in its fighter characteristics. Tactically, however, the Sabre pilots were far superior to their communist counterparts, and it was this factor that proved decisive in the battles that were to come.

The fact that they were operating over North Korean territory gave the communists great 'battlefield' advantages: the MiGs were operating from airfields just over the Yalu, whereas the Sabres had

War in the skies over Korea was to add a new dimension to air combat. For the first time the newly-developed turbojet fighter aircraft of the East and West came into confrontation.

to fly up from Taengu or Kimpo, which reduced their patrol time to a maximum of only a few minutes. To conserve fuel the Sabres cruised at lower speeds than was advisable tactically, and this added to the MiG's advantage, for the latter could now make their diving firing pass and break away before the Sabres had accelerated to a reasonable combat speed. The Sabre pilots quickly worked out a counter, however: 'combat loiter' speed on patrol was reduced yet further, but four flights of Sabres were sent into the same area at five minute intervals at high speed to engage any MiGs that might have been tempted down onto the first flight. The tactics worked excellently, as was shown on 22 December, when six out of 15 MiGs were shot down in a brief battle with eight Sabres. By the end of the year the Sabre pilots had flown 234 missions, resulting in 76 combats during which eight MiGs had been destroyed, two probably destroyed and seven damaged.

The Russians and Chinese were not overly dismayed, though, and pressed on with the build-up of their forces in Manchuria. The objective of this was eventually to secure air superiority over northwestern Korea. The communists' chance came in January 1951, when the 4th Interceptor Wing was pulled back to Japan, the necessary maintenance facilities for its Sabres not being available in Korea. While the Sabres were absent the communists struck, and the 75 MiGs of a complete air division at Antung gained air superiority over north-west Korea. Even the reappearance of the Sabres, now operating from Suwon, did not entirely remove the communists' superiority. March was occasioned by several large-scale combats, but the Sabres scored no kills. Indeed, the MiGs fared better, at times being able to dive through the Sabre top cover to engage the bombers below, the close escort of Republic F-84 Thunderjets being unable to deal with the MiGs.

The scale of the communist effort is indicated by the battle of 12 April, when some 50 MiGs took on 39 B-29s escorted by Sabres and Thunderjets. Three B-29s were shot down and another six damaged, the Sabres claiming four MiGs destroyed and six damaged, and the Thunderjets three probables. The last seems an unduly optimistic claim. Quite apart from the MiGs' material successes, though, what amazed and concerned the American pilots was the determination of the MiGs' pilots. Only later was the reason revealed by a defector: in March large numbers of Russian, Polish and Czech pilots began to arrive for three-month tours of operational flying. As General O. P. Weyland, commanding the Far

East Air Force, pointed out, the main reason for this short stay in the front line (US pilots did eight months per year in the combat zone) was probably the Russian desire to give as many of their own and their satellites' pilots as possible experience of jet warfare. At the same time whole squadrons of Russian MiGs were attached to the Chinese air force, their aircraft having the Russian markings replaced by Chinese ones.

It was clear towards the end of April 1951 that the communists were preparing airfields in North Korea (i.e. south of the Yalu) in expectation of a major air offensive. Accordingly the US heavy bombers virtually destroyed these fields in a series of major raids at the end of the month, the Sabre escorts also scoring convincingly over the MiGs. The failure of the planned air offensive made the communists revert to the earlier diving attacks from across the Yalu. During May there was only one major clash, in which 36 Sabres knocked down three of 50 MiGs and damaged another five.

During the next month the squadrons (regiments) of the Red Air Force joined combat on the communist side. Although the reason for it was not known by the American pilots, the entry into combat of regular Soviet units was indicated by the general rise in flying standards on the communist side. Although the Sabres were still enjoying a highly favourable kill:loss ratio (only one Sabre having been lost in combat so far), they now started losing more of their number as the MiG pilots probed far south of the Yalu. On 18 and 19 July a Sabre was lost on each day, and others were lost thereafter. Yet these communist incursions into the south were of little practical value, for the pilots kept their aircraft at high altitude, where they enjoyed a performance advantage over the Sabres. This meant that for the most part the UN fighter-bombers could operate below the MiGs unmolested. On the few occasions when the MiGs did attack lower down, the F-80 and F-84 pilots were usually able to look after themselves, and also managed to down a few MiGs.

The communists' build-up of aircraft was continuing all the while, and by the middle of the year the three main MiG bases at Antung, Ta-ku-shan and Ta-tung-kou contained 300 MiGs or more, with another 150 in reserve in China. Despite the fact that the Americans had only 44 Sabres in Korea, with a further 45 in reserve in Japan, the MiGs were pushed back onto the defensive. This occurred despite Russian experimentation with a number of different tactical schemes designed to test out the Americans' combat efficiency. One of these, the 'Yoyo', consisted of a circle of MiGs at their service ceiling, from which sections would break off to make firing passes before climbing up to rejoin the safety of the parent circle. The very same tactic had been used by the Germans from 1944 onwards for attacks on the American bomber streams.

The pace of operations increased again in September, when formations of up to 75 MiGs were met. This gave the Russians and their allies a numerical advantage in combat of up to four-to-one on many occasions, yet for the loss of three Sabres, one Mustang, one Shooting Star and one Thunderjet, the Americans could claim the destruction of 14 MiGs. The pace increased yet again in October, when the communists flew some 2,500 missions, losing 32 MiGs to an American loss of 15 aircraft. The reason for this increase in the air fighting of October was that once again the communists were attempting to build up new airfield complexes south of the Yalu, and the Americans were determined to prevent them being brought into service, no matter what the cost. Despite this feeling, however, the vigour with which the communists launched their fighter attacks on the big American bombers had the effect of putting them off their balance, with the result that they could not neutralize fully all the new airfields. Consequently, for the first time in the war some MiGs, about 25 in number, were able to start operating from a base south of the Yalu, in this case Uiju.

This last fact, combined with the tactical lessons learned during the previous month, convinced the planners of the US Air Defense Command that a greater effort was needed in Korea. Within a fortnight, exemplary speed for a large organization, a further 75 Sabres were on their way to Korea via Japan. These aircraft were to be operated by the 51st Fighter Wing, and were an improved version of the Sabre, the F-86E. This featured an all-flying tail and powered controls, making the aircraft far more responsive. Taking over from the tired 4th Fighter Interceptor Wing in January, the 51st Fighter Wing made its presence felt immediately, its improved aircraft and fresh pilots downing 25 MiGs for no loss in January 1952.

That the communists appreciated this change in equipment and manpower became quickly apparent as large formations of MiGs virtually disappeared from the North Korean skies, their place being taken by small groups rarely numbering more than a dozen or so. Nevertheless serious air fighting continued. Then in May the Russians sprang a surprise on the Americans, when suddenly MiGs began to

appear by night. At last the Russians had finished their new GC1 (Ground-Controlled Interception) system, and the MiGs, despite their lack of radar gunsights and other sophisticated electronic equipment, began to take a toll of the American bombers, which had been forced to operate by night to avoid the attentions of the Russian day fighters.

The night-fighting MiGs drew their first blood on 10 June, when they shot down the alarming total of three B-29s out of a formation of four over North Korea. It seemed that the Americans had suppressed one threat only to raise another in its stead.

But despite such successes, the communist air forces failed to gain any real measure of air superiority. Yet they were at the same time continuing to build up considerable air power north of the Yalu. During December 1952, for example, estimates put the number of aircraft available to the units on the northern bank of the Yalu at some 2,100 aircraft, of which some 1,300 to 1,500 were jets. Of these jets, it was reckoned that about 950 were MiG-15s. More dangerously, perhaps, photographic reconnaissance revealed that the Russians had also deployed in the theatre some 100 jet bombers, the latest aircraft produced by Russia, Ilyushin Il-28 twin-jet medium bombers. Although they could not hope to operate by day, the Il-28s could have operated with relative safety in the night, for the Americans possessed no high performance night-fighter or even the GCI equipment to allow day fighters to be used at night with any chance of success. Aware of the possibility that the Americans might react with atomic weapons to such a provocation, however, these bombers were never used.

Sporadic air fighting continued during the closing months of the year, and on 18 November there occurred the only overt Russian effort to tackle American tactical support operations. While attacking Hoeryong, aircraft of the US Navy's Task Force 77 saw MiGs climbing from the Russian base at Vladivostok. Although they carried no national markings, the aircraft could only be Russian. On joining combat, however, they lost one of their number to a Grumman Panther and the rest then made off.

Although they felt that they had the measure of the MiGs by the end of 1952, the American pilots were surprised at the beginning of 1953 when the MiG pilots seemed to improve yet again. This was due, no doubt, to the introduction of the new model of the MiG, the MiG-15*bis* with uprated engine and improved electronics. Confident of the potential of their new machines, the Russian pilots, many of them flying in aircraft marked with the Russian star, at last stayed to fight. Unfortunately, for them, their new aircraft were not enough to turn the scales, and in January 1953 the Americans destroyed 37 MiGs for the loss of only one Sabre.

The war in Korea was gradually winding down at this time as the armistice negotiators went through the wearying round of talks that finally ended the war with the armistice of 27 July 1953. The war in general had been a costly one, but the performance of the Sabres had been one of the factors that had allowed the United States to prevail generally in the air war. For the loss in combat of only 68 of their own number, the North American F-86 Sabres of the United States Air Force had claimed the destruction of 810 aircraft: one Ilyushin Il-12, six Lavochkin La-9s, no less than 792 Mikoyan-Gurevich MiG-15s, nine Tupolev Tu-2s and two others. The grand total of communist aircraft destroyed by all the United Nations air units was only 1,050, so the importance of the Sabre may easily be gauged.

It should always be remembered, though, that to a great extent the USSR had achieved its aims in the air – it had tested its aircraft, personnel and other equipment to the full against the likely enemy, had acquired much American equipment, especially radar and other electronics, and was then able to digest this information at leisure and implement the conclusions to produce a far more formidable air force for the future.

The Vietnam War

It is difficult to fix an exact date for the real beginning of the 'Vietnam War'. A convenient year is 1965, when for the first time the Communist guerrillas in South Vietnam, the Viet Cong, began to receive overt support from the regular forces of the North Vietnamese Army. The guerrilla wars had begun in 1959. So far as aviation history is concerned, 1965 marks the time at which US air forces were freed to attack targets within North Vietnam, clear political proof that the US administration of President Lyndon B. Johnson had decided to enter the fray in Vietnam.

However, limited US air involvement in the Vietnam War dates back to 1961, when the first US Army helicopter squadrons arrived to bolster the troop lift of the South Vietnamese air force (RVNAF). US aircrew were already operating as advisers in South Vietnam, helping to improve the operational efficiency of the RVNAF, which was at

the time equipped with a relatively small number of aircraft suitable only for limited operations against guerrilla forces lacking effective AA armament. During 1961, though, the rules governing the conduct of US advisers was relaxed to allow them to fire on guerrillas in 'self defence,' that is when aircraft in which they were 'advising' came under hostile fire. Despite the small scale of Viet Cong activities at this time, generally limited to attacks on government outposts and convoys bringing supplies up to remoter areas, inefficiency and other factors dissuaded the RNVAF from taking adequate steps to crush the still weak Viet Cong. US political considerations prevented US aircraft from tackling the main supply route from North Vietnam, the 'Ho Chi Minh Trail,' running from North Vietnam south through Laos parallel with the Vietnamese border, with spurs running over the border into South Vietnam at intervals. Matters improved after the accession of Marshal Ky to the office of President of South Vietnam in 1963, but only marginally so. The net effect was that the Viet Cong were able to infiltrate deeper and deeper into South Vietnamese territory laying a foundation for the war that would soon break out with North Vietnamese assistance.

During May 1964, the US decided that the flow of

The development of helicopters in the Korean War had shown that VTOL aircraft were an important new component of any nation's fighting forces. A Boeing Vertol Chinook delivers supplies to a forward position covered by an armed helicopter.

traffic down the Ho Chi Minh Trail had reached such alarming proportions that it should be monitored more or less continuously by aircraft such as the RB-26, so that some estimate of the *materiel* reaching South Vietnam could be made. Within a month of the start of these operations two US reconnaissance aircraft had been shot down, probably by Laotian Communists, the Pathet Lao. Nevertheless, US aircraft continued to make reconnaissance sorties over the Ho Chi Minh Trail.

The open involvement of the US in the Vietnam War came a step closer in August 1964 when the destroyer *Maddox* was allegedly attacked by the North Vietnamese in the Gulf of Tonkin on 2 and 4 August, as was the destroyer *Turner Joy*, also on August 4. President Johnson immediately ordered retaliatory strikes by carrierborne aircraft from the carriers *Constellation* and *Ticonderoga* against the North Vietnamese attack craft and their bases in the Gulf of Tonkin. Johnson's actions were retrospectively endorsed by Congress in the 'Tonkin Resolution,' giving the president discretionary powers to safeguard the national interests of the US in the event of acts of aggression.

Thereafter matters continued as before until 7 February 1965, when the Viet Cong, with North Vietnamese support, attacked the US base at Pleiku, home of many of the fighter-bomber and medium bomber units deployed to South Vietnam by the US Air Force since 1964. President Johnson felt that only a sharp blow against the North Vietnamese could convince them of their supposed folly, and so sanctioned the start of an air offensive against the north, though this was to be strictly limited in scope. Hitherto, only the Viet Cong had been seen to be implicated in previous attacks on US bases – such as that in November 1963, when nine US aircraft were destroyed at a base just outside Saigon, the South Vietnamese capital. Also, a small, but steady, number of US aircraft had been shot down by AA fire over South Vietnam.

The area designated for the US retaliatory strikes was the 'panhandle' of southern North Vietnam which was raided by USN, USAF and RVNAF aircraft on 7 and 11 February 1965. Useful results were obtained, but no decisive advantage to the South Vietnamese or Americans was seen to accrue. The Johnson administration therefore decided to step up the air war against North Vietnam with operation 'Rolling Thunder,' which began on 18 March 1965: the rationale of the operation was that as major strikes by the US air forces moved slowly

north from the 'panhandle' region towards Hanoi, the North Vietnamese capital, the backers of the Viet Cong would realise the futility of opposing the US militarily, and call for the beginning of peace talks. President Johnson defined the campaign in an unrealistic way which was widely undermined by the awful practice of war: 'Our objective is the independence of South Vietnam and its freedom from attack . . . We have no desire to devastate that which the people of North Vietnam have built with toil and sacrifice. We will use our power with restraint and with all the wisdom that we can command.' In purely military terms, Johnson revealed on another occasion, that the bombing of North Vietnam had the following objectives: 'To back our fighting men and our fighting allies by demonstrating that the aggressor cannot bring hostile arms and men to bear against South Vietnam from the security of a sanctuary; to exact a penalty against North Vietnam for her flagrant violations of the Geneva Accords of 1954 and 1962; to limit the flow, or substantially increase the cost, of infiltration of men and *matériel* from North Vietnam.' The trouble lay in the political connotations of the offensive, summed up in the two words 'with restraint.' The US administration was acting with extreme caution to mitigate the repugnance that world opinion in general, and US opinion in particular, was bound to feel at the use of massive US air power against North Vietnam. The planners of the offensive were therefore hamstrung by the need to secure strategic results by the use of tactical methods. This was to be one of the great anomalies of the air war in Vietnam: strategic tasks over North Vietnam ('out country' missions) were generally undertaken by tactical aircraft of the Tactical Air Command, subordinated to the control of the Pacific Air Forces for Vietnamese operations, while many tactical tasks in South Vietnam ('in country' missions) were undertaken by the Boeing B-52s of the 7th Air Force.

Another notable factor which became apparent in the early stages of the war was that US aircraft were generally too sophisticated for their tasks: the generation of US aircraft in service in the mid-1960s had been designed for major operations in Europe in the context of an all-out war with the Warsaw Pact, and had therefore been designed for high-performance operations for a limited time, with the need to deliver nuclear weapons paramount. In Vietnam, on the other hand, the USAF was suddenly faced with the need to engage in a limited war of prolonged duration in a climate unsuitable for sophisticated aircraft. Greater manoeuvrability was

called for, as were long endurance, inbuilt cannon armament and the need to carry large quantities of conventional equipment instead of a limited number of nuclear weapons. It is greatly to the credit of US groundcrews that field modifications were undertaken so successfully, and that aircraft were kept so serviceable even in the most adverse circumstances.

Given these conditions, and the political turmoil which surrounded the whole campaign, it is hardly surprising that when the first 'bombing pause' was called in May to allow the North Vietnamese to indicate their willingness to come to the negotiating table, no such indication was received. Moreover, the first US aircraft had been shot down over North Vietnam: on 4 April 1965 four North Vietnamese air force (NVNAF) Mikoyan-Gurevich MiG-17 fighters had 'bounced' a formation of bomb-laden Republic F-105 Thunderchiefs, shooting down two of them. Thereafter the USAF was forced to provide McDonnell Douglas F-4C Phantom II escorts for Thunderchief missions. Missions over the 'demilitarised zone' along the 17th parallel, the border between North and South Vietnam, were at this time made safer by the deployment of US Marine Corps EF-10B Skyknight electronic reconnaissance and countermeasures aircraft in the 'panhandle' region to jam the North Vietnamese AA guns' fire-control radars. Nevertheless, AA fire was one of the most potent threats to be met by the aircraft of the USAF, USN and USMC, especially at low level. The problem was compounded by the North Vietnamese use of the Russian SA-2 surface-to-air missile (SAM) which the Soviet Union had announced it was supplying for the protection of the Hanoi–Haiphong region of North Vietnam during the first bombing pause. US reconnaissance aircraft had spotted such missiles as early as 5 April 1965, however. The first SA-2 success came on 24 July, when a USAF F-4C was lost just south of Hanoi, the bombing pause having lasted only one week. A joint USAF/USN strike to eliminate the missiles before they were fully operational had been proposed, but vetoed by the administration on political grounds. The whole episode typifies the problems that bedevilled the actual conduct of air operations, where the realistic use of the USAF's and USN's air strength in an effective way was prohibited by political restraints. After the loss of the F-4C to an SA-2, the ban was removed, and on 27 July the SAM site was attacked by 55 USAF fighter-bombers, which lost four of their number to AA fire.

The whole problem of the Russian-supplied SAMs gave the Americans every reason to press ahead with the development of means of countering the threat: special external pods were developed by the USAF and USN ('Wild Weasel' and 'Iron Hand' respectively) for fitting to fighter-bombers and attack aircraft operating as support planes. With the aid of these pods, the aircraft could detect the North Vietnamese radar 'illumination' of an intended victim, and fire a Shrike air-to-surface missile to home on the illuminating radar's emissions, thus rendering the missile 'blind.' The North Vietnamese countered by turning on the radar to draw the Shrike, and then turning it off, in turn blinding the US missile, before finally illuminating the target at the last minute. Once the SAM was on its way, the US aircraft found that the best defence was to outmanoeuvre the ponderous SA-2, or dive to low level, where the SA-2 was ineffective. At low level, however, there was the fatal threat of the AA guns.

These early operations revealed the shortcomings of the USAF aircraft, and the generally superior capabilities of the attack aircraft operated by the USN and USMC, the Douglas A-4 Skyhawk day attack bomber and the Grumman A-6 Intruder all-weather attack bomber. Lacking the sophistication of their USAF counterparts, and designed to operate from the 'climatically' adverse conditions on board aircraft-carriers, these aircraft could carry more conventional ordnance than the USAF types, and in many instances were lighter when carrying bombs than the USAF fighter-bombers without bombs. The aircrew of USN and USMC aircraft were also trained to hit small targets with conventional ordnance, time and time again proving superior to USAF crews in bombing accuracy.

The strange imbalance between the use of tactical aircraft for strategic operations in North Vietnam, and of strategic aircraft for tactical operations in South Vietnam was highlighted in June 1965 by the first use of SAC B-52 bombers from the island base of Guam to bomb targets in South Vietnam. The North Vietnamese were aware of the fallacy of the Americans' tactics (the use of 'limited force with limited means to gain limited results'), and believed that the Americans would eventually realise this fallacy themselves, and consequently dispersed their hitherto completely vulnerable industries and their fuel supply production centres.

The US bombing offensive continued to gain momentum until the second 'bombing pause' in December 1965. But when after six weeks the North Vietnamese had still showed no signs of real willingness to negotiate, the bombing was once again resumed.

Right through this period, the NVNAF played a small but important part in hindering the US air effort over North Vietnam, as much by its threat as by its actual air efforts. The force mustered only about 70 fighters at most, of the MiG-17, 19 and 21 types, but at times managed to inflict losses on the US raiders out of all proportion to their numbers. For the loss of about 110 of their number, the North Vietnamese fighters shot down some 48 US aircraft, a 'kill' ratio in the Americans' favour of some 2·3:1. When the value of the American aircraft is compared with the Russian planes flown by the North Vietnamese, the balance is about equal. However, the Americans were more disturbed by their lack of widescale success. At first it was thought that it was only aircraft limitations, such as the lack of fixed cannon armament, that was responsible, but then it became clear that it was the US aircrews' training that had to take the primary blame. Between January and May 1968 the USN lost more aircraft than it shot down, and its pilots failed to score a single hit with 50 Sidewinder and Sparrow air-to-air missiles. In September of that year the USN therefore instituted the 'Top Gun' pilot training scheme, which soon began to produce pilots with the abilities needed to defeat the North Vietnamese pilots. By 1971 the USN was enjoying a victory: loss ratio of 11·5:1, compared with the USAF's 2:1. The USAF ratio improved only with the introduction of a scheme similar to the USN's in October 1972.

The cost of the 'Rolling Thunder' campaign was high: by the end of 1967 the US forces had lost about 3,000 aircraft in Vietnam, 1,400 in operations, and 1,550 from other operational causes. More than 670 of these losses had been over North Vietnam, with the USN suffering slightly higher losses (52 per cent), understandable in view of the fact that the more efficient carrierborne aircraft were flying more than 65 per cent of all missions over North Vietnam. During 1968, however, it was realised that the extensive use of aircraft-carriers off Vietnam was no longer cost effective, there now being large numbers of airfields ashore. Accordingly, carrier operations were cut back. In many ways the decision was an odd one, for it had been the experience of the USAF that a major cause of aircraft loss was the distance a damaged aircraft had to fly to return to its base. Here the carriers were ideal bases, able to operate only short distances from the target area. The range of USAF bases from targets in North Vietnam also necessitated the use of tanker aircraft, fighter-bombers such as the F-105 having to refuel both on the way to and on the way back from strikes in northern Vietnam. US aircrew losses over North Vietnam were reduced considerably, it should be noted, by the widespread use of rescue helicopters, operating either from destroyers in the Gulf of Tonkin, or from bases in Laos and Thailand. Such helicopters, swiftly alerted to the loss of an aircraft over North Vietnam, could reach the area without delay, and pluck downed aircrew out of the hands of certain captivity. The first such rescue occurred on 20 September 1965, only 32km (20 miles) from Hanoi. Until the helicopter arrived, the downed aircraft's companions usually remained in the area to deter the North Vietnamese, these being replaced when the helicopter arrived by the Douglas A-1 Skyraiders which usually escorted the rescue aircraft. The helicopter was also able to supply some of its own protection by means of its extremely effective machine-guns.

In its basic task, though, the bombing of North Vietnam failed: the Communist government was not forced into defeat and the movement of men and supplies south into battle zones of South Vietnam was not halted. In this context it must be admitted that such movement was hindered, but in a war in which the Communists made only limited use of motor transport, the elimination of bridges and roads could only hinder not halt movement that was more dependent on bicycles and manpower. B-52s were now used for raids on North Vietnam, but even this failed to improve matters materially. Operating from bases in Guam, Okinawa and Thailand, the B-52s were committed to attacks only on the 'Route Package Areas' of North Vietnam, not on the 'Iron Triangle' formed by Hanoi, Haiphong and Thanh Hoa, the political and economic centre of the country.

This 'Iron Triangle' had become the US primary target in North Vietnam during mid-1966, and was soon a devastated area. North Vietnam was in dire straits, particularly from the virtual elimination of electricity generating stations in the triangle, and the destruction of oil storage and handling facilities, especially in Haiphong, North Vietnam's main port. Had it not been for the warning given by the Americans' earlier offensive, the situation would have been far worse. The trouble now lay in the fact that there was no rational control of the bombing, and in an inexplicable assumption on the part of Washington planners that all bomb tonnage intended for a target actually hit it. During 1967 the situation improved, and although political necessities made the bombing of Hanoi and Haiphong an impossibility, the US embarked on a programme to

isolate the two cities from the rest of the country. This they did by cutting all lines of communication, and closing the harbour at Haiphong by the sinking of the dredgers needed to keep the harbour mouth deep enough for the Russian ships that brought in North Vietnam's weapons and oil. Late in August 1967 there were signs that the campaign was working, for nonessential civilians were ordered to clear the area, and stockpiling of stores in Haiphong whenever a ship got through indicated that communications with the rest of the country had been cut. This evidence was borne out by the gradually decreasing number of missiles fired at attacking aircraft, until a respite occasioned by a spell of bad weather gave the North Vietnamese the long-awaited chance to resupply.

Then just as the Americans seemed to be getting on top of the situation, political pressures called the 1967 'holiday stand-down' of the bombing during Christmas, New Year and the Vietnamese New Year. Inevitably, the Vietnamese called no halt themselves, and moved massive quantities of *matériel*. The general failure of the bombing campaign, and the stupidity of the 'holiday stand-down,' were signalled by the devastating success of the Communists' 'Tet offensive' in January 1968, which was finally halted only by a determined use of huge numbers of US aircraft.

At last, though, negotiations had started, and to show their goodwill, the Americans ordered a halt to bombing north of the 'panhandle' from 1 April 1968. Talks continued in a desultory fashion, and from 1 November 1968 all bombing of North Vietnam was halted, after the loss of 915 fixed-wing aircraft. In South Vietnam, however, the war continued unabated, while the North Vietnamese took the opportunity to strengthen their defences and start on the reconstruction of their country. It is noticeable that the fighters were now all MiG-21s, and that the SA-2 'Guideline' SAM was supplemented by the SA-3 'Goa,' a more advanced weapon with faster acceleration and better performance at low level.

Finally the US, unable to reach a coherent and decisive plan at the Paris peace talks, ordered a resumption of the bombing of North Vietnam on 26 December 1971. The targets were the same as before, but losses were at first higher as a result of North Vietnam's new weapons. Gradually the Americans countered these, and the damage in North Vietnam rapidly approached a decisive level. To try to divert attention elsewhere, the North Vietnamese Army was launched on an invasion of South Vietnam, slowly driving back the South Vietnamese Army until the deployment of massive US air strength restored the situation slightly. Nevertheless, the situation in the south was critical, and President Richard M. Nixon finally allowed Haiphong harbour to be mined to prevent the arrival of Russian arms ships. The harbour was mined on 9 May 1972 by US carrierborne aircraft, and so successful was the operation that the harbour was closed for the rest of the war.

On 23 October the Americans yet again halted the bombing in response to a North Vietnamese offer to reopen negotiations, and with the exception of one massive raid by B-52s and strike aircraft on 22 November, this final 'pause' lasted until 18 December 1972, when American patience with North Vietnamese delays in getting down to serious negotiations finally ran out: President Nixon ordered an all-out air offensive against North Vietnam's economy and war-making potential, with B-52s to be used as necessary. In this 'Linebacker II' offensive, some 17 B-52s were brought down by SAMs, while the Mig-21s failed to score. The Americans called a halt over Christmas, but then resumed with overwhelming strength on 28 December 1972. By the end of the year it seemed as though North Vietnam was finished as an effective base for operations in South Vietnam. That the earlier irresolution of the American political and military leadership had been at fault had been demonstrated by the success of the campaign.

Nevertheless, the USA finally lost the war, though this is no reflection on the capabilities of the air forces involved. Pressed by fierce anti-war opinion both at home and abroad the American administration finally withdrew from South Vietnam and following the Paris peace agreement in January 1973, the last US troops left in April. The North Vietnamese, regaining confidence as the American effort was scaled down, were able in November 1974 to launch the final offensive that took them into Saigon early in 1975.

Over South Vietnam tactical air power reached a new peak, although only one fundamental tactical innovation appeared: the use of the helicopter as a basic means of improving ground force mobility, and as a means of fire support using machine-guns, cannon, rockets, missiles, grenade-launchers and even bombs. Thus while the air war over South Vietnam reached huge proportions in 'conventional' terms, it is for the emergence of the armed helicopter that the Vietnam War will be remembered as regards innovations. Although the type at first appeared to be too vulnerable to ground fire to make

an effective weapon, rapid strides were taken in the development of helicopters with fewer vulnerable points, better performance and agility, and considerable firepower. Indeed, the helicopter 'gunship' has become a potent symbol and reminder of this tragically confusing period in US history.

The Six Day War

Since her establishment in 1947 Israel has been plagued by both overt and covert Arab attempts to eliminate her as a nation. The Arab-Israeli conflict, however, is too complex to be treated satisfactorily here, but it must be pointed out that although Israel and Egypt have now reached a peaceful accommodation, the anomosity between the Israelis and the rest of the Arab world continues unabated. Now, however, the conflict is waged with economic and political, rather than military, weapons.

In the 30 years since her formation Israel has fought four major wars against her Arab neighbours, and in the intervals between these large-scale conflicts has survived a fairly continuous war of attrition in conventional terms and continuous harassment from Palestinian guerilla movements.

The Israeli Defence Force/Air Force
This semi-permanent state of war has turned the Israeli Defence Force into one of the most efficient and sophisticated armed forces in the world. The economic situation of the country will not permit large standing forces to be maintained, however, and the main strength of the Israeli forces thus rests on well-trained reservists, who can in general join the full-time cadres of their formations within 72 hours of the mobilization signal. One primary exception to this rule, though, is the Israeli Defence Force/Air Force, for the nature and sophistication of modern air war precludes in general the use of reservists as combat pilots. The IDF/AF is therefore based on a large cadre of permanent aircrew, with reservists joining in secondary capacities. Fairly continuous combat experience has helped the IDF/AF develop into a first-class force, well suited to the operational requirements it is designed to undertake. All in all, therefore, the IDF/AF may in most respects be considered the elite formation of the whole IDF.

At the time of the Six-Day War in 1967, most of the IDF/AF's equipment was of French origin, a legacy of the support and encouragement the new state had enjoyed from France since its formation.

Dassault Mystère and Super-Mystère fighter-bombers, and Mirage delta-wing interceptor/ground-attack aircraft of the Israeli Defence Force/Air Force strafe enemy positions during the Six-Day War of 1967.

(This aid was to be cut off peremptorily, shortly after the Six-Day War, when Israeli commandos raided Beirut airport in the Lebanon, destroying several Arab airliners in retaliation for a Palestinian attack. Thereafter the United States became the Israelis' prime supplier of weapons as well as of finance.) Geared to fight only relatively short wars, the Israeli economy being unable to sustain any prolonged large-scale conflict, the IDF/AF is relatively 'tooth' heavy (i.e. it has a relatively large combat element together with a small administrative and logistic 'tail'). Thus for a total mobilized strength of some 20,000 men, the IDF/AF could put into the air a strength of some 450 aircraft, of which about 350 could be classified as combat aircraft suitable for first line duties. These combat aircraft were, as noted above, mostly of French origin, and were in some cases of relatively venerable vintage: the best aircraft fielded by the IDF/AF were some 73 Dassault Mirage III Mach 2+ delta-winged interceptor and ground-attack machines; the principal fighter-bombers were 20 Daussault Super Mystères and about 50 of the older Dassault Mystère IVA; for less arduous combat the IDF/AF could also call on 48 obsolescent Dassault Ouragan fighter-bombers, 60

Fouga Magister training aircraft modified as light ground-attack aircraft, and about 25 obsolescent Société Nationale de Constructions Aéronautiques du Sud-Ouest (SNCASO) Vautour II ground-attack bombers. To support this offensive force the Israelis had about 20 Nord Noratlas and Boeing Stratocruiser transports and some 25 helicopters of various types. In world terms, the only two really first-line aircraft deployed by the Israelis were the Mirage and Super Mystère, both supersonic, and armed with inbuilt 30-mm cannon and provision for air-to-air missiles. With the exception of the Vautours, the combat radius of the Israeli aircraft was in the order of some 650 km (400 miles), which would apparently limit the IDF/AF's ability to strike deep into Arab territory.

The organization of the IDF/AF was into 13 squadrons: four of interceptors, five of fighter-bombers and ground-attack aircraft, two of transports and two of helicopters. In organization and equipment, therefore, there was little remarkable in the IDF/AF. What was remarkable, however, was the efficiency and ability of the pilots and other aircrew, the sophistication of the tactics, and the general superiority of the groundcrew, who man-

aged to make a very high proportion of the IDF/AF's overall strength serviceable before the beginning of the war, and thereafter kept a very high percentage serviceable despite combat damage and other adverse factors.

The Arab Air Forces

The main opponent for the IDF/AF in this war, as in the other Arab-Israeli conflicts, would be the Egyptian Air Force. After World War II it had used British equipment and methods for the most part, but the disaster of the 1956 war, in which the Israelis had won total superiority and plunged through the Sinai to the Suez Canal, led to a re-arming of the whole of the Egyptian armed forces along Russian lines with Russian support and equipment. With the aid of generous Russian terms, the Egyptians had thus been able to turn their armed forces between 1957 and 1967 into the apparently most formidable forces in the Middle East. Although their air force was only about the same in size as the IDF/AF, the Egyptians possessed some of the very latest Russian aircraft and missiles, supplied in some cases even before the Warsaw Pact forces had received such equipment. Front-line equipment thus consisted of

about 130 Mikoyan-Gurevich MiG-21 fighters, 80 MiG-19 fighter-bombers, 180 MiG-17 and MiG-15 fighter-bombers, 20 Sukhoi Su-7 fighter-bombers, 30 Tupolev Tu-16 reconnaissance bombers and 40 Ilyushin Il-28 bombers. In support the Egyptians had some 90 transport aircraft (Ilyushin Il-14s and Antonov An-12s) and 60 Russian helicopters. Training had been conducted by Russian advisers and some combat experience, albeit against minimal opposition, had been gained in the Yemen. All in all, though, the Egyptian Air Force seemed formidable, especially as the combination of its bombers and MiG fighter cover gave it the ability to deliver strategic strikes against Israel's main cities. Unfortunately for them, the Israelis possessed no similar capability, and the threat to their civilians weighed heavily on the politico-military leadership. Like the Israeli fighters, moreover, the Russian MiGs had inbuilt 30-mm cannon as well as an air-to-air missile capability, and this gave them a material equality with the Israelis. Under Russian tutelage, though, the Egyptian pilots had developed extreme confidence in their overall superiority to their Israeli counterparts. The confidence was based on propaganda rather than fact, as the war that was about to break out would show to the world at large.

One of the main tactical and strategic problems faced by Israel in any confrontation with the Arab powers lies in the fact that she is surrounded on three sides by her opponents: to the north lies the largely neutral Lebanon, which did, however, harbour large Palestinian guerrilla elements; to the north-east is Syria, with Iraq behind her but not contiguous with Israel; to the east is Jordan; to the south-east is Saudi Arabia, a non-combatant but provider of considerable economic aid to the combatants; and to the south-west, Egypt. The strategic implications of this encirclement are obvious. Slightly less obvious are the tactical problems of Israel's position: she is a long, thin country running along a north-south axis, and Arab forces stood a strong chance of cutting the country in two across this axis. The Arab air forces, moreover, could attack any part of Israel with their tactical aircraft, let alone their larger bomber formations, whereas the converse most certainly did not apply to Israel, possessing the ability to attack only the areas of the Arab opponents close to her own frontiers.

Although not individually comparable with the Egyptian or the Israeli Air Forces in numbers, equipment or training, the air forces that could be deployed by the Lebanon, Syria, Iraq, Jordan and possibly Saudi Arabia posed a great threat to Israel. The Lebanon could put some 35 aircraft into the air, mostly Hawker Hunters of limited combat use. Syria was better off, with an air force numbering about 9,000 men and 120 aircraft, all of them Russian. Organization and equipment were on Russian lines, with one MiG-21 and one MiG-19 squadron, and two MiG-17 squadrons. The problem with the Syrian air force, though, was lack of trained aircrew and groundcrew, and although the Russian advisers in Syria were making good progress, the war caught the Syrian air force in a difficult position. Iraq was also able to field an air force on Russian lines, in this case of about 220 aircraft: 1 Tu-16, 1 Il-28, 3 MiG-15 and MiG-17, and two MiG-21 squadrons. There were also transport and helicopter squadrons, but as with the Syrian air force, trained personnel were in scanty supply and this was seriously to hamper the Iraqi air force's war effort. Jordan was poorly placed with her air force, having only 22 Hawker Hunter ground-attack fighters in two squadrons, plus a few elderly transport aircraft and only three helicopters. As with Syria and Iraq, trained personnel were wholly inadequate – there were only 16 pilots for the 22 Hunters, for example. Saudi Arabia, finally, had only about 40 aircraft, and these in the event

remained non-combatant. In all, however, the Arab states who could become involved in a war with Israel possessed some 895 aircraft, whereas Israel could muster only some 350 first-line machines against them. In the war to come, though, this material imbalance was to be more than offset by the Israelis' superior training and tactical expertise.

The Israeli plan

The exact cause and responsibility for the Six-Day War, which began in the early hours of 5 June, 1967, will in all probability never be known. Clearly the Arab nations, especially Egypt and Syria, were threatening Israel militarily. But whether their preparations were in earnest, or merely 'sabre-rattling', remains an enigma. Considering the overwhelming strength deployed against them (2,790 tanks against 800; 540,000 men against 264,000; and the aircraft figures already quoted), the Israelis had decided that a pre-emptive strike against the Arabs was essential as soon as war appeared imminent. And it was this pre-emptive strike, launched originally by the IDF/AF, which was to decide the course of the war. On this both Major-General Moshe Dayan, the Defence Minister, and Major-General Yitzhak Rabin, the Chief-of-Staff, were agreed. In the short term the threat posed by the Arab air forces, especially the strategic bombers, was paramount, and considerable thought had been exerted on means of defeating this menace before it could be used. At the same time it was realized that even the elimination of the bomber menace would leave the Arab air forces with a considerable tactical superiority, and so means of reducing this imbalance were sought.

The plan finally decided upon was bold, and had the advantages of tactical simplicity allied with superior training, surprise, determination and excellent planning taking full advantage of the Israelis' abilities and weapons. That the threat posed by the bombers was a real one can be seen readily enough from the fact that the Egyptian Air Force alone could deliver some 300 tons of bombs in a single strike, and the other Arab air forces another 200. And whereas Cairo was nearly 30 minutes flying time from the nearest Israeli air base, Tel Aviv was only 4½ minutes from the Egyptian forward base at El Arish just to the west of the Gaza Strip. Further to increase this imbalance, the Arabs deployed fairly considerable numbers of the Russian SA-2 'Guideline' surface-to-air missiles (SAMs). The Israelis, on the other hand, had to a certain extent discounted the effectiveness of SAMs, and thus had only a few American Hawk missiles, enough to provide a scanty defence of Tel Aviv and more substantial protection for the Israeli nuclear research establishment in the Negev desert.

The Israeli air forces, therefore, had long prepared for an all-out aerial pre-emptive strike against the Arab airfields at the first sign of hostilities, the object being to 'take out' a large proportion of the Arab aircraft in 'sitting duck' positions on the ground. The plan was simple enough in essence, but was complex to execute. Split-second timing was needed to catch the Arab aircraft in just the right positions, and the runways would also have to be destroyed to prevent, or at least to hamper, their use by surviving Arab aircraft. The problem in such an operation was how to secure surprise, always a difficult matter in modern war, when radar can warn the defence while the attackers are still many miles distant. The plan worked out by the IDF/AF's staff was shrewd and took full advantage of known Arab routines, especially those of Egypt: knowing that the Israeli aircraft were short-ranged for the most part, the Arabs would expect any attack to come straight in; the Israeli staff, therefore, planned a raid to come in from the Arabs' rear, where less comprehensive radar watch was kept. Flying a circuitous approach to the target, the Israeli fighters would come in fast and low, underneath the primary radar cover. They would thus be close to their objectives before they were spotted.

While the planners worked out the schedules, offensive loads to be carried and the other logistical aspects of the operation, in a remote area of the Negev desert the Israeli pilots were undergoing intensive training in low-level operations and navigation, while groundcrew practised the rearming and refuelling of their aircraft so that returning fighters and fighter-bombers could be 'turned round' for another mission with minimum delay. In the event, many aircraft were turned round in well under five minutes, an extraordinarily short time for a sophisticated modern combat aircraft.

The first Israeli strikes against Egypt

At 0745 on 5 June, Israeli aircraft struck at nine Egyptian airfields in simultaneous raids. Their timing was impeccable – approaching the Egyptian airfields from their rear, the Israeli fighter-bombers arrived over their targets at just the right moment. Dawn, about three hours earlier, had seen considerable Egyptian activity, for this was the time that an Israeli raid might be expected, and standing patrols of MiG fighters had been up in the air; at 0745, though, visibility was excellent, and the standing patrols had been brought down, the conditions being seemingly against an Israeli attack. Duty controllers had relaxed, but senior commanders had yet to arrive for the day's work, when all of a sudden radar operators picked up the 'blips' of fast aircraft approaching from the rear. These were the Israeli fighter-bombers, which had approached 'on the deck' and then climbed sharply through the Egyptian radar cover at a nicely determined range from their target airfields. The height gained in the climb put the attackers at just the right altitude for their strikes, and also 'warned' the Egyptians of the impending raid. Pilots rushed from their ready stations to man their aircraft while mechanics started up the engines. Just at this moment the Israeli aircraft appeared and released their French Nord AS-30 air-to-surface missiles. These infra-red (heat-seeking) missiles then had perfect targets in the exhaust gases of the MiG fighters; just as importantly, the aircraft were caught in the open preparing for take-off, so their destruction blocked the runway and usually killed the pilots.

In the first strike, eight of the nine airfields designated (El Arish, Jebel Libni, Bir Gifgafa and Bir Thamada in the Sinai; Kabrit on the Suez Canal; Abu Sueir and Cairo West in the Nile delta; and Beni Sueif south of Cairo) were eliminated within 15 minutes. Only Fayid on the Suez Canal was taken out late – because the Israeli pilots had difficulty finding it.

The planning has been masterly, and was in fact aided by the fact that Field-Marshal Ali Amer, Egypt's commander-in-chief, and General Mahmoud Sidky, Egyptian air force commander, were flying across the area in an Ilyushin to visit bases in the Sinai. The Egyptian AA gunners had thus been instructed not to fire, which in itself aided the Israelis, and at the same time the fact meant that the two most relevant Egyptian commanders were unavailable for 90 minutes while they flew back to Cairo to take command.

These first strikes had been launched by the Israeli Mystères, with a top cover of some 40 Mirages. The only Egyptian aircraft in the air at the time of the first strike were four trainers, which quickly fell to the Israeli fighters. Each wave of Israeli aircraft spent about ten minutes over the target, in this time making about four attacking passes before ammunition, rockets and bombs were exhausted. In the first three passes the runways and dispersal areas were engaged with cannon and missile fire, and during the last pass the Mirage fighter-bombers sprang a last surprise on the Egyptians – the newly

Seizing the initiative in the first hours of the Six-Day War, Israeli air strikes virtually eliminated Egypt's air force.

developed 'concrete dibber' bomb designed to destroy runways. Dropped from low altitude, these bombs were retarded after dropping to allow the parent aircraft to get away from the area, then turned vertical and accelerated into the concrete runways by a rocket motor in the tail. Their speed driving them through the concrete, the bombs then detonated, ripping large craters in the runways and thus rendering them useless. After ten minutes the first wave broke off to return home, refuel and rearm, its place over the target being taken by the second wave, which had taken off ten minutes later than the first wave. A third wave took over from the second, and a fourth from the third. Meanwhile the first wave had returned to base, where the groundcrew got quickly to work, and flew back to take over from the fourth wave. Mission after mission was flown until 1200, by which time Egypt's air force had virtually ceased to exist. Only eight MiGs had managed to get off the ground, and these were shot down for the loss of two Mirages.

The pattern was repeated over other Egyptian airfields, and by 1200 17 major airfields, including Cairo international airport, had been attacked. The cost to Egypt was devastating – in the order of 300 aircraft, including 30 Tu-16 and 27 Il-28 bombers, 70 MiG-19 and 90 MiG-21 fighters and fighter-bombers, 12 Su-7 fighter-bombers, and 32 transports and helicopters.

Egyptian repair work was hampered in the morning by the fact that Israeli aircraft were overhead the whole time, and by the fact that the dibber bombs (which were not used in the Sinai so that the advancing Israeli forces would take intact airfields) had delayed action fuses. Israeli losses had been minimal, despite the spirited defence put up by the Egyptian AA gunners later in the morning. But without fighters the Egyptians were at a severe disadvantage, and the SAM-2 then proved useless at low altitude, its take-off and acceleration being far too slow to enable it to deal with the low-flying, fast Israeli aircraft. For virtually no losses, therefore, the IDF/AF had during the morning of 5 June wiped out about six-sevenths of the Egyptian Air Force's serviceable strength. During the afternoon Israeli aircraft resumed their attacks, revisiting the scenes of their morning triumph to deal with any 'stragglers', and destroying some 23 Egyptian radar sites.

The victory was complete, and ranks as the single most crushing blow ever inflicted by one air force on another of comparable size and equipment.

Attacks against Jordan and Syria
Yet this victory was only one part of the Israelis' effort for the day. Convinced by the Egyptians and Iraqis that they were striking into Israel, King Hussein of Jordan decided to throw in his lot with the other Arab nations during the morning of 5 June, and ordered his Hunters to attack Israeli airfields, where four aircraft were claimed destroyed. On their return home, however, the Jordanian aircraft were caught on the ground by an Israeli strike of eight Mirages shortly after noon. Eighteen Hunters were destroyed, and the Jordanian air force's runways destroyed. None of the pilots was killed, and Hussein sent them off to fly Iraqi Hunters for the rest of the war. Yet Jordan had lost her air force, and on the ground her army, fighting desperately and often brilliantly, was being forced slowly backwards out of the Jordanian part of Jerusalem. At the same time the Israeli armoured and mechanized forces were thrusting deep into the Egyptian positions in the Sinai, unhindered by Egyptian airpower.

It was then the turn of Syria. At 1215 some 16 Mystères arrived over the complex of four airfields south of the Syrian capital, Damascus; only 20 minutes later the Syrian air force had virtually ceased to exist, its aircraft being nothing more than twisted piles of burning metal on cratered runways. Some 45 of the 120 available Syrian aircraft had been destroyed, many of the remnants were unserviceable, and many pilots had been killed.

There was only one brief, and in the event inconclusive, clash between Israel and Lebanon before the latter opted for a hostile neutrality. The other main enemy to be faced by Israel was Iraq, but the latter's only real air effort was restricted to one minor attack by a single Tu-16. During the early afternoon, though, Israeli aircraft struck at an Iraqi airfield, as a warning as much as a raid, destroying some seven aircraft. Thereafter the Israelis were content to leave the Syrian, Iraqi and Jordanian air forces alone, and concentrated their efforts on eliminating the last vestiges of Egyptian airpower and on supporting the fast moving spearheads of their ground forces, now deep in the Sinai *en route* to the Suez Canal, which was reached in the north during the 7th. On the same day, in the centre of the Sinai front, Israeli aircraft blocked the western exit of the Mitla pass, thereby denying any escape route to the Egyptians who suddenly found themselves trapped to the east of it.

During 8 June the remnants of the Egyptian air force struck three times at the Israeli forces in the Sinai, but could make no real impression, and were severely handled by the exuberant Israeli fighter pilots. As the main Israeli forces had by now reached the Suez Canal, the IDG/AF was able to turn its attentions to the Syrians, whose heavy artillery had proved particularly troublesome to the Israeli positions on the Golan heights. These artillery positions were quickly subdued, but the air force then had to turn its attention to the Syrian positions and bunkers on the top of the heights. Egypt had agreed to a ceasefire during the evening of 8 June, but the stubborn Syrians finally ceased fighting only at 1930 on 10 June, during which time the IDF/AF had inflicted severe losses on them on the Golan heights and on the retreat back towards Damascus.

The Israeli victory was a great one in overall terms, but an absolutely crushing one in the air. More than 1,000 sorties were flown by the Israelis on each of the first two days of the Six-Day War, with some pilots making as many as eight sorties. Some 26 out of 240 aircraft used were lost in this two-day period, the Arabs in turn losing about 415 aircraft.

1973 Arab-Israeli War

In 1978 the state of Israel celebrated its 30th birthday: and during those 30 years the state has weathered four-and-a-half wars to become one of the most powerful states in the Middle East. In 1948, the new state faced attacks from her Arab neighbours in the 'War of Liberation,' and despite the novelty of the Israeli armed forces, prevailed over superior forces to maintain and even enlarge her territory by the end of the war in 1949; in 1956 Israel cooperated with France and Great Britain in the Suez operation, launching a devastating attack across the Sinai desert; in 1967, after a massive preemptive air strike against the Arab nations preparing for an all-out war of destruction, Israel's armed forces crushed all opposition in the Six-Day War to seize the Sinai – that portion of Jordan to the west of the Jordan river, and the Golan area of Syria, whose mountains and foothills dominate northern Israel; between 1968 and 1970 Israel dominated the half-war known as the War of Attrition, which involved artillery battles and air combat for the most part, as each side attempted to probe for the other's weaknesses; and then in 1973 came the Yom Kippur War, so called because the Egyptians and Syrians launched their offensives into Sinai and the Golan on the Jewish Day of Atonement – Yom Kippur.

On 6 October 1973, the armed forces of Egypt and Syria launched a carefully-planned and coordinated offensive into Israel's buffer zones in the south-west and north-east, both seized in the Six-Day War of 1967. The Israelis had been expecting war, but the timing of the Arab attack was a tactical surprise for the Israelis, many of whose units in front-line areas were temporarily depleted since men had returned to their homes for the holy day. At first the Arabs, their ground forces covered by massive air 'umbrellas,' swept all before them, while the Israeli authorities ordered all units to hold their positions until the mobilisation orders immediately issued brought the Israeli Defence Forces up to strength for a counter-offensive.

In October 1973 Israel mustered some 1,700 tanks to her opponents' 5,500, 480 aircraft to 1,160, an unrevealed number of guns against 3,200. and 300,000 men against 1,200,000. Outnumbered in every important respect, Israel had now to rely on the quality of her fighting men, the determination of her people, and the capabilities of her weapons to stem and then throw back the twofold Arab advance.

Recently, Israel had relied very greatly on the superiority of her air force, the Israeli Defence Force/Air Force or IDF/AF to protect the ground forces from attacks by Arab air formations, and then attack her enemies with bombs, missiles and guns, paving the way for the advance by Israel's powerful armour/artillery/infantry ground combat teams. However, as the Arab air forces swept into battle in the opening stages of the war, and Israeli commanders called for defensive air support, it was soon appreciated that all was not going to go the way the Israelis hoped it would. The Yom Kippur War was going to prove far more difficult than the Six-Day War.

First, however, it is necessary to outline the course of the war, so that the air activity may be seen in its context. In the first stage, between 6 and 10 October, Israel's buffer zones in Sinai and the Golan were invaded by Egypt and Syria, the Egyptians pushing forward some 16km (10 miles) into Sinai after crossing the Suez Canal and breaking through the Bar-Lev Line, and the Syrians nearly clearing the Israelis off the Golan Heights by 7 October before being driven back themselves, in a series of costly battles, by 10 October. In the second stage, between 11 and 15 October, the Egyptians started to push on towards the strategically vital Giddi Pass, but made little progress, while in the north the Israelis managed to push the Syrians about 10km (6 miles) back into Syria, inflicting enormous casualties on them. In the third and final stage between 16 October and the ceasefire on 24 October, the Israelis eased pressure on the Syrians and diverted their main effort to the south-west, where they broke through the Egyptian bridgehead to the east of the Suez Canal, crossed the canal just to the north of the Great Bitter Lake in the 'Battle of Chinese Farm' and then cut off the Egyptian 3rd Army after taking the town of Suez. The deteriorating Egyptian position prompted the USA and Russia to organise a ceasefire, which came into effect on 22 October, failed, and finally became effective on 24 October, bringing the war to an end.

Of Israel's 480 aircraft, about 330 could be considered to be first-line, including some 130 supersonic machines (McDonnell Douglas F-4 Phantom multi-role aircraft and Dassault Mirage interceptor/ground-attack aircraft, plus some Israeli-produced IAI *Nesher* versions of the Mirage) and a number of McDonnell Douglas A-4 Skyhawk attack bombers. Of the Arabs' 1,160 aircraft, the most important were the 520 supersonic Mikoyan–Gurevich MiG-21 interceptors and Sukhoi Su-7 ground-attack aircraft, and 180 subsonic Mikoyan–Gurevich MiG-17 ground-attack aircraft.

But in the opening stages of the war, as Israel's aircraft arrived to try to stem the Arab advance, it was not the Arab aircraft that dominated, but the arsenal of Soviet surface-to-air missiles deployed in a vast aerial umbrella over the advancing Arab forces. There were four basic types: the SA-2 'Guideline,' an elderly weapon designed for high-altitude interception from fixed sites (32km [20 miles] range), with command radar guidance; the SA-3 'Goa' low-to-medium altitude (100-4,500m [328-14,763ft]) weapon for use in fixed sites, with command radar guidance; the SA-6 'Gainful' mobile system, mounted on tracked carriers, for low to medium altitude interception, using radar command guidance with infra-red terminal homing; and the SA-7 'Grail' shoulder-fired low-level pursuit-course infra-red homing missile for infantry use. As the Israeli aircraft arrived over the battlefield, relatively large numbers of them fell to the Russian missiles, especially the SA-3 and SA-6. Seeing the fate of the first arrivals, Israeli pilots subsequently dived only to fall to the SA-7 and ZSU-23-4 self-propelled 23mm quadruple AA cannons used by both the Egyptians and the Syrians. In the first day of the war, which did not start until 1400 hours, the IDF/AF lost 30 Skyhawks and ten Phantoms, most of them in the Golan area. In the first week of the war, Israel suffered heavy losses to the AA defences: some 80 aircraft, about 55 of them over the Golan battlefield. The IDF/AF had electronic jamming equipment, but as the frequencies on which the SA-6s radar worked were unknown, there was no quick way to tune the electronic countermeasures (ECM) pods to the right jamming frequency. Widespread use of chaff, tinfoil of varying lengths to reflect a number of radar wavelengths, proved a palliative, but the Israelis finally had to settle in the short term, for a risky suppression method and a no less risky evasion method. For suppression, the Israelis decided, the best method involved acute diving attacks with unguided rockets, bombs or cannon fire against the missile carrier: it was known that the SA-6s trajectory at launch was low, and if the pilot timed his attack well, he could make his attack before the missile had climbed to meet him. For evasion, the Israelis made extensive use of spotter helicopters behind the front line. Spotting the launch of a missile, the observer in the helicopter warned the target aircraft, which immediately dived straight towards the climbing missile, thus presenting the 'cold' nose of the aircraft, rather than the 'hot' exhausts of the engines, to the missile's infra-red sensors. At the same time flares were launched to distract the missile's sensors. However, the tactic had been foreseen by the Russians, and filters were added to the sensors to discriminate between flares and jet exhausts, leaving the Israelis with the only answer of turning towards the missile so sharply that they could pass it in a turn that could not be matched by the unmanoeuvrable missile. However, the main disadvantage of both tactics was that the dive usually brought the aircraft within range of the deadly radar-directed ZSU-23-4 guns and the SA-7 shoulder-fired missile. The only other method of using attack aircraft, the Israelis discovered on the Golan front right at the end of the first day's fighting, was to fly 'on the deck' at maximum speed, to get under the Arab's radar and missile umbrella, and surprise the ZSU-23-4s and SA-7s, then disposing of ordnance before escaping 'on the deck.'

The Israelis made the defence of the Golan their first priority, and on 7 October the air situation began to look slightly less bleak, partially because of the new tactics, and partially because the efficiency of the SA-6s radar was reduced by having to look down into the valley of the upper Jordan from the Golan Heights. During the day, therefore, the IDF/AF began to play its customary part in the land battle, destroying large numbers of Syrian armoured fighting vehicles and pieces of artillery, striking at Syrian dumps, and even beginning to probe the SAM umbrella over Syria proper.

On the next two days the IDF/AF continued to play a more decisive part in operations, despite further losses. However, on 8–9 October the Syrians fired a number of FROG surface-to-surface missiles, possibly ten, towards an Israeli air base. Unfortunately, several landed in *kibbutzes*, and Israel immediately retaliated with raids against Syria, for earlier probings against the SAM umbrella revealed that it 'protected' only the field army. The IDF/AF therefore launched a series of raids against key Syrian installations: the ports at Latakia and Tartus; the oil port and refinery at Banias and Homs; various airfields and powerplants within Syria; and even the Defence Ministry buildings within Damascus, the Syrian capital.

However, the main task still remained the need to provide the ground forces with air support and the combination of a number of tactics finally allowed the IDF/AF to operate over the narrow axis of 10km (6 miles) along which the Israeli army was driving a powerful armoured column. Starting from Qneitra,

by 12 October this had reached a position only 35km (22 miles), from Damascus, despite the Syrians' determined and courageous efforts to halt it.

In the south the battle was still dominated by the Egyptians' SAM umbrella along the line of the Suez Canal, with the aircraft of the Egyptian and other Arab air forces operating on each flank, over Port Said in the north, and over the Red Sea in the south. Major air battles developed over Port Said with a smaller number over the Red Sea. The IDF/AF has admitted to the loss of only four fighters in air combat, and although this figure is almost certainly an understatement, the losses of 450 aircraft to AA fire and in air combat by the Arabs shows that the IDF/AF still maintained an enormous superiority over its Arab opponents.

The Egyptians had at first seemed content with their first position in Sinai, but then desperate pleas for assistance from the Syrians on 13 October persuaded them to push forward and so compel the Israelis to divert forces from their advance in the Golan (little did they know that the Israelis had already done just this in preparation for an offensive in Sinai). Major air battles, usually ending in Israeli victories, occurred as the Egyptian air force attempted to fly reconnaissance missions towards the Giddi Pass along the 24km (15 mile) front across which the Egyptians hoped to push an armoured column supported by infantry and artillery, the latter two being intended for the seizure of the pass itself.

However, the advance played right into Israeli hands: the attack had been foreseen, the IDF/AF had planned for such a battle, the Israelis were more familiar with the terrain, the Egyptians had the insufficient superiority of only 2:1 in armour, and most important for the air force, the advance would take the Egyptians beyond the cover of their SAM umbrella. The result was almost inevitable: the Egyptian forces were decimated by the Israeli armour, artillery and aircraft.

The initiative on the southern front then passed to the Israelis, whose advance towards the canal, not altogether what the high command had planned, eventually proved a masterstroke. The Egyptians and Syrians still fought on under their SAM umbrellas, which prevented the IDF/AF from playing as decisive a part as it had in the Six-Day War, until the great airlift of American *matériel* finally brought in ECM pods that neutralised the Arab SAM radars. Thereafter the fighter-bombers and attack aircraft of the IDF/AF once again played their customary part in aiding the ground forces.

Airlines and Airliners

In the first decade of this century, to be precise on 16 November 1909, the world's first airline was founded. This was Deutsche Luftschiffahrts AG, generally known as Delag, with headquarters at Frankfurt-am-Main, Germany, and its purpose was to operate passenger flights with Zeppelin airships and to train Zeppelin crews. Count Ferdinand von Zeppelin had flown his first rigid airship, LZ1, at Manzell on Bodensee (Lake Constance) on 2 July 1900, but it was not a complete success. But on 19 June 1910, his seventh airship, the 147·97m (485ft 6in) long LZ7 *Deutschland* (*Germany*) made its first flight. It had been ordered by Delag and, although short-lived, was the first powered aircraft to carry passengers.

Delag had plans for a system of Zeppelin-operated air services, built a number of airship stations with sheds, and even published maps of the routes, but activities were limited to voyages from these stations and no scheduled services were operated. By the time World War I began Delag had operated seven passenger Zeppelins, made 1,588 flights covering 172,535km (107,211 miles) and carried 33,722 passengers and crew without injury.

It was on the other side of the Atlantic, in Florida, that the only pre-World War I scheduled air services were operated. On 4 December 1913, the St Petersburg–Tampa Airboat Line was established, and at 10·00 on New Year's Day 1914 a single-engined 75hp Benoist biplane flying boat, piloted by Tony Jannus, left St Petersburg for the crossing of Tampa Bay and alighted at Tampa 23 minutes later to complete the world's first scheduled airline service. The city of St Petersburg subsidized the service, and traffic demand made necessary the use of a second flying boat. Operations continued until the end of March, 1,024 passengers being carried.

The next important step in establishing air services was taken in Britain, when the Royal Air Force set up a Communication Wing to provide fast transport, mainly between London and Paris, for members of the Government and other officials attending the Peace Conference. Using mostly D.H.4s, 4As and, later, Handley Page O/400 twin-engined bombers, regular London–Paris services began on 10 January 1919. This was the start of cross-Channel air services, and it is claimed that the modified O/400 *Silver Star* carried the first non-military cross-Channel passengers and also made the first passenger night flight across the Channel. The RAF services continued until September 1919; 749 flights were made with 91 per cent regularity and 934 passengers and 1,008 bags of mail were carried.

The start of commercial air transport

Aircraft Transport and Travel (AT and T), although founded in 1916, had to wait for the return of peace and for government permission before it could begin commercial operations. However, all was eventually ready, and a civil Customs airport was established at Hounslow, near the present London Airport at Heathrow, for the start of scheduled services on 25 August 1919. At 12·40 that day Major Cyril Patteson took off with four passengers in a de Havilland 16 on the first scheduled London–Paris service. On that same morning at 09·05 Lieut E. H. 'Bill' Lawford had left Hounslow for Paris in an AT and T D.H.4A with one passenger and some goods; Lawford's flight has often been reported as the inaugural service, but this is not true.

A second British airline, Handley Page Transport, had been formed on 14 June that year, and on 25 August one of the company's O/400s flew from London to Paris with journalists. Regular Paris services did not begin until 2 September, when Lt-Col W. Sholto Douglas (later Lord Douglas of Kirtleside) flew an O/7 from Cricklewood to Le Bourget.

Handley Page opened a London–Brussels service on 23 September 1919, and a London–Amsterdam service on 6 July 1920, although AT and T had already begun a London–Amsterdam service on 17 May in conjunction with the Dutch KLM.

A third British airline was that operated by the Aerial Department of the shipowners S. Instone and Co. This company had begun a private Cardiff–Hounslow–Paris service on 13 October 1919, for the carriage of staff and shipping documents, and on 18 February 1920 began a public London–Paris service when Capt F. L. Barnard flew a D.H.4A on the inaugural service. From 12 December 1921, operations were under the title The Instone Air Line.

By the end of 1920 the British airlines were in financial trouble. French companies had begun Paris–London services in September 1919, and the Belgian SNETA had begun flying between Brussels and London in June 1920. These British companies were now competing with each other and with subsidized continental companies for very limited traffic. At the end of October Handley Page ceased operating the Amsterdam route, in November gave up regular operation of the Paris route, and then on 17 December AT and T stopped all its operations.

On 28 February 1921, all British air services came to an end through lack of finance. As a result, small government subsidies were granted to Handley

The de Havilland D.H.4A, originally a two-seat day bomber for the RAF, was used after the war as an airliner.

Page and Instone and they resumed London–Paris services in the second half of March at the same £6 6s single fare as the French airlines. The situation was not helped when the Daimler Airway began London–Paris services on 2 April 1922, so that three British airlines and two French competitors were all fighting for the same traffic; in the year ending 31 March 1922, a total of 11,042 passengers had crossed the Channel by air, 5,692 flying in British aircraft and 4,258 in French.

From April 1922 government subsidies were granted to Daimler, Instone and Handley Page for the Paris route and to Instone for a London–Brussels route which was opened on 8 May. However, it soon became obvious that the competitive operations could not continue, and in October a revised scheme was introduced and routes allocated. Daimler was allocated £55,000 for a Manchester–London–Amsterdam route with a connection to Berlin; Instone £25,000 for London–Brussels–Cologne; Handley Page £15,000 for London–Paris; and British Marine Air Navigation (not then formed) £10,000 for a Southampton–Channel Islands–Cherbourg flying boat service. The Southampton–Guernsey service opened on 25 September 1923, but no regular service was flown to Cherbourg.

Operations on these lines, with some changes, continued, and Handley Page opened a London–Paris–Basle–Zürich service in August 1923; but the financial situation was still unsatisfactory and the government decided to set up a national airline. This came into existence on 31 March 1924, as Imperial Airways with £1 million capital and guarantee of a subsidy of £1 million spread over ten years. Imperial Airways took over the fleets, staff and operations of the four earlier airlines and was due to begin operation on 1 April but unfortunately a pilots' strike prevented the working of any services until 26 April.

Most of the aircraft used by the pioneer British airlines were of two families: Airco and Handley Page. AT and T, having close associations with Airco, chose de Havilland designs. The RAF Communication Squadrons had used single-engined D.H.4 two-seat bombers, and some of these were converted to D.H.4As with two passenger seats over which there was a hinged cover with windows. AT and T had four of these aircraft and eight D.H.16s, which were four-passenger adaptations of the D.H.9 bomber design. Both types were of wooden construction, powered by 350/365hp Rolls-Royce Eagle engines, and cruised at about 160km/h (100mph).

Following the D.H.4A and D.H.16, de Havilland designed the purely civil D.H.18, with 450hp Napier Lion engine and eight seats. This type appeared in 1920 and was used by AT and T and Instone. A much improved development was the Lion-powered eight-passenger D.H.34 which first flew on 26 March 1922, and went into service with Daimler Airway on 2 April! It also served with Instone and one was exported to Russia.

The RAF Communication Squadrons had also used the large Handley Page O/400 with two Eagle engines and a span of 30·48m (100ft). A few had passenger cabins in place of the bomb cells, while others only had the military equipment removed and seats installed. Handley Page Transport used O/400s with austere cabins, but developed a number of sub-types with improved accommodation. These included the O/7, O/10 and the O/11 freighter. In all of these aircraft two passengers could be carried in the open nose cockpit. The maximum weight was 5,470kg (12,050lb), cruising speed about 112km/h (70mph) and passenger accommodation was 14 in the O/7, 12 in the O/10 and 5 in the cargo O/11.

Based on the O/400 series, Handley Page designed and built the W.8 with reduced span, better passenger accommodation and single fin and rudder in place of the O/400's box-like structure. The sole W.8, with 12 to 14 seats and Lion engines, first flew on 4 December 1919, saw limited service with Handley Page Transport and was followed in 1922 by three 12/14-seat Eagle-powered W.8bs.

On 30 April 1920, Instone took delivery of an aeroplane that was destined to become the best known of all in those pioneering days. It was the Vickers Vimy Commercial named *City of London*. This was basically the same as the modified Vimy bombers which in 1919 made the first nonstop transatlantic flight and the first England–Australia flight, but incorporated a new fuselage having seats for as many as ten fare-paying passengers.

British Marine Air Navigation, for its cross-Channel services, used three wooden Supermarine Sea Eagle amphibian flying boats. These had a Lion or Eagle engine with pusher propeller and a small six-seat cabin in the bows. Only the three Sea Eagles were built, and two passed to Imperial Airways along with the Vimy Commercial, the three H.P.W.8bs, seven D.H.34s and, unairworthy, a D.H.4A and an O/10.

France quickly set about the development of air services after the war, formed a number of airlines and made plans for trunk routes to South America and the Far East. On 8 February 1919, a Farman Goliath flew from Paris to Kenley, near London, with 11 military passengers. It is often claimed that this was the start of French cross-Channel services, but, although this is untrue, France did open the first regular scheduled international passenger service as early as 22 March 1919, when Farman began a weekly Paris–Brussels service.

Early in 1919 an association of French aircraft manufacturers established Compagnie des Messageries Aériennes (CMA), and this airline developed a system of air services beginning with a daily link between Paris and Lille from 1 May. Breguet 14 single-engined biplanes were used and it was this type with which CMA began Paris–London services on 16 September, later working in pool with Handley Page Transport.

Compagnie des Grands Express Aériens was also founded in 1919, but did not begin operations until the following year, starting a Paris–London service with Farman Goliaths on 29 March.

The network of French air routes rapidly expanded and services were begun between France and Corsica and France and North Africa. But the great French dream was an air service to South America, and Pierre Latécoère, the aircraft manufacturer, set about its achievement. The route through Spain to West Africa was opened in stages for the carriage of mail and operated as far as Casablanca by April 1920. Services were initially worked under the title Lignes Aériennes Latécoère, but in April 1921 the name was changed to Cie Générale d'Entreprises Aéronautiques, to be succeeded by Cie Générale Aéropostale in 1927. Operating under severe climatic conditions and crossing stretches of desert where a forced landing could mean death or torture for the crew at the hands of hostile tribesmen, the line was pushed forward through Agadir, Cap Juby, Villa Cisneros, Port Étienne and St Louis to Dakar – the full route to West Africa being opened in June 1925.

French airlines were also looking towards the Orient, and the first step was the founding, on 23 April 1920, of Cie Franco–Roumaine de Navigation Aérienne. This airline opened the first sector of its eastward route, from Paris to Strasbourg, on 20 September 1920. Prague was reached in that October and Warsaw in the following summer. In May 1922 the Paris–Strasbourg–Prague–Vienna–Budapest route was opened, Bucharest was reached in September, and by 3 October the entire route was open to Constantinople (now Istanbul). This route across Europe involved flying in mountainous terrain and through some of the continent's worst weather, and therefore represented a great achievement. Up to 1923 all flights were made with single-engined aircraft and confined to daylight, but in that year three-engined Caudron C.61s were introduced and night flying was pioneered with the first such service between Strasbourg and Paris on 2 September 1923, and between Bucharest and Belgrade on 20 September. On the first day of 1925 the company's name was changed to CIDNA, and by the end of 1927 the line was operating no less than 76 aircraft.

Night flying was also pioneered on the Paris–London route in June 1922 when Grands Express made a return night flight with a Goliath, but it was not until April 1929 that regular night services were flown over the route – by Air Union, which in 1923 had been created by the merger of CMA and Grands Express.

The numerous French pioneer airlines employed a very wide range of aircraft. Wartime Salmson 2-A.2 and Breguet 14 single-engined reconnaissance and bomber biplanes were used in large numbers, Latécoère employing well over 100 of the latter on the route to West Africa.

Franco-Roumaine used Salmsons and some single-engined Potez biplanes before acquiring a main fleet of Blériot Spad cabin biplanes. The first of

these was the Spad 33 of 1920, and it was developed into a whole family of similar aircraft with seats for four to six passengers and a variety of engines including, in later models, the air-cooled Jupiter. More than 100 of these attractive little biplanes were built, and some could cruise at 170km/h (106mph).

Widely used by the Farman Line, CMA, Grands Express and some non-French airlines, was the Farman Goliath. This was designed as a twin-engined bomber but converted to have two passenger cabins with seats for up to 12 passengers. The wing span was 26·5m (86ft 10in), most had two 260hp Salmson water-cooled radial engines, and the cruising speed was about 120km/h (75mph).

Germany also made an early start in establishing regular services. Numerous companies began operating, but the two most important were Deutsche Luft-Reederei (DLR) and Deutscher Aero Lloyd. DLR began Europe's first sustained regular civil daily passenger services when it opened the Berlin–Weimar route on 5 February 1919. The operation grew rapidly and employed a large fleet of ex-military single-engined biplanes, mostly the L.V.G. C VI, with room for two passengers in the open rear cockpit, but there were also considerable numbers of A.E.G. J IIs and some twin-engined

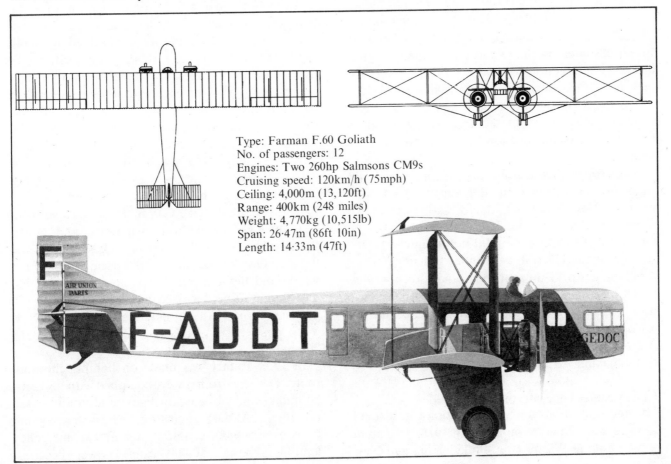

Type: Farman F.60 Goliath
No. of passengers: 12
Engines: Two 260hp Salmsons CM9s
Cruising speed: 120km/h (75mph)
Ceiling: 4,000m (13,120ft)
Range: 400km (248 miles)
Weight: 4,770kg (10,515lb)
Span: 26·47m (86ft 10in)
Length: 14·33m (47ft)

A.E.G. G Vs and Friedrichshafen G IIIas.

Junkers began flying a Dessau–Weimar service in March 1919 with a modified J 10 two-seat attack aircraft. This was a low-wing cantilever monoplane of all-metal construction with corrugated metal skin, and is almost certainly the first all-metal aeroplane to have operated an air service. In 1921 Junkers set up Junkers-Luftverkehr to operate air services and promote its F 13 cabin monoplane. The F 13 was the first all-metal aeroplane designed and built as a transport, having an enclosed cabin for four passengers, and made its first flight on 25 June 1919. More than 300 were built, in many versions.

In January 1926 Deutsche Luft Hansa was created by the merging of Aero Lloyd and Junkers-Luftverkehr and operations began on 6 April. Thereafter Luft Hansa (written as one word from the beginning of 1934) was the German national airline, but numbers of small airlines continued to exist.

Although short-lived, there was one German airline operation which must be mentioned. This was the postwar revival of Delag, with ambitious plans for airship services within Germany and on international routes, and two new Zeppelins, the LZ 120 *Bodensee* (*Lake Constance*) and LZ 121 *Nordstern* (*North Star*) were built. The *Bodensee* flew on 20 August 1919, and with accommodation for 21 to 27 passengers began working a Friedrichshafen–Berlin service on 24 August, flying in opposite directions on alternate days. *Bodensee* flew until December, made 103 flights (including one from Berlin to Stockholm) and carried 2,253 revenue passengers, but the Inter-Allied Control Commission would not allow the service to restart in 1920.

Numerous other European countries began air transport operations. In Belgium SNETA did pioneering work leading to the formation of Sabena in 1923. KLM Royal Dutch Airlines was founded in October 1919. Although it did not initially operate its own services, KLM was to play a major part in developing world air transport and still operates under its original title. DDL – Danish Air Lines – was another very early European airline, beginning a Copenhagen–Warnemünde seaplane service on 7 August 1920. Today DDL is a constituent of SAS – Scandinavian Airlines System.

United States concentrates on the mail

The first stage in opening up US nationwide airmail services was the start on 15 May 1918, of a mail service linking Washington, Philadelphia and New York. Curtiss JN-4 biplanes were used, flown by Army pilots.

The Post Office took over the US Aerial Mail Service on 12 August 1918. Specially built Standard JR-1B mailplanes were introduced and by the end of the year the service had achieved an average 91 per cent regularity. The Washington–New York mail service was closed down at the end of May 1921, by which time the entire transcontinental mail service was in being.

Late in 1918 the Post Office acquired a large number of war-surplus aircraft, including more than 100 US-built DH-4Bs with 400hp Liberty engines. This fleet made it possible to start establishing the coast-to-coast service and on 15 May 1919, the Chicago–Cleveland sector was opened, saving 16hr on the Chicago–New York journey. On 1 July the New York–Cleveland sector was opened, with through New York–Chicago flights from September, and the San Francisco–Sacramento section was opened on 31 July. On 15 May 1920, the mail route was opened from Chicago to Omaha via Iowa City and Des Moines, and the full route came into operation on 8 September with the opening of the Sacramento–Salt Lake City and Salt Lake City–Omaha stages. Branch lines were also opened between Chicago and St Louis and Chicago and the twin cities Minneapolis/St Paul.

In order to save time it was decided to make experimental flights in each direction, with night flying over some stages. On 22 February 1921, two aircraft took off from each end of the route and so began a saga that is now part of American history. One of the eastbound aircraft crashed in Nevada, killing the pilot, and only one of the west-bound aircraft managed to reach Chicago because of extremely bad weather, and at that point the flight was abandoned. The surviving eastbound mail was taken over at Salt Lake City by Frank Yeager, who flew through the night to Cheyenne and North Platte, where he handed over to Jack Knight for the flight to Omaha. Because of the termination of the westbound flight there was no aircraft at Omaha, and Knight secured a place in history by flying on to Chicago. Ernest Allison flew the last stage to New York, and the total coast-to-coast time was 33hr 20min.

In 1922 a start was made on lighting the mail route. The aerodromes were equipped with beacons, boundary and obstruction lights and landing-area floodlights. At the regular stops there were revolving beacons of 500,000 candle-power, and at emergency landing grounds 50,000 candle-power beacons.

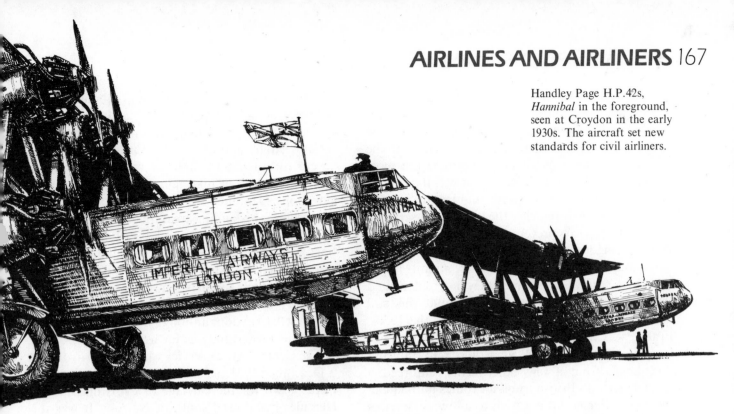

Handley Page H.P.42s, *Hannibal* in the foreground, seen at Croydon in the early 1930s. The aircraft set new standards for civil airliners.

Flashing gas beacons were installed every three miles along the route, and by the end of 1925 the entire 3,860km (2,400 miles) of route had been lit at a cost of some $550,000.

During 1926/27 the Post Office Department turned over its airmail services to private contractors. When the Post Office's own operations ceased it had flown more than 22 million km (13¾ million miles) with 93 per cent regularity and carried more than 300 million letters; but there had been 4,437 forced landings due to weather and 2,095 because of mechanical trouble. Worse still, there had been 200 crashes with 32 pilots killed, 11 other fatalities and 37 seriously injured.

Pioneering the trunk routes

In spite of sparse ground facilities, harsh climatic conditions, and low-performance aircraft, Britain, Belgium, France and the Netherlands were all eager to establish air communication with their overseas territories, and this was to lead to the opening of trunk routes which ultimately grew into our present globe-encircling system uniting what were remote corners of the world.

The first attempt to operate commercial air services in the tropics was that by the Belgian SNETA in the Congo (now Zaïre) when it opened the first stage of the Ligne Aérienne Roi Albert (King Albert Air Line) between Kinshasa and N'Gombé on 1 July 1920, using a three-seat Lévy-Lepen flying boat with a 300hp Renault engine. The N'Gombé–Lisala sector was inaugurated on 3 March 1921, and by that July the entire Congo River route was open between Kinshasa and Kisangani.

In Britain, at the end of October 1919, the Advisory Committee on Civil Aviation had recommended the establishment of trunk air routes linking the Commonwealth with the United Kingdom. They stated that 'the proper place for initial action' was the route to India and ultimately thence to Australia, 'to be followed by a service to South Africa . . .' and in October 1926 there was an agreement between Air Ministry and Imperial Airways for operation of an Egypt–India service. But things, unfortunately, did not quite work out like that.

In March 1921 there was a Cairo Conference to examine administration and control of mandated territory, and Sir Hugh Trenchard, Chief of the Air Staff, proposed policing Mesopotamia (now Iraq) with air forces instead of the orthodox ground forces. The proposal was adopted and this in turn led to the establishment of the Desert, of Baghdad, Air Mail to speed communication between the United Kingdom and Baghdad. In June a survey was made of the route, landing grounds were marked out in the desert at intervals of 24 to 48km (15 to 30 miles) and a furrow was ploughed over some sectors as a navigational aid to the pilots. The route between Cairo and Baghdad was opened by the RAF on 23 June 1921, and maintained until commercial operations began in January 1927.

Ground and air surveys were made of the route between Cairo and India, and a fleet of special aircraft was ordered for Imperial Airways so that commercial service could replace the Desert Air Mail and implement the plan for a service to India. The route from Cairo involved long desert stages, several mountain crossings and the use of comparatively high-elevation aerodromes in extreme temperatures. The limited range and low speed of the aircraft also meant that aerodromes had to be provided at frequent intervals. Perhaps the most serious obstacle was political, for it was the task of obtaining permission to overfly and land in various countries that was to delay the opening of services, and for many years prevent through flights from the United Kingdom.

It should be realized that the early aircraft found the Alps a formidable barrier; they had to fly round them over France or the Balkans, and this still applied to many transport aircraft into the 1950s. France was in dispute with Italy, and would not allow British aircraft to fly via France into Italy and this, apart from a few experiments, enforced a special sector in British trunk air routes until late in the 1930s. Persia did not wish to allow air services through her territory, and so the first British trunk air route was confined to the Cairo–Baghdad–Basra sectors.

Five de Havilland 66 Hercules biplanes were specially built for the route. These had three 420hp Bristol Jupiter air-cooled engines, accommodation for eight passengers, cruised at 177km/h (110mph) and had a range of about 640km (400 miles).

On 7 January 1927, one of the Hercules left Basra to open the fortnightly service to Cairo via Baghdad and Gaza. It arrived in Cairo on 9 January and three days later the first eastbound service left Cairo.

Eventually Persian permission was granted for a coastal route, and this enabled Imperial Airways to open its long awaited England–India service. The Armstrong Whitworth Argosy *City of Glasgow* took off from Croydon on 30 March 1929, and carried its passengers and mail to Basle. From there they travelled by train to Genoa, then in a Short Calcutta flying boat to Alexandria and on to Karachi in a Hercules. The Egypt–India sector involved eight scheduled intermediate stops, the total journey took seven days and the single fare was £130.

The service was extended to Jodhpur and Delhi in December 1929, and in association with Indian Trans-Continental Airways to Calcutta in July 1923, to Rangoon that October and to Singapore by the end of the year. Because of political and operational difficulties the route was transferred from the Persian to the Arabian shore of the Gulf in October 1932.

The setting up of a route through Africa was much more difficult. The continent rises slowly from the coast and then steeply from southern Sudan to over 1,525m (5,000ft) at Nairobi and about 1,830m (6,000ft) at Johannesburg. Temperatures can be extremely high, and tropical storms of great intensity led to flooding of aerodromes as well as severe turbulence in flight. Aircraft performance falls off sharply in high temperatures and at high elevations. Termites were also apt to build solid obstructions on landing areas. In the 1920s and early 1930s Africa posed a formidable challenge to the operation of air services. However, all the problems were tackled, aerodromes constructed, rest houses built and fuel supplies laid down, and on 28 February 1931, the inaugural service to Central Africa left Croydon, although at that stage passengers were carried only as far as Khartoum.

The Central Africa service which left Croydon on 9 December 1931, was extended experimentally to Cape Town for the carriage of Christmas mail, a Hercules being used south of Nairobi. It was this type which operated the Nairobi–Cape Town sector when the entire route was brought into operation in the following January.

In 1925, while employed with the French company CIDNA, Maurice Noguès made a Paris–Teheran survey flight with a Blériot Spad 56, and in 1927 CIDNA began negotiations for the extension of its services to Beirut, Damascus, Aleppo and Baghdad. Noguès later transferred to Air Union's Ligne d'Orient which was founded in 1927, that organization joined with Air Asie to form Air Orient, a regular Marseilles–Beirut mail service was started in June 1929 and the route was extended to Baghdad at the end of the year. On 17 January 1931, the entire line to Saigon was opened with a ten-day schedule, although at first passengers were carried only as far as Baghdad. In 1938 the route was extended to Hong Kong, and in 1939 260km/h (161mph) three-engined Dewoitine D.338 monoplanes were working the entire route, by then operated by Air France.

On 1 October 1931, KLM inaugurated a regular Amsterdam–Batavia (now Jakarta) passenger service using Fokker F.XIIs. These aircraft had luxury-type seats for 4, although in Europe they could carry 16 passengers. The scheduled journey time was ten days and the flying time 81hr for what was almost certainly at that time the world's longest air route. In 1932 Fokker built five slightly larger F.XVIIIs to replace the F.XIIs, and the new aircraft had sleeper seats. In 1934 a KLM Douglas DC-2 took part in the England–Australia race and, carrying three passengers, covered the 19,795km (12,300 miles) to Melbourne in just over 90hr. In June 1935 DC-2s

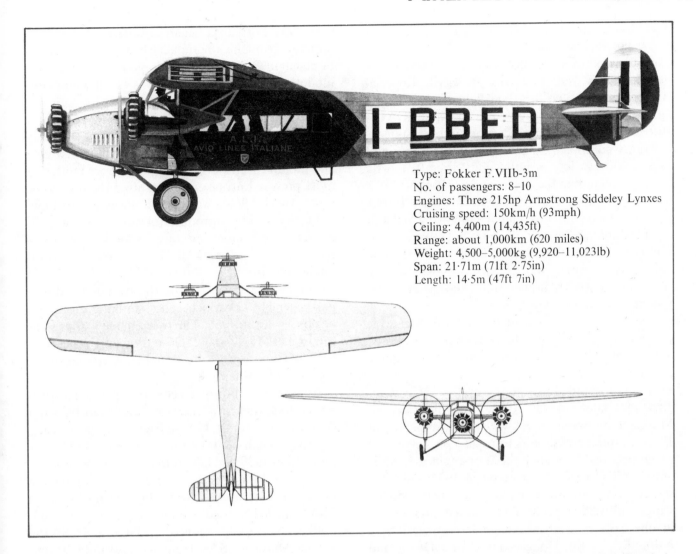

Type: Fokker F.VIIb-3m
No. of passengers: 8–10
Engines: Three 215hp Armstrong Siddeley Lynxes
Cruising speed: 150km/h (93mph)
Ceiling: 4,400m (14,435ft)
Range: about 1,000km (620 miles)
Weight: 4,500–5,000kg (9,920–11,023lb)
Span: 21·71m (71ft 2·75in)
Length: 14·5m (47ft 7in)

were introduced on the Far East service with a six-day schedule.

Consolidation in Europe

In Europe the pioneer years of air transport continued through the 1920s and, according to definition, into the 1930s. For most of this time many routes were operated only in the summer, and night services were only gradually introduced. There was steady development of aircraft, particularly in Germany and the Netherlands, and the air-cooled engine largely replaced the water-cooled engine, with its attendant plumbing problems and heavy radiators. Radio came into increasing use, airport lighting was developed, and Germany, in particular, gave considerable attention to developing ways of navigating and landing in bad weather.

The British national airline operated a very small European route system. Nevertheless, Imperial Airways did improve its fleet, at first with developments of the early Handley Pages but from 1926 with a small number of three-engined Armstrong

Whitworth Argosy 18/20-seat biplanes. Unlike their British predecessors, these had metal structures although they still employed fabric covering.

A much greater improvement in standards came in 1931 when Imperial Airways began using a fleet of Handley Page H.P.42 Hannibal and H.P.45 Heracles biplanes. These were large four-engined aircraft with Jupiter engines, 38 seats on European services and 16 to 24 on the Egypt–India and Egypt–Central Africa routes. On Heracles two stewards were carried, full meals were served, and during peacetime this fleet operated with perfect safety. The only problem was their low speed of around 160km/h (100mph).

The Handley Pages remained in operation until World War II although some smaller and faster aircraft had been added to the fleet by this time, including the beautiful de Havilland Albatross monoplanes, known as the Frobisher class by Imperial Airways, which cruised at 338km/h (210mph).

In 1937 a second British operator, British

Airways, began using fast Lockheed Electras on European services and later added Lockheed 14s. They flew to Germany, Poland and Scandinavia as well as Paris and, in terms of speed, Imperial Airways was shown in a bad light. It was finally agreed to amalgamate the two airlines and BOAC – British Overseas Airways Corporation – resulted.

France continued expansion of its air routes and used a wide variety of indigenous aircraft, mostly biplanes, including the single-engined Breguet 280T and twin-engined Lioré et Olivier 21. But CIDNA had introduced the three-engined low-wing Wibault 280T monoplane on its Paris–Istanbul route and Air Union had at least two on the London–Paris route before Air France was formed in 1933 as the successor to these airlines and Aéropostale, Air Orient and the Farman Line.

The new organization introduced a further wide range of types including the twin-engined Potez 62 and Bloch 220 and the three-engined Dewoitine series.

Sabena expanded its routes with Fokker F.VIIb-3ms, and later introduced three-engined Savoia-Marchetti monoplanes. KLM used a whole series of Fokker monoplanes for its European passenger and cargo network, including the four-engined F.XXII and F.XXXVI which appeared in 1934 and 1935 respectively. But the Fokkers, with their wooden wings and welded steel-frame fuselages and, except in one case, non-retractable undercarriages, were outmoded by the Douglas DC-2 and DC-3, and KLM became the first European operator of these advanced aeroplanes.

It was the German airline Lufthansa, which dominated the air transport map of Europe. When the airline was founded it took over 162 aircraft of 19 different types. The largest batches of standard, or reasonably standard, aeroplanes were 46 Junkers F 13s, 19 Fokker-Grulich F.IIs and 13 F.IIIs. There were also numbers of Dornier Komets, and it was one of these that operated the airline's first service. Three-engined Junkers G 24s were used for the first night service on 1 May 1926, and three-engined Rohrbach Rolands pioneered trans-Alpine services from 1928.

Lufthansa expanded rapidly and in its third year had a European network of more than 33,000km (20,500 miles) over which it flew 10 million km (6,214,000 miles) and carried 85,833 passengers and 1,300 tons of cargo, mail and baggage. The airline had the biggest domestic network of services in Europe, and also served most of the continent's main cities.

In 1932 Lufthansa began operation of that very famous aeroplane, the Junkers Ju 52/3m. This was a three-engined low-wing monoplane with accommodation for 17 passengers, was of all-metal construction with corrugated skin and, although not particularly fast, was extremely reliable. Lufthansa was to use about 230 with a maximum of 78 at any one time. The Ju 52/3m was widely used as a civil and military transport; nearly 5,000 were built, and most prewar European airlines used them at some time – two on floats working in Norway until 1956.

Lufthansa also introduced a number of limited-capacity high-speed aircraft including the four-passenger single-engined Heinkel He 70, commissioned on *Blitz* (*Lightning* or *Express*) services in Germany in June 1934, and the twin-engined ten-passenger He 111 which had a smoking cabin and began service in 1936. The four-engined 40-passenger Junkers Ju 90 and 25/26-passenger Focke-Wulf Fw 200 Condor were prevented by war from showing their full potential.

Almost every European country operated its own air services. Austria's Ölag was founded in 1923 and flew Junkers aircraft. Poland began very early, using mostly French and German aircraft; LOT was founded in 1929 and flew Junkers and Fokker types and a few Polish designs, and later adopted the Lockheed Electra and Model 14. Czechoslovakia's CSA and CLS used Czechoslovak aircraft and Fokkers before CLS adopted the DC-2 and CSA the Savoia-Marchetti S.73. Hungary used Fokkers and Junkers; Greece used Junkers types; Rumania used a mixture of British, Czechoslovak, French and German aircraft; and Yugoslavia used mostly British and French aircraft. Most of these airlines operated domestic services and services to other European destinations, although LOT had a route to Palestine.

Spain began mail services in October 1921, when CETA began operating D.H.9s between Seville and Spanish Morocco. In 1925 UAE was formed and, using Junkers aircraft, opened services linking Madrid with Seville and Lisbon. A Madrid–Barcelona route was opened in 1927 by the original Iberia using Rohrbach Rolands, but in March 1929 the three pioneer Spanish companies were amalgamated to form CLASSA, in turn superseded in 1932 by LAPE, which operated domestic and international services with Fokker F.VIIb-3ms, Ford Trimotors and, later, Douglas DC-2s.

A number of small Swiss companies were founded and at the end of 1919 one of these was renamed Ad Astra Aero. Initially it operated a number of small

A Short Calcutta flying boat, one of the family of waterborne craft used by Imperial Airways to establish worldwide routes.

flying boats from the Swiss lakes, but did not begin regular scheduled operations until June 1922 when it began Geneva–Zürich–Nuremberg services with Junkers F 13s. This airline expanded its operations, mostly with Junkers aircraft, and, in 1931, amalgamated with Balair to form Swissair. Balair had begun services in 1926 and was equipped with Fokkers. Since its inception Swissair has been one of Europe's most technically progressive airlines, has introduced a range of advanced aircraft and was among the first to adopt such types as the DC-2 and DC-3. It was Swissair's introduction of the Lockheed Orion monoplane with retractable undercarriage in May 1932 that led Germany to produce the high-performance Heinkel He 70 and Junkers Ju 60 and Ju 160.

Scandinavian air transport differed from that in other parts of Europe because of terrain and the difficulty of providing aerodromes. DDL, formed in Denmark in 1918, began operation with a seaplane, but otherwise was able to develop its routes with landplanes; Sweden and Norway, however, had to rely mainly on seaplanes.

AB Aerotransport – Swedish Air Lines – began services in June 1924 between Stockholm and Helsinki using Junkers F 13 floatplanes. As traffic grew a three-engined Junkers G 24 was added and in August 1932 a Ju 52/3m floatplane – almost certainly the first regular airline operations of the Ju 52/3m, even before Lufthansa. Lack of a land aerodrome at Stockholm forced seaplane operation to continue until May 1936, when Bromma Airport was opened. However, ABA had established services to Denmark, Germany, Amsterdam and London from the airport at Malmö on the west coast. ABA was an early operator of DC-3s, acquiring three in 1937. Like Swissair, ABA was a technically advanced airline, as its fleet composition shows, and it also opened the first experimental night mail service in Europe when, on the night of 18/19 June 1928, an F 13 with onboard sorting facilities flew from a military aerodrome near Stockholm to London in four stages via Malmö, Hamburg and Amsterdam.

The real beginning of air transport in Norway came with the founding in 1932 of DNL. A coastal route from Oslo through Kristiansand, Stavanger and Bergen was established into Arctic Norway. It was mainly operated by the Ju 52/3m on floats, although a few other types were used, including Short Sandringham flying boats after World War II.

Aero O/Y, in Finland, began its services in 1924 with a Helsinki-Reval route. F 13 seaplanes were used, followed by a G 24 and Ju 52/3ms, and pre-

World War II operations were mainly confined to routes across the Baltic and the Gulf of Finland. In winter many of the aircraft had to be fitted with skis because ice prevented the use of seaplanes. Aero O/Y now operates as Finnair.

Most of the pioneer Italian services, too, were operated by marine aircraft, but Transadriatica, based in Venice, began operation in 1926 with a fleet of F 13 landplanes and, with three-engined G 24s, opened a Venice–Munich service in 1931.

Avio Linee Italiane was founded in 1926, used Fokkers on its early services, and had established a route to Munich by 1928 and Berlin by 1931. It also opened a Venice–Milan–Paris service and in June 1938 extended this to London, with Fiat G.18s.

Ala Littoria was formed in 1934 to take over most of the Italian airlines and it operated numbers of Savoias and other Italian aircraft but long remained a large operator of flying boats and seaplanes, although flying three-engined Caproni landplanes on its services to Italian East Africa.

Marine aircraft in Europe
Much use of marine aircraft was made in Europe in the pioneer years of air transport. There was an erroneous belief that seaplanes and flying boats offered greater security on overwater crossings.

Britain used flying boats on the trans-Mediterranean section of the Empire routes. France and Italy, with many Mediterranean and Adriatic routes, were Europe's biggest users of commercial seaplanes and flying boats, although Germany used such aircraft on coastal resort services and in the Baltic.

When the England–India route was opened in 1929, Imperial Airways used Short Calcutta flying boats on the trans-Mediterranean section. They were 12-passenger metal-hulled biplanes with three 540hp Bristol Jupiter engines, and the first example flew in February 1928. The Calcutta was followed by the Short Kent, of similar configuration but with four 555hp Jupiter engines and improved accommodation for 16 passengers. There were only three Kents, they comprised the Scipio class and entered service in May 1931.

In July 1936 Short Brothers launched the first of the S.23 C-class flying boats, the *Canopus*. These had been designed for implementation of the Empire Air Mail Programme and were very advanced high-wing all-metal monoplanes powered by four 920hp Bristol Pegasus engines, had accommodation for 16 to 24 passengers and a top speed of just on 320km/h (200mph). *Canopus* made the first scheduled flight of the type, from Alexandria to Brindisi on 30 October

1936. Eventually the S.23-Cs worked the entire routes between Southampton and Sydney and Southampton and Durban, and some remained in service until after World War II.

Sandringham and Solent developments of the C class operated some of BOAC's postwar routes, and the last British flying boat services were those operated by Aquila Airways' Solents to Madeira until the end of September 1958.

The German resort services were mostly flown by small seaplanes and the Baltic services by Dornier Wals. The first Dornier Wal flew in November 1922 and, because of Allied restrictions on German aircraft production, most early Wals were built in Italy and Italian airlines employed them in considerable numbers. The Wal was an all-metal monoplane with the wing strut mounted above the hull and braced to the stabilizing sponsons, or sea wings. Two engines were mounted back to back on top of the wing, and the passengers occupied a cabin in the forward part of the hull. There were many versions of the Wal, with different engines, weights, performance and accommodation, but the early examples had seats for eight to ten passengers and were usually powered by Rolls-Royce Eagle or Hispano-Suiza engines. SANA was the biggest operator of passenger Wals, and put them into service in April 1926 on a Genoa–Rome–Naples–Palermo service.

There were three major flying boat constructors in Italy – Cant, Macchi and Savoia-Marchetti. The Cants were biplanes, mostly of wooden construction and SISA used the single-engined four-passenger Cant 10*ter* on its Trieste–Venice–Pavia–Turin service, which opened in April 1926, and on its Adriatic services. To supplement them SISA had a fleet of three-engined ten-passenger Cant 22s. Aero Espresso used a twin-engined eight-passenger Macchi M.24*bis* biplane flying boat on its Brindisi–Athens–Istanbul service.

Although the biplane served well into the 1930s, it was the Italian monoplanes that proved of most interest. In 1924 Savoia built a twin-hulled monoplane torpedo-bomber to the design of Alessandro Marchetti, and from this developed a passenger aircraft for Aero Espresso's then planned Brindisi–Istanbul service. This was the S.55C with thick-section 24m (78ft 9in) span wooden wing, accommodation for four or five passengers in each hull, triple rudders and two tandem-mounted 400/450hp Isotta-Freschini engines. These aircraft went into service in 1926 and were followed in 1928 by the more powerful S.55P version, of which SAM had 14 or 15 on trans-Mediterranean services.

Developed from the S.55 was the 33m (108ft 3¼in) span S.66 of 1932. This had three 550 or 750hp Fiat engines mounted side by side, and the twin hulls could accomodate up to 18 passengers. It is believed that 24 were built. They were operated by Aero Espresso, SAM and SANA, and some passed to Ala Littoria, working services between Italy and Tunis, Tripoli and Haifa. A number survived to be taken over for war service.

In a completely different category was the Cant Z.506 which saw widescale service with Ala Littoria on Mediterranean and Adriatic routes. This was a three-engined twin-float low-wing monoplane with seating for 12 to 16 passengers.

As early as 1923 L'Aéronavale started using twin-engined four-passenger Lioré et Olivier 13 wooden flying boats between Antibes and Ajaccio, and in May that year a Latécoère affiliate opened a Marseilles–Algiers service with them. More than 30 were built.

Also in 1923, CAMS had begun production of a series of civil and military flying boats designed by Maurice Hurel. One of them was the wooden CAMS 53 biplane with two tandem 500hp Hispano-Suiza engines and a small cabin for four passengers. It was introduced by Aéropostale on the Marseilles–Algiers route in October 1928.

The biggest French biplane flying boat in passenger service was the 35·06m (115ft) span Breguet 530 Saïgon, with three 785hp Hispano-Suiza engines, three-class accommodation for 19 to 20 passengers and a maximum weight of 15,000kg (33,069lb).

As a replacement for its CAMS 53s, Air Union had also ordered a fleet of four-engined 10 to 15 passenger Lioré et Olivier H 242 monoplane flying boats. These had their 350hp Gnome Rhône Titan Major engines in tandem pairs above the wing, and 14 went into service with Air France at the beginning of 1934.

The last trans-Mediterranean flying boat services of Air France were those operated to Algiers immediately after the war by two Lioré et Olivier H 246 four-engined 24 to 26 passenger monoplanes.

Mail and passenger services in the USA

On 1 March 1925, Ryan Airlines opened its Los Angeles–San Diego Air Line with modified Standard biplanes and a Douglas Cloudster, and this is claimed to have been the first regular passenger service wholly over the US mainland to be maintained throughout the year.

The first major legislative step in creating an airline industry was the passing of the Contract Air

Mail Act (known as the Kelly Act) in February 1925. This provided for the transfer of mail carriage to private operators, and it was followed in May 1926 by the Air Commerce Act, which instructed the Secretary of Commerce to designate and establish airways for mail and passengers, organize air navigation and licence aircraft and pilots. This Act came into force at midnight on 31 December 1926.

The history of United States air transport over the next few years was extremely complex, with numerous airlines competing for mail contracts. Some were small concerns, but others were backed by large financial organizations and closely linked with the aircraft manufacturing industry. Here only the briefest details of this intricate picture can be given.

The first five mail contracts were let on 7 October 1925, and 12 airlines began operating the services between February 1926 and April 1927 as feeders to the transcontinental mail route which was still being operated by the Post Office.

The Ford Motor Company had begun private daily services for express parcels between Detroit and Chicago on 3 April 1925, using single-engined all-metal Ford 2-AT monoplanes. Ford secured Contract Air Mail Routes (CAM) 6 and 7 covering Detroit–Chicago and Detroit–Cleveland and was the first to operate, beginning on 15 February 1926. Passengers were carried from August that year.

Next to start was Varney Air Lines, on CAM 5, between Pasco, Washington State, and Elko, Nevada, via Boise, Idaho. Leon Cuddeback flew the first service on 6 April 1926, using one of the airline's fleet of Curtiss-powered Swallow biplanes, but the operation was immediately suspended until the Swallows could be re-engined with air-cooled Wright Whirlwinds.

Also in April 1926 Robertson Aircraft Corporation began flying the mail on CAM 2 between St Louis and Chicago, and this operation is claimed as the first step in the eventual creation of the present American Airlines. Western Air Express began working CAM 4 between Los Angeles and Salt Lake City with Douglas M-2 biplanes, and opened its first passenger service over the route on 23 May that year.

CAM 1, New York–Boston, was awarded to Colonial Air Transport, but the services did not start until June 1926 and passenger service, with Fokkers, began almost a year later in April 1927.

PRT – Philadelphia Rapid Transit Service – obtained CAM 13 for Philadelphia–Washington and operated three flights a day from 6 July 1926. Whereas most of the mail carriers used passengers as a fill-up load or ignored them entirely, PRT, with its Fokker F VIIa-3ms, catered for passengers and used mail as a fill-up.

The prize routes were those covered by CAM 17, from New York to Chicago, and CAM 18, San Francisco to Chicago. National Air Transport (NAT) was founded in May 1925, secured CAM 17, and began operating it with Curtiss Carrier Pigeons on 1 September 1927, having opened CAM 3, Chicago–Dallas, on 12 May the previous year. NAT was not very interested in passenger traffic and soon acquired a fleet of 18 ex-Post Office Douglas mailplanes; but, later, did buy Ford Trimotors.

Boeing, however, secured the San Francisco–Chicago route and was very interested in passengers. The company built a fleet of 24 Model 40A biplanes for the route. They were powered by 420hp Pratt & Whitney Wasp engines and could carry pilot, two passengers and 545kg (1,200lb) of mail, and were superior in payload and performance to the aircraft operated by NAT and most other mail carriers. By midnight on 30 June 1927, the Boeing 40s were deployed along the route and service began the following day.

Western Air Express (WAE) had hoped to get the San Francisco–Chicago mail contract, but Boeing's tender was much lower. Instead WAE developed a

Los Angeles–San Francisco passenger service with Fokkers and established a high reputation for reliability and safety.

In 1926 and 1927 two events took place which had a marked effect on the growth of air transport in the United States. The first was the appearance of the Ford Trimotor, which first flew in June 1926. This was an all-metal high-wing monoplane powered by three 200/300hp Wright Whirlwind engines, and had accommodation for 10 or 11 passengers in its original 4-AT version. Later came the 5-AT with 400/450hp Pratt & Whitney Wasps and seats for 13 to 15. The other event, in May 1927, was Lindbergh's flight from New York to Paris. This was the first solo flight from New York to Paris and it created enormous interest in aviation.

Brief mention must be made of three other airlines. Pacific Air Transport was organized by Vern Gorst in January 1926 and opened service over CAM 8, Seattle–San Francisco–Los Angeles, on 15 September; Maddux Air Lines opened a Los Angeles–San Diego passenger service with Ford Trimotors on 21 July 1927; and Standard Airlines under Jack Frye began a Los Angeles–Phoenix–Tucson passenger service with Fokker F.VIIs on 28 November 1927.

The Standard Airlines operation, by using a rail

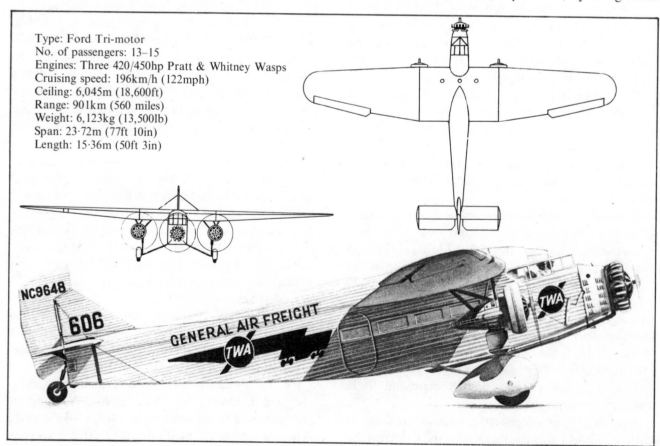

Type: Ford Tri-motor
No. of passengers: 13–15
Engines: Three 420/450hp Pratt & Whitney Wasps
Cruising speed: 196km/h (122mph)
Ceiling: 6,045m (18,600ft)
Range: 901km (560 miles)
Weight: 6,123kg (13,500lb)
Span: 23·72m (77ft 10in)
Length: 15·36m (50ft 3in)

connection, provided a 70-hour air-rail transcontinental service from February 1929 and further development cut the time to 43hr 40min. Boeing Air Transport acquired Pacific Air Transport in January 1928, and this gave Boeing a connection between the Chicago–San Francisco route and Seattle. Boeing's passenger traffic was developing, but its aircraft were unsuitable, so the company built the three-engined Model 80 with Pratt & Whitney Wasps and accommodation for 12 passengers. Four Boeing 80s were built and began service in October 1928, to be followed by ten more powerful Model 80As with 18 seats.

In May 1928 Transcontinental Air Transport (TAT) was formed, and began planning a transcontinental route using Ford Trimotors on which meals could be carried, although the first air stewardesses appeared on Boeing's Model 80s in 1930. TAT came to an agreement with the railways for an air-rail coast-to-coast service. It began on 7 July 1929. Westbound passengers left New York's Pennsylvania Station in the *Airway Limited* and travelled overnight to Port Columbus in Ohio, where a special combined air-rail terminal had been built. From there they flew to Waynoka in Okalahoma where they transferred to the Atchison, Topeka and Santa Fe Railroad for the night journey to Clovis in New Mexico. The final stage by air took them to Los Angeles, and from there they could continue free to San Francisco by train or by Maddux Air Lines. The New York–Los Angeles journey took exactly 48hr and the fare ranged from $337 to $403 one way. In November that year TAT acquired Maddux.

In the east, on 1 May 1928, Pitcairn Aviation began operating CAM 19 between New York and Atlanta with Pitcairn Mailwings and, using the lighted airway, did some night flying. That December the airline took over Florida Airways' Atlanta–Miami route, and the company became Eastern Air Transport in January 1930. The route was extended to Boston and passengers were carried over part of it from August. In December 1930 Curtiss Condors were introduced, and were also used by TAT. These were large twin-engined biplanes with 600hp Curtiss Conqueror engines and 18 seats. They led in 1933 to the T-32 model, which could carry 12 passengers in sleeping berths.

In 1929 Walter Folger Brown became Postmaster General, and he had strong views on how US airways should be developed. The Post Office contract for the central transcontinental route – CAM 34 New York–Los Angeles – was up for tender, but

before TAT could secure the contract Brown forced the merger of TAT with Western Air Express, thus forming Transcontinental and Western Air (TWA). The new airline began the first coast-to-coast all-air through service on 25 October 1930, using Ford Trimotors on the New York–Los Angeles route with a nightstop at Kansas City and an overall time of 36 hours.

American Airways, formed on 25 January 1930, obtained the southern route CAM 33 via Nashville and Dallas by a series of route extensions and company take-overs.

Then in July 1931 United Air Lines was organized to take over officially Boeing Air Transport, NAT, PAT and Varney, which for some time had been working under the United title.

Thus, by mid-1931, the Big Four had been created out of nearly 30 airlines of varying size. They were American Airways, Eastern Air Transport, TWA and United Air Lines.

Although this is but a brief summary of the early development of US air transport, mention must be made of the Ludington Line and of two more transport aircraft. Ludington was an airline which really believed in passengers, and on 1 September 1930, began a service 'every hour on the hour' from 08.00 to 17.00 over the Newark (New York)–Camden (Philadelphia)–Washington route, mainly with Stinson Trimotor monoplanes. In the two years before the company was taken over by Eastern it carried 124,000 passengers.

An event of great importance was Boeing's production of the Model 247 ten-passenger low-wing all-metal monoplane with two Pratt & Whitney Wasp engines and retractable undercarriage. This can be claimed as the prototype of the modern airliner. The Boeing 247 first flew on 8 February 1933, 60 were ordered for the United group, and they went into service with Boeing Air Transport, National Air Transport and Pacific Air Transport and became United Air Lines' standard equipment up to the end of 1936. The Boeing 247 was able to climb with one engine inoperative and cruise at 250km/h (155mph), and it made obsolete all other US airline equipment.

Boeing 247s were not available to non-United group airlines, and so TWA asked Douglas to produce a competitive aircraft. This was the DC-1, of similar layout to the Boeing. It first flew on 1 July 1933, and TWA ordered 20 of the improved production DC-2s.

But before the DC-2 could enter service, in July 1934, President Roosevelt, believing that some airlines had been unduly privileged in securing contracts, cancelled all the main contracts on 9 February 1934, thus ending an important chapter in US airline history. Under White House instructions the last mail flight had to be completed on 19 February. That day, as a dramatic gesture, Jack Frye of TWA and Eddie Rickenbacker flew the DC-1 from Los Angeles to Newark, via Kansas City and Columbus, in a transcontinental record time of 13hr 4min.

Development in the Commonwealth

After completing the first flight from Britain to Australia in December 1919, Ross and Keith Smith continued across the continent to Sydney, Melbourne and Adelaide. In preparation for the flight across Australia a route was surveyed, aerodromes prepared and fuel and oil provided. Responsible for this work on the Darwin–Longreach section were W. Hudson Fysh and P. J. McGuinness, and it was these men, with others, who founded Queensland and Northern Territory Aerial Services (QANTAS) on 16 November 1920.

It was to Norman Brearley's West Australian Airways that the honour went of starting the first subsidized air service in Australia. On 4 December 1921, a Bristol Tourer took off from Geraldton, the railhead north of Perth, to inaugurate a weekly mail service to Carnarvon, Onslow, Roebourne, Port Hedland, Broome and Derby. The route was extended to Perth in January 1924, and from Derby to Wyndham in July 1930. In 1929 a weekly Perth–Adelaide service began using de Havilland Hercules biplanes, and Vickers Viastra monoplanes were introduced in 1931, covering the journey in less than 24 hours.

On 2 November 1922, QANTAS opened its first scheduled service when P. J. McGinness flew an Armstrong Whitworth F.K.8 from Charleville to Longreach with mail, and on the next day Hudson Fysh flew on with mail and one passenger to Cloncurry.

QANTAS steadily expanded its operations in Queensland, and in 1931 took part in the first experimental England–Australia mail flights, carrying the mail from Darwin to Brisbane. QANTAS

was chosen as the partner to Imperial Airways to carry passengers and mail between Singapore and Brisbane when the England–Australia service opened in December 1934. A new company, Qantas Empire Airways (referred to hereafter as Qantas), was registered in January 1934 with Imperial Airways and QANTAS each holding half the share capital.

Qantas ordered a fleet of D.H.86 four-engined biplanes for the Brisbane–Singapore operation, but, when the first service left Brisbane on 10 December 1934, they had to use a single-engined D.H.61 and D.H.50 because of the late delivery, and Imperial Airways worked the Darwin–Singapore sector until February 1935. Passengers were carried over the entire England–Australia route from April.

In September 1932 the Holyman brothers began a Launceston–Flinders Island service with a Fox Moth, and soon after merged with another concern to form Tasmanian Aerial Services, which a year later became Holyman's. In October 1935 Adelaide Airways began operations, and sometime that year Airlines of Australia began services based on

Brisbane and Sydney. Then, in July 1936, Australian National Airways (ANA) was incorporated to include Holyman's, Adelaide Airways, Airlines of Australia and West Australian Airways.

Also in 1936 came an insignificant event which, years later, was to have a major influence on Australian air transport. Reginald Ansett had been refused a licence for a bus service between Hamilton and Melbourne, so, on 16 February 1936, he began an air service between these points, for which a licence was not required. After World War II Ansett was to get control of ANA and take over other companies to form Australia's largest private airline.

In Canada, during 1920, about 12,070km (7,500 miles) of landplane and seaplane routes were surveyed, the first flight to penetrate the Northwest Territories was made in the following year, and in 1923 the first air mail flight was made between Newfoundland and Labrador.

On 11 September 1924, Laurentide Air Service and Canadian Pacific Railway established an air service linking the railway at Angliers with the Quebec goldfields at Rouyn. This was the first regular air service introduced in Canada for the carriage of passengers, mail and freight.

Much of Canada's early air transport was on these lines, with air services linking remote areas with the nearest railhead. In summer the aircraft operated as landplanes and seaplanes, but the harsh winter climate enforced a change to skis.

Many concerns were involved in these early operations, mostly using single-engined aircraft. On 26 December 1926, Western Canada Airways began operations at Sioux Lookout and regular services were begun to Rolling Portage and Red Lake mining districts, and over the next few years a fairly extensive route network was established. On 3 March 1930, the airline began the nightly *Prairie Air Mail* service over the Winnipeg–Calgary and Regina–Edmonton routes. The initial service was flown by a Fokker Universal, and these were the first scheduled night flights in Canada.

On 25 November 1930, Canadian Airways Ltd began operations, having been formed by Canadian Pacific Railway, Canadian National Railway, Western Canada Airways and a group of airlines controlled by Aviation Corporation of Canada. The new concern established services in many parts of the country and undertook a lot of mail and freight

A United Air Lines Boeing 247. United ordered 60 following the maiden flight in February 1933.

carriage. But although it linked many pairs of cities, it did not provide a transcontinental service.

However, work went ahead on preparing airports and navigational services for a transcontinental route and on 10 April 1937, Trans-Canada Air Lines (TCA) was created by a Government Act. Survey flights over the route began from Vancouver in July 1937 and on 1 September that year TCA began operation with a Vancouver–Seattle service flown by Lockheed Electras. Early in 1938 Vancouver–Winnipeg mail and cargo services began, and that October the route was extended, for cargo only, to Toronto, Ottawa and Montreal, mail being carried from the beginning of December.

On 1 March 1939, the full-scale official transcontinental mail service was inaugurated, and passengers were carried from 1 April on this route and on the Lethbridge–Calgary–Edmonton route. TCA now operates as Air Canada.

The other major Canadian airline is CP Air, which was formed on 30 January 1942, as Canadian Pacific Air Lines. This was founded as a subsidiary of Canadian Pacific Railway, and represented the merger of Arrow Airways, Canadian Airways, Dominion Skyways, Ginger Coote Airways, Mackenzie Air Service, Prairie Airways, Quebec Airways, Starratt Airways, Wings Ltd and Yukon Southern Air Transport. The initial CPA fleet consisted of some 77 single- and twin-engined aircraft.

On 31 July 1929, Mrs F. K. Wilson founded Wilson Airways in Nairobi, and when Imperial Airways' Central Africa route was opened Wilson Airways operated connecting flights between Kisumu, on Lake Victoria, and Nairobi, using a Puss Moth. In August 1932 Wilson Airways began Nairobi–Mombasa–Tanga–Zanzibar–Dar-es-Salaam mail services and subsequently developed a network of services in the region.

With the introduction of the Empire Air Mail Programme in 1937, Imperial Airways introduced flying boats via a coastal route to Durban, and Wilson Airways opened a weekly Kisumu–Nairobi–Moshi–Dodoma–Mbeya–Mpika–Broken Hill–Lusaka mail service to connect with Imperial Airways at Kisumu.

In 1931 two airlines were formed in Central Africa, Rhodesian Aviation and Christowitz Air Services (in Nyasaland). The former opened a weekly Bulawayo–Salisbury service, subsidized by the Government of Southern Rhodesia and the Beit Trustees. The service was only operated as required and used South African-registered Puss Moths.

Christowitz opened a Blantyre–Beira service, also using a Puss Moth. In 1933 Rhodesian Aviation began a weekly passenger and goods service over the Salisbury–Gatooma–Que Que–Gwelo–Bulawayo–Johannesburg route with a Fox Moth, and Christowitz began a Salisbury–Blantyre service. Then in October 1933 Rhodesia and Nyasaland Airways (RANA) was formed. This acquired the assets of Rhodesian Aviation, taking over the Salisbury–Johannesburg service, which was terminated at Bulawayo, and in February 1934 the Christowitz Salisbury–Blantyre service. RANA also developed new routes and, like Wilson Airways, operated services connecting with Imperial Airways flying boat operations.

The first Government air mail flight in the world took place in India when, on 18 February 1911, Henri Pequet, in a Humber biplane, flew mail from Allahabad to Naini Junction about 8km (5 miles) away. The first actual mail service in India began on 24 January 1920, when the RAF opened a weekly service between Bombay and Karachi. This was maintained for only a few weeks, and there was no Indian air service until 15 October 1932, when Tata Sons opened a Karachi–Ahmedabad–Bombay–Bellary–Madras mail service to connect with Imperial Airways' England–Karachi flights. The Tata aircraft was a Puss Moth, and it was flown from Karachi to Bombay by J. R. D. Tata.

In May and June 1933 two airlines were formed in India. The first, Indian National Airways (INA), was established to participate as a shareholder in the second, Indian Trans-Continental Airways, and to develop services in northern India. Indian Trans-Continental was set up to operate the trans-India route in association with Imperial Airways, beginning the Karachi–Calcutta service on 7 July 1933, with Armstrong Whitworth Atlanta monoplanes.

On 1 December 1933, INA started a weekly Calcutta–Rangoon service with Dragons, and on the same day opened the first daily air service in India – between Calcutta and Dacca.

INA and Tata Air Lines (as successor to Tata Sons) continued to expand their operations, Tata opening a Bombay–Delhi service on 6 November 1937, and INA establishing a Karachi–Delhi service a year later.

A number of short-lived air services were operated in New Zealand during the 1920s, and in 1930–31 Dominion Airways operated a Desoutter monoplane on about 100 flights between Hastings and Gisborne before the loss of the Desoutter ended the undertaking. Air Travel (NZ) started a

Hokitika–Okuru service with a Fox Moth on the last day of 1934, East Coast Airways began a twice daily Napier–Gisborne service with Dragons in mid-May 1935, and at the end of that December Cook Strait Airways opened a Wellington–Blenheim–Nelson service with two Dragon Rapides.

The year 1935 saw the founding of Union Airways of New Zealand as an offshoot of Union Steam Ship Co, and this airline began a daily Palmerston North–Blenheim–Christchurch–Dunedin service on 16 January 1936, with D.H.86s. The airline commissioned three Lockheed Electras in June 1937 and put them on a daily Auckland–Wellington service. Union Airways purchased East Coast Airways in July 1938, and the following March opened a Palmerston North–Napier–Gisborne–Auckland service. After the war Air Travel (NZ), Cook Strait Airways and Union Airways were all absorbed by the newly founded New Zealand National Airways Corporation.

South Africa was slow in setting up any form of air transport, and it was only in March 1925 that the South African Air Force began a weekly experimental mail service over the Cape Town–East London–Port Elizabeth–Mossel Bay–Durban route, using D.H.9s. Only 32 flights were made, with 276 bags of mail, and it was not until more than four years later that a private air service began when Union Airways opened a subsidized Cape Town –Port Elizabeth service with extensions to Durban and Johannesburg.

During 1930 the administration of South West Africa concluded an agreement with Junkers to form South-West Africa Airways and operate a weekly passenger, freight and mail service between Windhoek and Kimberley. This and a number of other services were begun in 1931 using F 13s and Juniors.

A change in policy came with the founding, on 1 February 1934, of South African Airways (SAA), which began operation with aircraft and staff taken over from Union Airways. SAA introduced Junkers Ju 52/3ms, and these were the first multi-engined aircraft used by a South African airline.

At the beginning of February 1935 SAA took over South-West Africa Airways. The airline steadily expanded its network and on 1 April 1936, took over the Cape Town–Johannesburg sector of the England–South Africa route. Junkers Ju 86s were added to the fleet in 1937, and by the time all civil flying was suspended in May 1940, SAA had opened routes to Lusaka, Broken Hill, Nairobi, Kisumu and Lourenço Marques.

The South Atlantic

Having opened the air route between Toulouse and Dakar, France set about establishing services between Natal in Brazil and Buenos Aires. Airports were prepared and radio equipment installed, at an estimated cost of $1·5 million. Aeropostal Brasileira and Aeroposta Argentina were established as subsidiary of Aéropostale, and the Natal–Buenos Aires services was opened in November 1927. The dream of a service linking France and South America finally came true on 1 March 1928, when the entire route from Toulouse to Buenos Aires was opened for mail, with a transit time of eight days, but the ocean sector had to be operated by ships.

In 1928 Aeroposta Argentina opened services from Bahía Blanca, south of Buenos Aires, to the oil centre of Comodoro Rivadavia 950km (590 miles) further south, and from Buenos Aires to Asunción in Paraguay. In July 1929 the great barrier of the Andes was conquered with the opening of the Buenos Aires–Santiago service, and French air services had reached the Pacific coast.

For several years the services in South America were operated by a series of single-engined Latécoère monoplanes, the Laté 17, 25 and 26, and open-cockpit Potez 25 biplanes. Much of the flying was done at night and the weather was frequently appalling. On the route to Comodoro Rivadavia there were very strong winds to contend with, but it was the service across the Andes which called for a very high standard of flying and a great amount of bravery. One incident is enough to illustrate the character of the route. In June 1930 Henri Guillaumet left Santiago in a Potez 25 with the mail for Europe. He encountered blinding snowstorms and violent turbulence and was forced to land close to Laguna Diamante at an elevation of 3,500m (11,480ft). The little biplane overturned, but Guillaumet survived and it took him five days and four nights to struggle to safety.

Great as were the achievements of what was known as The Line, the South America route could not be regarded as satisfactory until the entire route could be covered by air. The first attempt was made in 1930, when on 12–13 May Jean Mermoz flew the ocean crossing from St Louis, Sénégal, to Natal in 21hr in the Latécoère 28 seaplane *Comte de la Vaulx*. But a single-engined floatplane was not a suitable aircraft for the ocean crossing, and the French Government ordered the three-engined Couzinet 70 *Arc-en-Ciel* landplane and the four-engined Blériot 5190 flying boat *Santos-Dumont*.

On 16 January 1933, Mermoz flew the *Arc-en-*

Ciel from St Louis to Natal in 14hr 27min. After modifications, as the Couzinet 71, it began regular South Atlantic mail flights at low frequency from 28 May 1934, completing eight ocean crossings by the end of the year. Flown by Lucien Bossoutrot, the *Santos-Dumont* made its first crossing, from Dakar to Natal, on 27 November 1934, and began regular service with Air France early in 1935. It made at least 22 ocean crossings and cut the Toulouse–Buenos Aires time to 3 days 20hr.

The construction of the big Blériot flying boat had been delayed by financial problems, and as a result it had been beaten into service by the four-engined Latécoère 300 flying boat *Croix du Sud* (*Southern Cross*). This boat made its first crossing on 3 January 1934, and continued as far as Rio de Janeiro. Thereafter it shared the route with the *Arc-en-Ciel* and made six crossings by the end of the year; but on 7 December 1936, radio contact was lost with the *Croix de Sud* about 4hr after it left Dakar. The flying boat, its famous commander Jean Mermoz, and his crew disappeared and were never found.

The overall success of the *Croix du Sud* led to construction of three similar Latécoère 301s, and these began service early in 1936.

Air France also employed a number of large four-engined Farman landplanes on the South Atlantic route, beginning with the F.220 *Le Centaure*, which made its first Dakar–Natal flight on 3 June 1935. In November 1937 the last of these, the F.2231 *Chef Pilote L. Guerrero*, owned by the French Government, flew from Paris to Santiago in 52hr 42min and made the Dakar–Natal crossing in 11hr 5min.

Although passengers were carried on the route to West Africa and in South America, none were carried on these prewar ocean crossings.

Germany used different methods in establishing its services to South America. Luftschiffbau Zeppelin decided to use the LZ 127 *Graf Zeppelin* on the route and on 18 May 1930, this airship left Friedrichshafen on a trial flight to Rio de Janeiro. This was followed by further trials in 1931. As a result the airship left Friedrichshafen on 20 March 1932, to open a regular service to Recife in Brazil, and this flight carried paying passengers – the first ever to fly on a transocean air service. Four flights were made that spring and five at fortnightly intervals in the autumn, with the last three continuing to Rio de Janeiro. In 1933 the *Graf Zeppelin* made 9 flights over the route, in 1934 twelve, in 1935 sixteen, and in 1936 nine, with an additional seven by the larger *Hindenburg*. From March 1935 this service was operated by Deutsche Zeppelin-Reederei, which was founded by Lufthansa and the Zeppelin company.

Germany's second method of flying to South America was to use landplanes between Germany and Africa and in South America and flying boats over the ocean. A system was devised for catapulting the flying boats from depot ships because they could not take off under their own power with sufficient fuel and while carrying a payload. The Dornier 8-ton Wal was used initially and the first experimental crossing made on 6 June 1933, when the *Monsun* was catapulted from the depot ship *Westfalen*. Regular mail services began in 1934, when *Taifun* made the first crossing on 7–8 February, and the entire route from Berlin to Buenos Aires was scheduled for four to five days. The 10-ton Wal was introduced on the route and, after trials over the North Atlantic in 1936, the much-improved Do 18s began working over the South Atlantic. Four-engined Do 26s made 18 mail crossings before war stopped the service and during their operations the Wals made 328 crossings.

The third country to begin South Atlantic air services was Italy. Ala Littoria Linee Atlantiche was set up as the Atlantic division of Ala Littoria, and had begun taking delivery of a special fleet of Savoia-Marchetti S.M.83 three-engined monoplanes before Linee Aeree Transcontinentali Italiane (LATI) was established to operate the services. After proving flights, a regular Rome–Rio de Janeiro service was inaugurated in December 1939. Avoiding British and French territory, the route was via Seville, Villa Cisneros, the Cape Verde Islands, Natal and Recife.

US air mail upheaval and a new start

After President Roosevelt's cancellation of the mail contracts and the last mail flight on 19 February 1934, the US Army Air Corps was given the task of flying the mail. The 43,450km (27,000 mile) network of mail routes was reduced to 25,750km (16,000 miles), and about 150 aircraft of various types were allocated to the operation. The weather was bad, the crews were inexperienced and ten pilots were killed before the last flight on 1 June.

The President admitted that he had been wrong, and in April 1934 the Postmaster General, James Farley, called in the airlines to find ways of salvaging what remained of the nation's airways. Temporary mail contracts were awarded, and among the conditions required to qualify for a contract was the stipulation that no contract carrier could be as-

The Blériot 5190 four-engine flying boat *Santos Dumont* with which Air France began regular services between Dakar and Natal in early 1935. Most of the long overwater routes were first pioneered and then established by this category of aircraft.

sociated with an aircraft manufacturer.

The Big Four were reorganized as American Airlines, Eastern Air Lines, TWA Inc and United Air Lines, and of the 32 new mail contracts they were awarded 15, with TWA and United both getting mail contracts for transcontinental routes – Newark–Los Angeles and Newark–Oakland respectively. Eastern got three important routes, Newark–New Orleans, Newark–Miami and Chicago–Jacksonville, while America's routes included Newark–Chicago and Newark–Boston. By combining Newark–Fort Worth and Fort Worth–Los Angeles they gained a transcontinental route, although it was not as direct as TWA's and United's.

Latin America

Most of the early airlines in South America were established by German nationals and with German capital, the first being SCADTA, which was founded in Colombia in 1919. This airline had a fleet of Junkers F-13 seaplanes and, after a period of experimental operation, opened a regular service in 1921 over the Magdalena River route linking the port of Barranquilla with Girardot, the railhead for Bogotá. The distance was 1,046km (650 miles) and the flight took 7hr, compared with a week or ten days by steamer. New routes were added and more modern equipment acquired until finally, in 1940, the airline was merged with another small company to form today's Avianca.

The next airline to be formed, again by Germans, was Lloyd Aéreo Boliviano (LAB). Equipped with F 13s, this company was founded in August 1925, and by the end of the year was running regular services between Cochabamba and Santa Cruz, taking three hours against the surface time of four days. This company was almost certainly the first to operate Junkers Ju 52/3ms, although it did so on military operations during the Gran Chaco war in 1932. LAB is still the Bolivian national airline.

In 1927 two German influenced airlines were founded in Brazil, Varig and Syndicato Condor. Condor opened a Rio de Janeiro–Pôrto Alegre–Rio Grande do Sul service in October, and during the year handed over the Pôrto Alegre–Rio Grande sector to Varig. Both companies used German aircraft, mostly Junkers, with floatplanes predominating, and built up a route system, initially in the coastal regions. In 1942 Condor was reorganized as Serviços Aéreos Cruzeiro do Sul and, although retaining its identity, was taken over in 1975 by Varig, which remains the principal Brazilian airline and now operates intercontinental services.

In Peru Elmer Faucett founded Compañía de Aviación Faucett in 1928, and by the following year had established air services extending the length of the country, from Ecuador in the north to Chile in the south. Faucett was unusual in that it built many of its own aircraft, based on Stinson single-engined monoplanes. Although now under different ownership, Faucett is still operating.

A military air service was introduced in Chile in 1929. This ran north from Santiago to Arica near the Peruvian border. It was taken over by Linea Aérea Nacional (LAN) in 1934, at first flying mail and then passengers, and since that time a considerable route system has been developed, including transatlantic services. LAN is the only airline to serve Easter

Island, which it includes on its Chile–Tahiti route.

The United States airline which became deeply involved in Latin America was Pan American Airways. This company, led by Juan Trippe for nearly 40 years, began regular contract mail services between Key West and Havana on 28 October 1927, using Fokker F.VIIa-3m monoplanes, and carried passengers from January 1928. PAA developed a network of services in the Caribbean, mostly using Sikorsky amphibians, and its routes reached as far as Christóbal in the Panama Canal Zone. Extension southward down the west coast of South America was blocked by the Grace shipping line; so in 1929 Panagra was formed, with Pan American and Grace each holding 50 per cent of the shares. Panagra secured a mail contract for the route from Cristóbal to Santiago, Chile, and across the Andes to Buenos Aires. The first mail left Miami on 14 May 1929, and from Cristóbal was flown by Sikorsky S-38 and then Fairchild FC-2 to Mollendo in Peru, where the route then terminated. The extension to Santiago was inaugurated on 21 July and, after the acquisition of Ford Trimotors, Buenos Aires was reached on 8 October and Montevideo on 30 November. A speeding up of the services in April 1930 enabled mail to travel from New York to Buenos Aires in 6½ days. Passengers were carried to Santiago from 15 August 1931, and through to Montevideo on 5 October.

Although Pan American had to settle for a half share in South American west coast operations, the company would not compromise on the east coast. In March 1929 NYRBA (New York, Rio and Buenos Aires Line) was founded, and that August it began in Buenos Aires–Montevideo service. The next month a mail and passenger service was started between Buenos Aires and Santiago with Ford Trimotors – five weeks ahead of Panagra's trans-Andes service.

Before the east coast route could be opened to the United States it was necessary to obtain suitable aircraft. There were long overwater sectors and few landing grounds, so a fleet of 14 Consolidated Commodore flying boats was ordered. These were large monoplanes powered by two 575hp Pratt & Whitney Hornet engines, and could carry 20 to 32 passengers. They cruised at 174km/h (108mph), and had a range of 1,600km (1,000 miles). Four had been delivered by the end of 1929 and on 18 February 1930, the entire route between Miami and Buenos Aires was opened, the 14,485km (9,000 miles) being flown in seven days. NYRBA also set up NYRBA do Brasil to undertake local operations in Brazil.

Pan American desperately wanted the east coast route, and finally managed to acquire NYRBA in September 1930, when it took over the Commodores – including those still on order – and changed NYRBA do Brasil to Panair do Brasil. Postmaster General Brown awarded Pan American the mail contract for the east coast route, FAM-10, on 24 September 1930.

Through NYRBA Pan American had acquired a fleet of Commodores, it urgently needed long-range aicraft with bigger payloads, and the need was met by three specially designed Sikorsky S-40s. These were large flying boats with four 575hp Pratt & Whitney Hornets, tail units carried on twin booms, and accommodation for 32 passengers.

A much more advanced and more important flying boat was the Sikorsky S-42, powered by four 700hp Hornets and having accommodation for up to 32 passengers. The S-42 cruised at 274km/h (170mph) and had a normal range of 1,930km (1,200 miles). Ten examples of three models were built, and the type was introduced on the Miami–Rio de Janeiro route on 16 August 1934.

It was also in 1934 that VASP came into being and today, with Cruzeiro, Transbrasil and Varig, this company operates the Ponte Aérea (Air Bridge) between Rio de Janeiro and São Paulo, with 370 flights a week in each direction.

In Central America, a New Zealander, Lowell Yerex, was responsible for much of the early air transport development. In 1931 he founded Transportes Aéreos Centro-Americanos (TACA) in Honduras, and built up a main airway linking British Honduras with the capitals of Costa Rica, Guatemala, Honduras, Nicaragua, Panama and Salvador. In 1933 he opened the first service between Tegucigalpa and San Salvador. Networks of routes were set up in the Central American republics, and eventually TACA associated companies were established in Colombia, Costa Rica, Guatemala, Honduras, Mexico, Nicaragua, Salvador and Venezuela.

Today this vast TACA empire has shrunk to the present TACA International, which in the spring of 1976 was operating three BAC One-Elevens and three DC-6As. In 1939 Yerex founded British West Indian Airways (BWIA), and the airline opened its first service, Trinidad to Barbados via Tobago, in November 1940, using a Lockheed Lodestar.

Mexico has had a large number of airlines, its first being Cia Mexicana de Aviación (CMA), which began work in August 1924, carrying wages to the oil fields near Tampico with Lincoln Standard biplanes.

With Martin M-130 flying boats, such as the one pictured here, Pan American Airways established the first regular passenger route across the Pacific Ocean on 21 October 1936.

This operation was devised to circumvent the activities of bandits. Under the title Mexicana, CMA is now a major international airline. Mexico's other major airline, Aeromexico, was created in 1934 as Aeronaves de Mexico. During the course of its long history this airline has absorbed a considerable number of other Mexican carriers.

Conquering the Pacific

The first air crossing of the Pacific was not made until 1928.

In July 1931 Charles Lindbergh and his wife made a survey of a northern route in a Lockheed Sirius single-engined seaplane, flying to Japan via Alaska, Siberia and the Kuriles, but political problems prevented the establishment of such a route.

The only alternative was to use island stepping stones which were United States territory, and bases were therefore prepared on Wake Island and Guam to enable a service to operate from San Francisco to Manila via Honolulu, Wake and Guam. This gave stages of 3,853km (2,394 miles) from San Francisco to Hawaii, 3,693km (2,295 miles) to Wake, 2,414km (1,500 miles) to Guam and 2,565km (1,594 miles) to Manila.

To obtain aircraft for the operation of a regular service, Pan American issued a specification for a flying boat capable of flying 4,023 km (2,500 miles) against a 48km/h (30mph) headwind while carrying a crew of four and at least 136kg (300lb) of mail. To this specification Martin built three M-130 flying boats, each powered by four 800/950hp Pratt & Whitney Twin Wasp engines. The boats had a span of 39·62m (130ft), weighed 23,700kg (52,252lb) fully loaded and could carry 41 passengers, although only 14 seats were installed for the Pacific route. Cruising speed was 253km/h (157mph) and the range

5,150km (3,200 miles), or 6,437km (4,000 miles) if the flying boat was only carrying mail.

The M-130s were named *China Clipper*, *Philippine Clipper* and *Hawaii Clipper*, and the first was delivered in October 1935. The *China Clipper*, under the command of Capt Edwin Musick, inaugurated the trans-Pacific mail service when it left Alameda on 22 November 1935, and it alighted at Manila 59hr 48min later. Paying passengers were carried from 21 October 1936.

Early in 1937 the Sikorsky S-42B *Hong Kong Clipper* was delivered, and that spring it made a survey of the southern Pacific route to Auckland, subsequently being used to extend the trans-Pacific operation from Manila to Hong Kong, the first service being on 27–28 April. This gave Pan American a direct link to China through the China National Aviation Corporation's Hong Kong–Canton–Shanghai service. From that time the Martin M-130s were completing the San Francisco–Manila out–and back–flights in 14 days.

On 23 December 1937, Pan American inaugurated a San Francisco–Auckland service via Hawaii, Kingman Reef and Samoa, but the S-42B, with its commander, Edwin Musick, and crew, was lost on the second flight and the service had to be suspended. But on 12 July 1940, a fortnightly service was opened via Hawaii, Canton Island and New Caledonia using one of the new Boeing 314s, with passengers being carried from 13 September.

The modern airliner

The principal passenger aircraft in use in the United States in the late 1920s and early 1930s were Fokker F-VIIs and Ford Trimotors. They had accommodation for 8 to 15 passengers, but cruised at only a little over 160km/h (100mph), and for much of the time very few of their seats were occupied. Some people believed there was a need for smaller and faster aeroplanes, and as a result there was a period when numerous airlines were operating fleets of small single-engined monoplanes capable of cruising at more than 241km/h (150mph).

John Northrop and Gerrard Vultee designed a superb four-passenger high-wing monoplane known as the Lockheed Vega. It was a wooden aeroplane, powered by a Wasp engine, and it first flew in July 1927. The Vega entered service with International Airlines on 17 September 1928, and subsequently there were several versions, including one with a metal fuselage. Cruising speed was 217–241km/h (135–150mph), and the type was used by a number of airlines, including Braniff and TWA. Very similar was the parasol-winged Lockheed Air Express which was produced for Western Air Express.

The last of Lockheed's single-engined high-speed transports was the six-passenger Orion, which had a low-mounted wing and retractable undercarriage. It went into service with Bowen Air Lines in May 1931. With either a Wasp or Cyclone engine the Orion cruised at 289–313km/h (180–195mph), and is claimed as the first transport aircraft capable of 320km/h (200mph). American Airways, Northwest Airways and Varney Speed Lines were among the Orion users, two were exported to Swissair, and Air Express Corporation operated them on a US transcontinental freight service, achieving coast-to-coast times of 16–17 hours.

The most advanced of these single-engined monoplanes was the Vultee V-1A, which could carry eight passengers and cruise at 340km/h (211mph). The Vultee had an 850hp Wright Cyclone engine, was of all-metal construction and had a retractable undercarriage. American Airlines introduced the type in September 1934.

There were also two Boeing prototypes. These were the Model 200 and 221, both named Monomail. They were all-metal low-wing monoplanes, each powered by a 575hp Pratt & Whitney Hornet. The Model 200 was originally a single-seat mail and cargo carrier. It first flew on 6 May 1930, and its semi-retractable undercarriage made possible a cruising speed of 217km/h (135mph). This first Monomail was followed in August 1930 by the slightly longer Model 221 with non-retractable undercarriage and a cabin for six passengers. It went into service with Boeing Air Transport, and like the Model 221A was later lengthened to provide two extra seats.

On 19 February 1934, the Douglas DC-1 had made its dramatic coast-to-coast flight, and TWA had already ordered 20 of the 14-passenger produc-tion DC-2s. These were each powered by two 720hp Wright Cyclones, giving a maximum cruising speed of 315km/h (196mph) and a range of just over 1,600km (1,000 miles). The first DC-2 was delivered to TWA on 14 May 1934, and four days later made a proving flight from Columbus to Pittsburgh and Newark. On 1 August DC-2s began transcontinental operation over the Newark–Chicago–Kansas City–Alburquerque–Los Angeles route to an 18hr schedule. Apart from providing much improved transcontinental services, the Newark–Chicago sector was the first nonstop operation over the route.

On 5 May 1934, American Airlines had begun transcontinental sleeper services with Curtiss Condors, and had also used them to build up frequency on its New York–Boston route. Condors were no match for the DC-2, and so American Airlines asked Douglas for a sleeper development of the DC-2.

Douglas enlarged the DC-2 by widening and lengthening its fuselage and added 3·04m (10ft) to its wing span. Powered by two 1,000hp Cyclone engines and having 14 sleeping berths, this new type was known as the DST – Douglas Sleeper Transport.

American Airlines had ordered 10 DSTs, but after its first flight on 17 December 1935, increased the order and changed it to cover eight DSTs and 12 dayplanes – DC-3s with 21 seats. The first DST was delivered at the beginning of June 1936 and, used as a dayplane, went into service on the New York–Chicago route on 25 June. American took delivery of its first DC-3 in August, and on 15 September was able to inaugurate its DST *American Mercury* skysleeper transcontinental service with an eastbound schedule of 16 hours.

Thus was launched one of the world's great transport aeroplanes, the DC-3, of which by far the biggest percentage was powered by 1,200hp Pratt & Whitney Twin Wasps. It was to be built in numerous civil and military versions, with a total of 10,655 produced in the United States and others being built in Japan and, under licence, in the Soviet Union. After the war large numbers of surplus military DC-3s became available to civil operators, and in the postwar years almost every airline operated them at some period. There are still several hundred flying.

Lockheed also embarked on production of a series of fast twin-engined monoplanes. The first was the ten-passenger Model 10A Electra with 450hp

A typical medium-size airliner of the 'sixties was the Handley Page Herald. Aircraft which belonged to this category were used to provide short/medium-range passenger/cargo services, worldwide.

The three-engine Ford Tri-motor, which first appeared in the US in 1926 is one of the 'old timers' of civil aviation. Rough and rugged, built to endure, it made use of the corrugated light alloy skin which had been pioneered in Germany. Examples of the Tri-motor, known as the 'Tin Goose', still fly in 1980.

Previous pages: A Boeing 747 (*top left*), a Lockheed TriStar (*top right*), and a Boeing 727 (*below*).

Below: The world's first international supersonic passenger services were initiated by Air France/British Airways on 21 January 1976, with the Anglo-French Concorde.

Pratt & Whitney Wasp Junior engines and a cruising speed close to 322km/h (200mph). The Electra was introduced on 11 August 1934, by Northwest Airlines. The original British Airways had seven, and also bought nine of the more powerful 12-passenger Model 14s, which were about 64km/h (40mph) faster.

The major United States airlines required an aeroplane with greater capacity and range than the DC-3 and, in March 1936, the Big Four and Pan American each came to an agreement to share the cost of developing the four-engined Douglas DC-4E, with a span of 42·13m (138ft 3in) and a maximum weight of 30,164kg (66,500lb). It was powered by 1,450hp Pratt & Whitney Twin Hornets, and cruised at 322km/h (200mph). The DC-4E had a nosewheel undercarriage, the first on a big transport, triple fins, and production models were to be pressurized. United Air Lines ordered six 52-passenger sleeper DC-4Es in July 1939, having put the prototype into experimental operation in the previous month. But the aeroplane was found to be unsuitable, the order was cancelled, and the only example was exported to Japan.

The last American transport landplane to go into service before the Japanese attack on Pearl Harbor was the Boeing 307 Stratoliner. Although only ten were built, it has an important place in history as the first pressurized aeroplane to go into airline service. The Stratoliner was a low-wing monoplane with a wing span of 32·69m (107ft 3in), powered by four 900hp Wright Cyclones and having a maximum weight of 19,051kg (42,000lb). It cruised at 354km/h (220mph) and had a range of 3,846km (2,390 miles). This design employed the wings, nacelles, power-plant and tail surfaces of the B-17 Flying Fortress bomber, but had a completely new circular-section fuselage with pressurized accommodation for 33 passengers and 5 crew. The first aircraft flew on the last day of 1938, but was lost before delivery to Pan American. PAA had three, TWA five, and Howard Hughes had a modified aircraft for record breaking.

TWA introduced the Stratoliner on its transcontinental route on 8 July 1940, and cut eastbound times to 13hr 40min; Pan American based its Stratoliners in Miami for Latin American operations; but at the end of 1941 these aircraft were ordered into war service with TWA's fleet being used over the North Atlantic. After the war the TWA

The Hawker Siddeley three-engine Trident short-haul airliner *left* which entered revenue service in 1964 adopted the rear-engine layout pioneered by the French Aérospatiale Caravelle.

aircraft, much modified, were put back into civil operation with 38 seats and increased take-off weight. Some Stratoliners were sold to Aigle Azur in 1951, and some saw service in Latin America and the Far East for several years, a few remaining in use until the mid-1960s.

United Kingdom domestic airlines

When sustained air services were established in the United Kingdom, most that were to prove successful were based on routes which involved a water crossing. In the spring of 1932 British Amphibious Air Lines began irregular operation of such a route, between Blackpool and the Isle of Man, using a Saunders-Roe Cutty Sark amphibian. From June until the end of September the service operated regularly, and there was a bus connection between Blackpool and a number of towns in Yorkshire.

In April 1932 Hillman's Airways opened a service between Romford and Clacton, and by June had achieved a frequency of one every three hours from 0900 until dusk.

June 1932 saw the start of a ferry service between Portsmouth and Ryde. This was operated, at very low fares, by Portsmouth, Southsea and Isle of Wight Aviation, using a three-engined Westland Wessex monoplane. The company was to work up a network of services in the south of England, use a wide variety of aircraft and carry a thousand passengers a day until World War II.

Two other events of importance in 1932 were the opening of a twice-daily Bristol–Cardiff service which began on 26 September and was operated by Norman Edgar with a Fox Moth; and the first flight of the de Havilland D.H.84 Dragon on 24 November. Norman Edgar's service was to develop into a sizeable operation under the title Western Airways, and the Dragon was to make possible economic short-haul airline operations in the United Kingdom and many other parts of the world. The Dragon was a six-passenger biplane with two 130hp de Havilland Gipsy Major engines, and 115 were built in the United Kingdom and 87 in Australia. The first Dragon was delivered to Hillman's on 20 December 1932.

The entry of the railways into UK airline operation took place in 1933, and the same year witnessed the start of air transport in Scotland and the beginning of large-scale air services to the Channel Islands. On 12 April the Great Western Railway (GWR) began a public service, twice each way on weekdays, between Cardiff and Plymouth, with a stop at Haldon to serve Torquay and Teignmouth.

The aircraft used was a Westland Wessex chartered from Imperial Airways. The service was extended to Birmingham in May, the frequency cut to once a day, and at the end of September closed for the winter.

On the day that the GWR began its services, Spartan Air Lines began a service between Cowes and Heston with a Spartan Cruiser three-engined monoplane, and when it was withdrawn at the end of the summer 1,459 passengers had been carried. This operation was continued in 1934, grew considerably, and became part of the railway group's operations.

Of much greater importance was the founding of Highland Airways by capt E. E. Fresson on 3 April 1933. On 8 May this airline opened a regular service linking Inverness and Kirkwall in Orkney, via Wick. A Monospar S.T.4 monoplane was used and, in spite of some appalling weather, set such a high standard of regularity that a year later the airline was awarded the first domestic mail contract. The airline extended its routes to include Aberdeen and Shetland.

In Glasgow, John Sword had established Midland and Scottish Air Ferries (M&SAF), and on 8 May it made its first recorded ambulance flight, from Islay to Glasgow. The air ambulance is still an essential element of Scottish air transport. Regular passenger services were begun on 1 June over the Glasgow–Campbeltown–Islay route, with Dragons, and M&SAF were to develop a number of routes before closing down at the end of September 1934.

On 18 December 1933, Jersey Airways began a daily Portsmouth–Jersey service. There was no airport on Jersey until 1937, so the fleet of Dragons operated from the beach at St Aubin's Bay near St Helier. Sometimes the whole fleet was on the beach at the same time, because schedules were governed by the tides and the fleet tended to fly in a loose formation.

Having suffered as a result of increasing road competition, the four mainline railways had obtained rights to operate domestic air services. On 21 March 1934, they registered Railway Air Services, with the London Midland and Scottish Railway,

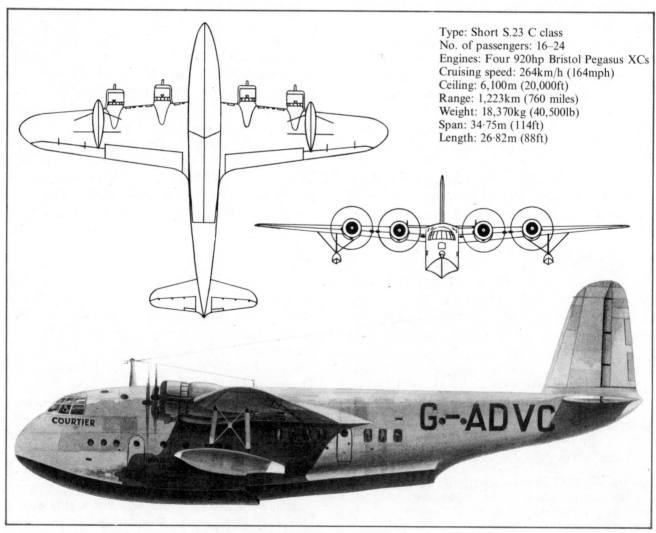

Type: Short S.23 C class
No. of passengers: 16–24
Engines: Four 920hp Bristol Pegasus XCs
Cruising speed: 264km/h (164mph)
Ceiling: 6,100m (20,000ft)
Range: 1,223km (760 miles)
Weight: 18,370kg (40,500lb)
Span: 34·75m (114ft)
Length: 26·82m (88ft)

COURTIER

G-ADVC

London and North Eastern Railway, Great Western Railway, Southern Railway and Imperial Airways as the shareholders and Imperial Airways responsible for undertaking flying operations. The first RAS service was opened on 7 May. This was the previous year's GWR operation, and the route was extended to Liverpool and operated by a Dragon. At the end of July RAS began a Birmingham–Bristol–Southampton–Cowes summer service, but the main RAS operation was to be the *Royal Mail* trunk route linking London, Birmingham, Manchester, Belfast and Glasgow. This was to be operated with D.H.86s and the opening was set for 20 August, but the weather was atrocious and only part of the route could be flown. Full working began the next day. This route, with modifications, continued throughout the years up until the war, but RAS's other routes were mainly confined to summer months, during which periods they catered for holiday traffic.

Hillman's Airways had opened a London–Liverpool–Isle of Man–Belfast service on 16 July 1934, and this was extended to Glasgow on 1 December, when the company got the mail contract instead of RAS.

Following the ending of Midland and Scottish Air Ferries' operations in September, George Nicholson gave up his short-lived Newcastle–Carlisle–Isle of Man service, moved to Glasgow and founded Northern and Scottish Airways. Operations began on 1 December with twice-weekly Glasgow–Cambeltown–Islay services, and the company steadily developed, with Skye being served from 5 December 1935. By July 1936 a circular route was linking Glasgow with Skye, North Uist and South Uist. Most services were flown by Dragons or Spartan Cruisers, and in August 1937 the company amalgamated with Highland Airways to form Scottish Airways, the link between the two networks coming in May 1938, when a Glasgow–Perth–Inverness–Wick–Kirkwall–Shetland service began.

On 30 September 1935, three domestic airlines, Hillman's Airways, Spartan Air Lines and United Airways, were merged to form Allied British Airways. The name was changed to British Airways on 29 October, and in the following August the company absorbed British Continental Airways. British Airways operated some domestic routes and controlled a number of United Kingdom domestic airlines, but its main effort was concentrated on developing fast and efficient services to the continent.

E. Gandar Dower founded Aberdeen Airways, later renamed Allied Airways (Gandar Dower) Ltd, and operated services in Scotland, including routes to Orkney and Shetland, and on 12 July 1937, started a service between Newcastle and Stavanger. Between March 1935 and the summer of 1936 Crilly Airways operated a number of domestic services; North Eastern Airways developed routes down the eastern side of the country between London and Scotland; Blackpool and West Coast Air Services operated over the Irish Sea; and there were others which operated for varying periods.

Although Aer Lingus is the Irish national airline, its beginning was closely involved with the United Kingdom domestic operations. The airline was founded on 22 May 1936, and a week later began a daily Dublin–Bristol service. In the same month it started working between Dublin and the Isle of Man, and in September opened a Dublin–Liverpool service and extended its Bristol service to London. All these operations were in association with Blackpool and West Coast Air Services, and they went under the title 'Irish Sea Airways'. Dragons were used on the first three routes, and D.H.86s flew to London.

A variety of aircraft served the United Kingdom routes, but the biggest contribution was made by the de Havilland biplanes – the D.H.84 Dragon, D.H.86 and D.H.89 Dragon Rapide. The Dragon Rapide had first flown in April 1934; it was a much-improved Dragon with two 200hp Gipsy Six engines, there were seats for six to eight passengers, and the cruising speed was about 210km/h (130mph). Several hundred Dragon Rapides were built; they served airlines in many parts of the world and a few are still flying on short routes today.

The Empire air mail programme

In December 1934 HM Government announced that from 1937 all letters despatched from the United Kingdom for delivery, along what were then the Empire routes, would, as far as practicable, be carried by air without surcharge. The carrier was to be Imperial Airways, with certain sectors to be covered by Commonwealth airlines.

A fleet of 28 Short S.23 C-class flying boats was ordered, and 12 Armstrong Whitworth A.W.27 Ensign landplanes, although some of the latter were for European operation. The first C-class flying boat was launched at Rochester on 2 July 1936, making its first flight on 4 July, and the first Ensign flew on 24 January 1938.

The Ensigns were large, high-wing monoplanes powered by four 850hp Armstrong Siddeley Tiger engines, and had accommodation for 27 passengers on Empire routes and 40 on European. Cruising speed was 274km/h (170mph) and range 1,287km

(800 miles). The Ensigns suffered numerous troubles and, although introduced in Europe towards the end of 1938, played little part in Empire operations until they had been re-engined with Wright Cyclones during World War II.

The C-class flying boats began operation in October 1936, and the first stage of the Empire Air Mail Programme was inaugurated on 29 June 1937, when the *Centurion* left Southampton with 1,588kg (3,500lb) of unsurcharged mail for the Sudan, East and South Africa.

On 23 February 1938, the Mail Programme was extended to cover Egypt, Palestine, India, Burma and Malaya when the *Centurion* and the Qantas 'boat *Coolangatta* left Southampton. By early April the schedules had been improved to give a Southampton–Karachi time of 3 days, a Singapore time of $5\frac{1}{2}$ days and Sydney time of $9\frac{1}{2}$ days. At that time the all-up mail did not apply to Australia, and the flying boats did not work through to Australia until June.

The Mail Programme was extended to Australia, New Zealand, Tasmania, Fiji, Papua, Norfolk Island, Lord Howe Island, Nauru, Western Samoa and certain other Western Pacific territories with the departure of *Calypso* from Southampton on 28 July 1938. Not all these places were served by air, the mail sometimes being sent over its last stages by sea.

The C-class flying boat also played a pioneer role in pre-war crossings of the North Atlantic.

The North Atlantic

Today there are some 30 airlines operating scheduled passenger services across the North Atlantic, and in 1975 the airlines made more than 80,000 scheduled flights over the ocean, carrying 8,782,176 passengers and more than half a million tons of cargo. In 1947 the airlines carried 209,000 passengers across the North Atlantic, whereas 415,000 travelled by sea. In 1957 the numbers using sea and air services were just about equal, at slightly over one million by each type of transport, and in 1958 the airlines carried 1,292,000, while sea passengers had dropped to 964,000. In 1970 the airlines carried 10,038,000 passengers on scheduled and charter services while sea traffic had fallen even further, to 252,000. Total North Atlantic airline passengers in 1975 numbered nearly $12\frac{1}{2}$ million.

It is not surprising that from the earliest days of aviation there were dreams of North Atlantic services, but the problems of making these dreams reality were formidable. The Great Circle distance between the present Shannon Airport, in Ireland, and Gander, in Newfoundland, is 3,177km (1,974 miles), and that is the shortest direct ocean crossing. North Atlantic weather is notoriously bad, with frequent fog in the Newfoundland area and very strong westerly prevailing winds.

Long range with worthwhile payload was difficult to achieve in the years prior to World War II. One solution was offered by the Short-Mayo flying boat/seaplane composite aircraft.

Pioneer west–east crossings were made in 1919. However, it was not until April 1928 that an aeroplane crossed the Atlantic Ocean from east to west, when Baron von Hünefeld, Cmdt J. Fitzmaurice and Hermann Köhl flew from Ireland to Greenly Island off Labrador in the Junkers W 33 *Bremen*.

As early as 1928 Pan American Airways began investigating possible North Atlantic routes. Numerous surveys were made, including one by Lindbergh, who studied a northern route via Greenland and Iceland, making his flight in a Lockheed Sirius seaplane. The very big problem was aircraft range, and Pan American and Imperial Airways did not begin trial flights until 1937.

France began designing a large transatlantic flying boat in 1930. During the design stage it had to be considerably modified, and emerged in January 1935 as the Latécoère 521, to be named *Lieutenant de Vaisseau Paris*. The Laté 521 was a 42-tonne monoplane powered by six 800/860hp Hispano-Suiza engines. It was designed to carry 30 passengers on Atlantic routes or 70 over the Mediterranean, and its range was 4,000km (2,485 miles). Its first successful North Atlantic flight, from Biscarosse to New York via Lisbon and the Azores, was made in August 1938, and it subsequently made trial flights over various North Atlantic routes.

Germany undertook North Atlantic trial flights during 1936 using two Dornier Do 18 twin-engined flying boats – *Zephir* and *Aeolus*. These were catapulted from the depot ship *Schwabenland* near the Azores, and from it the *Zephir* flew to New York in 22hr 12min on 11 September, arriving with 10 hours' reserve fuel. By 20 October, when the trials ceased, the two Do 18s had flown 37,637km (23,386 miles) on eight flights over various North Atlantic routes. In the periods August–November 1937 and July–October 1938 a further 20 flights were made by three Blohm und Voss Ha 139 four-engined seaplanes operating from the *Schwabenland* and the *Friesenland*.

But the honour of operating regular North Atlantic air services still went to Germany, when the Zeppelin LZ 129 *Hindenburg* began operation in 1936. Operated by Deutsche Zeppelin-Reederei, the *Hindenburg* left Friedrichshafen for New York with the first paying passengers on 6 May 1936, and on its return flight landed at Frankfurt, which became the regular European terminal. Ten return flights were made before the service was suspended for the winter, and so great was the demand that the airship's passenger accommodation had to be in-

creased from the original 50. During the first season 1,309 passengers were carried, and the fastest flight from Lakehurst, New Jersey, to Frankfurt was made in 42hr 53min.

A programme of 18 return services was announced for 1937, and on 3 May the airship left Frankfurt with 97 passengers and crew. But on 6 May (local time) fire broke out as the ship was landing at Lakehurst, 35 people lost their lives and German Zeppelin services came to an end, although three new Zeppelins had been ordered and the LZ 130 was test flown in September 1938.

Germany did not re-establish North Atlantic air services until 1955, but in August 1938 made a spectacular non-stop flight from Berlin to New York in 24hr 56min and back in 19hr 55min. This flight was made by a special Focke-Wulf Condor, and was the first across the ocean by a four-engined landplane.

The first regular service by heavier-than-air craft to be operated over part of the North Atlantic began on 16 June 1937, when Imperial Airways opened a Bermuda–New York service with the C class flying boat *Cavalier*, and Pan American started working a reciprocal service with the Sikorsky S-42 *Bermuda Clipper*.

Then on 5–6 July that year the two airlines made their first North Atlantic survey flights. The special long-range C class 'boat *Caledonia*, commanded by Capt A. S. Wilcockson, flew from Foynes on the Shannon to Botwood in Newfoundland in 15hr 3min, and the Sikorsky S-42 *Clipper III* (Capt H. E. Gray) flew in the opposite direction. *Caledonia* continued to Montreal and New York, and the Sikorsky flew on to Southampton.

Neither the C class boats nor the S-42 were suitable for commercial operation because they did not have the ability to carry sufficient fuel and a payload. Pan American ordered the large Boeing 314 for its services, and Imperial Airways undertook experiments designed to increase the range of its aircraft. One experiment involved refuelling in the air. An aircraft can carry a greater load than it can lift off the ground or water, so modified C class boats were built which could take off with a payload and then receive their main fuel supply from tanker aircraft via a flexible hose.

The other experiment involved launching a small aircraft from the back of a larger one. The Short-Mayo Composite Aircraft was therefore built. This consisted of a modified C class 'boat, on top of which was mounted a small floatplane. Take-off could be achieved with the power of all eight engines

and the lift of both sets of wings, and then at a safe height the aircraft would separate, the smaller mailplane flying across the ocean and the flying boat returning to its base. The launch aircraft was the Short S.21 *Maia*; the mailplane the S.20 *Mercury*, powered by four 340hp Napier Rapier engines and having a crew of two. The first separation took place successfully near Rochester on 6 February 1938, and on 20–21 July the *Mercury* made the first commercial crossing of the North Atlantic by a heavier-than-air craft when, commanded by Capt D. C. T. Bennett, it flew from Foynes non-stop to Montreal in 20hr 20min carrying mail and newspapers. From Montreal *Mercury* flew on to Port Washington, New York, but it played no further part in North Atlantic air transport.

On 4 April 1939, Pan American's Boeing 314 *Yankee Clipper* arrived at Southampton on its first proving flight from New York, and on 20 May the same aircraft left New York on the inaugural mail service. Flying via the Azores, Lisbon and Marseilles it arrived at Southampton on 23 May and left on the first westbound service the next day.

The northern mail route was opened on 24 June by the same aircraft, and on 28 June the *Dixie Clipper* left Port Washington on the inaugural southern-route passenger service – the first by a heavier-than-air craft. On 8 July, with 17 passengers, the *Yankee Clipper* left Port Washington on the inaugural service over the northern route, via Shediac, Botwood and Foynes. The single fare was $375.

On 4 August Britain began a weekly experimental mail service between Southampton and New York via Foynes, Botwood and Montreal, using S.30 C class flying boats which were refuelled in flight from Handley Page Harrow tankers based at Shannon and Botwood. The first service was flown by the *Caribou*, commanded by Capt J. C. Kelly Rogers. The full programme of 16 flights was completed on 30 September in spite of war having started. Because of the war Pan American terminated its services at Foynes, and then withdrew altogether on 3 October.

BOAC, as successor to Imperial Airways, operated the C class flying boat *Clare* on four round trips between Poole, in Dorset, and New York via

The Hawker Siddely Trident was designed around British European Airways' requirement for a short haul airliner. It came into operation in the 1960s.

Botwood and Montreal during the period 3 August–23 September 1940, and one round trip by the *Clyde* during October. These flights, most of which were made while the Battle of Britain was in progress, carried mail, despatches and official passengers, but the first truly commercial British North Atlantic services did not effectively begin until 1 July 1946.

Airlines at war

BOAC had the wartime task of maintaining communication between the United Kingdom and the Commonwealth. This was largely achieved in spite of the German occupation of Europe and Axis control of much of the Mediterranean. The C class flying boats operated the *Horseshoe* route between Durban and Sydney via the Middle East, and connection to this route was provided by services between the United Kingdom and West Africa and thence via a trans-African route to Khartoum. The *Horseshoe* services were maintained until Japan cut the route at the beginning of 1942, after which they terminated at Calcutta. But by a brilliant opera-

tional feat Qantas reopened the route by introducing a service between Perth and Ceylon. This began on 10–11 July 1943, and was operated by Consolidated Catalina flying boats, which had to maintain radio silence while flying the 5,654km (3,513 miles) overwater route. On the inaugural flight the *Altair Star* took 28hr 9min from Koggala Lake to the Swan River, but the longest crossing made took 31hr 35min. Qantas also played a very important part in the fighting around New Guinea and suffered heavy casualties.

In May 1941 the Atlantic Ferry Organization of the Ministry of Aircraft Production began operating the North Atlantic Return Ferry service. This was flown with Consolidated Liberators, and had the task of returning crews to North America after they had delivered military aircraft to the United Kingdom. In September that year BOAC took over the operation.

BOAC also maintained North Atlantic services over various routes with three ex-Pan American Boeing 314s, operated a considerable number of vital routes in the Middle East, and flew regularly between Scotland and Sweden – flying unarmed aircraft across occupied Norway to maintain essential communication and import much-needed ball bearings to Britain. It was on this last operation that the airline used the all-wood de Havilland Mosquito. An outstanding BOAC operation was the evacuation of 469 British troops from Crete to Alexandria in April and May 1941 by the flying boats *Coorong* and *Cambria* in the course of 13 return flights.

The war had a major effect on many of the United States airlines. Within the US they had to maintain vital services with what fleets they were allowed to keep – mostly DC-3s – and they had to provide a wide range of services for the government. Until that time Pan American had been the only US airline to operate outside North America, and it undertook global operations. But many others undertook long-haul trans-oceanic flying for the armed services, mostly with military Douglas DC-4s (the C-54 Skymaster). TWA operated North Atlantic flights with Boeing Stratoliners, and American Export Airlines began a New York–Foynes service on 26 May 1942, with Sikorsky VS-44 flying boats.

One of the epic operations of the war was that over the 'Hump'. This was the supply route from India across the mountains into China from 1942 until 1945. This was really a military operation, but was flown by the crews of several US airlines, and of US Air Transport Command and China National Aviation Corporation. DC-3s were used as well as some Consolidated C-87 Liberator transports, but the bulk of the work was done by Curtiss C-46 Commandos. The scale of the 'Hump' operation can be appreciated from the fact that 5,000 flights, carrying 44,000 tons, were made in one month. At some stages of the operation aircraft were taking off at two-minute intervals.

Finally, mention must be made of the New Zealand–Australia link. Before the war three C class flying boats had been ordered for Tasman Empire Airways (TEAL), although the airline was not founded until April 1940, with New Zealand, Australia and the United Kingdom holding the shares. Only two boats, *Aotearoa* and *Awarua*, were delivered, but not until March 1940, and on 30 April *Aotearoa* flew the first Auckland–Sydney service. In June 1944 the 1,000th Tasman crossing was completed, and throughout the war these two flying boats provided the only passenger service of any kind between the two countries.

The return to peace

The war had brought about three major changes in air transport. These were the design and construction of higher-capacity longer-range aircraft capable of transoceanic operation, the large-scale construction of land airports, and worldwide operations by United States airlines. A further change was to follow shortly, as the overseas territories of European countries gained their independence and set up their own international airlines.

Most of the immediate postwar airline operations was undertaken with the DC-3, most of them war surplus, but the four-engined Douglas DC-4, mostly ex-military Skymasters, began to appear in some numbers and to be used to inaugurate transatlantic services, although with intermediate fuelling stops.

In Britain the decision had been taken to nationalize the air transport industry, with three corporations each having its own area of responsibility. BOAC already existed and, on 1 August 1946, Britain European Airways Corporation (BEA) and British South American Airways Corporation (BSAA) were established under the Civil Aviation Act which was passed in 1946.

BOAC was to be responsible for all long-distance services except those to South America, BEA's sphere was mainly Europe and the United Kingdom, and BSAA was to operate all services to Latin America. BOAC had already opened a number of European routes and set up a BEA Division, and these were handed over on 1 August. But the domestic airlines retained their identities until early 1947, working the routes on behalf of the new BEA. Before the war there had been no British air services to South America, but in January 1944 British Latin-American Air Lines had been founded by a number of shipping companies. The name was changed to British South American Airways in October 1945 and, following a series of proving flights, BSAA opened regular London–Buenos Aires services with Avro Lancastrians on 15 March 1946 – the first scheduled operations from the new Heathrow Airport. A route had been opened down the west coast of South America to Santiago in Chile, and the east coast route was also extended to that city by the time the company was changed to a State corporation.

BSAA was later absorbed into BOAC, and over the years numerous private companies were to establish regular air services. Finally, in 1972, BEA and BOAC were to be amalgamated to form British Airways, while British Caledonian Airways – resulting from the mergers of several private airlines – was to become the second major British airline, with responsibility for South American operations from October 1976.

Throughout Europe the airlines were rebuilding, mostly with DC-3s on short routes and DC-4s on intercontinental services. Air France, Sabena, KLM, Swissair and others established transatlantic services and other long-haul routes into Africa or the Far East. Sweden established SILA to operate Atlantic services and then eventually formed SAS – Scandinavian Airlines System – as a consortium of Danish, Norwegian and Swedish airlines. Some airlines were established with outside help – Alitalia was set up in Italy with BEA assistance, and TWA was to be involved in assisting airlines in Greece, Ethiopia and Saudi Arabia.

In Eastern Europe the prewar CSA in Czechoslovakia and LOT in Poland restarted services, and in Bulgaria, Hungary, Rumania and Yugoslavia airlines were set up with Soviet participation – BVS in Bulgaria, Maszovlet in Hungary, TARS in Rumania and JUSTA in Yugoslavia. JAT was also established in Yugoslavia. Later these East European airlines became completely nationally owned as TABSO (later Balkan Bulgarian Airlines), Malév (Hungary) and Tarom (Rumania). JUSTA ceased to exist in 1948 and Tarom was renamed LAR at the end of 1976. These airlines

developed domestic and international services and in the cases of CSA, JAT, LAR and LOT have North Atlantic services.

The year 1955 saw the rebirth of Lufthansa, and it has now developed into one of the world's major airlines. In East Germany a separate Deutsche Lufthansa had been established the previous year, but it was forced to abandon this name and since 1963 has worked as Interflug, which had been set up in 1958 to work services between East Germany and Western Europe.

The United States rebuilt its domestic networks and started developing worldwide services on a big scale. Pan American resumed commercial North Atlantic services, and was joined by TWA and American Overseas Airlines (AOA), successor to American Export Airlines. TWA opened up routes to the Middle East and Far East, and AOA was absorbed by Pan American which eventually inaugurated round-the-world services. United Air Lines and Pan American found lucrative traffic between the USA and Hawaii, and Northwest Airlines established a northern Pacific route to the Orient. Braniff opened services to South America and in more recent times National Airlines opened a Miami–London route.

In Canada, Trans–Canada Air Lines and Canadian Pacific Air Lines made rapid progress, both building up services to Europe and the latter

Douglas DC-3s, such as this example in service with Swedish Air Lines, first entered service in 1936. They were built in thousands during World War II for service as troop and cargo transports. Many war surplus machines are still flying.

establishing routes to Australia and Japan. They are Canada's biggest airlines and now trade as Air Canada and CP Air respectively.

Qantas, as Australia's international airline, developed its system, opened up networks of services in the New Guinea area, introduced round-the-world services and pioneered the Indian Ocean route to South Africa. ANA was the main Australian domestic operator and, on behalf of the newly formed British Commonwealth Pacific Air Lines, worked a Sydney–San Francisco–Vancouver service until BCPA could undertake its own operations. But a change in Australian civil aviation policy in 1945 was to have a profound effect on ANA. The Australian National Airlines Act of 1945 led to the founding of the State-owned Trans-Australia Airlines (TAA), and all main domestic air services had to be shared equally between TAA and ANA. Over the years Ansett had been acquiring various Australian airlines, and in October 1957 acquired ANA to form Ansett-ANA, now renamed Ansett Airlines of Australia. These two major operators, each carrying more than four million passengers a year, give a very high standard of service and have, with Qantas, outstanding safety records.

The Douglas DC-4E was the prototype of a new-generation airliner that was before its time. It had such features as a tricycle type landing gear, power-boosted flying controls and reversible propellers to shorten the landing run.

In New Zealand the newly created New Zealand National Airways ran the domestic services, and TEAL (now Air New Zealand) developed its services between New Zealand and Australia, retaining flying boats until June 1954. TEAL also opened a number of South Pacific routes and a trans-Pacific service to the United States.

India's airlines rapidly developed after the war, new companies were formed, and in July 1946 Air-India was founded as the successor to Tata Air Lines. In March 1948 Air-India and the Indian Government formed Air-India International, and this airline began a Bombay–Cairo–Geneva–London service on 8–9 June 1948, using Lockheed Constellations. Now State-owned as Air-India, this airline enjoys a remarkable reputation for efficient operation and has a jet-operated network stretching from New York through India to Japan and Australia. By 1953 India had eight domestic airlines operating an extensive route system with a total fleet of nearly 100 aircraft, but on 1 August that year they were amalgamated to form the State-owned Indian Airlines Corporation which now carries in excess of three million passengers a year.

With the creation of Pakistan in August 1947, that country rapidly had to develop its air transport in order to provide communication between its widely separated west and east wings (the latter is now Bangladesh). The first airline was Orient Airways, which had been founded in India the previous year on the initiative of Jinnah, the founder of Pakistan. This airline established vital services including the supply routes through the Indus Valley and the world's highest mountains to Gilgit and Skardu, Pakistan International Airlines (PIA) was set up, and with three Lockheed Super Constellations began Karachi–Dacca nonstop services in June 1954 and Karachi–Cairo–London services on 1 February 1955. The airline was reorganized as a corporation with the majority government holding in March 1955.

South African Airways greatly expanded operations after the war, worked the *Springbok* services to London in association with BOAC, opened a trans-Indian Ocean route to Australia, another to Hong Kong and one to New York via Ilha do Sul.

In January 1947 the Governments of Kenya, Tanganyika, Uganda and Zanzibar established East African Airways, initially with a fleet of six de Havilland Dragon Rapides. Owned by Kenya,

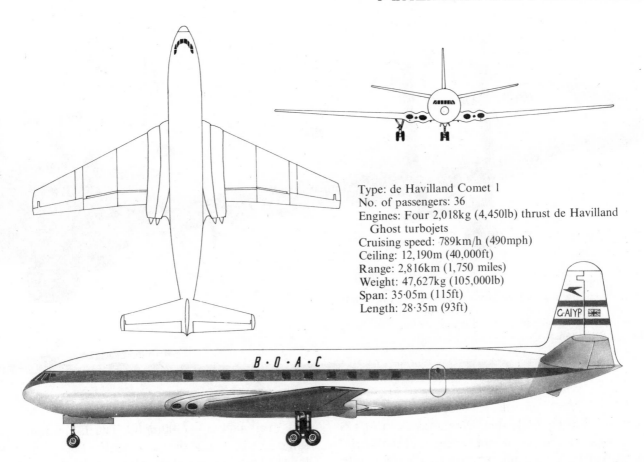

Type: de Havilland Comet 1
No. of passengers: 36
Engines: Four 2,018kg (4,450lb) thrust de Havilland Ghost turbojets
Cruising speed: 789km/h (490mph)
Ceiling: 12,190m (40,000ft)
Range: 2,816km (1,750 miles)
Weight: 47,627kg (105,000lb)
Span: 35·05m (115ft)
Length: 28·35m (93ft)

Tanzania and Uganda, East African Airways was a jet-equipped airline flying regional services, and routes to Europe, India and Pakistan. It was on the point of dissolution in early 1977.

Starting in 1951, Japan Air Lines began in a small way with the help of Northwest Airlines, which supplied aircraft and crews – Japanese nationals still being forbidden to fly as pilots. This airline rapidly established itself, and is now widely regarded as a major international carrier, having a very extensive route system and carrying about nine million passengers a year.

Today there is an enormous volume of domestic air traffic in Japan, and this is shared between JAL, All Nipon Airways, TOA Domestic Airlines and a few small companies. Japan Asia Airways was formed to operate services to Taiwan after JAL began operating to China.

A new generation of transport aircraft
In the first full year of peace after World War II, the scheduled services of the world's airlines carried 18 million passengers, double the 1945 total, and in 1949 the total for the year was 27 million. For some time to come the passenger total was to double every five years, and after a while cargo tonnage was to grow even faster.

These enormous traffic totals were largely due to the production of a remarkable series of transport aeroplanes – mostly designed and manufactured in the USA. Ever since the war the DC-3 has played a major role, although its numbers have declined; but even in the early 1940s it was obvious that larger, longer-range aircraft were required.

The four-engined Douglas DC-4E prototype had been abandoned as unsuitable in 1938, but Douglas designed a smaller, unpressurized DC-4. Before any could be delivered the United States was at war, and DC-4s were put to military use, as C-54 or R5D Skymasters. US airlines gained experience with these aircraft, flying them on military duties, but had to wait for peace before they could acquire surplus ex-military aircraft and some of the 79 civil examples built after the war.

The DC-4 was a low-wing monoplane with nose-wheel undercarriage and four 1,450hp Pratt & Whitney R-2000 engines. Initially it had 44 seats, but high-density seating for up to 86 was later installed. American Overseas Airlines introduced the DC-4 on New York–Hurn (for London) services at the end of October 1945. There were two intermediate stops and the scheduled time was 23hr 48min. On 7 March 1946, DC-4s went into US domestic service on the New York–Los Angeles route with American

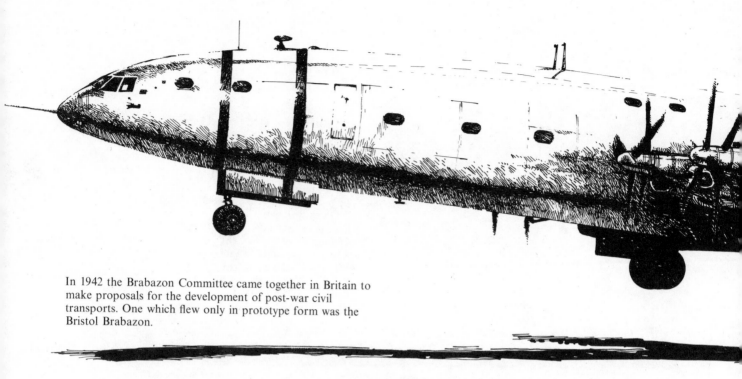

In 1942 the Brabazon Committee came together in Britain to make proposals for the development of post-war civil transports. One which flew only in prototype form was the Bristol Brabazon.

Airlines. Subsequently they were used by most US carriers and by many airlines throughout the world. A total of 1,242 civil and military DC-4s were built.

The DC-4 was a magnificent aeroplane, but it was unpressurized and its cruising speed of 352km/h (219mph) was no match for the other US wartime transport, the Lockheed Constellation. This had been designed for TWA as a long-range aircraft, but it did not fly until January 1943. TWA and Pan American had each ordered Constellations, but the small number built went to the United States Army Air Force as C-69s, and they entered service in April 1944.

After the war a civil production was resumed and the military aircraft were brought up to civil standard. Pan American began a Constellation service between New York and Bermuda on 3 February 1946, and in the same month TWA introduced Constellations on its New York–Paris and New York–Los Angeles services. The original Constellation had four 2,200/2,500hp Wright Cyclone R-3350 engines and 44 seats – later increased to as many as 81 in some configurations. It had a pressurized cabin, and at 480km/h (298mph) its cruising speed was at least 127km/h (79mph) higher than the DC-4's. Several improved models were built, and total production amounted to 233.

The superior performance of the Constellation led Douglas to build the comparable DC-6, with pressurized cabins and a maximum cruising speed of just over 482km/h (300mph). The DC-6 had the same 35·81m (117ft 6in) span as the DC-4 but was longer, had accommodation for 50 to 86 passengers, and was powered by four 2,400hp Pratt & Whitney R-2800 engines.

American Airlines and United Air Lines were the first to operate DC-6s, introducing them on 27 April 1947, on the New York–Chicago and transcontinental routes respectively. Its eastbound transcontinental scheduled time was 10hr. It was used by many major airlines and 170 were built.

In Montreal Canadair produced versions of the DC-4 and DC-6 modified to take Rolls-Royce Merlin liquid-cooled engines.

Douglas followed the DC-6 with the longer fuselage DC-6A and DC-6B, which had increased range. The DC-6A was a cargo aircraft mainly used by the US Air Force and Navy, but the DC-6B proved to be one of the finest piston-engined passenger aircraft ever produced. It entered service on 29 April 1951, on American Airlines' transcontinental route, began service with 54 seats but, later, in high-density configuration could seat 102 passengers. The DC-6B cruised at 507km/h (315mph) and had a range of just over 4,830km (3,000 miles). It was used by a large number of airlines, a total of 288 being built.

Lockheed developed the Constellation into the L.1049 Super Constellation, with 5·48m (18ft) increase in fuselage length and 2,700hp Cyclone

engines. It went into service as a 66-seat aircraft with Eastern Air Lines on 17 December 1951.

At about that time Curtiss-Wright's 3,250/3,400hp Turbo-Compound engine became available, and both Douglas and Lockheed developed their aircraft to make use of the new engine. Douglas slightly lengthened the DC-6B, but retained the same wing and produced the longer range DC-7, while Lockheed retained the basic Super Constellation airframe to produce the Turbo-Compound-powered L.1049C, D, E, G and H.

The DC-7 was the first aircraft capable of operating US transcontinental services in both directions without stops, and American Airlines introduced it on non-stop New York–Los Angeles services on 29 November 1953, scheduled to take 8hr 45min on the westbound flight, and 8hr eastbound – although it did not always make these times. Only 110 DC-7s were built, all of which were initially purchased by US airlines. Some later served as DC-7F freighters.

TWA had begun non-stop coast-to-coast services with L.1049C Super Constellations between Los Angeles and New York on 19 October 1953, but these had to stop at Chicago for fuel on westbound flights. However, the L.1049G, with wing tip fuel tanks, enabled TWA to match American Airlines' non-stop performance in both directions.

The DC-7 was developed into the longer-range DC-7B. This entered service in 1955, but only 112 were built – 108 for US airlines and four for South African Airways.

The DC-7 and L.1049G had made US transcontinental non-stop operation a practical undertaking, but at that time non-stop North Atlantic operation in both directions was rarely possible. In order to attain this desirable goal, Douglas added 3m (10ft) to the span of the DC-7 and increased its fuel tankage. It thus became the DC-7C Seven Seas with 7,412km (4,606 miles) range and North Atlantic non-stop capability. Pan American introduced them on the route on 1 June 1956 – many other airlines also adopted the type.

Lockheed went further than Douglas and designed a completely new 45.72m (150ft) span one-piece wing, containing fuel for more than 9,655km (6,000 miles), and mated this with the Super Constellation fuselage to form the L.1649A Starliner, with up to 99 seats. It entered service with TWA on the New York–London service on 1 June 1957.

These two very advanced piston-engined aircraft also made possible one-stop services between Europe and Japan over the Polar route, with a call at Anchorage in Alaska. SAS was first with the DC-7C on the Copenhagen–Tokyo route on 24 February 1957, and Air France followed about a year later, with L.1649As flying between Paris and Tokyo.

In addition to the Douglas and Lockheed types, there was one other large American four-engined airliner – the Boeing Stratocruiser. This was bigger than the other types, and had a lounge and bar on a lower deck. It had seats for up to 100 passengers, was powered by four 3,500hp Pratt & Whitney Wasp Major engines, and entered service over the North Atlantic with Pan American in 1949. Only 55 civil Stratocruisers were built, and they earned a poor reputation because of engine and propeller problems. In spite of this they were popular with crews and passengers, and maintained BOAC and Pan American first-class North Atlantic services until the introduction of jet aircraft in 1958.

In the immediate postwar years there was an urgent need for a twin-engined short-haul aeroplane to replace the DC-3, and several countries designed aircraft to meet this requirement. In the United States, Martin and Convair both produced aircraft in this category. They were low-wing monoplanes with nosewheel undercarriages and 2,500hp Pratt & Whitney R-2800 engines.

The 42-seat Martin 2-0-2 was unpressurized and entered service in autumn 1947 with Northwest Airlines in the United States and LAN in Chile. The type was involved in a number of serious accidents, and one suffered a wing failure which led to the type being withdrawn. After strengthening, the 2-0-2 returned to service in 1950 as the Martin 2-0-2A. It was followed by the 48 to 52-seat pressurized Martin 4-0-4, which entered service with TWA in October 1951. A total of 149 aircraft of both models was built for several airlines.

Much geater success was enjoyed by the Convair-Liner. The CV-240, with 40 seats and pressurized cabin, entered service in June 1948 with American Airlines. It quickly found favour, and 176 civil examples were built. The improved 44-passenger CV-340 followed in 1952, and in 1956 the 52 to 56 seat CV-440 Metropolitan made its debut. More than 1,000 civil and military Convair-Liners were built, of which more than 240 were later re-engined with propeller-turbines.

In Britain in the early postwar years BOAC was using DC-3s, Avro Lancastrians and Yorks developed from the Lancaster bomber, and Short flying boats. But plans had been made for new aircraft, and two of them flew before the end of 1945. They were the Vickers-Armstrong Viking, developed from the Wellington bomber, and the Bristol 170 Freighter/Wayfarer.

The Viking was a mid-wing monoplane with two Bristol Hercules engines, unpressurized cabin, and a tailwheel undercarriage. It cruised at about 320km/h (200 mph), and on entering service with BEA on 1

September 1946, had 27 seats. For several years it was the airline's principal type. It also served with many other airlines, and 163 were built.

The Bristol 170 was very different. It was a high-wing monoplane with a slab-sided fuselage, non-retractable undercarriage and two Hercules engines. The Wayfarer was a passenger aircraft, but the Freighter undertook the duties which its name implied – the best known operation being Silver City Airways' cross-Channel vehicle ferries, which it and the longer-fuselage Superfreighter undertook for many years. The type also operated cargo services between North and South Island in New Zealand.

A very handsome British aeroplane was the high-wing Airspeed Ambassador, with two 2,625hp Bristol Centaurus engines and pressurized cabin. It had accommodation for 47 to 49 passengers and was introduced into scheduled service as the Elizabethan by BEA in March 1952. Only 20 were built.

France's first postwar transport aircraft to enter airline service was the 33-seat Sud-Est SE.161 Languedoc which, as the Bloch 161, had made its first flight in September 1939, but could not be produced until after the war. It was originally powered by four Gnome Rhône engines, but most had Pratt & Whitney R-1830s. A total of 100 Languedocs was built, entering service with Air France on the Paris–Algiers route in May 1946 and serving a number of airlines in Europe and the Middle East.

Italy continued to produce three-engined Fiats and four-engined Savoia-Marchetti S.M.95s. One of the latter inaugurated Alitalia's Rome–London service on 3 April 1948, but these types made no significant contribution to air transport outside Italy.

There is one other European transport aircraft of the postwar era worthy of mention. This was the Swedish Saab Scandia, a low-wing monoplane with 32 to 36 seats and two 1,800hp Pratt & Whitney R-2180 engines. It flew in November 1946 and 18 were built, including six by Fokker.

Turbine power

In Britain the Brabazon Committee was set up in December 1942 to make recommendations for the development of postwar transport aircraft. Among its recommendations was the design and production of a North Atlantic turbojet aeroplane, Brabazon Type IV, and a short-haul propeller-turbine type, Brabazon Type IIB.

The Type IV evolved through a number of designs to become the de Havilland Comet – the world's first

The Stratocruiser was the first of Boeing's civil airliners to enter service after World War II. It maintained luxury first-class services across the North Atlantic until the advent of the jet-airliner.

turbojet airliner. Two types were designed to meet the Type IIB recommendation: the Armstrong Whitworth Apollo, of which only two prototypes were built, and the Vickers-Armstrong Viscount. Turbine power was to change the whole standard of airline flight by reducing vibration and interior noise levels. In the case of the turbo-jet, flight times were to be virtually halved.

The prototype Viscount, the V.630, made its first flight on 16 July 1948. It was a low-wing monoplane with a pressurized cabin for 32 passengers, and was powered by four 1,380hp Rolls-Royce Dart turbines driving four-blade propellers. Although the V.630 was considered too small to be economic, BEA did use it to operate the first ever services by a turbine-powered aeroplane. On 29 July 1950, it flew from Northolt, west of London, to Le Bourget, Paris, carrying 14 revenue passengers and 12 guests. The V.630 stayed on the Paris route for two weeks and then, from 13 August until 23 August, operated over the London–Edinburgh route on the world's first turbine-powered domestic services.

Rolls-Royce increased the power of the Dart and enabled Vickers to stretch the Viscount to become a 47 to 60 seat aircraft as the V.700. BEA placed an order for 20 and regular Viscount services began on 18 April 1953, when RMA *Sir Ernest Shackleton* operated the London–Rome–Athens–Nicosia service. The Viscount immediately proved a success, being produced in several versions and sold in many parts of the world, including the USA and China. A total of 438 were built.

The Comet 1 was a very clean low-wing monoplane with four 2,270kg (5,000lb) thrust de Havilland Ghost turbojets buried in the wing roots. There was accommodation for 36 passengers in two cabins and pressurization enabled it to fly at levels over 12,190m (40,000ft). The first prototype flew on 27 July 1949, and soon made a number of spectacular overseas flights. BOAC took delivery of ten Comet 1s and on 2 May 1952, operated the world's first jet service – over the London–Johannesburg route. With their cruising speed of 788km/h (490mph) Comets covered the 10,821km (6,724 miles) in less than 24hr. On the London to Singapore route they cut the time from $2\frac{1}{2}$ days to 25hr, and they reduced the London–Tokyo time from 86hr to $33\frac{1}{4}$hr.

Air France and UAT introduced Comets, and they were ordered by several other airlines. But exactly a year after their introduction a Comet broke up in flight near Calcutta, and in January 1954 another disintegrated and fell into the sea near Elba. After modifications the Comet was put back into service, but less than three weeks later, on 8 April, a third Comet broke up, and the type was withdrawn from service.

Pressure failure of the cabin, specifically bursting of the square windows in the original model, was said to have caused the failures, and some fuselage redesign resulted. Comet 2s, already under construction, were modified and went to the RAF. Work went ahead on the Rolls-Royce Avon-powered Comet 4 with longer fuselage, seats for up to 81, and additional, wing-mounted, fuel tanks.

BOAC ordered 19 Comet 4s and on 4 October 1958, operated the first ever North Atlantic jet services, the London–New York flight being made in 10hr 22min with a fuel stop at Gander. The eastbound flight was made nonstop in 6hr 11min. A shorter-span longer-fuselage Comet 4B, with seats for up to 101, was introduced by BEA on 1 April 1960, and in the same year the Comet 4C was commissioned – this combined the Comet 4 wings with the Comet 4B fuselage. A total of 74 Comet 4-series aircraft were completed.

The second European turbojet airliner to enter service was the French Sud-Aviation SE.210 Caravelle. It was a twin-engined aircraft and the first to have its turbojets mounted on either side of the rear fuselage. The Caravelle was designed primarily for operation between France and North Africa. It first flew on 27 May 1955, was powered by two 4,762kg (10,500lb) thrust Rolls-Royce Avons and, as the Caravelle I, had seats for up to 80 passengers. The Caravelle I entered service with Air France in May 1959 on the Paris–Rome–Istanbul route, and with SAS between Copenhagen and Cairo.

Numerous versions of the Caravelle were produced, up to the 128-passenger Caravelle 12B, and they were operated in many parts of the world, including the United States, where United Air Lines had 20. The Caravelle was an exceptionally fine aeroplane, and of the 280 built nearly 200 are still in service.

Although BOAC introduced the pure-jet Comet as early as 1952, the airline was to commission a large propeller-turbine aircraft nearly five years later, and BEA began using a similar type several months after it had introduced Comets. BOAC issued a specification early in 1947 for a Medium Range Empire (MRE) transport, and Bristol designed the Type 175 with four Centaurus piston engines to meet the requirements. Subsequently it was agreed that Bristol Proteus propeller-turbines would be used and in this form, as the Britannia, the prototype flew on 16 August 1952. The Britannia was a large low-wing monoplane with seats for up to 90 passengers and a take-off weight of 70,305kg (155,000lb).

BOAC introduced this Britannia 102 on the London–Johannesburg route on 1 February 1957 and the longer-range Britannia 312 on London–New York services on 19 December 1957. The latter were the first North Atlantic services by turbine-powered aircraft. The Britannia was an extremely good aircraft, but it appeared much too late and only 85 were built.

Canadair produced a number of Britannia variants, one of which, the CL-44, went into airline service. These aircraft generally resembled the Britannia, but had 5,730hp Rolls-Royce Tyne propeller-turbines and lengthened fuselages. Known as the CL-44D, and provided with swing-tail fuselage for cargo loading, the type entered service with the US cargo carrier Seaboard World Airlines in July 1961. Then came the CL-44J, with even longer fuselage and accommodation for up to 214 passengers, for the cheap-fare North Atlantic services of the Icelandic airline Loftleidir.

The large propeller-turbine aircraft built for BEA was the Vickers-Armstrong Vanguard, also powered by Tynes. It could carry 139 passengers and had considerable underfloor cargo capacity. The Vanguard first flew on 20 January 1959, and began operation with BEA on 17 December 1960, with regular operation from March 1961. About a month earlier it had entered service with Trans-Canada Air Lines, the only other customers for new Vanguards. Only 43 were built, and both airlines converted some into freighters.

Only one type of large civil propeller-turbine aeroplane, the Lockheed L.188 Electra, was built in the United States. This was smaller than the Vanguard, having accommodation for up to 99 passengers, and was powered by four 3,750hp Allison 501 engines. The Electra first flew on 6 December 1957, and entered service with American Airlines and Eastern Air Lines in January 1959.

The most successful of all the western propeller-turbine transports has been the Fokker F.27 Friendship. This is a high-wing monoplane powered by two Rolls-Royce Darts and normally accommodating about 48 passengers. The first prototype flew on 24 November 1955, and the aircraft went into production in the Netherlands and also in the United States as the F-27, built by Fairchild (later Fairchild Hiller).

The type first entered service in the United States, with West Coast Airlines, on 28 September 1958. In Europe Aer Lingus began operating the Dutch-built aircraft on Dublin–Glasgow services on 15 December 1958. Numerous versions have been built, including the 56-passenger FH-227, more than 650 have been ordered and the type is still in production at Amsterdam.

Britain's attempts to break into this market have been less successful. Handley Page produced the twin-Dart Herald, which was very similar to the F.27, but only 48 were completed. More successful was the low-wing Avro (later Hawker Siddeley) 748, also Dart-powered, which entered service in 1962 and is still in production. More than 300 civil and military examples have been built, about 50 of these being assembled or constructed in India.

There has also been one Japanese turbine-powered transport, the NAMC YS-11. This is a 46 to 60 passenger low-wing monoplane powered by two 3,060hp Dart engines. It first flew on 30 August 1962, and entered airline service in Japan in April 1965. More than 180 have been built and, apart from serving with several Asian airlines, they have been exported to the USA, South American, Canada, Europe and Africa.

The first 'big' jets

Not until the Comet 1 had been withdrawn from service did the first United States turbojet transport make its maiden flight. At Seattle, in Washington State, on 15 July 1954, a large brown and yellow swept-wing monoplane with four pod-mounted engines took to the air and began a new era in air transport. It was the Boeing 367-80, known affectionately as the Dash 80, and was in fact the prototype Boeing 707. This was a much more advanced aeroplane than the Comet: its wings were swept back 35°, and it embodied the experience gained with Boeing B-47 and B-52 jet bombers.

Pan American placed the first order, and the type went into production as the Model 707-100 with the customer designation 707-121. The production aeroplane had a span of 40·18m (131ft 10in), a length of 44·04m (144ft 6in), accommodation for up to 179 passengers, and a maximum weight of 116,818kg (257,000lb). It was powered by four 5,670kg (12,500lb) thrust Pratt & Whitney JT3C turbojets, cruised at 917km/h (570mph) and had a maximum range of 4,949km (3,075 miles) with full payload. The 707-121 entered service with Pan American on the North Atlantic in October 1958.

This original Boeing 707 was followed by a whole family of 707 passenger and cargo variants, with different lengths and weights, and turbofan power. The passenger-carrying 707-320B and passenger cargo 707-320C models are still in production. Also developed was the short-to-medium-range version known as the Boeing 720 or, with turbofans, 720B. By the end of October 1976, 920 Boeing 707s and 720s of all models had been ordered, and they had flown more than 30 million hours and carried just under 522 million passengers.

The second United States jet transport was the Douglas DC-8. It closely resembles the Boeing 707, and has likewise been produced in several versions. The first DC-8 flew on 30 May 1958, and the JT3C-powered DC-8-10 entered service with Delta Air Lines and United Air Lines on 18 September 1959. Then followed the -20 with more powerful JT4A engines, the -30 long-range aeroplane with the same engines, the -40 with Rolls-Royce Conway bypass engines and the -50 with JT3D turbofans. There were also freight, convertible, and mixed passenger/cargo versions. All had a span of 43·41m (142ft 5in) and a length of 45·87m (150ft 6in). Maximum seating was 177 and maximum weight

Lockheed's Model 14 Super Electra, seen here in service with the original British Airways in 1937, introduced such advanced features as Fowler-type area-increasing flaps, two-speed superchargers, and fully-feathering propellers.

varied between 120,200kg (265,000lb) and 147,415kg (325,000lb).

In 1966 the Series 60 DC-8 appeared. This was built in three main versions, the -61 with the same span as earlier models but a length of 57·09m (187ft 4in), seating for up to 257 and take-off weight of 147,415kg (325,000lb); the -62 with 45·23m (148ft 5in) span, 47·98m (157ft 5in) length, seats for 201 and take-off weight of 151,952kg (335,000lb); and the -63, combining the wing of the -62 with the fuselage of the -61. This had a maximum weight of 158,755kg (350,000lb) and seats for up to 269.

The DC-8-61 was put into service between Los Angeles and Honolulu by United Air Lines on 25 February 1967; the DC-8-62 entered service on the Copenhagen–Los Angeles route with SAS on 22 May 1967; and the DC-8-63 was introduced by KLM on the Amsterdam–New York route in July 1967. There were cargo and convertible versions of the Series 60 aircraft, and a total of 556 DC-8s of all models was built before production ceased in 1972.

The third United States constructor of large commercial jet transports was the Convair Division of General Dynamics, which produced two types. The first was the CV-880. It was smaller than the Boeing and Douglas types, its narrower fuselage having a maximum seating capacity of 130. It was powered by four 5,080/5,285kg (11,200/11,650lb) thrust General Electric CJ-805 engines, had a maximum weight of 87,770kg (193,500lb) and cruised at up to 989km/h (615mph). It was designed for TWA, but first went into service with Delta Air Lines on 15 May 1960, over the Houston–New York, New York–New Orleans and New York–Atlanta routes. Only 65 were built.

The second Convair was the CV-990 (named Coronado by Swissair). It was designed for American Airlines and closely resembled the CV-880, but had 7,257kg (16,000lb) thrust General Electric CJ-805-23B turbofans, seating for up to 158 and a maximum cruising speed of 1,005km/h (625mph). Distinctive external features were the four canoe-like shock bodies which extended aft of the wing. The CV-990 entered service with American Airlines and Swissair in March 1962. Total production was 37.

Britain made an attempt to break into the so-called 'big' jet market, but with little success. The aircraft was the Vickers-Armstrongs VC10, which was in the same category as the Boeing 707 but had its four 9,525kg (21,000lb) thrust Rolls-Royce Conway bypass engines mounted in pairs on each side of the rear fuselage, and a high-mounted

tailplane and elevators. The VC10 was designed for BOAC and was required to operate from restricted, hot and high runways which demanded specially good take-off and landing performance. Indeed, the VC10 could undertake some operations which were impossible for the early Boeings.

The Standard VC10 could carry up to 139 passengers, had a maximum weight of 141,520kg (312,000lb), and a maximum cruising speed of 933km/h (580mph). The VC10 went into service on the London–Lagos route on 29 April 1964, and was followed by the Super VC10 which was 3.96m (13ft) longer, had more powerful Conways, seating for up to 163 and a maximum take-off weight of 151,952kg (335,000lb). Super VC10s were introduced on the London–New York route on 1 April 1965. Only 54 of both were produced.

Medium- and short-range jets

In the summer of 1956 BEA issued a specification for a 965km/h (600mph) aircraft to meet its short- to medium-range route requirements. The de Havilland 121 Trident was selected, and the Aircraft Manufacturing Co, comprising de Havilland, Fairey and Hunting, set up to build the aircraft, but de Havilland then became part of the Hawker Siddeley Group. The design was changed to meet differing BEA demands before the first aircraft made its initial flight on 9 January 1962.

The Trident is a low-wing monoplane with high T-tail and three rear-mounted Rolls-Royce Spey bypass engines. BEA ordered 24 Trident 1s with 4,477kg (9,850lb) thrust engines, 88 seats, a maximum weight of 52,162kg (115,000lb), cruising speed of 941kg/h (585mph) and typical full-payload range of 1,448km (900 miles) with full fuel reserves. The first revenue flight by a Trident was made on 11 March 1964, and full scheduled services with Tridents began on 1 April 1964.

The Tridents have given good service and played a major role in the development of automatic landing, using a triplex system to provide a very high standard of reliability and safety. The first automatic landing by an aircraft on a passenger service was made on 10 June 1965, by a Trident landing at Heathrow on arrival from Paris.

The Trident 1E with uprated engines, leading-edge slats and seating for up to 139 passengers in high-density layout, flew in November 1964. Fifteen were built, and customers included Air Ceylon, Iraqi Airways, Kuwait Airways and PIA. Then came the Trident 2E of which 15 were built for BEA, 33 for China and two for Cyprus Airways. This had

5,411kg (11,930lb) thrust Speys, increased span, seats for 97 in BEA service, a take-off weight of 64,863kg (143,000lb) and increased range. This version entered service on 18 April 1968.

The last major version of the series was the Trident 3B. This has a tail-mounted 2,385kg (5,250lb) thrust Rolls-Royce RB.162 booster engine to improve take-off performance, is 5·0m (16ft 5in) longer than the earlier Tridents, has a maximum weight of 70,305kg (155,000lb) and accommodation for up to 180 passengers. BEA ordered 26, and began using them in March 1971. Finally came the Super Trident 3B with 3,628kg (8,000lb) weight increase and maximum-payload range of 3,380km (2,100 miles) and seats for 152 passengers. Two Super Tridents were ordered by CAAC in China, bringing the Trident total to 117.

In the same category as the Trident, and of very similar layout, is the Boeing 727. Design began in June 1959, but Boeing regarded the decision to go into production as a major gamble. However, the decision has been completely justified by the fact that 727 sales exceed those for any other Western jet transport by a very large margin, 1,352 having been ordered by October 1976.

The Model 727 has the same fuselage cross-section as the Boeing 707, 32 degrees of sweep on the wing, which is fitted with triple-slotted flaps and leading-edge slats, and much better take-off and landing performance than the original Tridents. The 727-100, powered by three 6,350kg (14,000lb) thrust Pratt & Whitney JT8D-1 turbofans and having accommodation for a maximum of 131 passengers, made its first flight on 9 February 1963, and entered service with Eastern Air Lines and United Air Lines at the beginning of February 1964.

Numerous versions of the Boeing 727 have been produced, the latest being the Advanced 727-200, which is 6·09m (20ft) longer. This is powered by 6,577/7,257kg (14,500/16,000lb) thrust JT8Ds and can carry up to 189 passengers. Maximum take-off weight is now 94,318kg (207,500lb) compared with 72,570kg (160,000lb) for the original aeroplane. Up to the end of October 1976 Boeing 727s of all versions had flown more than 22 million hours and carried more than 873 million passengers while serving over 80 airlines.

In addition to the Boeing 707s, 720s, 727s, Douglas DC-8s, Convairs and Tridents, there was considered to be a need for about 1,000 smaller twin-jet aircraft. The first designed to meet this requirement was the British Aircraft Corporation's BAC One-Eleven, powered by two rear-mounted Speys and having accommodation for up to 89 passengers. The first announcement about the BAC One-Eleven was made in May 1961, when it was stated that British United Airways had ordered ten. The first aircraft flew on 20 August 1963, but was lost in a deep stall, and it was not until April 1965 that the production Series 200 entered service with BUA and Braniff International Airways. The Series 200 was followed first by the increased-range Series 300 and then by the similar Series 400, which was specially adapted to meet US conditions and entered service with American Airlines on 6 march 1966.

Then came the Series 500, with more powerful Speys, 1·52m (5ft) increase in span, 4·21m (13ft 10in) increase in length, seating for up to 109 passengers and a take-off weight of 44,450kg (98,000lb) compared with 35,605kg (78,500lb) for the Series 200. BEA ordered 18 of the bigger version and introduced the type on 17 November 1968. There is also

The McDonnell Douglas DC-10 wide-body civil transport which is in large-scale use by the world's major airlines. Power comes from three large, economical-to-operate, turbofan engines.

The Fokker F.27 Friendship turboprop-powered short-haul transport, one of the most successful aircraft in this category to be developed in Europe, is in worldwide service.

the short-field Series 475, which has been ordered in small numbers, bringing the total sales to more than 220.

Douglas also produced an aircraft in this class and of similar appearance. This was the DC-9, which first flew on 25 February 1965, and entered service with Delta Air Lines on 8 December the same year. As the DC-9-10, powered by two 6,350kg (14,000lb) thrust Pratt & Whitney JT8D-1 turbofans, the aircraft has a span of 27·25m (89ft 5 in), a length of 31·82m (104ft 4¾in), accommodation for 80 passengers and a maximum weight of 41,140kg (90,700lb).

As with most previous Douglas types, success was assured by producing a family of aeroplanes to suit varying requirements, and the DC-9 was stretched by nearly 4.57m (15ft) to produce the 105-seat 28·47m (93ft 5in) span DC-9-30. This larger type flew on 8 August 1966, and entered service on 1 February 1967, with Eastern Air Lines on its shuttle operations. The correctness of the decision to stretch the DC-9 is shown by the fact that about 600 of the more than 840 ordered have been DC-9-30s or DC-9-30F or CF freighters or passenger/cargo aircraft.

Two special versions of the DC-9 were produced to meet the needs of SAS. These were the -40 with just over 1·82m (6ft) longer fuselage, 6,803kg (15,000lb) thrust JT8D-11 engines, 115 seats and a maximum weight of 51,710kg (114,000lb); and the -20, which combined the -10 fuselage with the -30 wing to achieve short-field performance. SAS bought 45 DC-9-40s and introduced them in March

1968, followed by ten DC-9-20s, which were commissioned in January 1969.

The latest version of the DC-9 is now the Super-80 much bigger than the -50, which first flew on 17 December 1974, and entered service with Swissair on 24 August 1975. This has a length of 40.7m (133ft 7in), accommodation for up to 139 passengers, and a maximum take-off weight of 54,884kg (121,000lb).

When the DC-9 was announced, BAC and Douglas expected to share a market for 1,000 aircraft. This total has been exceeded, although BAC got only a low share. But in spite of this Boeing produced a competitor which up to the end of October 1976 had itself attracted 489 orders. The decision to build the Boeing 737 was taken in February 1965, the type first flew on 9 April 1967, and it entered service early in 1968 with Lufthansa and United Air Lines.

Because the Model 737 retains the fuselage cross-section of the Boeing 707 and 727, it appears to be very stubby, having a length of only 28·57m (93ft 9in). It also differs in having its two 6,350kg (14,000lb) thrust Pratt & Whitney JT8D turbofans attached close beneath the wings. The tailplane is attached to the fuselage, and there is a tall single fin and rudder. Maximum weight is 42,410kg (93,500lb) and maximum seating 99. Like the other Boeings, the 737 has been developed into a family, and the Advanced 737-200 has a length of 30·48m (100ft), is powered by 7,257kg (16,000lb) thrust JT8D-17 engines, weighs up to 53,070 kg (117,000lb) and can accommodate a maximum of 135 passengers. By the end of October 1976, 474 Boeing 737s had been delivered and they had flown more than $5\frac{1}{4}$ million hours and carried 356,890,000 passengers with more than 60 operators.

In the Netherlands, Fokker (now Fokker-VFW) designed a jet successor to the F.27 Friendship. This was the F.28 Fellowship, which flew on 9 May 1967, and entered service with the Norwegian airline Braathens SAFE on 28 March 1969. The F.28 is a low-wing monoplane powered by two rear-mounted Rolls-Royce Speys. It was designed to operate from short rough runways and, to achieve good low-speed handling, the wing sweep was restricted to 16 degrees.

The original, or basic, version, known as the Mk 1000, had a span of 23·58m (77ft $4\frac{1}{4}$in), a length of 27·4m (89ft $10\frac{3}{4}$in), a maximum weight of 28,125kg (62,000lb) and accommodation for 65 passengers when operating from a 1,220m (4,000ft) runway. The Mk 2000 is 2·26m (7ft 5in) longer and has 79 seats; the Mk 3000 and Mk 4000 have 1·49m (4ft 11in) greater span and uprated engines, and will replace the Mks 1000 and 2000 on the production line; the Mk 6000 has the increased span and, like the Mk 4000 a longer fuselage with up to 85 seats. F.28s are truly international aeroplanes, components being built in the Netherlands, Germany and Britain. By October 1976 a total of 116 F.28s had been ordered by 33 operators in 23 countries.

Different from other jet transports is the VFW-Fokker VFW 614, designed for low-density traffic, short-haul routes, and to have the ability to operate from runways of under 1,220m (4,000ft). It is a low-wing monoplane with modest sweepback, and its two 3,527kg (7,760lb) thrust Rolls-Royce/Snecma M 45H bypass engines are mounted on pylons above the wing. The VFW 614 has a span of 21·5m (70ft 6in), a length of 20·6m (67ft 6in), a maximum take-off weight of 19,950kg (43,982lb) and accommodation for 40 to 44 passengers. Maximum cruising speed is 702km/h (436mph) and range with 40 passengers and fuel reserves 1,200km (745 miles). The VFW 614 first flew on 14 July 1971, was sold in small numbers and went into service with Cimber Air in Denmark late in 1975. It was dropped in 1977.

More independence creates new airlines

As related earlier, the granting of independence to India and other countries led to the creation of new airlines or the expansion of existing ones. In 1949 the Netherlands East Indies became Indonesia and Garuda Indonesian Airways was created in the following year, to become a major airline with a wide network of domestic and regional services, services to Europe and the Far East, and an annual passenger volume in excess of two million. In more recent times numerous other Indonesian airlines have been founded, including Merpati Nusantara and Bouraq Indonesia Airlines.

In 1957 Malaya became independent and formed the Federation of Malaysia, which later included Borneo, Sabah and Sarawak. Malayan Airways, founded in 1947, gave way to Malaysia–Singapore Airlines, but Singapore withdrew from the Federation and two separate airlines were formed: Malaysian Airline System, mainly equipped with Boeing 737s and Fokker F.27s, and Singapore Airlines, which has established an outstanding reputation with its Boeing 747 services to Europe.

Most of the countries which became independent in the period 1956–1965 were in Africa, and many of these had been French territories. In the north, Morocco and Tunisia became sovereign states in 1956 and Algeria in 1962, but these countries already had well-established airlines – Royal Air Maroc, Tunis Air and Air Algérie.

French Equatorial Africa in 1960 became the Central African Republic, Chad, Congo (Brazzaville) and Gabon, and together with Cameroun, Dahomey (now Benin), Ivory Coast, Mauritania, Niger, Sénégal and Upper Volta, formed Air Afrique as a joint airline. This has an extensive route network, including services between Dakar and New York, and operates DC-8s, DC-10s and Caravelles. Cameroun withdrew from Air Afrique in 1971 and set up Cameroon Airlines with Boeing equipment, but Togo later joined the Air Afrique group.

Of the British territories, the Gold Coast became Ghana in 1957 and formed Ghana Airways; Nigeria became independent in 1960 and replaced WAAC (Nigeria) with Nigeria Airways; Sierra Leone became independent in 1961 and Sierra Leone Airways is operated with assistance from British Caledonian Airways; Gambia became independent in 1965 and its Gambia Airways is also connected with BCAL; Air Malawi and Zambia Airways were created after the break-up of the Central African Federation.

The Belgian Congo became independent in 1960 and in the following year founded Air Congo as the national airline. The country's name was changed to Zaïre in 1971. The airline has a large network of domestic and regional services and operates to Europe. Its jet fleet includes Caravelles, DC-8s, DC-10s and Boeing 737s.

Although Libya became an independent state in 1951 it did not form an airline until 1964. It was founded as Kingdom of Libya Airlines and began flying Caravelles between Benghazi and Tripoli and to several neighbouring countries and Europe. F.27s were used on some domestic services. In 1969 the carrier was renamed Libyan Arab Airlines, and now operates a fleet which includes several Boeing 727s.

Various Australian airlines operated in New Guinea and Papua and between those territories and Australia, and when Air Niugini was founded in 1973 as the national carrier of Papua New Guinea, Ansett, Qantas and TAA all became shareholders. Services began on 1 November 1973, with DC-3s and F.27s.

One of the latest of the independent national airlines is Royal Brunei Airlines, which was founded in November 1974 and began services to Hong Kong, Kota Kinabalu, Kuching and Singapore on 14 May 1975.

Wide-bodied fleets

In 1960 the world's airlines, excluding those of China and the Soviet Union, carried 106 million passengers, and in 1966 this figure had nearly doubled, standing at 200 million. This rapid increase in traffic called for larger fleets of aircraft, and these added to airport and airway congestion. One way of absorbing the traffic growth without increasing aircraft movements was to build much bigger aircraft. At the same time these aircraft would reduce seat-mile costs and ease the noise problems in the vicinity of airports – an important aspect of air transport which was beginning to reach generally intolerable levels.

The first of this new generation was made known in April 1966, when Boeing announced that it would build the Model 747 and that Pan American had placed an order for 25. In one step the Boeing was doubling the capacity, power and weight of the transport aeroplane.

In general appearance the Boeing 747 was similar to the Boeing 707, but was scaled up to have a wing span of 59·63m (195ft 8in), a length of 70·51m (231ft 4in) and a height from the ground to the top of the fin of 19·32m (63ft 5in). The maximum interior width is just over 6m (20ft), and the ceiling height 2·43m (8ft). Seating on the main deck can be nine- or ten-abreast (in tourist class) with two fore and aft aisles, and this feature has caused the term 'wide-body' to be applied to this category of airliner. The flight deck is on an upper level, and behind this is a passenger cabin which serves as a first-class lounge

in most 747s. A stairway connects the two levels. Three cargo and baggage holds have a total volume of 175·28m³ (6,190cu ft), which is about equal to the entire volume of a cargo Boeing 707.

The original engines in the Boeing 747 were four 18,597kg (41,000lb) thrust Pratt & Whitney JT9D-1 turbofans, with a diameter of just under 2·43m (8ft). The first quoted maximum take-off weight was 308,440kg (680,000lb), and landing weight 255,825kg (564,000lb). The first Boeing 747 flew on 9 February 1969, and when the type entered service with Pan American on the New York–London route on 22 January 1970, the brake-release weight had already risen to 322,956kg (712,000lb).

Pan American's aircraft had 58 first-class and 304 economy-class seats, the maximum payload was 56,245kg (124,000lb) and the maximum cruising altitude 13,715m (45,000ft). In spite of its size the aircraft was completely orthodox in appearance, except that it had four four-wheel main undercarriage units in order to spread the load on runways, taxiways and airport aprons. Maximum seating was originally quoted as 490.

The Model 747 underwent rapid development, with increased power and consequently higher permissible weights, and the 747-200B with 22,680kg (50,000lb) thrust Rolls-Royce RB211s, 23,585kg (52,000 lb) Pratt & Whitney JT9D-70s or 23,815kg (52,500lb) General Electric CF6-50Es has a brake-release weight of up to 371,943kg (820,000lb) and has taken off at about ten tons above this weight.

The Boeing 747 has also been built as the all-cargo 747F, with upward-swinging nose for front-end loading and a payload of up to 113 tons, as the 747C, which can have all-cargo or all-passenger configuration or a combination of both.

In September 1973 Boeing flew the 747SR with structural reinforcement to allow high-frequency operation over short routes, which imposes greater stress on the structure – particularly the wings and undercarriage. This model can carry 500 passengers, 16 on the upper deck, but its take-off weight is limited to 272,154kg (600,000lb). Seven Boeing 747SRs were built for Japan Air Lines' domestic services, and they began operation on 10 October 1973.

All these Boeing 747s had the same external dimensions, but, on 4 July 1975, Boeing flew the first Model 747SP (Special Performance). This has a shortened fuselage, measuring 14·73m (48ft 4in) less in length than the other 747s, but its fin is 1·52m (5ft) higher. It has been designed for very long-range operation over routes where traffic volume does not require the larger-capacity 747s. The 747SP can carry up to 321 passengers in mixed-class configuration, with 32 on the upper deck, and can fly 11,105km (6,900 miles) with full passenger payload. Pratt & Whitney, General Electric or Rolls-Royce engines can be used and maximum brake-release weight is 299,370kg (660,000lb).

The enormous range potential of the 747SP was shown during a world tour before entering service. It flew non-stop from New York to Tokyo with 200 on board, a distance of 11,152km (6,930 miles) in 13hr 33min; non-stop from Sydney to Santiago, Chile, 11,495km (7,143 miles) in 12hr 14min; and non-stop from Mexico City to Belgrade, 11,595km (7,205 miles) in 12hr 56min.

Pan American took delivery of its first 747SP on 5 March 1976, and at the end of April introduced the type on non-stop services between Los Angeles and Tokyo and New York and Tokyo. The latter was the longest scheduled non-stop service until Pan American began a non-stop San Francisco–Sydney service in December 1976 – a distance of 11,582km (7,197 miles). On a delivery flight in March 1976 a South African Airways 747SP flew non-stop from Paine Field, near Seattle, to Cape Town – 16,560km (10,290 miles) in 17hr 22min.

To the end of October 1976, 313 Boeing 747s had been ordered, 291 had been delivered and they had flown just under 4 million hours and carried more than 127 million passengers. In $6\frac{3}{4}$ years of service only one had been involved in a fatal accident.

The Boeing 747 was too big for some airlines which required large-capacity aircraft, and McDonnell Douglas and Lockheed both built very similar three-engined aircraft to meet this need. In layout both were wide-bodied aircraft with two wing-mounted engines and one tail-mounted, although the two companies adopted different methods of mounting the rear engine. In the Douglas DC-10 the rear engine is built into the fin structure, but in the Lockheed L.1011 Tri-Star it is within the fuselage with the air intake above the fuselage forward of the fin.

The first DC-10 flew on 29 August 1970. This was the Series 10 version with 18,145kg (40,000lb) thrust General Electric CF6-6D turbofans, seating for 270 passengers in basic mixed-class configuration or a maximum of 345 in economy class, a maximum brake-release weight of 195,043kg (430,000lb) and a still-air range of 6,727km (4,180 miles). Span of the -10 is 47·34m (155ft 4in), and length of 55·55m (182ft 3in). This version entered service on 5 August 1971, on American Airlines' Los Angeles–Chicago route,

and on 14 August on United Air Lines' San Francisco–Washington route.

On 28 February 1972, the DC-10-20 made its first flight. This version, later redesignated -40, has a 1·82m (6ft) increase in span, 22,680kg (50,000lb) thrust Pratt & Whitney JT9D-59 engines, a brake-release weight of 251,742kg (555,000lb), and a still-air range of 9,060km (5,630 miles). The DC-10-40 was built for Northwest Airlines, and entered service on 16 December 1972.

The most used DC-10 is the -30 model which first flew on 21 June 1972. This has the increased span of the -20/40, is powered by 23,133kg (51,000lb) thrust CF6-50C engines, has the same weight as the -20/40 but a range of 9,768km (6,070 miles). The DC-10-30 first went into service with Swissair on North Atlantic services on 15 December 1972. There are also cargo and convertible versions of the DC-10, and by autumn 1976 a total of 246 had been ordered.

The Lockheed L.1011 TriStar, which first flew on 16 November 1970, has exactly the same span as the DC-10-10 but is slightly shorter at 54·43m (178ft 7½in). It was the first aircraft to be powered by the Rolls-Royce RB.211, which in its -22B form develops 19,050kg (42,000lb) of take-off thrust. The TriStar can carry 272 passengers in mixed-class configuration, 330 in coach class or up to 400 in economy class. Maximum take-off weight is 195,043kg (430,000lb), and the range with 272 passengers and full fuel reserves is 6,275km (3,900 miles). Maximum payload is 38,782kg (85,500lb). The TriStar entered service on 26 April 1972, with Eastern Air Lines and 159 had been ordered by the autumn of 1976, with options held on a further 50. Included in these orders are Saudi Arabian Airlines' L.1011-200s with 22,680kg (50,000lb) thrust RB.211-524 engines, and six L.1011-500s ordered by British Airways. These last six are to be longer range aircraft with 21,772kg (48,000lb) thrust RB.211-524B engines, 24,493kg (54,000lb) additional fuel, a take-off weight of 224,980kg (496,000lb) and a range of 9,817km (6,100 miles) with full passenger payload of 246. The L.1011-500 will be 50m (164ft 2in) in length. Flight trials are due to begin in 1978 with entry into service in 1979.

The fourth type of wide-bodied airliner is the Airbus Industrie A300B, which is slightly smaller than the DC-10 and TriStar, with a span of 44·84m (147ft 1in) and a length of 53·62m (175ft 11in). Unlike the other wide-bodied aircraft, the A300 is twin-engined, being powered by two wing-mounted 23,133kg (51,000lb) thrust General Electric CF6-50C turbofans. Maximum cabin width is 5·35m (17ft

A Tupolev Tu-114 four-turboprop airliner in service with the Soviet airline Aeroflot. For some years it was the world's largest airliner, based on the Tupolev Tu-20 bomber aircraft.

7in) and seating ranges from 251 in mixed class to 336 in high-density nine-abreast layout.

The Airbus is an international product with French, German, Dutch and Spanish partners, involving government backing. Hawker Siddeley designs and builds the wings in Britain, but without government support. The wing is claimed to be of the most advanced design in service.

The A300 first flew on 28 October 1972, and the B2 version entered service on the Paris–London route with Air France on 23 May 1974. It has a maximum take-off weight of 142,000kg (313,000lb), a cruising speed of 870km/h (541mph) and a range of up to 3,860km (2,400 miles). The B4 model has increased fuel capacity, a maximum take-off weight of 157,500kg (347,100lb) and a range of 6,000km (3,728 miles).

By autumn 1976 a total of 34 A300s had been ordered and options taken on a further 23, and Airbus had announced its intension of offering airlines the B4FC freighter conversion with a large cargo door and a payload of more than 40 tonnes over distances of up to 3,500km (2,175 miles).

In the Middle East there has been fantastic growth in air transport. This is due to large-scale development of Middle East countries made possible by oil revenue, but tourism has also played a part, particularly in Egypt and Lebanon. Most of the Middle East airlines have grown in the past 20–30 years from small local or regional carriers into major international airlines.

Soviet air transport

In 1921 or 1922 there were a few experimental air services in the Soviet Union, and in 1923 three air transport undertakings were formed. But before the formation of Soviet concerns, a Königsberg (now Kaliningrad)–Kowno–Smolensk–Moscow service had been opened, on 1 May 1922, by the joint Soviet–German airline Deruluft, using Fokker F.III single-engined monoplanes. This company, with various types of aircraft and with varying routes, was to operate between Germany and the USSR until 1937.

The three undertakings set up in 1923 were Dobrolet, Ukrvozdukhput in the Ukraine, and Zakavia. These organizations developed a number of routes, mainly with German aircraft, and in 1930 the first two were reorganized as Dobroflot, Zakavia already having been taken over by Ukrvozdukhput in 1925. In 1932 Aeroflot was created when all Soviet air services came under the Chief Administration of the Civil Air Fleet.

Until the German invasion of the Soviet Union in June 1941, Aeroflot gradually built up its network, and by 1940 had a route system measuring 146,300km (90,906 miles) and that year carried nearly 359,000 passengers and about 45,000 tons of cargo and mail.

During the war Aeroflot was engaged on essential

tasks, and when peace returned, faced the job of re-establishing and expanding its nationwide operations and building up an international route system. Its success is illustrated by the fact that Aeroflot now serves some 3,500 places in the USSR, operates to 65 other countries and, in 1975, carried more than 100 million passengers.

During the war the Soviet Union had received considerable numbers of military DC-3s from the United States, and in 1939 had already begun licence production of these under the designation PS-84, changed in September 1942 to Lisunov Li-2. In the early post-war years these Li-2s and ex-military DC-3s formed the backbone of the Aeroflot fleet, and some were supplied to neighbouring Communist countries.

But even while the war was being fought, Sergei Ilyushin's design bureau began work on the USSR's first postwar transport aeroplane – the Il-12. This was a low-wing 21/32-seat monoplane powered by two 1,650/1,775hp Shvetsov ASh-82FN engines. It had a nosewheel undercarriage, but was unpressurized. The type was introduced by Aeroflot on 22 August 1947, large numbers were built, and on the last day of November 1954 the Il-12s were joined by the improved but very similar Il-14.

In the mid-1950s the Il-12s were taking 33hr to fly from Moscow to Vladivostok, with nine intermediate stops, and on the Moscow–Sverdlovsk–Novosibirsk–Irkutsk route they took 17hr 50min with 14hr 35min flying time. These aeroplanes were obviously not acceptable for the distances which had to be covered and in 1953, as part of a major modernization plan, the first Soviet jet transport was designed. This was the Tupolev Tu-104, which first flew in 17 June 1955, and went into service on the Moscow–Omsk–Irkutsk route on 15 September 1956, with a schedule of under seven hours.

The Tu-104 had a low-mounted swept-back wing, swept tail surfaces and two 6,750kg (14,881lb) thrust Mikulin RD-3 or AM-3 turbojets close in to the fuselage. The pressurized cabins seated 50 passengers. The Tu-104 had a span of 34·54m (113ft 4in), a loaded weight of 71,000kg (156,528lb) and a cruising speed of 750/800km/h (466/497mph).

The Tu-104 transformed Soviet long-distance air services, but it was not particularly economic and was therefore followed by the 70-passenger Tu-104A and 100-passenger Tu-104B. It is believed that between 200 and 250 examples were built and many are still in service after 20 years.

The Soviet aircraft industry next designed three types of propeller-turbine transport aircraft. Two of these were for heavy-traffic routes and the third was a long-range aircraft. Ilyushin produced the Il-18, which was to prove of major importance to Aeroflot, about 600 being put into service. This was a low-wing monoplane powered by four 4,000hp Ivchenko AI-20 engines, and initially having accommodation for 80 passengers. The Il-18 made its first flight on 4 July 1957, when it bore the name Moskva (Moscow), and it went into service on the Moscow–Alma Ata and Moscow–Adler/Sochi routes on 20 April 1959. Several versions of Il-18 were built, including the Il-18V which has accommodation for up to 110, a maximum take-off weight of 61,200kg (134,922lb), a cruising speed of 625/650km/h (388/404mph) and a maximum-payload range of 2,500km (1,552 miles). About 100 Il-18s were sold to non-Soviet airlines.

The Antonov An-10 was of similar size to the Il-18, but was a high-wing monoplane designed to operate from poor aerodromes. Like the Il-18, the An-10 had four 4,000hp Ivchenko AI-20 engines, and when it entered service, on the Moscow–Simferopol route, on 22 July 1959, had 85 seats. The improved An-10A which followed had 100–110 seats. The number of An-10s and An-10As built is not known, but they did carry 10 million passengers in just under seven years, until all were withdrawn following a fatal accident at Kharkov on 18 May 1972. A freighter version, with rear under-fuselage loading was developed as the An-12. Large numbers were produced for both civil and military use and the type operates many of Aeroflot's cargo services at home and abroad.

One long-range aeroplane was unlike any other to be used in airline service. It was the Tupolev Tu-114, developed from the Tu-95 bomber. The 51·1m (167ft 7¾in) wing was low-mounted, swept-back 35 degrees and carried four massive Kuznetsov NK-12M turbines, each of 12,000hp and driving 5·6m (18ft 4½in) diameter eight-blade contrarotating propellers. Later the 15,000hp NK-12MV was installed. Fully loaded, the Tu-114 had a take-off weight of 175,000kg (385,809lb), and for nearly a decade it was the biggest aeroplane in airline service. It was also the fastest propeller-driven airliner ever in service, having a maximum speed of 870km/h (540mph).

The Tu-114 entered service on the Moscow–Khabarovsk route on 24 April 1961; went onto the Moscow–Delhi route in March 1963; opened the first Soviet transatlantic service, to Havana, on 7 January 1963; and the first to North America when it began operating to Montreal on 4 November 1966.

Tu-114s began working joint Aeroflot-Japan Air Lines Moscow–Tokyo services on 17 April 1967, and bore JAL's name and badge as well as Aeroflot's. This large aircraft could carry 220 passengers, but standard seating was for 170 or, on intercontinental routes, 120. Only about 30 Tu-114s were built and most have now been replaced.

Having re-equipped with medium- and long-range turbine-powered aircraft, Aeroflot set about the task of modernizing its short-haul fleet and introduced two new types in October 1962. One was the Tu-124, which was virtually a three-quarter-scale Tu-104, and the other was the Antonov An-24, which resembled the Fokker Friendship. The Tu-124, with two 5,400kg (11,905lb) thrust Soloviev D-20P turbofans, had accommodation for 44 to 56 passengers and entered service on the Moscow–Tallinn route. There was much criticism of the Tu-124, only about 100 were built, and it was replaced on the production lines by the similar, but rear-engined, 64–80-seat Tu-134 and Tu-134A, which are now used in quite large numbers, some having been exported to East European airlines.

The An-24, with two 2,100hp Ivchenko AI-24 propeller-turbines, is a high-wing monoplane. It has been built in several versions and in very large numbers, and it has accommodation for up to 50 passengers. There are cargo versions, including the An-24TV and An-26, and the An-30 survey aircraft. Numbers of An-24s have been exported.

As Aeroflot began developing its long-haul inter-national and intercontinental routes, it required a jet transport to replace the Tu-114. To meet this requirement Ilyushin designed the Il-62, which closely resembles the British BAC VC10. The Il-62 has four rear-mounted Kuznetsov or, in the Il-62M, Soloviev turbofans, and normal seating for 186. The Il-62 first flew in January 1963, but its development was prolonged and it did not enter service until March 1967, on domestic routes. On 15 September 1967, Il-62s began working the Moscow–Montreal services and on 15 June 1968 the extension to New York was opened.

Aeroflot has a vast number of local services, and a modern aeroplane was required to replace the Li-2s, Il-14s and smaller aircraft on these operations. The type chosen was the Yakovlev Yak-40, a unique tri-jet with the ability to operate from small, rough fields and carry up to 32 passengers. The Yak-40 flew in October 1966, went into service in September 1968 and has been exported to several countries including Afghanistan, Germany and Italy. The engines are 1,500kg (3,306lb) thrust Ivchenko AI-25 turbofans. It is reported that 2,000 Yak-40s are being built, and it is believed that nearly half that number have been delivered.

On 4 October 1968, a new Soviet airliner, also a tri-jet, made its first flight. This was the Tu-154, which has been designed to replace the large numbers of An-10s, Il-18s and Tu-104s. It has been reported that more than 600 were ordered for Aeroflot, and the type entered service on the

Left: A Soviet Antonov An-22 long-range very heavy transport. Having obvious military applications it holds several payload-to-height records. The Tupolev Tu-104A transport (*above*) was Aeroflot's first jet airliner.

Moscow–Simferopol and Moscow–Mineral'nye Vody routes on 15 November 1971.

For future re-equipment two other Soviet types are known to be under development: the 120-passenger Yak-42, which first flew in March 1975, and the 350-passenger Il-86.

Supersonic transport

In Britain and France the decision was taken to produce jointly an airliner capable of cruising at more than Mach 2, about 2,143km/h (1,332mph). This was the enormously costly Concorde project, with Aérospatiale and the British Aircraft Corporation being responsible for the airframe and Rolls-Royce and Snecma for the engines. The Concorde is a slim delta-winged aircraft with four 17,260kg (38,050lb) thrust Olympus 593 turbojets, a span of 25·37m (83ft 10in), a length of 61·94m (203ft 9in) and a maximum take-off weight of 181,436kg (400,000lb).

The first Concorde flew on 2 March 1969, and on 21 January 1976, the first supersonic passenger services were inaugurated – Air France flying the Paris–Dakar–Rio de Janeiro route and British Airways the London–Bahrein route. Air France also put Concordes on the Paris–Caracas route, and on 24 May 1976, Concordes of Air France and

British Airways made simultaneous arrivals at Dulles Airport, Washington, inaugurating services from Paris and London respectively. On 9 December in the following year services began to run to New York. Only nine Concordes have been ordered for airline service, five for British Airways and four for Air France. However, the transatlantic services (for which Concorde was designed) have proved popular, the aircraft consistently carrying 80–85 per cent of their capacity and by mid-1978 over 100,000 people had travelled supersonic.

The Soviet Union was actually first to fly a supersonic transport, its Tupolev Tu-144 making its first flight on the last day of 1968. Superficially the Tu-144 resembles the Concorde, but has so far not proved to be so successful. One was lost in an accident at the 1973 Paris Air Show and, although the type began operating a Moscow–Alma Ata cargo service on 26 December 1975, passenger service did not begin until 1977.

The United States held a design competition for a supersonic airliner, and selected the 298-passenger variable-geometry Boeing 2707 with four General Electric engines. This aircraft would have been 96·92m (318ft) long, spanning 53·08m (174ft 2in) with the wing swept forward and 32·23m (105ft 9in) with the wing swept back 72 degrees for supersonic flight. This design was then replaced by a pure-delta type, but finally the whole supersonic project was considered to be uneconomic and, after a long series of design hold-ups, it was cancelled.

Light Planes and Aerial Sport

The barnstorming era

The aeroplane 'came to town' in the 1920s and 1930s. While the international air routes were being slowly opened up by airlines, aviation was brought to the masses by the barnstormers. They were like an Elizabethan troupe of strolling players. They moved from town to town, landing on any suitable open space, and gave daring air displays.

Barnstorming developed in the wake of World War I but its origins were, like those of aeroplane flight itself, American. The impetus came in Great Britain from the number of war-surplus aeroplanes. Young men, in many cases faced with unemployment, put their few available pennies together and bought an old Avro 504K or D.H.9. Then they travelled and flew, learning as they went how to make a sensational show for the spectators.

There was a strong element of competition among the barnstormers as they tried to develop their acts. They flew low and fast along the front of the crowd; they learned to turn their aircraft upside down by the loop or the slow roll. Slowly one manoeuvre was joined to another, which led to individual aerobatic displays. Sadly, some people were hurt, for safety standards were minimal.

When not being used for a performance the aeroplane would do a spell of joyriding. A five-minute trip around the flying field usually cost five shillings – 25p in today's money, though more like £5 at today's values. The price was not enough to pay heavy overheads so the aircraft stayed out at night and the ground crew often stayed out as well, often in tents beneath the wings. Maintenance was adequate rather than lavish and the aim was just to keep the aircraft fit to fly. The ground engineer had to be a good all-rounder to survive and would as likely as not find himself posting bills among his other chores to announce the aircraft's arrival in the district. Certainly, once the performance was over and joyriding was under way, it would be the engineer who would be drumming up the trade and taking the money.

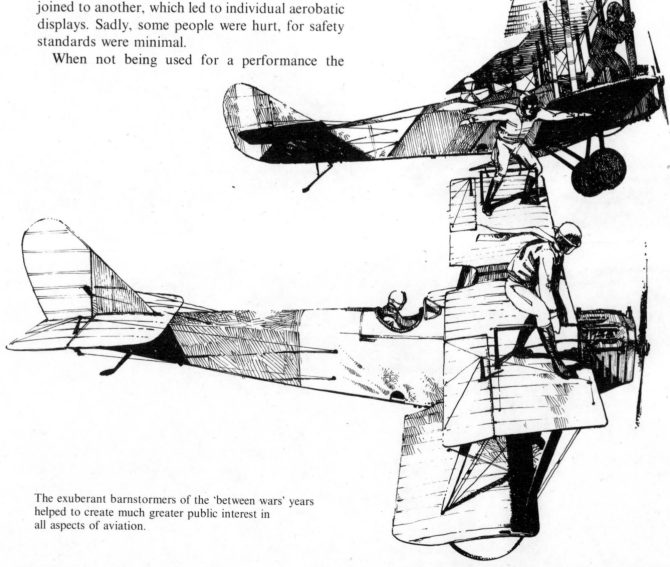

The exuberant barnstormers of the 'between wars' years helped to create much greater public interest in all aspects of aviation.

Many barnstormers started out as one-pilot-one-aircraft outfits but those that prospered added more aircraft and developed speciality acts. There was always a 'daredevil' appeal about wing walking and even today many a flying display is enlivened by the appearance of someone standing on the top wing of a Tiger Moth. Nowadays such a passenger is firmly strapped to the aircraft but this would have been called 'chicken' by the pioneers. One popular American trick was not only literally wing walking but stepping or dropping from the wing of one aircraft to the wing of another just below.

As individual pilots developed aerobatic skills and as tricks were worked out that involved more than one aircraft, the barnstorming act grew nearer to a circus performance. Indeed, Sir Alan Cobham used the title 'aerial circus' to describe his show. The performance outstripped the potential of the faithful Avro 504s. Their strength had been in their suitability both for aerobatics and for joyriding. Most were modified to squeeze in a second passenger in the rear cockpit and one Avro is on record as having accommodated four passengers, which cannot have done much for its handling characteristics.

Strictly speaking, the aerial circus pilots should have held the professional (B) licence because the operation was, in legal terms, 'for hire and reward' but there were ways of getting round the law. Instead of charging the passenger for his flight the barnstormer would give him a free flight but charge him an entry fee to the flying field.

The growth of civil aviation in general gradually stifled the activity of the barnstormers though the last of the flying circuses only finished under the increasing threat of war. More aviation meant more regulation and by the mid-1930s it was no longer practical to land on any convenient open space at the edge of town. The law required landing sites to be notified to the Air Ministry, which might not necessarily approve them. Only one typical barnstorming strip survives – the sandy beach. Once the most favoured strips, because they were unlikely to be obstructed, beaches are still used daily by Loganair in Scotland and for joyriding at Blackpool. But they are closely supervised operations and, as such, are a far cry from the carefree days when the beach was the temporary home of the barnstormers.

Between the wars

Twenty-one years of peace between the two great wars in Europe saw light aviation climb to a peak of popularity, as flying became increasingly available to the man and woman in the street. In the beginning flying had a remoteness and a glamour that made earthbound mortals stop and stare. Newspapers put up prizes for aeronautical feats – a race around the British Isles, a flight to India, South Africa or even Australia. The first air services began by using converted First World War light bombers and those that travelled in them were thought very daring as well as very rich.

There were men of vision in the British Isles who recognised that many people wanted to fly. To do so they needed simple, cheap and reliable aeroplanes and it was just such machines that emerged from the drawing boards of de Havilland, Miles and Percival. In the 1930s a handbuilt light aircraft, such as one of the early Moths, cost a little more than a family house and perhaps five times as much as a motor car. The modern £20,000 mass-produced machine is still in the same relationship to the car and is not far from maintaining parity with house costs.

The present-day aircraft is surrounded by an infrastructure of airfields, radios, air-traffic control towers, beacons, none of which was available to the pioneers. Jean Batten, on her first epic flight to Australia in a Gipsy Moth, arrived over a landing field in Cyprus to find the specially ordered windsock had been carefully tied to its mast lest the loose end should blow in the wind and be damaged. Alan Cobham, the Mollisons, Hinkler, Chichester – these were among the pioneering names who flew beyond Europe, documenting their routes as they went and so opening the way for the airliners to follow.

There were innumerable designers whose efforts resulted in unlikely prototypes that never caught on. For example, G. T. R. Hill and the Granger brothers were convinced that a light aeroplane could be built more cheaply without a tail but they only just succeeded in adapting the natural laws of stability and control, and the handling of light tailless types nowadays is generally regarded as demanding. Slightly less so, and thus something of a trap for the unwary, was the Flying Flea, a genuine attempt to bring flying to the masses by offering a cheap kit. Unfortunately there was no rigorous inspection system to ensure that the parts were assembled correctly and eventually the accident rate became so bad that the Air Ministry was forced to withdraw the Flea's certificate of airworthiness.

Light aviation received the patronage of royalty, and the King's Cup became the premier air-racing trophy. Nowadays it is contested over short courses, but it used to mark long races which needed navigational skill just as much as piloting ability.

The races became a challenge to designers who produced such beautiful classics as the Mew Gull and the Hawk Speed Six. One of the most remarkable and challenging of races was that held in 1934 from Mildenhall in Suffolk to Melbourne. For this de Havilland designed the original Comet, taking only nine months from the first idea to handover at the starting line, despite such comparatively new features as stressed-skin wooden construction and a simple retractable undercarriage.

The Comet won the race by covering 18,825km (11,700 miles) in just under 71 hours and despite having an unenviable reputation for being difficult to fly it went on to achieve other successes. It competed for the King's Cup and was flown in another of the typical long-range races of the era, the 1937 event from London to Damascus and back to Paris.

Flying had never been subsidized in England and those who learnt to fly were either rich enough to afford it or enthusiastic enough to spend long hours at flying clubs cleaning aeroplanes, refuelling them, holding the wingtips when the wind got up and generally earning their time at the controls. As the political situation in Europe deteriorated, the Air League encouraged the formation of a pool of aviators who would be immediately available to fly in the Royal Air Force and so the Civil Air Guard was borne. The idea was sound but it came too late and in 1939 the Moths and Hawks were made ready for battle. From their club beginnings the two famous designs had fathered the Tiger Moth and the Magister, which were to teach a generation of military pilots the first principles of flying.

Light planes in modern times
Talk of light aviation to the average town dweller in the western world and he will probably think of aeroplanes puttering overhead on a Sunday afternoon flown by pilots with more money than sense. Discuss it, however, with a crofter on a remote Scottish island, with a doctor in the Australian outback or with a farmer in the American Midwest

Right: Homebuilt lightplanes, such as this Aerosport Rail, enable enthusiasts to fly in an excitingly different way.

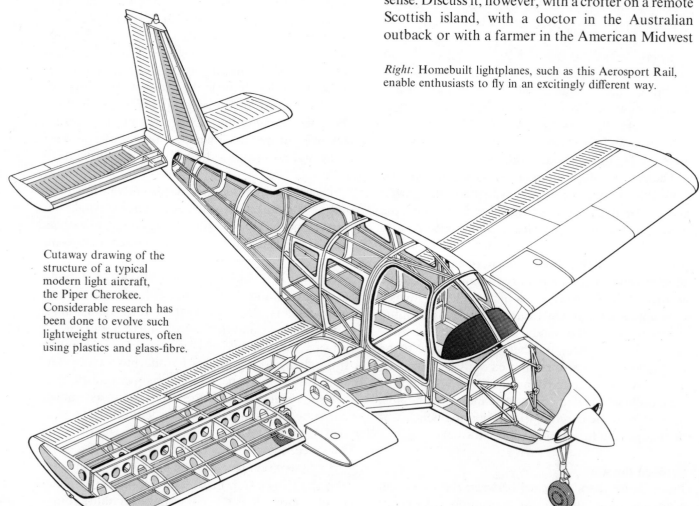

Cutaway drawing of the structure of a typical modern light aircraft, the Piper Cherokee. Considerable research has been done to evolve such lightweight structures, often using plastics and glass-fibre.

Far left, top: The British Aerospace HS 125 is typical of that type of lightweight business aircraft which comes between the lightplane and the 'jumbo jet'.

Far left, below: The B. 206 five/eight-seat twin-engine plane is a good example of a light executive plane.

This page: Hang-gliding, a well established sporting activity, makes one appreciate the courage of the men who first dared to fly.

Far right, top: A Bell Model 47 helicopter, which remained in production for more than 25 years.

Far right, below: Sikorsky's 30-passenger S-61, which has proved invaluable for commuter services close to city centres.

Below: A light naval helicopter can provide far-seeing 'eyes' for both small and large ships.

Left: Boeing Vertol Chinooks were built to carry men, their weapons and supplies into battle.

Right: The twin-rotor Mil Mi-12, the world's largest helicopter.

Below: An advanced naval helicopter, equipped to seek and destroy enemy submarines

Above: Vertical take-off
capability is not confined to
helicopters; this Dassault-
Breguet Mirage 111-V
prototype relies upon direct
jet-lift for take-off
and landing.

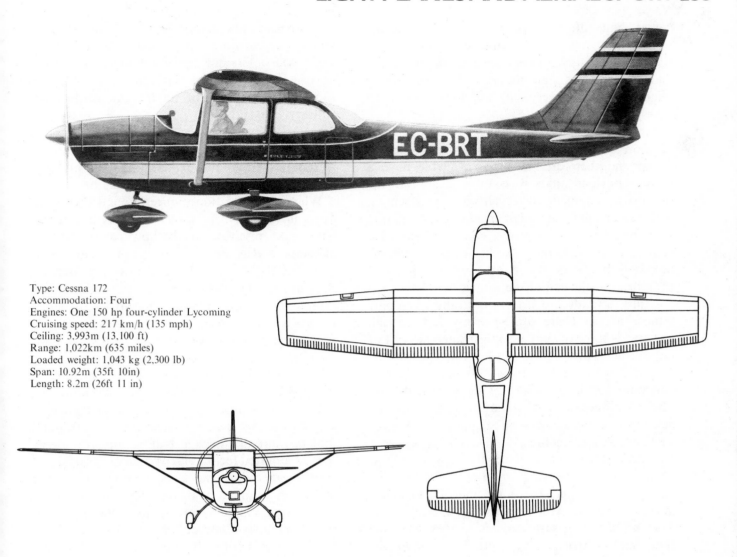

Type: Cessna 172
Accommodation: Four
Engines: One 150 hp four-cylinder Lycoming
Cruising speed: 217 km/h (135 mph)
Ceiling: 3,993m (13,100 ft)
Range: 1,022km (635 miles)
Loaded weight: 1,043 kg (2,300 lb)
Span: 10.92m (35ft 10in)
Length: 8.2m (26ft 11 in)

and in all probability you will receive an extremely different reaction.

The home of the post-war light aircraft is America, from where about 17,000 new machines emerge every year. The USA has an aircraft population exceeding 180,000, overwhelmingly consisting of light aircraft. The majority of these are two or four-seat touring types.

The European light-aircraft owner tends to be more of an enthusiast. While many do use their aircraft for business journeys – and who more than the English with their need to cross the Channel before they can travel almost anywhere – most fly for recreation and relaxation. A well-developed road and rail network eases the lot of the European business traveller and the unreliable weather imposes restraints on the would-be business pilot. The professionals learn to come to grips with the weather, equip their aircraft accordingly and fly on; the amateurs declare the winter a close season.

Five per cent of American aircraft work is in agriculture – the 'crop dusters' who have helped to increase world food production by enabling large areas of crops to be treated with pesticides in the shortest time possible. Agricultural aviation calls for manoeuvrable aircraft that can lift large chemical loads and spread them evenly and quickly. They must be equally able to cope with solids and liquids and they must be built so simply that they can be taken apart quickly and cleaned to prevent corrosion setting in. Both America and Russia have enormous expanses of open wheatfields, and are in the forefront of agricultural-aircraft use, but those produced in the East tend to be much larger and more rugged than the Western designs.

Poland is the design centre for East Europe's agricultural aviation and it has turned out several thousand 1,000hp Antonov An-2s. Present developments include adaptation of America's Thrush Commander, again using a 1,000hp radial piston engine, and initial service trials of the M-15, the world's first production jet biplane. Two very large hoppers are installed in fairings between the upper and lower wings and the design is apparently intended to allow a very wide path to be sprayed on each pass.

American designs tend to adapt proven general-aviation technology to aircraft of around 300hp to 400hp and therefore the types available have a smaller payload than Russian models. But the American manufacturers are actively experimenting with new engines, varying from adapted car motors to turboprops, and there is a constant search for economy and reliability as this class of aircraft becomes more and more an everyday item of equipment at the call of the farmer.

Utility aircraft are used universally to supply remote areas. There are probably not enough passengers to justify either sophisticated aircraft or large airfields but even so surface transport may be well-nigh impossible and the only way to keep a community in being is by serving it from the air. This is how the mail reaches the Orkney islander or a new drilling bit reaches an engineer in Alaska. The aircraft are functional – Twin Otters, Beavers, Islanders, twin Cessnas, all tough designs with large cabins that can be carrying passengers one day and stores the next. There are now 800 Islanders in service and many of them live up to their name. They serve islands just as much as they serve the hinterland of Africa or the South American jungle. In Papua, New Guinea, they have cut travelling time between districts from weeks to hours by flying across mountain ranges which otherwise could be crossed only by tortuous tracks over inhospitable terrain.

In remote districts the light aircraft provides, literally, the lifeline, enabling sick people to be transported to the cities or doctors to visit the remote townships. Australia's flying doctor service has been built up on the use of light aircraft. Some doctors fly their own; others use larger types in which there is room for a stretcher and a nursing attendant. In Scotland an aircraft is always available at short notice to fly from Glasgow with a medical team to cope with a climbing casualty or even a premature baby.

Versatile as the conventional light aircraft is, there are areas in which it cannot operate, and here the helicopter often comes into its own. The entire North Sea oil-drilling operation is supported by the fleet of Sikorsky helicopters based in Scotland, Holland, Germany and Norway. They carry the personnel for crew changes on the drilling platforms and any urgent stores; they perform in all weathers. Light helicopters are used by many police forces, hovering above a road traffic-jam or looking for someone thought to be on the run across open country.

Aerobatics

Aerobatics is the most exhilarating form of flying, whether for the pilot or for the spectator.

When early aviation pioneers found themselves flying upside-down by mischance they had to find out for themselves exactly how to rectify their dilemma. A Russian pilot is generally credited with flying the first loop in public during an appearance at Kiev in 1913, but it was the years of the Great War that taught pilots to use manoeuvres to gain a tactical advantage over their opponents. Perhaps the best remembered of the wartime manoeuvres was the Immelman turn, nowadays usually referred to as a half roll at the top of a loop. It gave, literally, an invaluable twist to the battle. An aircraft being pursued would dive to gain speed and then pull up to convert the excess energy to height. A vertical pull took the aircraft through half a loop so it gained several hundred feet but arrived there upside down. Height generally being a most sought after commodity in aerial combat, the pilot who was being chased would hope to reverse the order of the hunt by rolling upright and pulling round so as to be on his adversary's tail by the time that he, in turn, had followed.

Airforces have always set great store by aerobatic training, airline schools less so. Before the age of the missile made aerial combat possible without even seeing the target aircraft, it was the skill and determination of the pilot in flying his aircraft to the limit that usually won the day. So the military trainee would spend many hours of his training learning to fly a loop, a level roll or a stall turn that sliced accurately through 180 degrees.

Few aeroplanes are designed just to perform aerobatics but since the setting up in 1960 of world aerobatic championships, Russia, Czechoslovakia, France, the United States and lately Great Britain have all put a lot of design effort into producing aircraft which are competitive. An aerobatic aircraft has to be a compromise. It must, above all, have powerful and sensitive controls so that the pilot can begin a movement just when he wants, and, even more importantly, overcome the momentum to end it with equal precision. Lightness is an advantage since a surplus of power overcomes the constant

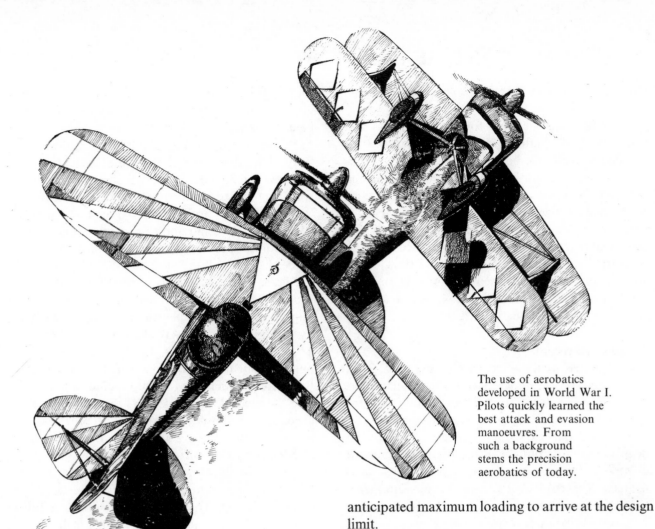

The use of aerobatics developed in World War I. Pilots quickly learned the best attack and evasion manoeuvres. From such a background stems the precision aerobatics of today.

tendency to lose height during a sequence of aerobatics. However, lightness is a luxury not often attained because the aerobatic aircraft has to be strong.

America's Pitts Special epitomizes the ideal aerobatic aircraft. It is a biplane which allows the mass to be concentrated and the span to be reduced. Less span – the Pitts measures only 6m (20ft) across – means less resistance to rolling, and when this is coupled with the ability of a biplane design to offer four ailerons it is not difficult to imagine the fingertip lightness of control available to the pilot.

The longer span and greater wing area of the monoplanes leads to some very strong structures, a fact quickly realized by students of the College of Aeronautics at Cranfield where Britain's first aerobatic design, the A1, has been developed. Although few pilots enjoy pulling more than about 5G – exerting five times their own weight on themselves – during sequences, the aircraft must withstand more and even then a safety factor must be applied to the anticipated maximum loading to arrive at the design limit.

Competition aerobatics has developed a unique character which tends to make it more angular than the flowing sequences which delight a display audience. The display pilot tries to make his manoeuvres flow from one to another, turning every gain of height or airspeed to advantage for the next move. Reversals of direction must be worked into the sequence so as to keep the aircraft in front of the crowd. While he may work to a pre-arranged sequence for the benefit of a commentator, the display pilot usually has more choice of what to do next than the competition pilot. He is expected to follow a set order and it was to make an easy kneepad crib sheet possible that Spaniard Jose Aresti compiled his dictionary of aerobatic manoeuvres. A simple symbol acts as a shorthand note and the Aresti method of recording aerobatics has become used universally.

One of the hardest aspects of flying aerobatics before spectators, whether they are competition judges or members of the public at an air display, is to fly in a straight line before the crowd. Easy enough on a calm day, this calls for a lot of concentration when there is a strong wind and results in manoeuvres which, if apparently straight to the spectator, are anything but straight to the pilot. The trick is not to apply too much corrective bank which would give the game away to the onlooker.

Constant aerobatics calls for extreme fitness and dedication, both qualities which Russia and the East Europeans seem to be able to call on at will. Czechoslovakians took the first three places in the first championships, appropriately enough since they hosted the meeting at Bratislava. Two years later prizes were awarded for teams as well as for individual performances and the top places went to Hungary, Russia and Czechoslovakia. Prizes for the ladies were introduced when the third championships were held at Bilbao in 1964, and again the Russians dominated the results.

The airspeed record

Less than ten years after man's very first flight in an aeroplane, the Fédération Aéronautique Internationale was established as the official body for recognising air speed records. Records could be set over short distances, from point to point or round measured circuits but the absolute airspeed record has always been one to be measured over a short distance. The FAI works in metric measurements and so the basic distance for a straight record is 3km (1·86 miles) there and back.

It was not always so and the first speed record to be recognized by the FAI was flown over two 10km (6.21 mile) laps. This record, set by Glenn Curtiss at 69km/h (43mph) in 1909, was the first to be documented. Santos Dumont is credited with providing enough proof to justify a record claim when his pusher biplane achieved 40km/h (25mph) in November 1906.

Speeds were rising fast as pilots and designers – often the same person – came to grips with aviation and the record book soon contained entries of 100km/h (62mph). The last record speed to be established before the start of the war in 1914 was 203km/h (126mph), achieved by a Deperdussin monoplane near Reims.

The chase for speed continued after the war but on two levels. Landplane development slowed and for a time there were not many attempts at the straight line record. Seaplanes, however, had received a great stimulus when Jacques Schneider set up a trophy especially for them. He had put up his challenge in 1912, but only two contests were flown before World War I. Restarted in 1919, races for the Schneider Trophy were flown every year, provided that enough entrants could be found, and they centred on four principal contenders: the United States, Italy, France and Great Britain. The prestige was so great

that individual entries soon ceased to matter as government-funded national teams began to compete. America took the 1923 and 1925 races at 285km/h (177mph) and 374km/h (233mph) respectively. This threw down the gauntlet to the other nations, for Schneider had decreed that a country winning three times in a row would keep the trophy.

In 1926 Italy broke the American run with the Macchi M.39 design and this aircraft was used for an absolute speed record attempt the same year. It succeeded at 454km/h (282mph). Italy staged the 1927 race in Venice and was the hot favourite but the Macchis were beaten by Royal Air Force Supermarine S-5s. When the race was flown again two years later the RAF again took the honours. The aircraft used was the S-6, another sleek Supermarine design, but one which substituted the Rolls-Royce R engine for the familiar Napier Lion. The successful team leader, Squadron Leader Orlebar, was allowed what proved to be a successful attempt at the outright world record and was timed at 576km/h (358mph).

Britain finally won the Schneider trophy outright in 1931 using a refined S-6 with its engine developed to give 2,350hp. The Italians had still more power and for a few years there was a tussle for the record. Britain was first to enter the record book with a speed above 644km/h (400mph) but the Italians were the fastest ever with a seaplane when, in 1934, a Macchi MC-72 reached 710km/h (441mph).

As governments chased the Schneider Trophy it was left to private individuals to compete for the land plane speed record. One of those who was successful was American industrialist Howard Hughes. He designed his own aircraft which, with a powerful Pratt and Whitney engine and retractable undercarriage, proved to be very fast. At the end of 1935 he flew the Racer over a measured course to return a new landplane record speed of 550km/h (342mph).

Tension was increasing in Europe and in Britain the Spitfire was soon to join Fighter Command. Clearly capable of development to well above the existing record speed, the Spitfire ultimately did not compete; instead, a secretly produced German design raised the official record to 747km/h (464mph).

By the end of the 1939–45 war the jets had arrived and there was only one more increase notched up by a piston-engined aircraft. This happened long after the jets had put the speed beyond 1,000km/h (600mph). A piston-engined class was established and in one great blaze of glory American Darryl Greenamyer flew a Grumman Bearcat over a

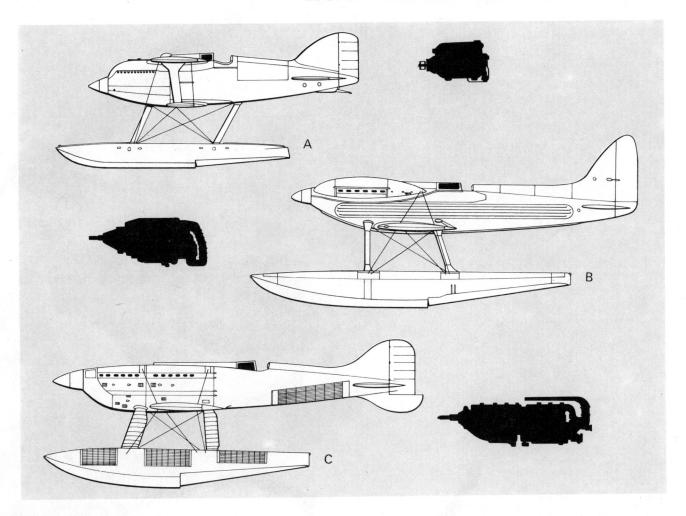

The results of the Schneider Trophy Contests were that engine designers concentrated their efforts on extra speed and on streamlining nose structures. Shown here are (A) the Curtiss V-1500 (D-12) engine in the Curtiss R3C, (B) the Rolls-Royce R engine in the Supermarine S6B and (C) the Fiat engine AS6 engine in the Macchi 72.

measured course at 776km/h (482mph).

Once the jets began to compete the timing equipment became extremely complex. The Gloster Meteors that were flown along the South coast of England in 1945 and 1946 ultimately achieved 991km/h (616mph), flew low in the traditional manner and were timed by stopwatches and theodolites. Not much more than 161km/h (100mph) was added before the discomfort of flying in high temperatures at high speed caused the Ministry of Defence to abandon further attempts by Hunters and Swifts. Both aircraft had made runs and that by the Swift above a Libyan road was the last British attempt at low level. It achieved 1,186km/h (737mph) but only by flying over the hot desert to minimize the drag rise as the aircraft approached the speed of sound. Thermals led to turbulence, which was uncomfortable, and American contenders were facing a similar problem during flights along their measured course

over the Californian desert. The FAI was asked to approve the use of radar to time flights made at high level.

The last low run and the first high one were flown by the same aircraft, a United States Air Force F-100, which put the speed up to 1,323km/h (822mph). One year later Britain's Fairey Delta FD2 flew the familiar coastal track, but at high level, and the record book took a giant leap to 1,822km/h (1,132mph).

Concorde's routine transatlantic flights make nonsense of such speeds now, but for sheer streaking across the ground the flight of the Blackbird takes some beating. To make a dramatic arrival at the 1974 Farnborough Air Show a Lockheed SR-71 Blackbird crossed the Atlantic in less than two hours from a flying start that originated over California. Radar offsets laid a track from south of Long Island to the middle of the English Channel, exactly the distance from central New York to central London. This was the fastest speed ever recorded. After such a speed, flown at a height of about 24,320m (80,000ft), it seems that future attempts must be left to spacemen.

Rotorcraft
and V/STOL

The development of rotorcraft – the term includes both helicopters and autogiros, or giroplanes as they are sometimes called – has always been in a world of its own. Apart from ballooning, it was mainly with moving wings, designed to emulate the flapping of bird flight, that man made his first attempts to fly. From the legendary experiments of Icarus and the sketches of Leonardo da Vinci, right through to the notable contributions of Sir George Cayley in the 19th century, most of the proposed flying machines had some means of flapping or rotating their wings to produce lift.

Initially, the practical problems of mechanical complexity were quite insuperable, and it is only within the past 100 years that any significant attempts have been made to depart from the moving-wing concept. Then, as it came to be more clearly discerned that so many of the finest natural fliers – such as the eagle and the albatross – seldom flapped their wings, the fixed-wing philosophy began to make great strides. But there were still some adherents of rotating wings who were not deterred in their search for vertical take-off and landing. To them, the aeroplane was a highly dangerous form of flight and, although they were habitually ridiculed, there was a great deal of evidence to support their contentions. Most of the frequent aeroplane crashes of the early 20th century occurred during the critical manoeuvres of take-off and landing at comparatively high speeds.

The group of rotorcraft pioneers contracted as many gave up the complexities and turned to the much simpler fixed-wing aeroplane, but during the first 20 years of this century a handful of designers did meet with some limited success. Among the helicopters to fly successfully in that period, the

Paul Cornu's tandem-rotor helicopter (*above*) was the first to achieve unrestrained free flight. The Cierva Autogiro of 1925 (*right*) introduced the first really practical rotary wing.

Breguet-Richet No. 1 was credited, on 29 September 1907, with being the world's first man-carrying helicopter to become airborne, albeit under restraint from tethering ropes controlled by the ground crew. This machine had a gross weight of 578kg (1,274lb) and was lifted by four biplane rotors of 8m (26ft 3in) diameter each and driven by a single Antoinette engine of 45bhp. It remained airborne for only about one minute outside the Breguet aircraft works at Douai, France, and reached a hight of only about 0·6m (2ft).

On 13 November of that same year and also in France, at Lisieux, the Cornu tandem-rotor helicopter became the world's first rotating-wing aircraft to achieve unrestrained free flight, with the designer himself at the controls. This maiden flight deserves its place in history even though, at less than 30 seconds, it was of even shorter duration than Breguet's. The machine weighed 260kg (573lb) and was powered by a 24hp Antoinette engine which drove the two twin-bladed 'paddle-wheel' rotors through a belt-and-pulley transmission system. To outward appearances, it gave the impression of having been built from parts of several bicycles and its short flight ended, regrettably, in the collapse of its frail tububular framework following a heavy landing. Still, it was a world first.

World War I brought a temporary halt to experiments but in the period that followed there were other limited successes. In France, the quadruple-rotor Oemichen helicopter, which was additionally fitted with five small variable-pitch propellers for control purposes, actually achieved a world helicopter distance record with a flight of just over 500m (1,640ft) in 1924. Another quadruple-rotor machine, the de Bothezat helicopter, was also being flight-tested in the USA by the US Army Air Force and achieved a maximum duration of 1min 40sec. Both were incredibly complicated mechanical contraptions. Hardly less complicated was the Pescara helicopter which was lifted by twin co-axial, contrarotating biplane rotors. This machine, built in France by a Spanish designer, proved itself superior to the Oemichen by increasing the world helicopter distance record to just about 800m (2,640ft). It was no wonder that the helicopter proponents were ridiculed by fixed-wing aeroplane pioneers. The Atlantic had already been flown non-stop by Alcock and Brown in a Vickers Vimy bomber.

The Cierva autogiro

The breakthrough in rotating-wing design came with the invention by the Spaniard Juan de la Cierva of the autogiro. By 1925, he had proved the validity of his concept beyond all doubt and was invited to England to continue his research and development work with British sponsorship. During the next ten years, the English company he formed produced a succession of autogiro prototypes which established all the design theory and data on which the helicopter industries of the world were subsequently founded. Tragically, Cierva himself was killed at Croydon in 1936, at the premature age of 41, without seeing the extent to which his invention would be developed. Ironically, he was a passenger in a fixed-wing airliner when he lost his life. He had geared the greater part of his working life toward eliminating the possibility of such an accident. Nevertheless, his ideas were a vital contribution to air safety.

Cierva's invention was based on a quite novel design philosophy which emerged from the mathematical analysis he made of contemporary helicop-

ter designers' problems. They were then all using their various forms of rotor system with the blades set at a high positive pitch angle, rotated by the application of power to the drive shafts on which they were mounted. In effect, they were large propellers which screwed themselves upward against the resistance of the air. There were two principal problems. On the one hand, the application of engine power to turn the rotor drive shaft produced an equal and opposite reaction which would turn the fuselage in the opposite direction when airborne. Known as torque reaction, this is a phenomenon familiar to all engineers. In order to avoid it, rotors on early helicopters were designed so that they could be used in pairs, turning in opposite directions to counteract the torque.

Following from this need to use pairs of rotors came the requirement for complex mechanical transmission systems to transmit the engine power and then elaborate structures to support the systems. Cierva's discovery was that none of this was necessary. His calculations of the aerodynamic forces at work in a simple rotor system satisfied him that, if the blades were to be set at a low positive pitch angle and started in rotational motion, the rotating system could be towed through the air in such a way that the blades would maintain a constant rotational speed in a state of equilibrium, without any application of power.

Autorotation

He termed this phenomenon autorotation and applied it in practice by building a single rotor and mounting it on a pylon above a conventional aeroplane fuselage. Instead of a complex transmission system there was simply one bearing at the head of the pylon which allowed the rotor to turn freely. There were a few teething troubles initially but he soon developed an arrangement that worked satisfactorily. Take-off in the early machines was achieved by hand-starting the rotation of the blades and then accelerating across the aerodrome, pulled by a conventional engine-driven propeller. As forward speed increased on the take-off run, rotor rpm also increased until the state of equilibrium was reached. At this stage, which was normally after a run of a few hundred metres to reach a speed of some 48km/h (30mph), sufficient lift was being generated by the spinning blades for the machine to become airborne.

It was an elegant solution to the problem of flight with rotating wings. A simple, single rotor system could be used because, as it was freely rotating, there was no torque to counteract. The extended flight-test and development work which the first practical autogiro made possible yielded, in due course, a vast accumulation of technical data on which future improved designs could be based. Subsequent refinements to the system included mechanical means to start the rotor before take-off and means of precise control of its tip-path plane in order to ensure a safe landing.

The latest forms of autogiro under development just before Cierva's untimely death had progressed still further. Methods were developed to enable the mechanical drive system to overspeed the rotor on the ground before take-off so that a sudden increase of blade pitch under the pilot's control would cause the machine to jump vertically without any forward run. The vertical jump was sustained for just long enough to allow the machine to gain normal forward flying speed while under the influence of propeller thrust.

The autogiro was never able to hover but it could fly much more slowly than a fixed-wing aircraft, down to less than 40km/h (25mph) without losing height, and could land with virtually no forward speed. If forward speed were reduced below this value in the air, the machine would begin to sink gently towards the ground. Rotor speed remained constant during the gliding descent, akin to the spinning descent of a sycamore seed. Since the control system governed the rotor's angle of tilt, and not that of the fuselage, the pilot could retain precise control right to the point of touchdown.

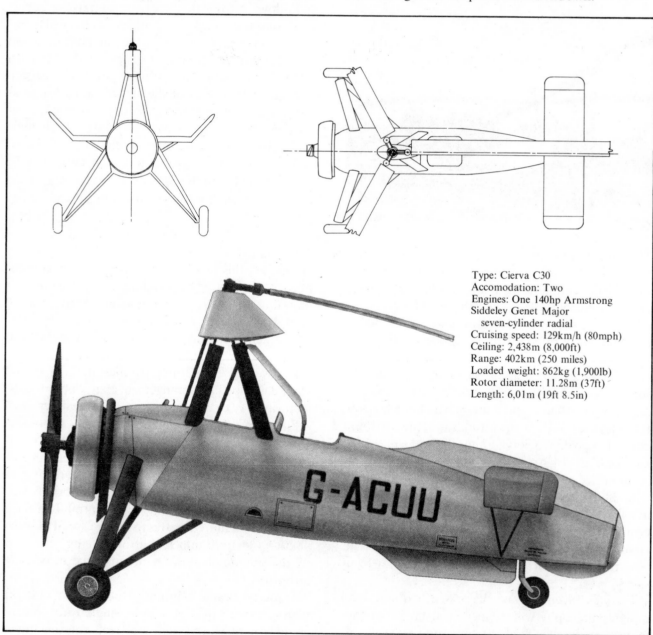

Type: Cierva C30
Accomodation: Two
Engines: One 140hp Armstrong
Siddeley Genet Major
    seven-cylinder radial
Cruising speed: 129km/h (80mph)
Ceiling: 2,438m (8,000ft)
Range: 402km (250 miles)
Loaded weight: 862kg (1,900lb)
Rotor diameter: 11.28m (37ft)
Length: 6,01m (19ft 8.5in)

Cierva C-30

The most effective of the many autogiros developed during the 1930s was the Cierva C-30, of which several hundred were produced by manufacturing licensees of the Cierva Autogiro Company in the UK and Europe. Modified versions of the type were also produced in the USA. The C-30 was a two-seat machine of some 862kg (1,900lb) gross weight and was powered by a 140bhp Armstrong Siddeley Genet Major radial engine. A power take-off from the rear of the engine crankshaft was used as a mechanical starting system for the three-bladed, 11·28m (37ft) diameter rotor. The two seats were arranged in tandem as open cockpits, with the pilot at the rear. The forward cockpit was located immediately below the rotorhead with its supporting pylon straddling the cockpit coaming. In this way, the weight of the passenger was exactly on the centre of gravity so that flying trim did not alter when the machine was flown solo.

The fuselage was a fabric-covered, tubular-steel structure, substantially the same as that of the Avro Cadet biplane, a successor to the famous 504K. The C-30 was, in fact, designed around this fuselage as Avro was one of Cierva's UK licensed manufacturers. From a production viewpoint it was advantageous to use an existing fuselage, although it was of course the rotor system rather than the fuselage which led to the C-30's success.

The fully articulated rotor was mounted universally on the pylon so that it could be tilted in any direction in response to movement of the pilot's hanging control column. Rotor blades were of aerofoil section and their construction was based on a tubular-steel spar with spruce ribs and plywood covering. Blade chord was 25cm (10in) and the pitch angle was fixed at 2° positive. Normal rotational speed in the autorotative condition was 200rpm.

Design modifications to the standard fuselage included the fitting of a long-travel, soft-oleo landing gear, to accommodate the unique autogiro landing characteristics, and a steerable tailwheel operated by the pilot's rudder pedals. The tailplane was also of unique design. Fitted with upswept tips, the aerofoil camber was positive on the starboard side and negative on the port side to provide a movement about the longitudinal axis (anticlockwise when viewed from the rear) and thus counteract propeller torque in forward flight.

Design maximum speed was over 160km/h (100mph). It was not always possible to achieve this, particularly at full load, but the type had a useful cruising speed in the order of 130km/h (80mph) and

carried fuel for just over two hours' endurance. The Cierva C-30 was sold widely for private and club flying and a number were also bought by the military agencies of several countries, mainly for army reconnaissance evaluation. Another military role, for which their slow-flying capabilities were found to be singularly well suited during World War II, was the calibration of ground radar stations.

If the war had not erupted when it did, it is more than likely that the pattern of progress might have been very different. In the years immediately preceding the outbreak of war, however, there was a small handful of designers who, having derived benefit from the accumulated knowledge of autogiro rotor design, were beginning to show promising results with new helicopter projects. Prominent among them were Heinrich Focke in Germany, James Weir in Scotland, who had earlier been instrumental in sponsoring Cierva's developments in England, and Igor Sikorsky in the USA. Sikorsky had, in fact, built two unsuccessful helicopters in Europe in 1909, at the tender age of 19 years, and had then changed to fixed-wing aeroplane design after emigrating to America. He was prompted to look at the possibilities of helicopter design again by Cierva's autogiro successes which he enthusiastically acknowledged.

Igor Sikorsky in his VS-300. It used a tail-rotor to stop the airframe from rotating in opposition to the main rotor.

Pioneer helicopters

Three projects – the Focke F61, the Weir W-5 and W-6, and the Sikorsky VS-300 – began to attract the interest of their respective military authorities in the late 1930s. They were all more complex than the autogiros, which by then were flying in quite large numbers, but the ability to sustain hovering flight was of special value in a variety of military roles. Even the vertical-jump take-off facility of the latest autogiros then flying was not considered by the military to be a suitable substitute for a hovering capability. Nor was the subsequent wartime use of autogiros on a scale sufficient to deflect their interest. So it was that the helicopter, when it finally began to emerge as a practical flying machine, was initially developed largely under military sponsorship and specifically as a military vehicle. This factor has influenced all its subsequent progress.

The demands made by the military upon the pioneer constructors were unbridled. There was a war to be won and the fledgling helicopter was seen to have enormous potential as a reconnaissance vehicle for swiftly moving ground forces. Initially, the interest came mainly from the US Army and the British Royal Navy. Government money was poured into development contracts with the result that by 1945 Sikorsky's main and tail rotor (MTR) design had become firmly established as the classic configuration. Following the VS-300 prototype, three new designs were developed and put into limited production by the same company to meet military orders. These were the Sikorsky R-4, R-5 and R-6. The first and last were two-seaters, supplied to special units for pilot training and operational evaluation, while the R-5 was a larger helicopter with a lifting capacity of about 508kg

(1,120lb). It was the world's first helicopter designed for a specific military role, having been ordered off the drawing board by the US Navy to operate from the decks of independently routed merchant vessels on anti-submarine patrol duties.

Other American companies had also been engaged in experimental work. Notable among them was Bell Aircraft, which was developing its own design of MTR helicopter. Its principal unique feature was a giro-stabilized two-bladed rotor having one common flapping hinge for the two blades. Bell termed it a teetering or see-saw rotor, and it has become well known throughout the world as a feature of most Bell helicopters.

Right: Sikorsky's R-4 two-seat helicopter trainer of which several hundred were built. The Focke-Achgelis Fa 61 *below* is a contra-rotating twin-rotor design free of torque problems, thus making it the first to achieve precision flights.

Post-war developments

After World War II, the established helicopter constructors turned their attention towards a possible commercial market. Bell was first to be awarded a commercial helicopter certificate of airworthiness, in 1946. This was for the Model 47, which remained in production for 27 years in many two- and three-seat variants. Sikorsky was not far behind, with a four-seat civil version of the R-5 designated S-51. It quickly became apparent, though, that the helicopter was anything but a motor car for the man in the street. Such was the cost structure established by the initial pressure of military procurement that manufacturers found the civil versions they were producing could be operated economically only in a limited number of highly specialized roles. Commercial sales were thus few and, with military interest becoming less intense, the crucial question of the hour was: 'Is the helicopter here to stay?'

In agricultural roles such as crop-spraying, for example, the high cost of the helicopter put it at a considerable disadvantage compared with fixed-wing aircraft. Commercial helicopter operators found themselves unable to penetrate more than about 20 per cent of the market, which comprised mainly those areas in which it was too difficult for the fixed-wing crop-sprayers to operate. The same applied in other aerial-work roles, particularly in survey and construction engineering support. In some applications, however, mainly those concerned with operations in remote areas or mountainous terrain, the helicopter was able to perform time-saving miracles. In such uses its apparently high cost was immaterial since to do the job any other way would have cost even more.

Unfortunately, the opportunities to engage in such specialized work were few, and progress in developing commercial uses for the helicopter was painfully slow for the first few years after World War II. There was, nevertheless, still something of a helicopter euphoria during this period, with new projects making their appearance in countries all over the world. In Europe, the pre-war French and British pioneers applied themselves to further development, seeking to make up for the time lost during the war years.

In France, during this period, three groups in the nationalized aircraft industry were engaged in rival helicopter projects. SNCA du Nord began in 1947 but its interest survived only a few years. SNCA du Sud Ouest had acquired as part of German war reparations the services for a few years from 1946 of members of a wartime Austrian design team. They had successfully built and flown the world's first jet-rotor helicopter, designed by Friedrich von Doblhoff in 1943. The Sud Ouest projects were thus all in the single jet-driven rotor (SJR) category and led to the development of the So-1221 Djinn helicopter, the only jet-rotor helicopter to go beyond the prototype stage into quantity production.

Other members of von Doblhoff's team went to Fairey Aviation in Britain, which led to the development of a Gyrodyne derivative with a jet-driven rotor and, from this, to the Rotodyne. Von Doblhoff himself went to the USA and joined McDonnell Aircraft Corporation to develop a ram-jet helicopter. No jet-rotor helicopter, however, has ever achieved any marked degree of commercial operating success.

Even the So-1221 Djinn helicopter was not a great success commercially. Production was discontinued after just over 100 had been built. A lightweight two-seater, the type was used mainly for pilot training and crop spraying. Its rotor was driven by what is known as 'cold' jets at the blade tips. This was, simply, compressed air bled from the compression chamber of the Turboméca Palouste turbine engine. The air was ducted through a rotating seal at the rotorhead and thence through hollow spars to the blade tip nozzles.

Other forms of jet driven rotor have made use of pulse jets, pressure jets and ram jets. These all involve the ducting of fuel through the head and blades, to be burnt in small combustion chambers at the blade tips. Hence the contrasting term, 'hot' jets. Many such experimental helicopters have been built and some have flown extremely well; but the fuel consumption has always been prohibitive.

The third French company, SNCA du Sud Est, followed more conventional lines in the development of MTR designs, and produced the series of helicopters which have dominated French involvement in the field. The SNCA du Sud Est and Sud Ouest were later amalgamated in Sud Aviation, which itself was subsequently regrouped under what is now the single French nationalized aircraft constructor, Aérospatiale. Among the wide range of helicopters Aérospatiale produce, three, the Puma, Gazelle and Lynx, are manufactured jointly with Westland Aircraft Ltd under the Anglo-French collaboration agreement.

In Britain after World War II, the original Cierva Autogiro Company reformed with many of its former key engineering staff and began the development of two new designs. The first to fly, the Cierva

W-9, was a two-seat single-rotor helicopter but, instead of following exactly the classic MTR configuration, it used a laterally deflected jet at the tail in place of the tail rotor for torque compensation. Concurrently, the company was also building a much larger machine, the triple-rotor Cierva W-11 Air Horse, which flew successfully at its design gross weight of 7,938kg (17,500lb).

This, in its day, was the largest helicopter in the world, with a cargo compartment 5·79m (19ft) in length capable of carrying wheeled vehicles. Volumetric capacity was in the region of 22·65m³ (800cu ft) and entry was by means of a ramp through clam-shell doors at the tail. The Air Horse was powered by a single water-cooled Rolls-Royce Merlin engine of 1,620bhp, the same engine that powered the famous Spitfire fighter. That the Air Horse did not mature to a successful conclusion was due mainly to lack of appreciation by the sponsoring British government of the need for a much higher level of funding to support so sophisticated a design.

Three other British aircraft constructors also entered the rotating wing field at the end of World

Royal Navy helicopters display their lifting capability at the Farnborough Air Show, demonstrating a vital aspect of the military helicopter's potential. Not only can they carry in troops and supplies, but they can also evacuate casualties.

War II: Westland Aircraft negotiated a manufacturing licence with Sikorsky to build the S-51 in Britain; Bristol Aeroplane Company began the development of a new MTR design with a cabin for five passengers, and Fairey Aviation Company developed a novel compound helicopter (CMP) project, also able to carry five passengers, named the Gyrodyne. All three were powered by the same type of engine, the Alvis Leonides nine-cylinder radial of 525bhp. The Fairey Gyrodyne, with its novel compound design for superior cruising speed, was the first rotorcraft to take the world helicopter speed record above the 200km/h (124mph) mark. Over a 3km (1·86 miles) course, it flew at 201km/h (124·9mph).

American initiative

In spite of inevitable setbacks in the comparatively early stages, helicopter projects grew both in number and variety, and nowhere was the profusion of new ideas so great as in the United States. At one time, just before 1950, there were more than 70 different active helicopter projects in America. Many were being built by small engineering companies or by individuals in private garages. Of these, only a few ever left the ground; many never progressed beyond the stage of being a gleam in their hopeful inventors' eyes.

One notable exception was the project of a young Californian graduate, Stanley Hiller Jr, who built his own back-yard co-axial, contra-rotating rotor (CXR) helicopter in 1944. He was fortunate in having links with a major industrial corporation which helped him to surmount the initial hurdles. He was later to develop his own servo-paddle rotor control system. Hiller Helicopters Inc, which he formed, produced more than 1,000 helicopters based on the design feature during the ensuing two decades. Among other American companies, the two which had been predominant in pre-war autogiro development were also involved with new helicopter projects. Pitcairn Autogiro Company, which had been taken over by the giant Firestone Tire & Rubber Company, produced a conventional MTR helicopter, but with a novel rotorspeed governor, while Kellett Aircraft Corporation concentrated its studies on what was then the unusual twin intermeshing-rotor (TIR) configuration. The same design was also favoured by Kaman Aircraft Corporation, which developed it to build several hundred helicopters for the USAF.

The tandem-rotor helicopter was introduced because it was thought that limitations in feasible rotor diameters necessitated multiple rotors to lift heavier payloads. Piasecki Helicopter Corporation was the first to produce a practical twin tandem-rotor (TTR) design, and derivatives of its first tandem-rotor machine, the XHRP-1 (jocularly known as the 'Flying Banana'), were still in production by Boeing Vertol in the mid-1970s. Among other configurations developed in the USA during the 1945 to 1950 period, SJR designs with rotors driven variously by pulse jets, pressure jets and ram jets all made their appearance – and, as often as not, were followed by their disappearance shortly afterwards.

By 1950, most of the weaker brethren had disappeared, leaving about 10 or 12 companies to form the nucleus of what was then a Cinderella industry struggling for recognition in a highly competitive aviation market.

Korean War

It was the outbreak of war in Korea, in the midsummer of 1950, that provided the next major impetus to transform this small group of manufacturers into the thriving helicopter industry which now exists. The US Air Force and US Navy units, equipped with the few hundred helicopters which had been delivered for evaluation, were despatched to Korea for trials under active service conditions. Their performance was far beyond the wildest expectations of the most optimistic military strategists.

The series of helicopter types principally involved were, initially, the Sikorsky H-5, derived from the S-51, Bell H-13 and Hiller H-23. Later, these were supplemented by the Sikorsky H-19 series, a larger machine capable of lifting ten men or a load of almost a ton, the Piasecki HUP-1, a TTR design based aboard US Navy aircraft carriers and used for ship-to-shore work. These helicopters, in their performance of otherwise impossible rescue missions, confirmed beyond any doubt that rotating-wing aircraft were here to stay. With their unique versatility and independence of prepared landing strips, they were able to save the lives of thousands of soldiers, wounded or stranded behind enemy positions. This in itself was enough to ensure their unreserved acceptance as a new ancillary to fighting armies. Perhaps even more significant was the potential that the success of such rescue operations revealed.

Military tacticians were quick to realize that if the helicopter could quite easily infiltrate behind enemy lines to rescue wounded soldiers, it could equally penetrate with offensive personnel, weapons and supplies to mount attacks from almost any unexpected quarter. By the time the Korean war had ended, American military strategy had been completely re-orientated in line with this new philosophy. The new US Army was to be composed largely of highly mobile task forces mounted, supplied and supported entirely from the air by fleets of helicopters designed specifically for the variety of operational tasks involved. At the same time, the helicopter's potential in marine warfare, as an anti-submarine weapon and to support marine commando raids ashore, was realized.

The new philosophy was pioneered by the American armed forces and quickly taken up by those of other nations throughout the world, including the Soviet Union where the history of rotating-wing development had had early origins but not the same degree of practical development as elsewhere.

With military procurement pressure once again calling the tune, the helicopter industry made rapid strides in the decade following the Korean war. Military agencies were then in a much better position to write operational specifications for helicopters to fulfil special duties for which active-service experience had shown the need. Thus, with specific targets to achieve and the money to spend in terms of development contracts, helicopter construction was suddenly transformed into a boom industry and the various new designs began to roll off the production lines in their hundreds.

The turbine engine

Another significant factor, which had a particularly important influence on the rapid rate of development from the mid-1950s onward, was the advent of the turbine engine and its application to helicopters. In fixed-wing aircraft, the shaft turbine as a power-plant to drive a conventional propeller – known as the turboprop – was not the sweeping success for which its designers had hoped. This was largely because it was so quickly superseded by the turbojet, which could produce much higher speeds and thereby justify its much higher costs.

For helicopters, though, the shaft turbine proved ideal. It had a much lower installed weight than the piston engine, which more than offset the greater fuel loads which at first had to be carried. Consequently, helicopter designers found for the first time that they had abundant power available. Moreover, the normal operating characteristics of the turbine, which permitted extended running without fear of damage at up to about 85 per cent power, was particularly well suited to a helicopter's requirement for long periods of continuous hover-

ing in certain military roles. The high initial cost and high fuel consumption were no drawbacks to the military, while the much shorter warm-up period before take-off and smoother operation in flight were considerable advantages.

So, with the power available, much bigger helicopters were soon found to be practicable. Earlier fears of limitations to the size of rotor systems proved groundless in practice, and the helicopter grew up. Before 1950, apart from the Cierva W-11 Air Horse prototype, of 7,938kg (17,500lb) gross weight, the biggest production helicopters powered by piston engines were in the region of 2,500kg (5,510lb) gross weight, with a lifting capability of some 680kg (1,500lb). Ten years later, at the outset of the Vietnam conflict in the early 1960s, transport helicopters like the Boeing Vertol CH-47A Chinook were in the air at gross weights of some 15,000kg (33,070lb). The Chinook, with twin Lycoming turbines providing 5,300shp total installed power, had a useful load capability of more than 7 tons.

Speeds too had risen. Before 1950, the world helicopter speed record stood at 201km/h (124·9mph). By 1963, a French MTR helicopter, the Sud Aviation Super Frelon, had taken it to 341km/h (211·9mph).

In addition to the technological progress, production quantities were also increasing substantially. Four major American constructors, Bell, Boeing Vertol, Hiller and Sikorsky, were all in the big league, each with more than 1,000 helicopters produced. Bell, in fact, was well in the lead with more than 3,000. Large numbers were also being built in Europe and the Soviet Union. Some indication of the extent to which the USA had by then developed the new strategic philosophy of helicopter mobility is that in the US Army alone there were some 5,000 qualified helicopter pilots. Their largest training school was equipped with more than 200 light helicopters, mainly of Bell and Hiller manufacture, and staffed by nearly 150 flying instructors.

Looking to the future at that time, the US Army sponsored a Light Observation Helicopter (LOH) design competition among the already extended industry, the prize for which was to be a production contract for 3,000 turbine-powered helicopters. With military contracts of such proportions in progress and in prospect, it was not surprising that helicopter constructors could spare but scant effort to meet the admittedly small needs of commercial operators. Progress was nevertheless made in commercial applications and a few constructors did set up small sales organizations to supply civil adap-

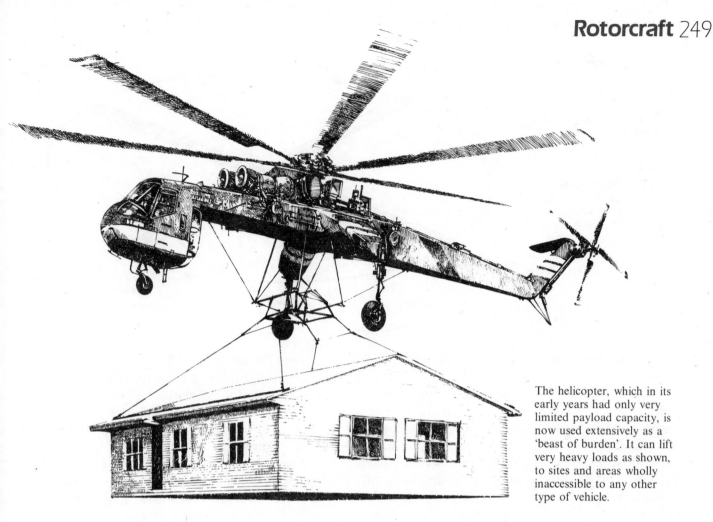

The helicopter, which in its early years had only very limited payload capacity, is now used extensively as a 'beast of burden'. It can lift very heavy loads as shown, to sites and areas wholly inaccessible to any other type of vehicle.

tations of the military helicopters which comprised their main production. Their selling price was high, but for specialized tasks they could be operated on a cost-effective basis. After some 15 years of consistent, if slow, growth, there were by 1960 about 1,000 helicopters operated commercially by some 300 companies throughout the world. More than half the companies, though, may well have been operating only one, two or possibly even three small helicopters.

Specialized tasks

The multiplicity of tasks undertaken by these civil operators impinged upon almost every sphere of industrial and commercial activity. The thousands of kilometres flown on such tasks as overhead power-line construction and patrol, or the innumerable hours flown on crop-spraying and other agricultural work, would rarely be noticed except by those directly concerned. Similarly, the constant aerial support work done for offshore oil prospectors became commonplace.

By its very nature, the greatest proportion of all such operations was performed in remote areas and received little publicity. Only when a helicopter was used for spectacular work in a populous area did its

unique attributes become more widely known. Typical of such a single feat was the placing of the cross on the summit of the rebuilt Coventry Cathedral by an RAF Belvedere. Following this remarkable demonstration of precision lifting, aerial crane work became more widely accepted as part of the commercial helicopter's repertoire. In particular, larger machines were used to lift air-conditioning equipment on to the flat roofs of 'high-rise' buildings, a job which helicopters still are not infrequently called upon to do.

The growth of this particular application, combined with military needs for heavy-lift helicopters, led to the development of aircraft designed specifically for aerial crane work. In the Sikorsky S-60 and S-64 crane helicopters, provision is made in the aft of the cabin for a third set of pilot's controls, with the seat facing rearwards. When the actual lifting is to be done, the helicopter is flown with the normal controls into the hover approximately above the load. The captain then leaves his usual seat and takes over the hover with the third, aft-facing, set of controls. In this position he is looking directly downward on to the load for the precision manoeuvre and can instruct the second pilot, over the intercom, to take over and resume normal flight as

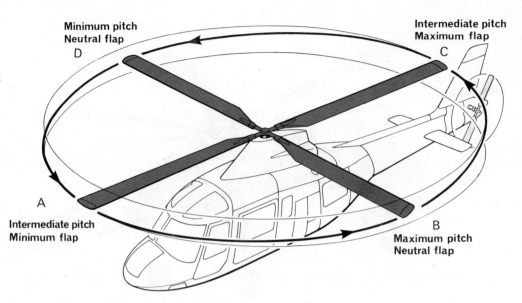

Blade rotation of helicopter in forward flight. (A) Blade at front produces virtually no lift, (B) retreating blade set to develop maximum lift to compensate for its relatively slower speed, (C) blade at rear produces virtually no lift, and (D) advancing blade travels relatively fast and so has a minimum pitch setting to equalise blade lift between the sides of the rotor disc.

Minimum pitch
Neutral flap
D

Intermediate pitch
Maximum flap
C

Intermediate pitch
Minimum flap
A

Maximum pitch
Neutral flap
B

soon as the load has been secured or released, as the case may be.

Mainly because of their high cost, these specialized helicopters have not yet come into general use with commercial operators, whose fleets are made up mostly of small and medium-sized machines with gross weights ranging between some 1,360kg (3,000lb) and 6,350kg (14,000lb). Some of the larger operators, with big oil-rig support contracts, may exceed this upper figure if they have Sikorsky S-61 series helicopters engaged on this work.

Passenger transport
The one area in which the civil helicopter had not made any significant headway up till the early 1960s was the carrying of passengers. There were a few meritorious attempts, mainly in the USA, to set up helicopter airline services and these met with some, if qualified, success. The routes flown provided direct links between the principal airports of the area served – Los Angeles, San Francisco, Chicago and New York – and the outlying suburbs. Their success was qualified in that, although the adapted military transport helicopters themselves were technically capable of providing the services, their operating economics were such that the companies needed some measure of government subsidy. There have been a few isolated exceptions, but most were forced eventually to close down for this reason. In Europe, a helicopter service centred on Brussels, operated by the Belgian airline SABENA, was discontinued but a British Airways service between Cornwall and the Scilly Isles has continued.

In its smaller sizes, too, the helicopter was as far from becoming an effective means of personal transport for the man in the street as it had ever been. In fact, by 1960 the dream of a back-garden flying machine for the masses had been completely abandoned. It was still hoped, though, that the unique vertical take-off and landing ability would enable executive transport and aerial taxi work to become an important addition to the helicopter's extensive range of applications.

One of the first helicopters to be developed specifically for this latter role was the Bell Model 47J, named the Ranger. This was not a new design but an adaptation of the earlier Model 47G, of which it used all the dynamic components, In other words, it was a 47G in all its rotating mechanisms but had a newly styled four-seat cabin in place of the earlier type's transparent plastic bubble which enclosed a bench seat for three, including pilot, sitting side-by-side. The bench was retained in the 47J for the three passengers, but the cabin was extended forward to provide space for a fourth, separate, seat for the pilot. The fuselage was more streamlined, with a stressed-skin monocoque structure instead of the open tubular-steel framework of the 47G.

With these improvements, the 47J Ranger was able to attain slightly higher speeds, up to a maximum of some 169km/h (105mph), and the type was used quite widely for passenger transport. In contrast, however, contemporary light fixed-wing aircraft of equivalent size and power could offer something like twice the speed at about one-third of the cost, so the helicopter was again operating in a strictly limited field. At that time, few more than 100 helicopters of this kind were in regular operation for company transport use and, in most cases, they were for a highly specialized requirement which virtually precluded the use of fixed-wing aeroplanes.

The position changed dramatically with the introduction of the small turbine engine, a by-product of the American LOH design competition previously mentioned. The turbine engines first used as helicopter power-plants produced some 1,000shp. Two types were used initially, the General Electric T58, with a power range between 900shp and 1,800shp and the Lycoming T53 of some 770shp to 1,450shp. Both were free-shaft turbines, more details of which can be found in the engine section of this volume. These two power units gave rise to the development of helicopters such as the Bell UH-1A Iroquois series, a ten-seat MTR helicopter of 3,856kg (8,500lb) gross weight; the Kaman H-43L, a TIR aircraft of similar size; and, with a twin turbine installation the 25-seat Vertol 107TTR transport

helicopter and the 28-seat Sikorsky S-61 series, the latter still in production in 1977. All these types were flying by 1960.

Although the 1,000shp turbines were regarded as small by contemporary fixed-wing standards, they were far too large for use in the four- or five-seat aircraft in the 1,300 to 1,400kg (2,850 to 3,085lb) gross-weight category. The French company Turboméca was developing a range of smaller fixed-shaft turbines producing some 400shp, but it was the Allison T63 free-shaft unit, designed specifically for the US Army LOH project, which provided the major breakthrough. It had a power-turbine spool no bigger than a two-litre oil can but produced 250shp (now 650shp) at its best operating speed.

All three finalists in the LOH competition, the Bell

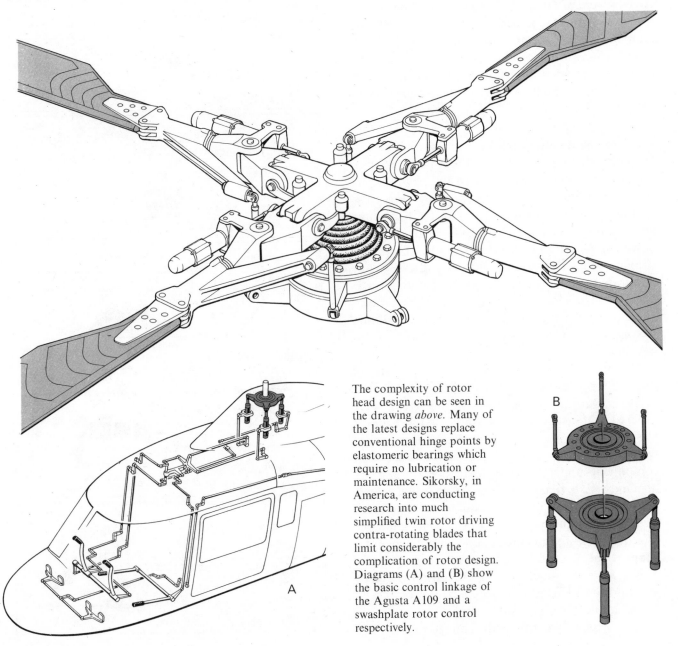

The complexity of rotor head design can be seen in the drawing *above*. Many of the latest designs replace conventional hinge points by elastomeric bearings which require no lubrication or maintenance. Sikorsky, in America, are conducting research into much simplified twin rotor driving contra-rotating blades that limit considerably the complication of rotor design. Diagrams (A) and (B) show the basic control linkage of the Agusta A109 and a swashplate rotor control respectively.

OH-4A, Hiller OH-5A and Hughes OH-6A, used this Allison T63 turbine. The declared winner was the Hughes OH-6A, a four-seater with a gross weight of 1,089kg (2,400lb) and a maximum speed of some 241km/h (150mph), and mass-production of the type was set in motion. Of the other two, Bell developed its OH-4A prototype, which had been built to full civil airworthiness standards as one element of the design competition conditions, into what was to become the Model 206A JetRanger. A refined version of this type now constitutes the mainstay of light executive helicopter fleets. The additional speed conferred by the turbine engine was enough to make helicopter transport worthwhile to a much wider field of prospective users. Against a stiff headwind, the earlier 160km/h (100mph) helicopters were often no faster than a good motor car for conventional journeys. The extra 80km/h (50mph) of the turbine helicopter placed it well ahead in any wind conditions. Later, Hughes was also to produce a civil version, known as the Hughes 500, of the company's own turbine-engined competition winner.

Vietnam and its aftermath

When the Vietnam War began, the American armed forces were fully reorientated to the new military strategy of helicopter mobility. They were also partially equipped with a variety of light, medium and heavy transport helicopters, all turbine-powered, to put the new tactics into effect. The strategy met with virtually instant success, to the extent that some historians have dubbed the war in Vietnam 'The Helicopter War'. That the final outcome did not bring with it a conventional military victory for the new tactics was no fault of the helicopters.

The effect on the American helicopter industry of the initial successes in active service was extremely rapid expansion. Helicopters of all types were ordered in thousands; at one stage, Bell had one of the Beechcraft aeroplane factories almost exclusively engaged in producing fuselage and other components under sub-contract to meet the military demands. Experience in Vietnam also had a corresponding effect in other countries as military strategists began to follow the American lead. The idea of an army going into action without helicopter support became totally redundant.

The advances made during the ensuing decade and into the 1970s were largely a matter of degree. For example, the UH-1A Iroquois, designed by Bell as the Model 204, originally entered service just

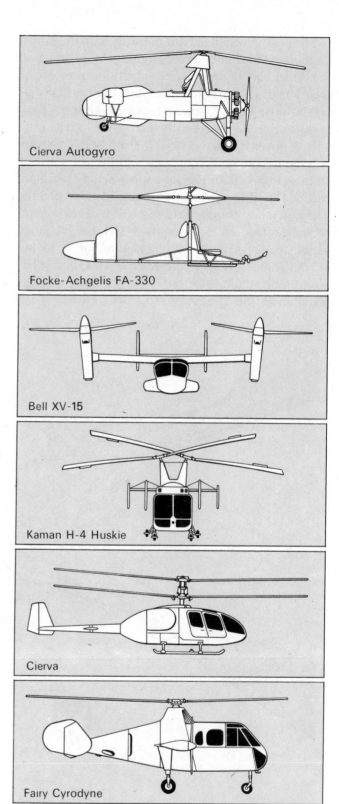

Cierva Autogyro

Focke-Achgelis FA-330

Bell XV-15

Kaman H-4 Huskie

Cierva

Fairy Cyrodyne

Canadair CL-84

before 1960 as a ten-seat utility helicopter powered by a 1,100shp Lycoming T54-L-11 turbine. Its gross weight was 3,856kg (8,500lb), rotor diameter 13·41m (44ft) and disposable load 1,805kg (3,980lb). Through the years, 15 variants of the design have been produced, amounting to some 10,000 helicopters of this one series. The latest derivative, Bell's Model 214, has a maximum gross weight of 7,258kg (16,000lb) and a lifting capacity for external loads of up to 3,629kg (8,000lb). The rotor diameter has gone up to 15·24m (50ft) and the powerplant is a 2,930shp Lycoming T-5508D turbine reduced to 2,050shp for normal sea-level operation. This leaves a substantial power reserve available for high-altitude operation, maintaining sea-level performance up to about 6,000m (19,685ft).

Another current variant of the series, the 15-seat Bell 212, has a twin turbine installation for greater safety. The powerplant in this variant is the Canadian Pratt & Whitney PT6T-3, rated at 1,800shp total, in two coupled turbines either of which will sustain level flight if the other fails. The Model 212 rotor diameter is slightly smaller than that of the 214 at 14·69m (48ft 2½in) and the gross weight is 5,080kg (11,200lb).

An indication of the extent of these advances can be gauged from the fact that the Lycoming T55 turbine in the model 214 is the same basic power unit used in a twin installation to power the 15-ton Boeing Vertol CH-47 Chinook helicopter. The Chinook itself is another typical example of the degree of advance made during the decade of the Vietnam War. The first prototype made its debut in 1961 as a 14,969kg (33,000lb) gross-weight TTR military transport helicopter. It was powered by twin Lycoming T55-L-7 turbines of 2,650shp each and had a disposable load capable of 6,804kg (15,000lb). It was the first big helicopter in which the rear loading door opened downwards to form a drive-on ramp for loading wheeled or tracked vehicles into the cabin, which was 9·14m (30ft) long; the diameter of the two tandem rotors was 17·98m (59ft).

In the latest Chinook, the CH-47C Model 234, the maximum gross weight has grown to 22,680kg (50,000lb) and the disposable load can now be more than 11,340kg (25,000lb). The rotor diameter has remained substantially the same but the systems and transmission have all been modified and strengthened to take the considerably greater power output from the twin Lycoming T55-L-11C turbines. These are rated at a maximum of 3,750shp each, with emergency reserve up to 4,500shp each.

Altogether, Boeing Vertol and the Piasecki Helicopter Corporation (which it absorbed) have produced more than 2,500 tandem-rotor helicopters.

Sikorsky Aircraft, which has also graduated mainly into the construction of large helicopters, has produced even more, though not of tandem-rotor design. This company pioneered the boat-hulled helicopter with its S-61 series and has consequently been concerned mainly with production of aircraft for maritime applications. One of the latest versions of the S-61, known as the SH-3D Sea King, is in service with many navies for anti-submarine and air-sea rescue operations. One of their much publicized applications has been the retrieval of American astronauts on their splash-down return from space missions in the Atlantic or Pacific oceans. The Sea King is built under licence in several countries, in England by Westland Aircraft. Their version has a gross weight of 9,525kg (20,995lb). Disposable load is 2,350kg (5,180lb). The machine is powered by twin Rolls-Royce Gnome H.1400-1 turbines, rated at 1,630shp each, driving an 17.98m (59ft) diameter 5-bladed rotor. The Gnome turbines are derived from the General Electric T58 turbines which power the American-built version. In addition to all-weather navigation systems, the Sea King's equipment can include full sonar detection apparatus plus four torpedoes and four depth charges. It is operated by a crew of four.

At the other end of the size range, the helicopter war in Vietnam brought out the light observation helicopters in their thousands. First the Hughes OH-6A Cayuse and, later, the Bell OH-58A Kiowa, derived from the original OH-4A which was itself the forerunner of the five-seat civil JetRanger. Bell's latest JetRanger variant, known as the Long-Ranger, has a passenger cabin 0·61m (2ft) longer than that of the JetRanger to provide space for two extra seats. The rotor is of 1·12m (3ft 8in) greater diameter and the power unit, an Allison 250-C20B turbine, has a maximum continuous rating of 370shp as opposed to the 270shp of its predecessor. Gross weight of the seven-seat LongRanger is up to 1,814kg (4,000lb) whereas the JetRanger's gross weight is 1,452kg (3,200lb). Both types are still in full production.

Helicopter gunships

One more significant concept emerged from the implementation of the new military strategy in Vietnam. This concerned a much faster, heavily armed helicopter which could give close support and protection to troop-transport and supply armadas

moving into forward battle zones. US Army commanders in the field found that aerial support available from conventional fixed-wing fighter squadrons was sometimes too remote and inflexible to be sufficiently effective for their specialized requirements. So the idea of the helicopter gunship was born, the best known example of which is probably the Bell AH-1 Huey Cobra. This design uses all the dynamic components of the latest Bell UH-1 utility helicopter in a slender, streamlined fuselage to give it a speed of up to 354km/h (220mph). It carries a crew of two, pilot and air gunner/observer, and can be armed with a range of missiles, rockets, rapid-firing machine-guns and other weapons. The type was in quantity production before the end of the Vietnam war and proved extremely effective on active service.

After Vietnam, attention was once more turned towards civil applications, the most significant current trend being the design of light and medium-sized twin-turbine helicopters able to operate in full Instrument Flight Rules (IFR) conditions. This has entailed the development of special instrumentation, autostabilizers and navigation aids and has resulted in the production of such sophisticated executive transport helicopters as the West German five-seat

Messerschmitt-Bölkow-Blohm (MBB) BO 105, the Italian eight-seat Augusta A 109 or the American ten-seat Bell 222. The twin-engine reliability common to all three is essential to meet the requirements of IFR operation in air traffic control zones and also to fly into heliports located at the centre of populous areas. All three follow the classic MTR configuration and all are capable of speeds of about 240km/h (150mph) or better.

Mil V-12

One helicopter which merits special mention, as the largest in the world, is the Russian V-12 heavy transport helicopter, designed by the Mil bureau. In the 1960s, Mikhail Mil produced what was then the world's largest helicopter, designated Mi-6. It was a conventional MTR helicopter with a rotor diameter of 35m (114ft 10in) and powered by twin TB-2BM turbines producing 5,500shp each. Its gross weight was about 40,642kg (89,600 lb), almost twice the weight of its nearest rival.

The V-12 is twice this size again, having been produced by combining two Mi-6 rotors, each with its twin-turbine power supply, in a giant twin side-by-side rotor (TSR) configuration. The rotors with their power units are mounted at the ends of two

The Bell Huey Cobra *above* fires a salvo of 2·75in rockets. Helicopter gunships like this are important battlefield weapons. *Left* is a CH-54A transport in USAF service.

inversely tapered outrigger booms, through which transmission shafts couple the two pairs of turbines through a common gearbox in the upper fuselage. Power can thus be transferred from either side in the event of partial failure. The capacious cargo compartment measures over 27·43m (90ft) in length, 4·27m (14ft) in height and width. These are the same dimensions as those of the Antonov An-22 cargo transport aircraft, so it seems that the helicopter is intended to carry some of the same outsize loads. Lifting capacity is more than 40,624kg (89,560lb).

Right at the other end of the size range, there should also be mentioned the many recent attempts to revive the autogiro, commonly in the form of ultra-light single-seaters designed principally for amateur flying. There has been some progress in this field, though not without its problems, and the popularity of these diminutive rotorcraft has been spasmodic.

There is no reason why the autogiro could not be developed to a standard comparable in its own way with that of the present-day helicopter. It would never, however, be capable of sustained hovering flight, and it has always been precisely this facility from which the helicopter has derived its operating superiority. Enormous sums of money have been spent in reaching the present stage of development. The millions of flying hours accumulated by the tens of thousands of helicopters in current operation have yielded knowledge which has elevated the helicopter engineer to a status of scientific parity with his fixed-wing conterpart.

Further design refinements, mainly to increase reliability, component life and comfort can be expected. Hingeless rotor systems such as in the new Westland Lynx and MBB BO105 are examples, as is the new nodal-beam suspension system being introduced by Bell, to reduce the effects of unavoidable vibration. Some marginal increases in cruising speeds may also be introduced to advantage. The use of 'exotic' materials will certainly increase. In the early days, rotor blades were manufactured on a tubular spar with wooden ribs and stringers, and covered in fabric. Today, extruded alloy spars are used, with stainless steel and glass-reinforced plastics also embodied. Titanium is beginning to replace steel for certain rotorhead components, and many more developments along these lines are on the way.

With its present-day sophisticated systems, and especially with its requirement for an extremely high degree of precision engineering, the helicopter is probably one of the most expensive means of transport yet devised by man. Its great saving grace is that, in its unique sphere, it is also, without doubt, one of the most effective.

V/STOL

The introduction of the turbojet engine, towards the end of World War II, came at a time when every available effort was being made by both sides to channel this new invention into a fighter aircraft. As events transpired, neither side was able to bring the first jet fighters into service until the war was virtually over.

After hostilities ceased, engine designers had more time to think of other future possibilities and vertical take-off, for both military and civil aircraft, was one obvious application. Even the early jet engines could develop a thrust of higher value than their own weight. By the time the Rolls-Royce Nene turbojet came into service, in the early 1950s, its maximum thrust of some 1,814kg (4,000lb) was roughly twice its own weight. There was thus ample margin for the additional weight of a supporting framework.

To evaluate the feasibility of controlling such a concept, Rolls-Royce mounted two Nenes horizontally in a tubular-steel engine test-bed and modified the engine tailpipes to direct the jet efflux vertically downwards. When it first flew, in 1953, the four-legged framework created a worldwide sensation and was promptly dubbed the 'Flying Bedstead'.

Gross take-off weight of the machine was 3,264kg (7,196lb). Maximum vertical thrust from the two Nenes was about 3,629kg (8,000lb) so there was no doubt about its vertical take-off capability. The main experimental purpose of the test rig was to evaluate the system devised for attitude control in hovering flight. Compressed air was bled from the two engine compressors, at 3·2kg (7lb) per second, into a common collector box and then ducted into four downward-facing nozzles, one positioned forward, one aft and one at either side.

The fore and aft nozzles each had a 10cm (4in) diameter orifice and produced some 132kg (290lb) average thrust at a 4·27m (14ft) arm from the centre of gravity. To control pitch attitude, their thrust could be varied differentially by diverter valves connected to the pilot's control column. The nozzles could also be swivelled differentially from side to side by the rudder pedals, to provide control in yaw. The lateral nozzles were approximately half the diameter and half the distance from the centre of gravity, producing some 15·9kg (35lb) thrust each. This, too, could be varied differentially by lateral movement of the pilot's control column. The aggregate of all the control thrust remained constant and continuous during hovering flight and so it contributed slightly to vertical lift. After a few inevitable teething troubles had been overcome, the system worked reasonably well. About 380 tethered flights and 120 free flights were made during the comprehensive tests that followed.

The success of the 'Flying Bedstead' conjured up futuristic visions of Vertical Take-Off and Landing (VTOL) airliners operating direct between the centres of the world's capital cities, without the need for aerodromes and all the travelling delays associated with them. It was the fixed-wing aircraft designers' answer to the helicopter, then still in the throes of demonstrating its basic practicability. The vision spurred Rolls-Royce into the development of a special series of light-weight turbojet engines with a remarkably high power:weight ratio; in the first of the series, the RB.108, it was 8:1. The RB.108 was rated at 1,002kg (2,210lb) maximum take-off thrust for a basic dry weight of only 122kg (269lb). It was designed specifically for a vertical-lift application with provision for deflecting the angle of the jet efflux by a few degrees to aid control when hovering in flight.

The first use made of this power unit was by Short Brothers & Harland in the SC-1 delta-wing research aircraft. Powered by five RB.108s, four mounted vertically in two pairs for lift and one mounted horizontally for propulsive thrust, the SC-1 became the world's first fixed-wing jet-powered VTOL aircraft to achieve vertical take-off, in 1958. The first full transition – from the hover to fully wingborne forward flight and then back to the hover sustained only by the lift jets – was made in 1960.

Provision was made in the SC-1 for the two pairs of lift engines to be swivelled fore and aft of the vertical. After take-off, they were swivelled aft a few degrees so that their thrust supplemented that of the propulsive engine in the transition to forward flight. On the approach to land, the lift engines could be swivelled forward a few degrees to provide a braking effect in the transition back to hovering flight. All five engines were fitted with a compressor bleed from which high-pressure air was fed into a common duct for hovering control. The duct system terminated in four small ejector nozzles, one at each wing tip, one at the nose and one at the tail, similar to the system used on the 'Flying Bedstead'.

The SC-1 was intended as the forerunner of a single-seat VTOL fighter which would use a more powerful version of the RB.108 engine for propulsive thrust. It was also intended as the scaled-down prototype of a much larger VTOL airliner. There was, however, another jet-lift system under concurrent development by Hawker and this, known as the vectored-thrust system eventually proved superior.

Based on the Bristol Siddeley BS.53 Pegasus turbofan engine, the vectored-thrust system was first tried in the Hawker P.1127. A developed version of this prototype was later named the Kestrel when it joined a trinational evaluation squadron of British, American and West German composition. Whereas in the Short SC-1 the greater part of thrust available was shut down during wing-borne flight, in the vectored-thrust system of the Kestrel it could all be used to attain much higher forward speeds. The efflux from the turbofan engine was ejected through four swivelling nozzles, two forward and two aft on either side of the fuselage under the wing. For vertical take-off and landing, the nozzles were swivelled to direct the jets vertically downward, while for forward flight the nozzles were swivelled to the rear to give horizontal thrust.

First hovering trials of the P.1127 began in 1960 and the first full transition from vertical take-off to conventional forward flight and then back to vertical landing was achieved in the following year. By 1964, the aircraft had attained supersonic speed in a shallow dive and more powerful, truly supersonic, versions were under development.

Meanwhile, in Europe, the French company Marcel Dassault was developing a VTOL adaptation of one of its Mirage fighters. By replacing the SNECMA Atar turbojet with a smaller Bristol Siddeley Orpheus turbojet, space was made in the fuselage for installing eight Rolls-Royce RB.108 lift jets, mounted vertically in four pairs. This aircraft, called the *Balzac*, began flight trials in 1962 and was

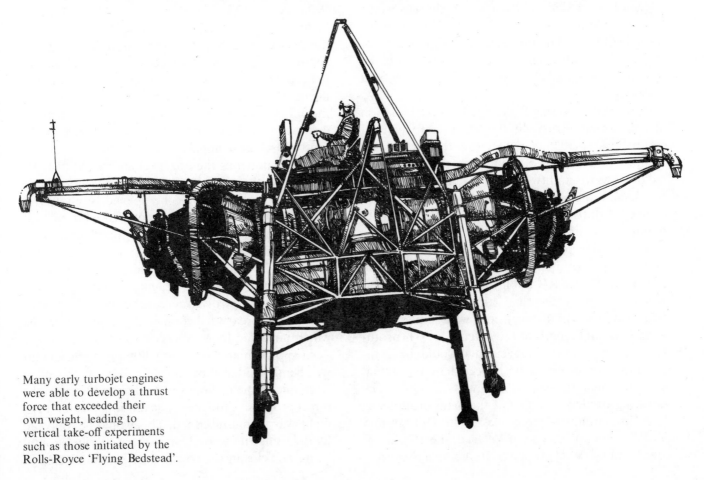

Many early turbojet engines were able to develop a thrust force that exceeded their own weight, leading to vertical take-off experiments such as those initiated by the Rolls-Royce 'Flying Bedstead'.

the forerunner of the Mirage III-V prototypes, which in turn were designed to lead on to a Mach 2 VTOL fighter-bomber. The production version of the Mirage III-V was planned to have a SNECMA-developed afterburning TF-306 (based on the Pratt & Whitney TF30) as its much more powerful main propulsion unit, and eight Rolls-Royce RB.162 jets for vertical lift. The RB.162 was a developed version of the RB.108 and produced just over twice the thrust 2,109kg (4,650lb). Development of the Mirage III-V was however discontinued after an unfortunate series of accidents, despite the prototypes having made successful transitions from hovering to forward flight and vice versa, and having achieved supersonic speed.

The more powerful Rolls-Royce lift jets were also chosen by the German company Dornier, which produced an ambitious prototype for a VTOL transport aircraft. It was designed as a high-wing monoplane with a cruising speed of 750km/h (466mph). Collaborators in the project were Vereinigte Flugtechnische Werke (VFW). The prototype, designated Do 31E, had two banks each of four RB.162s to provide a total lift thrust of some 16,000kg (35,274lb), mounted in wing-tip pods. This was supplemented at take-off by the vectored thrust of twin Bristol Siddeley Pegasus 5 turbofans which combined to add some 10,433kg (23,000lb) to the static lift thrust. The production version was planned to have still higher take-off thrust, with ten RB.162 lift jets, and its gross weight was to be about 23,500kg (51,808lb).

Control in hovering flight was by a combination of differential thrust on the lift jets, for lateral control, and small 'puffer' nozzles at the tail for fore-and-aft control. Two of the tail nozzles were directed downwards and two upwards, all four being fed by ducted high-pressure air bled from the lift engines. Control in yaw was by differential inclination of the lift-engine nozzles.

To test this proposed control system, Dornier built an open-framework hovering rig powered by four Rolls-Royce RB.108 lift jets. This test rig had much longer arms than the original 'Flying Bedstead'. Its overall dimensions were similar to the wing span and length of the actual Do 31E prototype so that control movements would be comparable with those to be expected in the actual aircraft when it flew, which was in 1967. The hovering rig flew in 1964 but the main project was later discontinued. Nevertheless, both Dornier and VFW retain an interest in Vertical/Short Take-off and Landing (V/STOL) aircraft. A similar hovering

rig was built by Fiat, in Italy, but was abandoned in 1966 before free flight was attempted.

Another German project, developed by the Entwicklungsring Süd research group formed by Bölkow, Heinkel and Messerschmitt in 1960, adopted yet another design configuration. In this project designated VJ-101C, a small high-wing monoplane of 6,010kg (13,250lb) gross weight had six Rolls-Royce RB.108 lift jets installed in three pairs. One pair was mounted vertically in the fuselage, immediately abaft the pilot's cockpit, while the second and third pairs were in swivelling wing tip pods. For vertical take-off, all six lift jets were used, with the wing tip pods swivelled into the vertical position. To make the transition into forward flight, the wing tip pods were swivelled forwards through a 90° arc until their efflux provided horizontal thrust. As forward speed increased, with the wing taking over the lift function, the forward pair of lift engines was shut down. The prototype VJ-101C started flight trials in 1963 but the project was abandoned in the following year.

Concurrently, a variety of different projects was in the course of development in the USA. In 1963, Lockheed produced and flew its XV-4A jet-lift VTOL fighter prototype, powered by twin Pratt & Whitney JT12A-3 turbojets rated at 1,497kg (3,300lb) static thrust each. Named Hummingbird, the XV-4A was a mid-wing monoplane. The two engines, mounted horizontally in nacelles alongside the centre fuselage, were arranged to provide either horizontal or vertical thrust. For vertical take-off, the efflux from both engines was diverted, through ducting into rows of downward-facing nozzles in the centre-fuselage compartment between the engine nacelles known as the nozzle chamber. Above and below the chamber, long doors in the upper and lower fuselage skin could be opened to allow free downward flow of the efflux, coupled with induced flow of ambient air through the upper doors to mix with and augment the jet efflux. The angle of the fixed nozzles was such that the aircraft hovered in a nose-up attitude. As in a helicopter, the application of forward control lowered the nose and thus introduced a rearward component of vertical thrust which imparted forward acceleration.

At approximately 145km/h (90mph), sufficient lift was being produced by the wing to allow the efflux from one of the turbojets to be diverted back to the direct propulsive function; conventional wing-borne flight was then established. This would be followed by diverting the second engine to direct horizontal thrust and closing the nozzle chamber doors. Design maximum speed was 837km/h (520mph). The system functioned reasonably well, but performance was not sufficient to warrant continuation of the project.

Ryan Aeronautical conceived a different way of achieving vertical take-off. Using what was known as the fan-in-wing system, the SV-5A prototype obtained its vertical thrust from two 1.59m (5ft 2½in) diameter, 36-blade lift fans mounted horizontally in the wings. A third, smaller, lift fan was mounted in the nose for control purposes. Around the periphery of each fan were fitted small turbine blades, or scoops, on to which the efflux from the twin General Electric J85-5 turbojets impinged to produce a fan speed of 2,640rpm at the full rated power of 1,206kg (2,658lb) static thrust on each engine.

Hinged semi-circular doors in the upper and lower wing surfaces above and below the fans were opened during the take-off. Below each fan there was also a series of transverse louvres, adjustable under pilot control. A thumb-wheel on the control column was used to rotate these louvres and so deflect the fan thrust rearwards to gain forward speed from the hovering position. Differential rotation of the louvres could be demanded through the pilot's normal controls to provide roll and yaw control in the hovering and slow-speed phases. Conventional wing-borne flight was established in this aircraft at approximately 225km/h (140mph), whereupon the turbojet efflux was transferred from the peripheral scoops on the fans to normal tail-pipe ejection and the fan doors were closed.

Flight trials of the Ryan XV-5A began in 1964 but, although the concept proved practicable, it too was later abandoned. Ryan was also involved in the experimental development of a 'tail-sitter' jet aircraft, the X-13, which was designed to take off vertically from a gantry. This, however, proved impracticable for a number of reasons. More successful was the company's contribution, with Hiller Helicopters, to the Ling-Temco-Vought XC-142A tilt-wing research military transport. Powered by four 2,850shp General Electric T64-1 shaft turbines, driving 4·72m (15ft 6in) diameter variable-pitch propellers, the XC-142A was designed to carry up to 7,112kg (15,680lb) of wheeled vehicles and other cargo in its 9·14m (30ft) long cabin. Its maiden flight was in 1964, but it never entered production.

Other North American companies concerned with tilt-wing projects have been Boeing Vertol and Canadair, the latter company's twin-engined CL-84 having been longest in development. The use of ducted fans or ducted propellers has been yet another concept in the USA's search for a vertical take-off capability. It has been proved feasible by Hiller, Piasecki and others, but is hardly practical for general service. Bell Aerosystems, an associate of the helicopter constructor, developed a project for an aircraft in which four ducted propellers mounted on two stub wings were capable of being tilted through 90° to provide either vertical or horizontal thrust. Powered by twin General Electric T58 shaft turbines of 1,250shp each, the machine was known as the X-22A. A similar aircraft, using four tilting propellers without the ducts, was built by Curtiss-Wright and designated X-19A.

Between 1950 and 1970, the US military authorities sponsored the development of almost any apparently feasible VTOL system for practical evaluation. None of the methods, however, was as reliable as the British vectored-thrust principle, so it has transpired that a developed version of the Hawker Siddeley Kestrel has been adopted for operational service with the US Marine Corps and Spanish Navy as well as with Britain's Royal Air Force (RAF).

Named the Harrier, this single-seat V/STOL close-support and reconnaissance aircraft is powered by a single Rolls-Royce Pegasus 11 turbofan developing 9,752kg (21,500lb) thrust. It is a shoulder-wing monoplane with pronounced anhedral both on the wings and tailplanes and with the leading edges of both swept back to an angle of about 40°. The first fixed-wing V/STOL aircraft to enter service resulted in some 105 Harriers being ordered for the RAF by the middle part of the 1970s, 110 for the US Marine Corps and 12 for the Spanish Navy.

Two new variants of the type are now under development. In Britain, Hawker Siddeley is building the Sea Harrier, substantially the same as its land-based progenitor but with a completely new raised cockpit and multi-role NAV-attack systems. The type is due to enter Royal Navy service in 1979 and will operate from a new class of anti-submarine cruisers. In the USA, Hawker Siddeley's licensee McDonnell Douglas is developing a variant designated AV-8B, which will retain the Harrier's fuselage and vectored-thrust engine but will have an entirely new and larger wing made substantially of composite materials. The AV-8B will also be endowed with strakes and other lift-improvement devices designed to eliminate the difficult 'suckdown' effects when the aircraft is hovering near the ground.

That the VTOL fighter is now firmly established

The 'Flying Bedstead', more correctly Thrust Measuring Rig, paved the way for sophisticated V/STOL military aircraft such as Hawker Siddeley's (now British Aerospace's) Harrier.

as a weapon with formidable potential is proved by the recent appearance in Russian naval service of the Yakovlev Yak-36 Forger, which uses a highly unusual lift- and lift/cruise-engine formula. A single lift and propulsion engine has two vectoring nozzles, one on each side of the fuselage beneath the trailing edge of the wing. This is supplemented for vertical take-off and landing by two lift jets mounted in tandem just aft of the cockpit. This layout, though demonstrably effective in vertical operations, is for a number of reasons operationally unsuitable for short take-offs.

Whether the vision of VTOL airliners will ever materialize must be left for the future to record. One of the greatest practical difficulties to be overcome may well be the noise generated at take-off and landing which, at current values, would be totally unacceptable in any city centre. It is by no means impossible, though, that some of the current principles may be applied to shorten the take-off runs of conventional jet or even supersonic airliners, in an endeavour to keep the size of tomorrow's international airports within reasonable bounds.

Stol Short Take-off and Landing Aircraft

For a wide range of aerial transport operations in which neither the vertical take-off capability nor high cruising speeds are essential, another quite specialized category of aircraft has evolved. This is the Short Take-off and Landing (STOL) aircraft which operates to the greatest advantage in the less well-developed regions of the world, often taking off from and landing on small, rough airstrips with few, if any, ground facilities.

Jet lift VTOL aircraft can, of course, be operated in the STOL mode. In other words, if space and a suitable surface are available for a running take-off to be made with the vectored thrust nozzles swivelled to a setting between the vertical and horizontal positions, this technique will permit greater loads to be carried. A typical application for such a technique could be for missions where longer range requires additional fuel to be carried. After its initial running take-off and flight to its operating area, the machine's VTOL capabilities will have been restored as the extra fuel load will, by then, have been consumed. The same technique can also be applied to many helicopters and some convertiplanes, such as tilt-wing aircraft, to enable them to supply forward bases at long range. If the vertical take off or hovering capability is never to be required, though, the STOL aircraft is much more economic.

Background to the development of this category of aircraft is to be found in the growth of 'bush flying' during the 1920s, mainly in the northern territories of Canada and the Australian outback. Here, the pricipal attributes required of an aircraft have always been ruggedness, reliability and economy of operation; high speed is not so important. For safety in these desolate regions, where any reasonably level open space has to serve as an airfield, short take-off runs and low landing speeds are essential.

As the rapid improvement of aircraft in the years following World War I began to result in progressively higher cruising speeds, landing speeds increased proportionately and airfields had to be made correspondingly larger. To meet the bush fliers' special needs, some aircraft manufacturers introduced a variety of aerodynamic devices, such as special flaps, slots and the like, to improve wing efficiency. Aircraft so fitted were found to possess greatly improved handling characteristics at low speeds and a better load carrying performance for an equivalent installed power.

Through the years, many such aircraft types have been produced and some have made substantial contributions to man's pioneering in the remote regions of the world. There has also been a good deal of military interest in STOL machines for general army transport and aerial reconnaissance duties. Today, sophisticated developments of the basic STOL theme, such as the augmentor wing concept designed by de Havilland Aircraft in Canada, are being applied to much larger aircraft which are designed in such a way that they have both civil and military functions.

Augmentor wing

The augmentor wing concept involves the use of full span slats at the leading edge, to delay the onset of turbulence over the wing's upper surface in an exaggerated nose-up attitude. This is combined with the use of full span double slotted flaps at the trailing edge, capable of downward deflection to an angle of 75°. The flaps are divided into three approximately equal length sections on each wing, the outboard sections serving also as ailerons. These are drooped to an angle of 30° when the inboard flaps are extended to 60° deflection and beyond. The drooped ailerons thus add considerably to the flap effect. They have a differential movement of plus or minus 17° on either side of the drooped datum position.

To supplement their effect still further, an air distribution duct system carries air bled from the Rolls-Royce Spey turbofan engines and expels it over and through the trailing edge flaps, to mix with and deflect downward the main airstream over the wing. Chokes in the airflow augmentor system can restrict the fan air outflow area at the flaps to control the lift of the flap system, the outboard chokes can be linked to the pilot's controls, to supplement the roll effect of the ailerons at slow forward speeds, and all four chokes can be used simultaneously as a 'lift dump' at the moment of touch down. In addition, the high pressure fan air is used for a boundary layer control system at the flaps, to delay airflow separation at slow speeds.

Various versions of the system, either in whole or in part, are embodied in current de Havilland Canada designs such as the DHC-5 Buffalo, DHC-6 Twin Otter and DHC-7 Dash-7. The full system, as described here, is being developed by the company in collaboration with the Canadian and US governments in a DHC-5D Buffalo procured by the US National Aeronautics and Space Administration. NASA's designation of test aircraft is the XC-8A.

STOL aircraft have progressed considerably during the last quarter century. In contrast to the earlier single engined machines which, typically, could carry up to 6 or 7 passengers or a payload of about 1 tonne, the Twin Otter has accommodation for a crew of up to 20 passengers (or a 2-tonne payload), while the Dash-7, with four engines, has accommodation for 50 passengers or over 5 tonnes payload. The military transport Buffalo will carry 41 equipped troops, at speeds up to 420km/h (261mph), or stores up to 8,164kg (18,000lb). It was designed to have a load capability compatible with the Boeing Vertol Chinook helicopter. Like the Chinook, the Buffalo has a rear loading ramp for wheeled vehicles.

The typical take-off distance these very big STOL aircraft need to become airborne is only about 300m (984ft) in still air. Landing runs for those types fitted with fully reversible pitch propellers may be as little as half this distance. While the concept may not be suitable for airliners flying the trunk air routes of the world, there can be little doubt that STOL aircraft will be coming into much more general use on short-haul routes as development proceeds.

Unusual
Aircraft

Research aircraft

The following subject is rather strange and is not usually included in discussions on the evolution and development of aircraft. Some aircraft, however, happen to fall into this category because, while appearing to be fairly conventional, they have been designed to fulfil a specific, but perhaps unusual, need such as fire-fighting.

To recap briefly: during World War II the first of the gas turbine engines were introduced into service. These had been developed independently by Pabst von Ohain in Germany and Frank Whittle in Britain, and the first fruits of their genius became operational at a time when the conventional piston-engine had almost attained the peak of its development.

Aircraft designers had done much to refine airframe structures so that the best possible performance could be achieved. It was clear, however, that the combination of piston-engine and propeller could not be taken much further. In terms of real performance, aircraft such as the British de Havilland Hornet, and Supermarine Spitfire XIX, the United States North American P-51H Mustang and Republic P-47M Thunderbolt, could attain speeds of approximately 760km/h (472mph), 740km/h (460mph), 784km/h (487mph), and 761km/h (473mph) respectively, in level flight at optimum altitude.

To match or better performance in this category, Germany had produced two unusual and interesting aircraft: the Dornier Do 335 *Pfeil* (Arrow), and Me 163 *Komet*. The former relied on a pull-push engine arrangement, with forward engine/tractor propeller, and rear engine/pusher propeller aft of a cruciform tail unit, to provide a maximum speed of 763km/h (474mph). The revolutionary rocket-powered tailless *Komet* could attain a speed of 950km/h (590mph) at optimum altitude, but performance in this category had as its penalty a powered endurance of about 12 minutes.

All of these aircraft could, if allowed, attain speeds in a dive where the airflow over the wing surface was approaching the speed of sound, when aerodynamic shock-waves began to build up as a result of air being compressed ahead of the wing leading-edge. Unless the resulting vibration or shuddering of the airframe structure was at once recognised by the pilot as being potentially dangerous, causing him to reduce airspeed, it was possible for wings or tail units to be torn away.

Extensive research into the aerodynamics of high speed flight, much of which had been conducted in

Germany, had suggested that a very thin-section aerofoil, a swept wing, or a combination of both, could do much to delay the onset of the buffeting, and so allow an aircraft to pass smoothly from subsonic to supersonic flight. In order to prove whether such ideas were practical, the USAAF, Bell Aircraft Corporation and NACA (predecessor of NASA) combined their talents to design an immensely strong aircraft, with very thin slightly swept wings, powered by a four-tube rocket motor that could develop a total thrust of 26·7kN (6,000lb st).

Built by Bell under the designation X-1, the aircraft was launched in unpowered flight on 19 January 1946, transported by a Boeing B-29 'mother plane.' Development was slow because everything was new, and it was not until December 1946 that it was possible to begin the first powered flights. Piloted by Captain Charles Yeager of the USAF, the X-1 was pushed slowly nearer and nearer the critical speed of sound. At Mach 0·94 (Mach 1 is the speed of sound) Yeager found the buffeting so severe that only such a specially constructed aircraft as his could have survived. Finally, on 14 October 1947, Yeager and the X-1 slipped through the 'sound barrier' into smooth supersonic flight.

Three X-1s had been built by Bell, and it was the first of these which was piloted into the pages of aviation history by Charles Yeager. The second was used for high-speed flight research by NACA, the third destroyed in a ground refuelling accident. They were followed by a generally similar X-1A, which had turbo-driven pumps for the rocket fuel instead of a nitrogen pressure system; and an X-1B which was instrumented to research the effects of kinetic heating in high-speed flight. The final version to complete a significant number of research flights was the X-1E, the second of the original X-1s fitted with wings of much reduced thickness/chord ratio, the turbo-driven fuel pumps of the X-1A, plus a knife-edge (special drag reducing) windscreen. The X-1A proved itself in December 1953, when flown at Mach 2·435, and in June of the following year it exceeded an altitude of 27,430m (90,000ft).

Next in the line of research aircraft evolved by Bell was the X-2, with stainless steel swept wings, intended to explore still further the problems of supersonic flight. It had a refined and much more powerful rocket motor (66·7kN: 15,000lb st) and, of considerable importance, the output of this new

A Canadair CL-215 amphibian, which is used as a water-bomber to help prevent or control forest fires in large timber areas.

The Ling Temco Vought XC-142A tilt-wing military aircraft is captured between vertical and horizontal flight. The low forward speed of the helicopter has lead to several different configurations which provide the advantages of vertical take-off and landing with a much higher forward speed. Though valuable for military purposes, few projects have developed beyond the prototype stage.

Left: Lockheed's P2V Neptune, an important maritime patrol aircraft, is equipped with ski landing gear and oversize wing-tip tanks. The Lake LA-4 (*below left*) has the Bell-developed air-cushion landing-gear system. The giant Aero Spacelines Guppy (*below*) is built to airlift bulky cargo loads.

The Lockheed C-130 Hercules cargo aircraft has been employed successfully in a wide range of roles, other than as the plain and simple cargo carrier for which it was originally designed. Here a KC-130R tanker version of the Hercules, in service with the US Marine Corps, fuels two USMC A-4 Skyhawks.

The North American X-15A-2 high speed research aircraft (*above*) has flown at speeds of up to 7,927 km/h (4,534 mph).

Curtiss-Wright engine could be controlled by a throttle. Two were built: the first was destroyed in an accident in May 1954, but the second of these aircraft, which flew for the first time on 18 November 1955, was to climb to 36,640m (120,200ft) and fly at Mach 3·2, before being destroyed in an accident on 27 September 1956.

While Bell had been working on the X-1 and X-2 programmes, the Douglas Aircraft Company had developed their D-558-2 Skyrocket with wings swept at 35 degrees, and a mixed power plant comprising a 13·3kN (3,000 lb st) Westinghouse turbojet engine, supplemented by a 26·7kN (6,000lb st) bi-propellant rocket motor. First flown on 4 February 1948, it was to become the world's first piloted aircraft to exceed Mach 2, on 20 November 1953. It was succeeded by the slim, needle-nosed Douglas X-3, powered by two Westinghouse turbojets which could each develop 18·7kN (4,200lb st) with afterburning. It was provided with instruments to record surface temperatures, airframe stresses and loads, and had more than 800 minute holes in the airframe skin with associated sensors to provide information on surface pressures throughout the flight regime. First flown on 20 October 1952, it was tested by

High-performance sailplanes, such as that pictured (*top left*), are often built entirely of smooth and resilient glassfibre. Dr Paul McReady's Gossamer Albatross (*left*) was used for the first man-powered flight across the English Channel in 1979.

Douglas before being turned over to NACA for advanced research in late 1953, and its test programme terminated in May 1956.

The most ambitious of the American high-speed research programmes had been initiated a year earlier, when North American Aviation received a joint USAF, USN and NACA contract to build three examples of an advanced high-speed research aircraft designated X-15. First powered flight, by the second aircraft, was made on 17 September 1959, and when this was involved in an accident, on 9 November 1962, the opportunity was taken to rebuild it to permit more advanced hypersonic research. It then became the X-15A-2, and was powered by a 253·5kN (57,000lb st) Thiokol rocket motor, coated with an ablative material to maintain the temperature of the structure well below the design maximum of 1,200 degrees F (649 degrees C), and had 12 jet nozzles to provide control during flight above the effective atmosphere. By the time that the X-15 programme terminated, in November 1968, the X-15A-2 had been flown to a height of 107·95km (67·08 miles) and had managed to achieve a speed of Mach 6·72 (7,297km/h; 4,534mph).

Research of this nature made it possible for North American Aviation to design an intercontinental-range strategic bomber (nuclear or conventional), able to travel the entire distance to and from its target at Mach 3. Changes in weapons policy brought an end to this project before it really got under way, but it was decided to build just two of these aircraft for research purposes under the designation XB-70A Valkyrie. This large aircraft had some very advanced features, including a very thin-section delta wing with hydraulically-folding wingtips to improve stability and manoeuvrability in low-altitude supersonic and high-altitude Mach 3 cruising flight. There was a large canard foreplane just aft of the flight deck, 12 elevons (composite elevators ailerons) to provide control in roll and pitch, and six 137·8kN (31,000lb st) General Electric turbojet engines with afterburning.

During this period of American high-speed research, designers in Britain had been devoting their efforts to achieving similar success, but had been rather limited by a penurious budget. Thus, the second and third examples of the de Havilland 108, a swept-wing research aircraft intended to provide aerodynamic data for the Comet 1 commercial transport, were intended specifically to investigate the potential of the wing design. It was while flying the second D. H. 108, on 27 September 1946, that Geoffrey de Havilland junior was killed. The third aircraft, built to continue this high-speed research, had a number of structural changes and was provided with powered controls. In this aircraft, on 6

The best high-speed high-altitude performer is the Lockheed SR-71 reconnaissance aircraft (*below and above right*) which is constructed mainly of titanium alloy to maintain structural integrity at high skin temperatures.

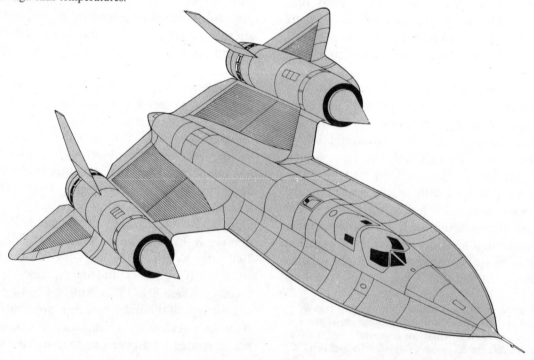

September 1948, pilot John Derry exceeded the speed of sound in a dive, and the D. H. 108 thus became the first gas turbine-powered aircraft to exceed a speed of Mach 1. In the following decade, the Fairey Aviation Company built two examples of a delta wing high-speed research aircraft which was known as the Fairey Delta 2.

The first of these took to the air on 6 October 1954, and in this aircraft on 10 March 1956 test pilot Peter Twiss set the world's first speed record in excess of 1,609km/h (1,000mph), averaging 1,822km/h (1,132mph) in two measured flights. A pioneering feature of the Delta 2 was a nose section which could be drooped to improve the pilot's view over the long needle-nose for taxiing, take-off and landing, and this innovation was used subsequently for the Anglo/French Concorde supersonic airliner. Before this however, the feature was retained when the Delta 2 was almost completely rebuilt by British Aircraft Corporations' Filton Division to produce the BAC 221, powered by a 62·3kN (14,000lb st) Rolls-Royce Avon RA-28R afterburning turbojet. Used from 1 May 1964 until 9 June 1973 for research

into a number of aspects of subsonic and supersonic flight, it was flown as fast as Mach 1·6 during this programme.

Supersonic speeds within the Mach 1 to Mach 3 range were found to be the most practical for operation within the earth's atmosphere. Even at the higher end of this scale the airframe must be constructed of materials such as titanium and stainless steel to enable it to withstand the result of sustained kinetic heating to temperatures as high as 427 degrees C (800 degrees F). This factor explains the choice of a Mach 2 to Mach 2·4 speed limitation for the Concorde and Russian Tupolev Tu-144, both of which have a structure which is basically conventional, with only a limited application of stainless steel and titanium. More costly and complicated construction in these latter metals and associated alloys is used only for special purpose military aircraft. Nevertheless, costly or otherwise, research aircraft such as the foregoing have banished the 'sound barrier,' making it possible for both civil and military aircraft to fly at speeds in excess of Mach 1, routinely and safely.

The space shuttle

Scientists, engineers and designers concerned with the American space programme appreciated at an early date the desirability of a reusable space shuttle, which would be able to ferry men and materials to and fro, between Earth and an orbiting space station. This required a vehicle which could be launched from Earth, fly and manoeuvre in space, and re-enter into the Earth's atmosphere to land conventionally. The first of the strange-looking lifting-body aircraft designed to investigate this concept was Martin Marietta's small unmanned X-23A. Launched by Atlas SLV-3 (Space Launch Vehicle), three of these wingless models were able to prove the feasibility of such a vehicle, which relies upon the shape of the airframe structure to supply the necessary lift for flight within the atmosphere.

First of the full-size test vehicles was the Northrop/NASA M2-F2, which in July 1966 was piloted on its first unpowered flight, released from beneath a B-52 'mother plane' at a height of 13,720m (45,000ft), making a successful landing after a gliding flight of about four minutes. Subsequently, and then designated M2-F3 after some reconstruction, this aircraft made its first powered flight, on 25 November 1970. On this occasion it was flown at a maximum speed of Mach 0·8 at 16,150m (53,000ft). At much the same period Northrop built and carried out 25 powered flights with a generally similar vehicle designated Northrop/NASA HL-10. The main difference between this and the M2-F2/3 was in the lifting-body aerodynamic section. In the HL-10 it was of D form, with the straight portion of the D forming the fuselage undersurface: the M2-F2/3 configuration had the straight portion forming the upper surface.

While the above-mentioned work was in progress, Martin Marietta was busy building for the USAF a lifting-body aircraft based on the small X-23A. This had the designation X-24A, and was of the same general configuration as the M2-F2. It was powered by a 35·6kN (8,000lb st) Thiokol XLRII turbo-rocket engine, and was launched into powered flight for the first time on 19 March 1970.. When its programme ended it had made a total of 28 flights, had attained a maximum speed of Mach 1·62, and had flown to a height of 21,770m (71,407ft). At that stage the X-24A was stripped down and rebuilt, appearing with a refined external shape, the most conspicuous change being replacement of the blunt fuselage nose by a thin tapering structure which extended the overall length by some 3·96m (13ft). A total of thirteen powered flights were made by this

vehicle during 1975, a maximum speed of Mach 1·72 being recorded and an altitude of 22,590m (74,132ft).

The research work carried out by the Martin Marietta and Northrop lifting-bodies has helped speed the design and construction of the first Space Shuttle Orbiter, *Enterprise*. This has already been launched in unpowered flight from heights of up to 7,620m (25,000ft), carried to that altitude by a specially modified Boeing 747 Shuttle Carrier Aircraft. The *Enterprise* is expected to be launched in orbital flight in 1980, and a second Orbiter is scheduled to make its first operational orbital mission in the latter half of the 1980s, or failing that, the early 1990s.

Research aircraft are used mainly to test new aerodynamic concepts. In the 34 years since World War 2 ended, there has been a constant stream of projects aimed at achieving far better performance without the need for ever-increasing output from power plants. Thus, the improved engines combined with aerodynamic advances, have made possible new generations of civil and military aircraft with performances which would have seemed unbelievable in the pre-war days of the piston-engine.

A Boeing 747 'jumbo jet' converted to serve as a transporter for NASA's Space Shuttle Orbiter *Enterprise*, is shown as it takes off during the early tests of this new space vehicle.

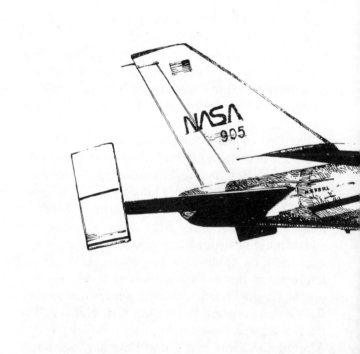

Ram wing

Among the concepts which have been or are being explored is the ram wing. In the late 1950s, the American Dr W. R. Bertelsen had noted that with most VTOL aircraft, and especially those of the deflected slipstream type, considerably more payload could be carried when the aircraft was operating in ground effect, i.e. when on the cushion of air compressed against the ground. To exploit this benefit he developed a series of ground-effect vehicles which he called Aeromobiles, typical of the air-cushion vehicles developed at about that period by many countries. All of these required special fans to provide lift, and Dr Bertelsen designed what he called a 'ram wing', a conventional aerofoil mounted at a high angle of attack between two long side walls, and powered by a conventional two-blade propeller just forward of this wing. The engine was mounted so that the propeller's thrust line of about 24 degrees was similar to the angle of attack of the aerofoil. Bertelsen demonstrated that with this system,

which he called an Arcopter, it was possible to become airborne with zero forward speed and that, as forward speed increased, so did hovering height.

Though an improvement on the then normal type of air-cushion vehicle in respect of payload, it has since become outdated as the air-cushion concept has developed. Far more significant for an aircraft in this class is the ground effect aircraft which was originated by the German aerodynamicist and designer Dr Alexander Lippisch. This was still undergoing development by Rhein-Flugzeugbau GmbH in late 1978, and the X-114 Aerofoilcraft, as it is known in its latest form, has a reversed delta wing with considerable anhedral to give stability in roll. Thus, when the aircraft is flown at low altitude, over land or water, the oncoming airflow creates a cushion of air between the aircraft and the surface. This air cushion not only provides a very considerable increase in lift, but also increases stability

and reduces drag, offering the benefit of lower operating costs. With the X-114, which is powered by a 149kW (200hp) Lycoming piston-engine, maximum cruising speed in ground effect is 150km/h (93mph), but the aircraft can also fly conventionally to avoid high obstacles, though with a reduced lift coefficient.

Flying Mattress

Another approach to low-altitude flight was provided by the Piasecki 'Flying Jeep' or 'Flying Mattress,' two examples of which were built under the designation VZ-8P for evaluation by the US Army. Each of these comprised a low fuselage or chassis structure, mounted on three wheels, with fore and aft horizontal rotors enclosed in ducts and shielded completely on all sides by the fuselage structure. Space was effectively provided between the rotor ducts for twin-engines (either of which could drive both rotors in emergency), as well as for a pilot and three passengers. Intended for operation at low level, it was claimed that the VZ-8P could be climbed high enough to avoid most natural obstacles in the US, and could attain speeds of up to 240km/h (150mph). Like many other low-altitude research vehicles, the 'Flying Mattress' was not built on a production basis.

Of far greater value, and likely to be of growing importance, are new applications of known techniques, to improve lift and low-speed control. They include various means of achieving Boundary Layer Control (BLC); NASA's XC-8A augmentor wing jet STOL research aircraft representing a current application of this, and other, advanced ideas. Developed in conjunction with Canada's Department of Industry, Trade and Commerce, the XC-8A uses BLC to ensure that the thin layer of air in contact with the fuselage skin continues to flow smoothly at all speeds, reducing drag to a minimum. In addition, air is blown at high pressure across the ailerons and trailing-edge flaps, maintaining a smooth airstream over their surfaces at all times, providing a very considerable increase in lift.

In the search to discover ways of reducing the take-off and landing run of military transport aircraft, the USAF initiated a prototype fly-off programme in which Boeing and McDonnell Douglas have taken part with aircraft designated YC-14 and YC-15 respectively. Each company adopted a different approach, Boeing's YC-14 using upper surface blowing to ensure that the airflow over the wing clings to the surface of wing and flaps, even when the latter are lowered. This exploits the long known

North American's XB-70A Mach 3 research aircraft. Only two were built after changes in US government policy ended the original plans to develop this design as a supersonic intercontinental strategic bomber.

Coanda effect, with two powerful turbofan engines mounted so that their efflux is directed over the surface of the wing/flap system and directed downwards to generate direct powered lift. The McDonnell Douglas YC-15 uses a simpler technique, in which the two-section trailing-edge flaps are lowered directly into the efflux of the four underwing-mounted turbofan engines, the high-velocity airflow passing through the slots in the flaps and acting in precisely the same way as the normal airstream. The fact that this jet efflux, or artificial airstream, is at a much higher velocity than that which would prevail at take-off and landing speeds, means that the lift of the entire wing system is enhanced.

Perhaps the strangest of all the research aircraft is one which has yet to fly. This is the AD-1 oblique-wing concept which has originated from NASA's Dryden Flight Research Center. In such an aircraft the wing is mounted conventionally for take-off, landing and all other low-speed operations. For high-speed flight the wing can be pivoted to an oblique angle of up to 60 degrees, and wind tunnel testing has indicated that reduced drag, with a resulting increase in speed and range, should result. It seems possible at this early stage that such a concept may enable civil transport aircraft to fly at speeds of up to 1,610km/h (1,000mph) at approximately half the fuel consumption of the Anglo-French and Russian supersonic civil transport aircraft currently in service. Even more advanced ideas are involved in another NASA research programme, for with a need to investigate alternative sources of fuel for aircraft propulsion, study is being made of an electrically-driven light aircraft which would derive its power from a microwave beam transmitted from the ground.

The history of aviation has seen many strange designs brought to fruition. Most were evolved in the early stages of development when, for example, an aircraft with sufficient fuel for a particular range was too heavy to lift itself into the air. This brought aircraft such as the British Short-Mayo composite, in which the *Maia* carrier lifted the fully-fuelled *Mercury* into the air, or the aerodynamic auxiliary fuel tank towed by some German aircraft to enable them to increase their range.

Giant transport

Modern airframe and engine design has eliminated 'oddballs' of that kind, but two strange specialised aircraft are still to be seen in the world's sky. The first was evolved by Aero Spacelines in America, where there was a requirement in the 1960s for the air transport of large booster stages and other specialised and outsize equipment used in the nation's space programme. Outsize cargo carriers were built by conversion of Boeing Stratocruisers or C-97 transport aircraft, the resulting Pregnant Guppy, Super Guppy and Mini Guppy aircraft being used for the carriage of cargo which could be as large as the third stage of a Saturn V launch vehicle. Largest was the Super Guppy, which had a cargo compartment 33·17m (108ft 10in) long and 7·62m (25ft) in diameter.

Fire fighters

The remaining 'aircraft' is the Canadair CL-215 twin-engined amphibian, designed primarily for fire-fighting operations. These aircraft have onboard tankage for 5,346 litres (1,176 Imp gallons) of water, or water plus fire retardant, and on occasions single aircraft operating as water-bombers have dropped as much as 545,520 litres (120,000 Imp gallons) in one day. Refuelling of the water tanks can be accomplished on the ground in approximately 90 seconds with the requisite equipment, but by dropping low to skim a suitable water surface the latest water scoop can fill the tanks in ten seconds. CL-215s are operated by Canada, France, Greece and Spain, and have proved invaluable in dealing with large-scale forest fires. More recent tests have shown that, with the use of a foaming retardant, these CL-215s can be equally effective against oil fires.

This 28·60m (93ft 10in) span amphibian, perhaps more than any other aircraft mentioned in this section, would be the one chosen by aviation pioneers as being most representative of their ideals. The role which they planned for their creation was a peaceful one: the knowledge that it could also fulfil some vital task would have made their work and sacrifices seem very worthwhile.

Gliders and Gliding

Origins

After the Montgolfier brothers' lighter-than-air successes in 1783, ballooning became an acceptable way to fly in the 19th century. However it fell to an Englishman, Sir George Cayley, to lay down sound principles of aerodynamics and especially 'downhill' flight or gliding. He built a glider that could be towed into the air, and in 1853 Sir George's coachman made the flight described in the first chapter of this book. Although this was a brief flight which resulted in a crash landing, it was a quite soundly designed and well built machine, for in the 1970s Derek Piggott, one of the best-known gliding instructors and authors, made several successful flights for a television film in a superb replica of Cayley's machine.

as a sport had caught on, one of the principal centres being a site at the Wasserkuppe in the Rhön district of Germany, the country which soon took the lead in world gliding developments. This is a lead it has maintained to this day, apart from a few years following World War II. Flights of a few seconds in 1920 had progressed to over three hours in 1922 – all involved flying parallel to the hillside and soaring in 'slope' or 'ridge' lift, which is the simplest means of keeping a glider up. In 1922 the *Daily Mail* sponsored a gliding contest at Itford, in Sussex. The prize went to a contestant who managed to stay in the air for over three hours.

Through the 1920s glider design and performance gradually improved, although gliding ambitions were thwarted by the belief that it was not possible to

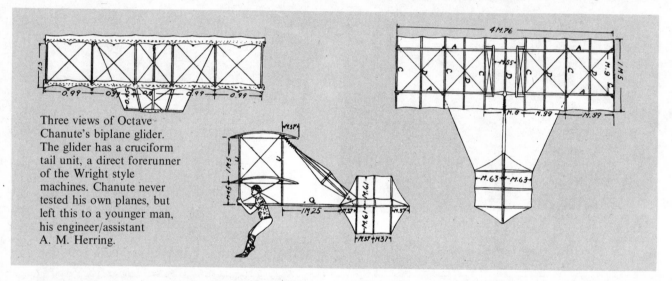

Three views of Octave Chanute's biplane glider. The glider has a cruciform tail unit, a direct forerunner of the Wright style machines. Chanute never tested his own planes, but left this to a younger man, his engineer/assistant A. M. Herring.

By the end of the 19th century at least six gliding pioneers had been launched from platforms, hills and sometimes balloons in foot-launched gliders that were akin to today's hang gliders. The German Otto Lilienthal controlled his glider by shifting his weight (just as a modern hang glider pilot does). At the beginning of the 20th century Montgomery used wing warping and weight shift. The Scot, Pilcher, the French-born American, Chanute, and the later pioneers, the Wright brothers, all used tail control surfaces similar in principle to those used today. These pioneers made thousands of successfully controlled glides, progressively improving control and performance. Montgomery even flew for 12.9km (8 miles) on one flight.

The Wright brothers' biplane gliders led to their historic first powered flight in America in 1903, but the rapid development of powered aircraft and then World War I hindered any significant progress in glider design. However, by the early 1920s gliding

stay up by any means other than slope lift. Pilots know that birds soared by using other upcurrents or 'thermals,' but it was thought impossible for a human-flown glider to use them.

Towards the end of the decade the variometer – the glider pilot's instrument which indicates whether he is in rising or sinking air – was invented, and it led to an understanding of vertical air movement which set pilots free from hills. In 1929 the well-known German pioneer Kronfeld climbed to over 1,830m (6,000ft), by venturing into a thunderstorm – a hazardous flight because his Wien glider had neither blind-flying instruments nor airbrakes to prevent the glider exceeding its maximum permitted speed.

The British Gliding Association was formed in 1929, as was the London Gliding Club which set up at a site near its present airfield, at the foot of Dunstable Downs. Pilots were soaring in thermals from Dunstable by the early 1930s, and on one historic day in 1934 three cross-country flights were

The German Otto Lilienthal made many flights with his gliders, and meticulously recorded data for the benefit of all aviators.

made from Dunstable. The longest, of 88.5km (55 miles), was the first of many records to be established by a great pioneer of the British gliding movement, Philip Wills.

Soon another form of lift was discovered: the standing wave phenomenon, in which wave-like motions of rising and sinking air are produced in the lee of hills or mountains roughly perpendicular to a reasonably strong wind. In 1933 two Germans soared in such a wave to nearly 1,524m (5,000ft) and in 1937 2,133m (7,000ft) was achieved near the later famous gliding site at the Long Mynd in Shropshire, England.

During the 1930s thermal skills and cloud flying steadily progressed, so much so that gliders could then fly for hundreds of miles, and reach heights of many thousands of feet. Blind-flying instruments were developed, and also airbrakes to limit speeds, should a pilot lose control in cloud and enter a steep dive. By the outbreak of World War II a popular top-class glider like the gull-winged German Minimoa of all wood-and-fabric construction, had a best glide ratio of 1:26 at a speed of 85km/h (53mph). Glide ratio is the basic criterion by which you start to assess a glider's performance. A ratio of 1:26 means it descends 0·3m (1ft) every 7·92m (26ft).

Gliding as a sport in Europe naturally took a back seat during the World War II but some benefits accrued from the war's busy technical developments, such as waterproof glues and increased knowledge about the low-drag aerofoil. Immediately after the war, the influence of pre-war German designs was still prevalent. The British manufacturer Elliotts of Newbury took the German Meise design and turned it into the Olympia, to be Britain's most popular single-seater for several years.

Other developments from the pre-war German Grunau Baby and Focke-Wulf Weihe were also produced in other countries. The British Slingsby Sky in which Philip Wills became world champion in 1952, and which boasted many other competition successes around that time, was very similar in appearance to the Weihe against which it often competed in the early 1950s.

Developments in aerofoil design and surface finish were a major step forward in the 1950s, with the Ross-Johnson RJ-5 achieving a then almost unbelievable glide ratio of nearly 1:40. An accurate, smooth shape on the finished wing, thereby giving less drag, became the main aim of many designs, the popular British Slingsby Skylark series achieved this by using thicker gaboon plywood as covering instead of the less effective thinner birch of earlier gliders.

In 1957 the world's first glider to be made of glassfibre was produced (glassfibre enables a superbly smooth surface finish to be achieved). Called the *Phoenix*, it was built by one of Germany's university groups (responsible for many advanced gliding designs) and had a best glide ratio of almost 1:40. Another modern development, that of a semi-reclined seating position to reduce the glider's frontal area, was seen in the Polish Zefir and Foka of 1960.

In 1964 another German university group produced the D-36 Circe, a forerunner of most modern top performers, with glassfibre construction, flaps as well as airbrakes, retractable landing wheel, T-tail and the sort of performance almost taken for granted today: a best glide ratio of 1:44 at 93km/h (58mph).

Early in the 1970s the two-seat German SB-10

Schirokko managed to achieve a glide ratio of 1:50, but with an enormous wingspan of over 26m (85ft), which is a little impractical for normal gliding operations on the ground and rather unwieldy in the air. The great compromise in gliding is to be able to fly slowly in weak lift, but also to fly fast at a flat glide ratio between thermals. Recent developments have featured variable geometry – changing the shape of the wings in flight. In 1975 Germany's FS-29 became the first glider in the world with telescopic wings, able to vary its wingspan from 13–19m (43–63ft) and at the 1978 world championships the German SB-11 (which won its class) caused a lot of worried faces amongst other competitors with its sophisticated Fowler flaps, which slide out from the wings to change their area in flight. When the flaps are out, it has a stalling speed of only 58km/h (36mph), which means it can turn more sharply than most conventional gliders and easily get into the strongest lift in the centre of a thermal, thereby out-climbing them. But it will probably be a long time before such expensive developments are seen on ordinary club gliders. The SB-11 is said to have taken 20,000 man hours to build.

Aerotowing, the launch of a sailplane by powered aircraft, is both the most effective and costly method of launch, but it can haul the sailplane into an area of 'lift' for release.

Launching methods

There are currently three widely used methods of launching gliders: winching, towing behind a car and towing by a powered aircraft. The fourth, 'bungee' or catapult launching, is still done at a few gliding sites situated on hill-tops, but is very rare today.

Winching is a popular method in the UK, partly because it is cheap, given good equipment, but largely because it can be done from a launching site with a rough or undulating surface. All you need is a reasonably smooth area just a few hundred feet long from which the glider can take off.

The best winches are specially designed for the purpose, and have engines of about 100hp driving drums of stranded steel cable with a breaking strain of about 127,000kg (2,500cwt). Two drums on one winch are preferable, since the car or truck which tows the cables out to the gliders can thus tow two out in one journey. One glider can then be launched immediately after the other. the winch-driver has a cage round him to protect him against the possibility of a cable snapping and catching him, and a guillotine mechanism for cutting the cable in an emergency, should the glider be unable to release it.

Winching is not an easy job, requiring judgement and a skilled touch on the throttle to achieve a good

launch. There are two standard airborne signals for use by the glider pilot: waggling the wings means 'too slow', yawing the nose from side to side means 'too fast.' The signals used at the launch point and relayed on the glider pilot's command are standard for all launching methods. Once the cable is hooked on, 'take up slack' is indicated by waving the arm from side to side below shoulder height, and 'all out' by moving the arm similarly above the head. Acceleration is fairly rapid on a winch launch, and as the glider becomes airborne it is held down in a level attitude close to the ground for just a few seconds to pick up speed, then the stick is eased back to lift the nose into the climb. The attitude during a good winch launch is quite a steep one. Towards the top of the launch the cable will start to pull the nose of the glider downwards, and it is then time to release the cable. On a good day with a reasonable wind blowing, a glider can reach a height of 300m (1,000ft) or about a third of the total length of the winch cable.

Towing behind a car, usually called auto-towing, needs a smooth surface such as a hard runway, since the cable is not wound in and the car has to accelerate to and maintain quite a high speed. It again requires quite a lot of skill from the driver, and a sturdy car which needs to be just as powerful as the winch. Piano wire is often used as a cable. In simple auto-towing the driver has to keep glancing backwards to see how the glider is doing, but a better system is reverse pulley auto-towing. In this method the car driver is able to drive towards the glider since he is pulling the cable round a pulley.

Being tugged into the air behind a powered aeroplane – aerotowing – is really the most efficient method of launching, since it requires the least effort from helpers on the ground, and gliders can be taken to a good height – usually 610m (2,000ft) – and to areas of known lift. But, like most, luxuries it has one drawback, expense. The tow rope is usually made of nylon, is about 45·72m (150ft) long, and like all launch cables has a weak link at either end so that it can break before an undue load is exerted on the glider. Ordinary light aeroplanes, of about 150hp, are used as glider tugs.

The glider is usually airborne several seconds before the tug, and once the tug takes off the aim is to fly behind and either above or below it to avoid the turbulence of the tug's slipstream. At the pre-arranged height the rope is released from the glider,

the tow to 610m (2,000ft) will probably have taken four to five minutes.

Bungee launching is about the quickest way to get airborne in a glider. A length of rubber rope is hooked on to the glider's nose, and at the end of the rope about three people walk or run down the side of the hill until the rope is sufficiently tight. On a signal from the person holding the glider's wingtip, another helper hanging on to the glider's tail releases it, and the glider shoots forward into the air. Since bungee-launching is only attempted from a hill into quite a strong wind, the glider can then turn to fly parallel to the hill and immediately start hill-soaring. It is the cheapest way to launch a glider, but the most strenuous.

Gliding and soaring

Most people who learn to glide do so by joining a club, and many of them get their introduction to the sport on a one or two-week holiday course run by the larger, full-time clubs. A course of this kind usually begins on a Monday morning after the ordinary gliding club members have finished their flying for the weekend. Up to a dozen course members will meet their instructor to see the glider or motorglider they will be flying, and will then be given an informal lecture on the principles of gliding. Out on the airfield they are shown how to park and handle gliders on the ground (a very important consideration, for gliders can be fairly easily blown about by the wind and damaged in other ways). They will then take it in turns to fly, and be introduced to one of the biggest factors in all forms of gliding, waiting. Any flying sport involves waiting for the weather to improve, and in gliding at club level there is also a lot of time spent in waiting for your turn to fly, since club gliders are hardly ever booked (unlike powered planes) and it is usually a case of 'first come, first served' on a list of names for a particular glider. On a gliding course one is usually guaranteed a minimum number of flights, weather permitting. In just a week's course it is unlikely that the average beginner will reach the solo-flying stage, although he might well do so in a fortnight if he has previous flying experience of some kind.

A full-time course, with daily flying, obviously leads to an efficient rate of progress since everything is constantly fresh in the memory from the previous flight. But many people have to learn to glide at the weekends, with perhaps just one to three flights a week, and it is usually a fairly long-drawn-out process, needing a lot of patience to sort out the determined, hard-working enthusiast from the

person who is just 'having a go.' A winch or car launch on a 'flat' day when the weather is not suitable for soaring will allow a flight of only four to five minutes, and the glider then has to be pushed back to the launching side of the airfield. Aerotowing behind an aeroplane will give a flight of perhaps 15 minutes on a non-soaring day. This is usually more than three times as expensive as a winch launch.

On the first flight a beginner is usually shown little more than simply what it is like to be up in the air in a glider, and how the local landmarks look from above. On the next flight, he will probably be shown how to control the glider. Soon he will learn to fly by attitude – keeping the nose in a constant position relative to the horizon, which keeps the airspeed constant. He also learns to coordinate stick and rudder movements either to fly the glider in a straight line or to turn it. This is a vital exercise in gliding, since soaring in thermals depends very much on really accurate turning. Another vital exercise is stalling (deliberately flying so slowly that the wings provide insufficient lift) for when the glider pilot is soaring he is often flying fairly close to his stalling speed.

By this time he will be handling the glider during the take-off and launch (the instructor, of course, has his own set of dual controls) and learning gradually to position himself for the approach and landing. Control of height on the approach to land is achieved with the spoilers or airbrakes mounted on the wing and controlled by a cockpit lever. When moved back towards the pilot, this opens the airbrakes and steepens the glide angle for a more rapid loss of height.

All the time spent on the ground waiting for and manhandling gliders leads to quite a high drop-out rate amongst people who start training, but this has decreased in recent years with the advent in some countries of the motor-glider. An aircraft like the German Falke can be started up, taxied out and taken-off just like an ordinary powered aircraft, and once in the air it handles similarly to a two-seat training glider. You can even stop the engine in the air, soar like an ordinary glider, and start it again when you wish. In the larger European clubs most students learn in motor-gliders first, and then go on to gliders, and the overall training process is much quicker.

On his early solo flights the new pilot gains the first of the international badges awarded by the FAI (Federation Aéronautique Internationale). These are the A and B certificates, followed by the C, which requires at least five minutes' soaring flight.

There is then a longish gap in the badge system to the first proper international soaring certificate, the Silver C. In Britain this gap is bridged by the Bronze C, requiring two soaring flights of at least 30 minutes each (60 minutes if launched by aerotow), flying tests and written examinations. The FAI Silver C requires a 50km (31 miles) cross-country flight, 1,000m (3,280ft) height gain and a five-hour duration flight; the Gold C calls for a 300km (186 miles) flight and 3,000m (9,845ft) height gain, to which the really accomplished pilot can add a diamond each for a 500km (310 miles) flight, 'goal' flight of 300km (186 miles) around a declared point and a 5,000m (16,400ft) height gain. There are also FAI badges for flights of 1,000km (621 miles).

The simplest soaring technique is, of course, ridge lift parallel to a hill, cliff or mountain, but this is not as simple as it seems, since it demands a high degree of skill. Although excellent for long flights, it limits the pilot to local soaring, unless he is over the kind of high ridges, sometimes hundreds of kilometres long found in parts of the USA.

Thermal soaring is much more difficult and the ability to contact thermals, find the strongest part and then climb as quickly as possible is a large part of the championship pilot's skill. As well as a good variometer (or two), he needs experience and the ability to shift his turns exactly so as to reach the strong core of a thermal, and not waste time circling, for the life of an active thermal upcurrent may be only a few minutes. Thermals frequently feed into cumulus (cauliflower-shaped) clouds but sometimes the thermals produce no clouds at all, and the glider pilot then finds it more difficult to find lift. The modern glassfibre sailplane is often flying so fast that on finding lift the pilot can zoom straight up, gaining height without stopping to circle. This is called 'dolphining'.

Another and rarer form of lift is standing wave. As mentioned previously this is found downwind of hills and mountains when they are roughly perpendicular to a reasonable wind and the upper air meets certain critical conditions. Sometimes marked by stationary lenticular (lens-shaped) clouds, the up-going part of each wave produces smooth lift, which has taken a glider to 14,020m (46,000ft), and produced climbs to over 9,144m (30,000ft), even from the comparatively low Scottish mountains. In simple terms, wave technique is similar to ridge soaring in that you fly parallel to the mountain along the line of lift, but in very strong winds you can sometimes point directly into the wind and climb almost vertically upwards as though in an elevator.

Wave systems also have corresponding down-going areas, of course, and the pilot sometimes has to go through these to get into the lift, as well as cope with severe turbulence at low levels, known as the 'rotor.'

The ambitious glider pilot will want to compete in gliding contests. Most countries have their own regional and national contests, culminating in a world championship every two years. There are now three contest classes: Open (with no limitations), Standard and 15m (49ft), (the last two being limited to gliders of 15 metres span with various other specifications). Most of the daily tasks set involve flying round a predetermined circular route (or sometimes a triangular route), or even an out-and-return, to and from a declared point. This way many competitors manage to get back to base and avoid landing in a field. The old task of 'free distance,' which resulted in top pilots landing hundreds of miles away is now almost unheard of, and for the sake of the ground crews who tow the gliders back in, this is probably a good thing. On one of the best days at the 1978 world championships, winning pilots averaged over 113km/h (70mph). If they had gone in a straight line at that speed all day, you can imagine what a long and expensive business it would have been to retrieve them afterwards.

High performance sailplanes
Single-seat sailplanes are currently grouped into three competition classes by the Fédération Aéronautique Internationale, the international sport aviation body, and these three classes are also used to describe glider specifications outside the comparatively small world of competitions. Taking them in ascending order of sophistication and expense, the first is the Standard Class, which calls for a maximum wingspan of 15 metres (49ft) and allows the carriage of water ballast that can be jettisoned in flight. Then comes the 15 Metre Class, which has the same span limitation but allows flaps as well as water ballast (the purpose of both flaps and water ballast will be described later). In the Open Class, the top performance category, there are no limitations on wing span width, but the span is rarely more than 20 metres (65ft) because beyond this size a sailplane is liable to become unwieldy both on the ground and in the air, as well as difficult and very expensive to construct. Even a new Standard Class sailplane costs some £10–12,000.

Top class sailplanes have very slender, smooth lines that enable them to slip through the air as fast as old biplanes. The slim fuselage curves elegantly around the cockpit area in a tadpole shape, and behind the wing it narrows to a very small diameter before rising to the fin, usually carrying the tailplane on top of it. 'T-tails' have certain aerodynamic advantages and also protect the tailplane from damage by long crops in the event of a field landing. The single wheel retracts completely into the fuselage after take off, so that only the little tailwheel or tailskid protrudes in flight – and on the latest British sailplane, the Vickers-Slingsby Vega, even the tailwheel retracts. As well as the conventional aircraft controls – ailerons, elevator and rudder – there are airbrakes which slide out at right angles from the top and bottom surfaces of the wings to create drag and produce a steep descent angle for landing. The rear portions of the wings take the form of flaps which can be raised or lowered by the pilot to alter the lifting characteristics of the wings in flight. For high-speed straight glides between thermals he raises the flaps, and then lowers them for a reduced stalling speed when he needs to circle slowly. When the flaps come down fully they, too, produce drag like the airbrakes for a steep landing approach into a small space. A few sailplanes have a parachute in the tail which is released to give extra drag on the landing approach.

Tanks are situated inside the wings and can hold over 90kg (200lb) of water ballast. This is carried to vary the wing loading. Extra weight, or a higher wing loading, enables the sailplane to fly faster for the same glide angle, thereby improving its cross-country speed. Later in the day, or at any other time when the pilot finds himself having to use weak lift, he can dump the water ballast, thereby decreasing his stalling speed, which will allow him to fly and circle much more slowly.

When you sit in a cockpit the first thing you notice is that you are lying back in a semi-reclining position. As well as reducing the shape of the fuselage for low-drag purposes, this makes for comfort, since on a good day you might be airborne for eight hours or so. The cushion or seat is adjustable, and so are the rudder pedals. The control column, or stick, is conveniently placed between your legs, with a little lever for the wheelbrake below the handgrip. A largish knob or handle, painted yellow by tradition, operates the cable release. Usually on the side of the cockpit is the lever which raises or lowers the undercarriage leg.

Two prominent controls, not too far apart, are the airbrake lever and flap lever. The former is moved back, towards you, to open the airbrakes, and forwards to close them. The flap lever moves forwards to raise the flaps to negative angles for high speeds, and back towards you to lower them for slow speed gliding. In the 'fully back' position the flaps are usually fully down for landing. There is also a small elevator trimmer knob in the cockpit which you adjust so as to relieve the load you feel on the stick when flying for any length of time at one speed. This makes accurate flying much easier.

The instrument panel contains an airspeed indicator, an altimeter to tell you how high you are, and a turn and slip indicator, the basic blind-flying instrument which tells you when and how much you are turning in cloud. Often there is also an artificial horizon, similar to that in most modern light aircraft, which makes instrument flying very much easier. There are probably two variometers to indicate when you are in rising or sinking air, and at what rate the air is going up or down. One of them is electrically operated, and linked to an audio system to produce a note which alters its pitch according to the strength of lift you are in.

And that is not the only sound that intrudes into the cockpit. The silent reverie of your flight is liable to be punctuated at any time by a voice over the radio. Most countries reserve certain radio frequencies which glider pilots and their ground crews can use to converse with each other, although an increasing number of sailplanes carry more sophisticated radios that can be tuned into any airfield or air traffic unit, just like powered aircraft.

Hang gliding

Hang gliding is about as close as man will ever get to the sensations of flying like a bird. It has become an extremely popular sport, and there are probably 60,000 enthusiasts throughout the world.

The first gliders flown by Lilienthal and other pioneers were hang gliders, but the modern kite, which in its basic delta-shaped form is known usually as a Rogallo, takes its name from the American Francis Rogallo, who developed a flexible kite originally as a means of recovering space capsules. Most pilots still train on the original conical type of Rogallo, but nowadays there are much more sophisticated developments with longer, slender wings, and the finest hang gliders are really very light editions of conventional gliders, often with a full set of flying controls.

The basic Rogallo is a flexible sail stretched over a frame, which needs a positive angle of attack to stay inflated by the airflow. The pilot controls it entirely by shifting his weight and by moving a control frame, or bar. To pitch down or descend, he moves the bar towards him, or pushes it away if he wants to

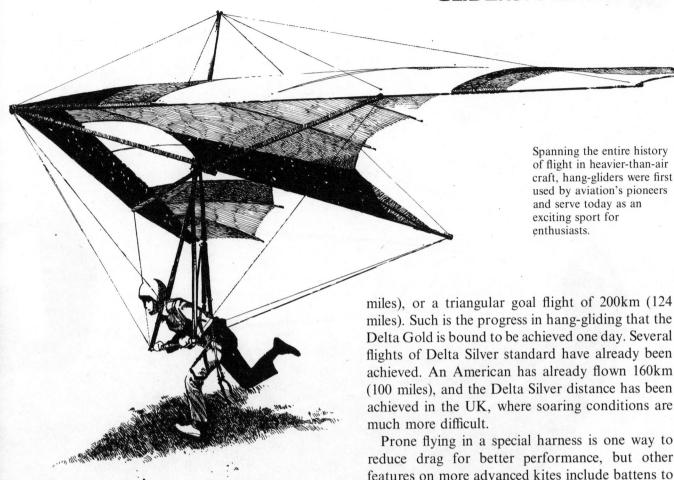

Spanning the entire history of flight in heavier-than-air craft, hang-gliders were first used by aviation's pioneers and serve today as an exciting sport for enthusiasts.

pitch the nose up and so flatten or decrease the glide path. Turns are made, just as with any other more conventional aircraft, by banking. To go left, the pilot pushes the bar to the right, and vice-versa.

Training is now done mostly at training schools, registered in the UK by the British Hang Gliding Association. Students usually begin on a slope that is only just steeper than the Rogallo's glide angle, early flights being made in a light wind, sometimes tethered and sometimes via radio contact through a receiver in the pilot's crash helmet.

The most basic method of soaring in a hang glider is of course by ridge or slope lift, but increasing use is being made of thermals. Although a hang glider's glide ratio is often no better than 1:8 – much worse than a conventional glider – its sink rate may be nearly as good, and in skilled hands it has a tighter turning circle.

An international badge system to mark soaring achievements is similar to that for conventional gliding. First comes the Delta Bronze, demanding five short flights and spot landings. Then comes the Delta Silver, with exactly the same requirements as conventional gliding's Silver C, and the Delta Gold, calling for an out-and-return flight of 300km (186

miles), or a triangular goal flight of 200km (124 miles). Such is the progress in hang-gliding that the Delta Gold is bound to be achieved one day. Several flights of Delta Silver standard have already been achieved. An American has already flown 160km (100 miles), and the Delta Silver distance has been achieved in the UK, where soaring conditions are much more difficult.

Prone flying in a special harness is one way to reduce drag for better performance, but other features on more advanced kites include battens to tauten and smooth the wings, plus fins, rudders or tip draggers on the wingtips for better control. There is even a class of rigid-wing gliders with pre-formed wing sections and a conventional aircraft's controls, although they are expensive, heavier to launch on foot, and much less portable. At this level it is difficult to say whether they are still hang-gliders, or lightweight gliders.

Competitions are gradually concentrating more on soaring and cross-country tasks than simple spot landings and precision flying. The safety record in such a rapidly growing sport is improving as training methods become much more refined and disciplined, and competence and technical know-how spreads across a very diverse activity.

The future will see some determined and exciting progress towards much better performances which should include more soaring and longer cross-country flights. Launching methods other than running down a slope will also be tried, and especially the addition of small engines enabling a take-off from a flat site, as with an ordinary motor-glider. But putting a small engine on a glider controlled entirely by weightshift raises some critical design factors, and this is a development that needs detailed attention.

Balloons and
Airships

It can be argued that the balloon is the most significant aircraft ever invented by man. Its original importance lay in the fact that for the very first time it enabled him to leave the surface of the Earth and travel freely in the air above, albeit at the whim of the breeze.

First demonstration of balloon flight

It has long been believed that the French brothers Etienne and Joseph Montgolfier were the originators of the hot-air balloon, the type of lighter-than-air craft with which man first achieved flight. But recent research has shown that in 1709 the Brazilian priest Bartolomeu de Gusmão demonstrated a model hot-air balloon at the court of King John V of Portugal. According to one Salvador Ferreira, the model balloon was constructed of thick paper and inflated by hot air. This came from 'fire material contained in an earthen bowl' which, as shown in a contemporary painting, was suspended beneath the open neck of the envelope.

On 8 August 1709, Gusmão presented his model balloon for examination by a distinguished and unimpeachable gathering which included the King, Queen Maria Anna, the Papal Nuncio and Cardinal Conti (later Pope Innocent III), together with princes, nobles, diplomats and other members of the court. It is recorded that the balloon rose to a height of 3·66m (12ft) before two valets, who feared it might set the curtains alight, terminated its flight by knocking it to the ground. Thus, almost three-quarters of a century before the Montgolfiers, the principle of the hot-air balloon had already been demonstrated.

The Montgolfier brothers

Many stories, doubtless fanciful, have been told to explain the way in which Etienne and Joseph Montgolfier discovered that hot air rises, and how they concluded that if contained in an envelope of sufficient size it would have enough 'lift' to raise passengers into the air.

The choice of paper for the envelope of their first balloon must have seemed natural to the brothers, who were paper-makers by trade. The balloon, fairly rigid, stood over a pit containing wool and straw which when ignited filled the balloon's envelope with hot air. On 25 April 1783, the first successful Montgolfière took to the air at Annonay near Lyons in France. It is reported to have risen to a height of about 305m (1,000ft) and travelled about 9,842·4m (3,000ft) horizontally before it fell to the ground as the air in the envelope cooled.

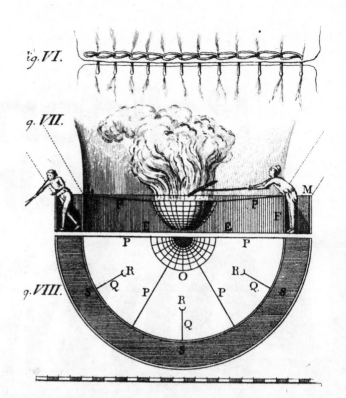

The brothers gave a public demonstration at Annonay on 4 June 1783, when a new balloon rose to about 1,830m (6,000ft). This success resulted in a summons to the capital, so that King Louis XVI could see the Montgolfiers' invention for himself. A balloon some 13m (42ft) in diameter was constructed especially for the event, and a basket hung beneath it to carry the world's first aerial voyagers: a cock, a duck and a sheep.

The balloon was launched at the Court of Versailles on 19 September 1783, climbing to approximately 550m (1,800ft) before the astonished gaze of King Louis, Marie Antoinette and their court. It landed about 3·2km (2 miles) away, and there was some concern when the cock was discovered to be a little the worse for his adventure. Had he been weakened by the great altitude at which the balloon had flown? Further investigation suggested that he was probably suffering from the effects of being trampled or sat on by the sheep.

The moment was fast approaching for manned flight, for which the Montgolfiers created a magnificent new balloon 15m (49ft) in diameter. Superbly decorated in a blue-and-gold colour scheme, it was emblazoned with the royal cipher, signs of the zodiac, eagles and smiling suns. Around its open neck was attached a wicker gallery capable of accommodating one or two men. In this vehicle the 26-year-old François Pilâtre de Rozier made a tethered flight to 26m (85ft) on 15 October 1783 and remained airborne for about 4½ minutes.

Just over a month later, on 21 November 1783, de Rozier and his passenger, the Marquis d'Arlandes, became the first men in the world to be carried in free flight by a balloon. They rose from the garden of the Chateau La Muette in the Bois de Boulogne, Paris, and were airborne for 25 minutes before managing to land about 8·5km (5·3 miles) away from their departure point.

The first hydrogen balloon

A couple of decades before the Montgolfier flights, in 1766, the British scientist Henry Cavendish had isolated a gas which he called Phlogiston. Weighing only 2·4kg (5·3lb) per 28·3m³ (1,000 cu ft) by comparison with 34·5kg (76lb) for the same volume of air at standard temperature and pressure, it seemed likely to prove invaluable for lighter-than-air flight.

By 1790 this gas had been named hydrogen by the French chemist Lavoisier. Seven years earlier on 27 August 1783, Professor Jacques A. C. Charles

Far left: A drawing of a Montgolfier balloon showing the layout of the crew station and central brazier; below it is a plan of the balloon and burner. *Above* is a contemporary impression of the first manned flight of a Montgolfière, from the Bois de Boulogne, Paris, on 21 November 1783. On the *left* is the world's first hydrogen balloon designed by Professor J. A. C. Charles, and flown unmanned from the Champs-de-Mars, Paris on 27 August 1783. Landing at Gonesse some 25 km (15.5 miles) away, it was attacked and destroyed by local residents, who were terrified by this visitor from an alien world.

had launched successfully a–small, unmanned Phlogiston-filled balloon. Flown from the Champs-de-Mars, Paris, it was airborne for about 45 minutes before coming to earth some 25km (15·5 miles) away at Gonesse. There it was attacked with pitchforks by panic-stricken villagers, who believed it to be some strange device of the devil and who were not satisfied until it had been torn to shreds.

To construct the envelope of the balloon, Professor Charles had sought the assistance of two brothers named Robert. They had devised a method of rubberizing silk to make it leak-proof. With their

help, following the successful flight of the model, Charles designed and built a man-carrying balloon. This balloon of 1783 was essentially the same as a modern gas-filled sporting design, and in it Professor Charles and one of the Robert brothers, Marie-Noel, became the first men in the world to fly in a hydrogen-filled balloon. Ascending in front of an estimated 400,000 people from the Tuileries Gardens in Paris on 1 December 1783, they completed successfully a flight of 43·5km (27 miles) in two hours. The balloon was 8·6m (28·2ft) in diameter and had a volume of 325·5m³ (11,500 cu ft).

The balloon goes to war

The success of the *Charlière*, as Charles's craft was named, generated tremendous interest in the balloon as a sporting vehicle, a money-spinner for showmen and daredevils, a possible means of exploration and, of course, as a weapon of war. Within days of de Rozier's first flight in the Montgolfier's hot-air balloon, he had taken up one André Giraud de Vilette as a passenger. Afterwards, in a letter to the *Journal de Paris* de Vilette commented on the ease with which he had viewed the environs of Paris. 'From this moment,' he wrote, 'I was convinced that this apparatus, costing but little, could be made very useful to an army for discovering the positions of its enemy, his movements, his advances, and his dispositions . . .'

Not surprisingly, the French were the first to take the balloon to war. Four military balloons – the *Entreprenant, Celeste, Hercule* and *Intrepide* – were constructed and a Company of Aérostiers formed as a component of the Artillery Service under the command of Captain Coutelle. On 26 June 1794, Coutelle ascended in the balloon *Entreprenant* during the battle of Fleurus in Belgium. Information signalled by Coutelle is believed to have played a significant part in the defeat of the Austrians. This flight represented the first operational use of an aircraft in war.

The spread of ballooning

Before the balloon's military debut, the first ascent by a woman had been recorded, as had first flights in Britain, Italy and America, and the first aeronaut, François de Rozier and a companion, Jules Romain, had been killed in a ballooning accident on 15 June 1785, while attempting a crossing of the English Channel from Boulogne in a combination hydrogen and hot-air balloon. The cause of the accident was unknown, but as the envelope was destroyed by flame it is reasonable to assume that hydrogen had ignited during the venting of gas, which was being supervised so as to control altitude.

The Channel had already been crossed, however, on 7 January 1785, by Jean-Pierre Blanchard of France and the American Dr John Jeffries. They took off from Dover, Kent, on a bitterly cold day and in a balloon which had only a small margin of lift when laden with the two men and their equipment. To avoid a ducking they had to throw overboard everything possible, including most of their clothes, but the two half-frozen gentlemen landed safely in the Forêt de Felmores, France, some two and a half hours later.

Balloons in America

Jean-Pierre Blanchard was well known as a balloonist in Europe before his crossing of the English Channel. Dr Jeffries must have then suggested that he visit America, because eight years later Blanchard arrived in Philadelphia, where arrangements were made for a flight from the yard of the old Walnut Street Prison. At the appointed time, 1000hrs on 9 January 1793, a large part of the city's population turned out to witness the historic event. They numbered among them President George Washington, as well as John Adams, Thomas Jefferson, James Madison and James Monroe, all four of whom were subsequently to become presidents of the United States.

The president handed Blanchard a letter asking 'all citizens of the United States, and others, that in his passage, descent or journeyings elsewhere, they oppose no hindrance or molestation to the said Mr Blanchard.' After wishing him a safe journey, the president joined the assembled thousands to watch the balloon rise easily into the sky. Some 46 minutes later Blanchard landed safety about 24km (15 miles) away.

Ballooning in America expanded rapidly after this first demonstration. But unlike the French government, those in authority in America could visualize no military employment for the balloon. One significant flight caused ballooning in the United States to be regarded in an entirely new light. On 2 July 1859, John Wise, John La Montain and O. A. Gager flew 1,770km (1,100 miles) from St Louis, Missouri, to Henderson, New York.

This achievement brought the realization that long-distance travel by balloon was possible, and a number of people discussed seriously and planned a transatlantic flight. Among them was a flamboyant showman named Thaddeus Sobieski Constantine Lowe. He made a balloon 39·6m (130ft) in diameter with a reported gross lift of 20 tons. Named the *Great Western*, this balloon was weakened during a trial flight on 28 June 1860, and finally burst when the gas pressure proved too much. Finding that the *Great Western* was beyond repair, Lowe, still intent on flying across the Atlantic, built a second balloon, which was named the *Enterprise*.

Balloons in the American Civil War

In April 1861 the first shots of the American Civil War were fired. On 15 April President Abraham Lincoln called for troops and four days later balloonist James Allen and a friend were among those who volunteered, taking along two balloons. They

After completing the first flight in a hydrogen balloon, on 1 December 1783, Professor J. A. C. Charles and Marie-Noel Robert land at Nesle after travelling 43·5km (27 miles).

were the spiritual fathers of today's enormous US Air Force, making their first military ascent on 9 June 1861. Other balloonists to join the Union Army included John Wise, John La Montain and Thaddeus Lowe.

On 18 June 1861, the latter demonstrated the potential of captive balloons for military reconnaissance. Climbing to a height of 152m (500ft) in the *Enterprise*, Lowe, a telegraph operator, and an official of the telegraph company, sent the first air telegraph through a wire trailing to the ground. In it he told President Lincoln: 'This point of observation commands an area nearly (80km) 50 miles in diameter. The city, with its girdle of encampments, presents a superb scene . . .'

The President was suitably impressed, but it was not until early August 1861 that Lowe received his first official instructions. Ordered to construct a balloon of 708m³ (25,000cu ft) capacity, he flew the *Union* within three weeks. On 24 September Lowe made military history. He observed the fall of artillery fire from the *Union* and transmitted correc-

tions by telegraph. The results were so good that Lowe was almost immediately required to provide four more balloons and crews to form the Balloon Corps of the Army of the Potomac.

But this first 'air force' was to last only 18 months. Though Lowe was a powerful driving force, the majority of his men were apathetic. When he resigned in May 1863 Lowe had made more than 3,000 ascents. His departure marked the end of the Balloon Corps, and almost 30 years were to pass before the US Army renewed its interest in military aviation.

Balloons in the Siege of Paris

When Paris was encircled by the Prussian army in September 1870 Chancellor Bismarck believed that the beleagured city would surrender quickly, being completely isolated from the rest of the world. But he had not counted on the balloon as a means of carrying important passengers and dispatches out of Paris.

Under the leadership of Gaspard-Felix Tournachon the balloonists of Paris formed the *Compagnie d'Aérostiers Militaires*, intending to use captive balloons to observe Prussian army movements. On

23 September 1870, the balloon *Neptune* took off from the Place St Pierre. Piloted by Jules Duruof, *Neptune* landed three hours later at Evreux, some 97km (60 miles) away. The 125kg (275lb) of mail and dispatches carried out of the city by Duruof was then sent on its way by more conventional means. The world's first airmail service had been inaugurated.

At the beginning of the siege there were only five or six balloons in the city, so arrangements were made to mass-produce further examples. 'Production' lines were started at the Gare d'Orléans and at a derelict music hall in the Elysées-Montmartre. New pilots were also needed, and circus acrobats and sailors – selected because they were likely to have a good head for heights – were trained to fly the balloons out of Paris.

The balloons offered only a one-way service, leaving Paris in the direction of the wind. Messages from outside were brought into Paris by carrier pigeons flown out in the balloons. The number of communications a single pigeon could carry was very limited, until a Monsieur Barreswil hit on the idea of photographing and reducing the letters and dispatches so that one bird could transport many more messages. Once delivered, these 'microfilms' could be enlarged for reading. During World War II this idea was to be developed into the Airgraph system which was designed to cater for the extremely

During the Siege of Paris (1870) the French enterprisingly used balloons to escape. Unfortunately for them several were captured by Prussian troops.

large volume of forces (and later civilian) airmail.

By the end of the siege, when a final sortie flown on 28 January 1871 carried news of the armistice. some 64 balloons had taken 155 passengers and crew to safety and 2½ to 3 million letters had reached destinations outside Paris. In the other direction, some 60,000 messages had been carried by 60 pigeons.

The siege gave rise to one other invention: the anti-aircraft gun. Not surprisingly, the Prussians were infuriated at the sight of important people and vital information floating to safety over their heads. Krupps, the famous armament manufacturers, responded by producing a special anti-balloon gun which could be elevated to fire at balloons passing above.

Military balloons in Britain

Although the first balloon flights in Britain took place in 1784, it was not until 1878 that the British Army showed any interest, allocating £150 for the construction of a balloon. Captains J. L. B. Templar and H. P. Lee were responsible for developing the *Pioneer*, which in 1879 became the first balloon to enter the Army's inventory. The first balloon to be actually used by the Army was Captain Templer's own *Crusader*, which entered service shortly afterwards.

On 24 June 1880, a balloon detachment was involved in military manoeuvres at Aldershot, Hampshire. Four years later the first British balloon detachment ordered to fly in support of military operations was sent to Bechuanaland. Arriving in Cape Town on 19 December 1884, this unit comprised two officers, 15 other ranks and three balloons. But order was restored without a shot having to be fired, and they never saw action.

In 1885 a detachment was sent to the Sudan, and in 1899 a balloon was used to keep an eye on the movements of the Boers besieging Ladysmith.

Development of military balloons

The conventional spherical balloon is quite at home when travelling freely with the wind. But when restrained by a rope anchored to the ground it can spin and buck so badly that the observation crew can soon lose interest in everything but a rapid return to the ground.

To overcome this problem the kite balloon was developed and brought close to perfection by the Germans August von Parseval and H. B. von Sigsfeld. They took the partly controllable sausage-shaped observation balloon of the time and added a

This partially inflated German kite balloon is typical of those used for observation by all nations during World War I.

tail fin to keep it pointing steadily into the wind. The French army officer Captain Caquot later developed this design by improving the streamlining of the basic shape and introducing three inflated tail fins disposed at 120° intervals, with one pointing vertically downward. Like the single fin of the von Parseval Drachen, they were inflated by an airscoop facing into the wind. The Caquot type was used extensively for aerial observation both on land and at sea during World War I. By the beginning of World War II the balloon had long since been superseded by heavier-than-air craft in this role. Instead, barrage balloons were used in large numbers to present a hazard to enemy aircraft and deter low-level attacks.

One unusual military application of the balloon came during World War II, when the Japanese tried to attack North America from the home islands. Launched into the jetstream winds, prevailing at a fixed altitude, bomb-carrying balloons proved capable of crossing the Pacific. They were equipped with an ingenious barometric device which worked in conjunction with an automatic gas-discharge valve to keep the balloon at the desired altitude of between 9,144m (30,000ft) and 11,582m (38,000ft). The higher daytime temperature would have caused the balloon, if unchecked, to ascend until it burst as the gas expanded. The valve discharged gas, and so reduced buoyancy and caused the balloon to descend. When the required altitude was reached the barometric device energized an electrical ballast-discharge circuit, causing the balloon to rise again. Enough ballast was carried for the 9,978km (6,200 miles) Pacific crossing. When the last of this had been jettisoned, the weapon load of two incendiary bombs and one 15kg (33lb) anti-personnel device was released during the next low-level cycle. A small explosive charge was detonated to destroy the balloon.

It was calculated that weapon-release would occur over the not insignificant target represented by the North American continent. Of about 9,000 balloons launched, between 11 and 12 per cent are believed to have completed the crossing. Despite this comparatively large number of weapons over the target, the damage was confined largely to fires in open areas and did not prove to be as damaging as they might otherwise have been.

Only six people were killed in the campaign, and the bomb-carrying balloons were finally countered by a complete ban on radio and newspaper reports. As a result, the Japanese were unaware of the operation's technical success and called it off. Had they known that one in eight of these weapons were reaching the target they might have launched many more, equipping them with a far more potent payload than the ones they carried.

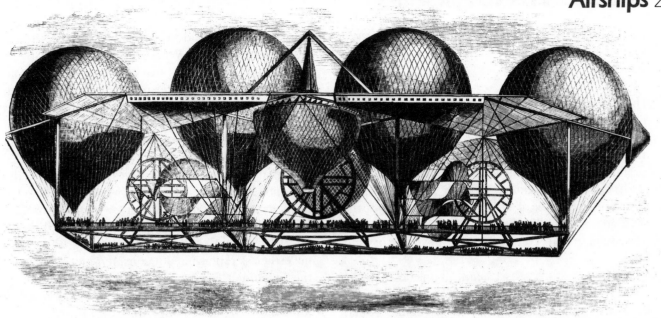

Scientific use of balloons

Balloons have also played a significant part in scientific research from the earliest days. J. A. C. Charles made recordings of air temperature and changes in barometric pressure during his first ascent in a hydrogen balloon on 1 December 1783. In 1784 Jean-Pierre Blanchard and Dr Jeffries made similar recordings.

In 1804 the French scientist Gay-Lussac made two flights to investigate the way in which the Earth's magnetic field varied with increasing altitude. In Britain James Glaisher made 28 ascents for scientific purposes during 1862 to 1866. On one occasion Glaisher became unconscious with the balloon still ascending at an altitude which he estimated at 8,839m (29,000ft). He survived, but the companions of Gaston Tissandier were not so lucky; they died during an ascent in 1875. In Germany, Professor Assman continued the work begun by Glaisher, achieving more significant results with his more highly developed instruments. By 1932 the Swiss physicist Auguste Piccard had climed 16,201m (53,153ft) in a sealed capsule suspended beneath his free balloon, having made the first flight into the stratosphere during the previous year. Some great heights have been reached, notably 34,668m (113,740ft) by US Navy reservist Commander Malcolm D. Ross on 4 May 1961. Unmanned balloons also have meteorological uses; some small ones indicate wind direction and speed, others carry instrument packages for transmitting data to earth.

Petain's giant 'airship' (*above*) built in 1850, linked several balloons for greater lift. Although an advanced concept it was never flown. World War II saw the kite balloon used as a barrage balloon (*left*) to deter low-level bombing.

Airships

In the 18th and 19th centuries the balloon had made many things possible, but there were some serious problems to be overcome if men were to travel easily by air from point to point. These shortcomings can be summarized briefly: the balloon could travel only in the direction of the wind; and at the end of each flight all the costly lifting gas had to be valved off to allow the balloon to be packed and transported back to its home base. Such disadvantages were acceptable to sportsmen, but they made the balloon completely impractical for most military or commercial tasks. What was needed was an independent source of power to render balloons navigable.

Steerable balloons: the first ideas

Once airborne the balloon is wedded to the wind, moving in the same direction and at the same speed. It has no independent motion within the airstream which would allow the rudders or sails proposed by many inventors to affect its direction of movement. So despite some ingenious plans to utilize muscle power, the evolution of a steerable balloon had to await the development of a compact power unit. By propelling the vehicle independently of the wind, such an engine would make it possible for rudders and elevators to control direction.

While balloonists waited for a suitable powerplant they came to realize that the spherical balloon was hardly ideal in shape for steering anywhere. They therefore began to streamline it into a spindle shape not so different from more modern solutions. But the early experimenters were disappointed to find that despite this 'ideal' shape, muscle-powered oars or propellers were incapable of steering their prototype airships.

The first practical airships

The autumn of 1852 saw French engineer, Henri Giffard, preparing a vehicle for flight which had the distinction of being the first powered and manned airship. He had constructed an envelope 43·9m (144ft) in length, pointed at each end and with a maximum diameter of 11·9m (39ft). Capacity of the envelope was 2,492m³ (88,000cu ft), and from it was slung a car containing the powerplant and pilot.

The power source was a steam engine designed by Giffard which weighed about 45kg (100lb) and developed approximately 3hp. The engine, boiler, three-blade propeller and an hour's supply of water and fuel weighed 210kg (463lb). The propeller, 3·35m (11ft) in diameter, was driven at a maximum speed of 110rpm. The airship was to be steered by what Giffard called a keel, a triangular vertical sail mounted at the aft end of the long horizontal pole to which the gondola was attached. It was effectively a rudimentary rudder.

On 24 September 1852, Henri Giffard piloted his airship from the Paris Hippodrome to Trappes, some 27km (16·8 miles) away at an average speed of 8km/h (5mph). This flight must have been made in virtually still air if it was in fact steered: control must have been marginal at such a low speed, and the airship would have been carried away if there had been any wind at all. Giffard's craft was nonetheless the first true airship.

The next important development came from the German Paul Hänlein, who in December 1872 flew a small airship at Brünn in Central Europe. Its 6hp gas engine was fuelled by the coal gas which filled the envelope. Designed by Etienne Lenoir, this power-plant was a development of his basic single-cylinder gas engine, first demonstrated in 1860. Hänlein's airship was not a success, but it is remembered for its powerplant, the first internal-combustion engine to take to the air. In 1883 the Tissandier brothers of France built and flew an airship powered by a 1·5hp electric motor. It was driven by primitive batteries which together with the motor weighed about 200kg (440lb), giving an improbably low power:weight ratio of 1hp:132.8kg (293lb). Despite this, the Tissandier airship was at that time marginally the most successful.

The following year saw the most important of the early airships, *La France*, designed by Captains Charles Renard and A. C. Krebs of the French Corps of Engineers. With an envelope of 1,869m³ (66,000cu ft) capacity, it was 50·3m (165ft) long and had a maximum diameter of 8·23m (27ft). Beneath the envelope was suspended a lightweight gondola which extended for about two-thirds of the length of the vessel. It comprised a framework of bamboo covered by silk, and in it was mounted the power-plant, a 9hp electric motor drawing on specially developed batteries. This drove a four-blade tractor propeller of 7m (23ft) diameter. A total powerplant weight of 857kg (1,890lb) gave a power:weight ratio of 1:210. Though still impractically low, this figure nevertheless represented a considerable advance on the Tissandier engine.

On 9 August 1884, Renard and Krebs lifted off in *La France* from Chalais-Meudon, France. They completed a circular course of about 8km (5 miles), reaching a maximum speed of 23·5km/h (14·6mph) en route. This is regarded as the first fully controlled powered flight by a manned airship anywhere in the world. *La France* made a further six flights later that year and in the following one.

Development of the airship

In spite of these achievements, electric motors powered by storage batteries were clearly impractical. Like the pioneers of heavier-than-air flight, those who believed that the airship represented the best method of air transport had to await the availability of a light and compact power source. Their patience was rewarded when an airship powered by a single-cylinder Daimler petrol engine was flown successfully on 12 August 1888. This was designed by the German Dr Karl Welfert, who was later to die in an airship. A flame from the petrol engine's exhaust is believed to have ignited gas being vented from the envelope.

Next on the scene with an airship powered by an internal-combustion engine was the Brazilian Alberto Santos-Dumont. He achieved a first successful flight in Paris on 20 September 1898, in a craft powered by a small motor-cycle engine. This little powerplant drove his *No 1* along quickly enough in the light breeze to enable him to steer the craft in the chosen direction.

But it was his *No 6* of 1901 that brought Santos-Dumont real fame. In it he won the 100,000 franc prize offered by M Deutsch de la Meurthe, a wealthy member of the Aéro Club, for a flight of about 11·3km (7 miles) from the *Parc d'Aérostation* at St Cloud, around the Eiffel Tower and back. *No 6* was pointed at both ends and had an overall length of

Count Ferdinand von Zeppelin's LZ3 (Z1) airship in flight over Munich in 1909. It was built to carry passengers but was used in World War I for long-range bombing.

33m (108ft 3in), a maximum diameter of 6m (19ft 8in) and a hydrogen capacity of 622m³ (21,966cu ft). Suspended from the envelope was a lightweight girder keel supporting the small control basket which accommodated the pilot, engine, tanks, pusher propeller and triangular rudder. Powerplant was a Buchet petrol engine based on the Daimler-Benz design and fitted with a water-cooling system. Developing about 15hp, it drove a crude two-blade pusher propeller.

The first Zeppelin

The airships discussed so far belonged to the classes known as non-rigids and semi-rigids. In the former the shape of the envelope was maintained by gas pressure alone; in the latter a rigid keel permitted a lower gas pressure. One of the drawbacks of these two types was that diminishing gas pressure brought a change in the shape of the vessel, and a serious pressure reduction could lead to the collapse of the forward end of the envelope. The resulting drag usually proved too much for the feeble powerplants of the day and was likely to thwart any attempts to steer the vehicle.

Though it was appreciated that a rigid airship would be far more practical, the weight penalty imposed by a rigid frame or envelope was unacceptable in the early years of airship development. The Austrian David Schwarz nevertheless designed an airship with a rigid aluminium-sheet envelope – reported to be 0·2mm (0·008in) thick – attached to a light-weight tubular framework braced by internal steel tension wires. Overall length was 47·55m (156ft) and gas capacity 3,681m³ (130,000cu ft). Power was provided by a 12hp Daimler petrol engine driving four propellers, two of which were for propulsion and two for directional control. When the airship flew for the first time, on 3 November 1897, the belt drives to the propellers were found to be useless. The ship drifted for about 6·5km (4 miles) before landing heavily and breaking up in the wind.

Meanwhile, a significant figure had taken the stage in Germany, Count Ferdinand von Zeppelin. Visiting America during the Civil War, he had taken to the air for the first time in one of the Army's balloons at St Paul, Minnesota. Impressed by this flight, influenced by a lecture on airship travel by the German Postmaster-General, and concerned at the comparative success of Renard and Krebs' *La France*, Count Zeppelin believed it was his duty to provide Germany with a fleet of military airships.

In 1899 he began the construction of his first rigid airship, *Luftschiff Zeppelin 1*, (LZ1) in a floating shed on Lake Constance. Like many pioneers before and after him, the Count seems to have believed erroneously that in the event of a heavy landing the surface of the lake would prove more forgiving than the ground.

Zeppelin's airship was truly a giant for its day, measuring 128m (420ft) in length and having a maximum diameter of 11·73m (38ft 6in). Its 11,327m³ (400,000cu ft) of gas was contained in 17 separate cells, an entirely new feature. Only its engines were less than huge: the two Daimlers together produced a mere 30hp for a weight of 771kg (1,700lb), giving a power:weight ratio of 1:57. Although this represented a great improvement on the airship *La France*, the enormous LZ1 needed large reserves of power if it was to be controllable in anything but still air. When the Zeppelin flew for the first time, on 2 July 1900, it was apparent immediately that it was grossly underpowered. Instead of flying at the expected 45km/h (28mph), the LZ1 had a maximum speed of only 26km/h (16mph). This would have given marginal controllability if the vessel had been equipped with efficient aerodynamic controls. Miniature rudders at bow and stern were expected to provide directional control, and a 250kg (551lb) weight could be slid fore and aft on a track between the two gondolas to take care of vertical control.

Not surprisingly, after two more test flights giving a total airborne time of two hours, LZ1 was scrapped in 1901. It was not until four years later that von Zeppelin began construction of the LZ 2, but by then an important airship had been built and flown in France.

The first controlled air journey

Under the guidance of an engineer, Henri Julliot, the brothers Paul and Pierre Lebaudy designed and completed an airship at Moisson, Seine-et-Oise, in 1902. It was strange in appearance, with an envelope 57·9m (190ft) in length and pointed sharply at each end. Beneath the envelope – made of two-ply rubberized material and with a capacity of 2,000m³ (70,629cu ft) – was a steel-tube structure carrying a basketwork gondola to contain the powerplant and crew. The German-manufactured engine developed about 35hp, driving two pusher propellers.

First flown early in 1903, it made a number of successful local flights and, on 12 November 1903, was flown 61km (38 miles) from Moisson to the Champs-de-Mars, Paris. This is considered to be the

Above: The steel tube structure of the Lebaudy airship. It was suspended below the envelope to carry gondola and powerplant. *Below:* A cutaway drawing of the Goodyear airship USS Akron, built for the US Navy and lost at sea in 1933.

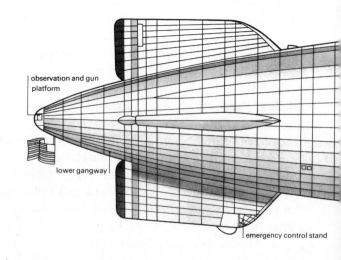

observation and gun platform

lower gangway

emergency control stand

first-ever controlled air journey by a practical dirigible. The Lebaudy airship subsequently suffered serious damage at Chalais Meudon. It was repaired and handed over in 1906 to the French government. Other examples were built for the French Army, Britain and Russia each acquired one, and a final example was built under licence in Austria.

Flight and control of an airship

The endurance of an unpowered hydrogen balloon of any size depends largely on the amount of ballast it can carry. With its envelope completely full of gas, such a balloon is ballasted so that the weight of the crew, equipment and ballast are just enough to keep it on the ground. The discharge of a small amount of ballast will cause the balloon to rise, and as atmospheric pressure decreases the gas will expand and the volume of the evelope increase. Gas is able to vent off to prevent undue stress on the envelope, and eventually a height will be reached at which the balloon is in equilibrium. Further ballast must then be dropped if the balloon is to climb. If it is to descend, more gas must be valved off, and a second climb will call for the dropping of more ballast. Finally, when all the ballast has been discharged, the flight must be terminated.

A dirigible – a navigable airship – is affected by very different problems arising from the powerplant which gives it airspeed and directional control. Like, the balloon, the dirigible can ascend vertically from the ground. Then, at a suitable height, the engines can be started and the ship can be steered in the required direction. The climb from ground level to the height at which the engines are started is made possible by the static lift provided by the gas within the envelope. Once the engines are started and the vessel is moving fast enough for its aerodynamic controls to be effective, the nose can be pitched up and dynamic lift created. In the same way, a negative dynamic force can be created by pitching the nose down.

An airship usually carries water ballast, as opposed to the sand carried normally by balloons. But just as with balloons, airships have to discharge ballast if they are to rise from the ground under static lift.

If the airship is of the non-rigid or semi-rigid type, the envelope will contain a number of ballonets which are filled with air by a pump or via scoops placed in the slipstream of the propellers. They maintain pressure within the envelope so that it retains its aerodynamic shape. An envelope pressurized with gas alone would vent it off throughout the ascent. Then, on descent, the remaining gas would be compressed under the increasing atmospheric pressure. The envelope's internal pressure would fall, possibly causing a drag-inducing change in its aerodynamic form.

To prevent this, about 95 per cent of the envelope's volume is taken up by lifting gas, with the

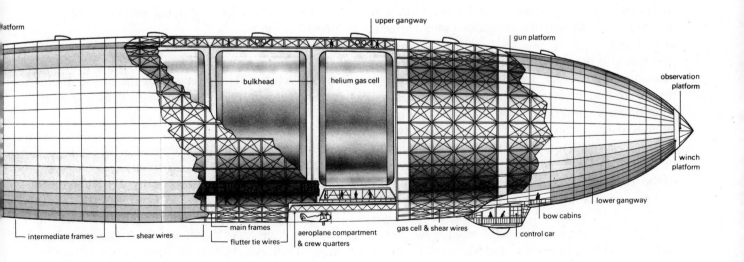

air-filled ballonets occupying the remaining 5 per cent. Air is vented off instead of gas as the airship climbs, and gas plus air maintain the aerodynamic shape. At optimum altitude the lifting gas will fill the envelope at the desired pressure, and the ballonets will be completely empty. The height to which an airship can climb without losing any lifting gas is known as its pressure altitude.

This simple arrangement becomes complicated when it is necessary to fly higher than the pressure altitude, and by the fact that the engines consume fuel. The airship will climb if ballast is discharged. Lifting gas will however have to be vented off as atmospheric pressure decreases. Secondly, the gross weight of the vessel will decrease as the engines consume fuel, and the airship will climb steadily unless gas is vented off. Changes in air temperature and temperature inversion present further control problems to the airship designer.

Here, very briefly, are some of the ways in which these problems can be overcome. If the lifting gas is heated on the ground, a volume of gas smaller than that normally required will expand to fill the envelope. The heating system can then be switched off as the airship starts its climb, and the falling gas temperature and external air pressure more or less cancel one another out. With such a system, lifting-gas pressure falls at about the same rate as external atmospheric pressure, making it unnecessary to vent gas during the climb above pressure altitude. An efficient means of condensing the water in the exhaust gas from the engines would provide additional ballast to keep the ship in equilibrium as fuel is consumed. Thus Paul Hänlein's airship, with an engine fuelled by lifting gas from the envelope, was not quite as shortsighted a design as it might appear at first sight.

It was also proposed that as the airship's engines consumed liquid fuel, reducing gross weight, the hydrogen that had to be valved off to maintain equilibrium should be mixed with liquid fuel in the engines' carburation system. The vented gas would not therefore be a total loss, being added instead to the fuel stock and so helping to increase range. One of the features of the German *Graf Zeppelin* of 1928 was the use of a fuel known as *blaugas*, a petroleum vapour some 20 per cent heavier than air. This gas was stored in separate cells occupying some 30,000m³ (1,059,434cu ft), and the vessel's gross weight changed little as fuel was consumed. The later *Hindenburg* was equipped to take in water when flying through heavy rain to increase her ballast reserves.

Airship pioneers in America

The first true powered dirigible to be built in the United States was Captain Thomas S. Baldwin's *California Arrow*. With an envelope 15·85m (52ft) in length and powered by a two-cylinder motor-cycle engine built by Glenn H. Curtiss, it was flown for the first time on 3 August 1904, at Oakland, California, piloted by Roy Knabenshue. When Brigadier General J. Allen of the US Army subsequently saw it in flight at St Louis, Missouri, he was so impressed that he persuaded the War Department to request tenders for the supply of a dirigible.

Baldwin's tender proved to be the lowest and on 20 July 1908, a 29·26m (96ft) dirigible powered by a 20hp Curtiss engine was delivered to Fort Myer, Virginia. When accepted on 28 August 1908, it was designated Dirigible No 1. The pilots trained to fly it were Lieutenants Frank P. Lahm, Benjamin D. Foulois and Thomas E. Selfridge. The Baldwin ship remained in service for three years before being scrapped, and was for all practical purposes the one and only US Army dirigible.

At about this time the wealthy and eccentric journalist Walter Wellman hired Melvin Vaniman and Louis Godard to design and build a dirigible for a flight to the North Pole. The resulting *America* was 69·49m (228ft) long, had a capacity of 9,911m³ (350,000cu ft) and was powered by a single 80hp engine. A canvas-covered hangar for the airship was completed at Dane's Island, Spitzbergen, in September 1906. A year later, a first attempt to reach the Pole ended when the *America* was forced down in a snowstorm. The airship was recovered and fitted with a second engine, and another polar flight was tried two years later, but unfortunately this also proved a failure.

A transatlantic attempt followed in 1910, with the *America* lifting off from Atlantic City, New Jersey, on 15 October. Three days later the five-man crew was rescued off New England by a British ship, the RMS *Trent*.

Airship developments in Britain

The earliest practical British airship was the first of five designed and constructed by Ernest Willows between 1905 and 1914. Willows' *No 1* was only 22·56m (74ft) in length and was powered by a 7hp Peugeot engine. *No 3* achieved the first airship flight across the English Channel, travelling from London to Paris. The largest Willows ship, *No 5*, had only 1,416m³ (50,000cu ft) of gas capacity, and they were

Right: A Tiros satellite, which is being used to take cloud-cover pictures, facillitates accurate weather forecasting.

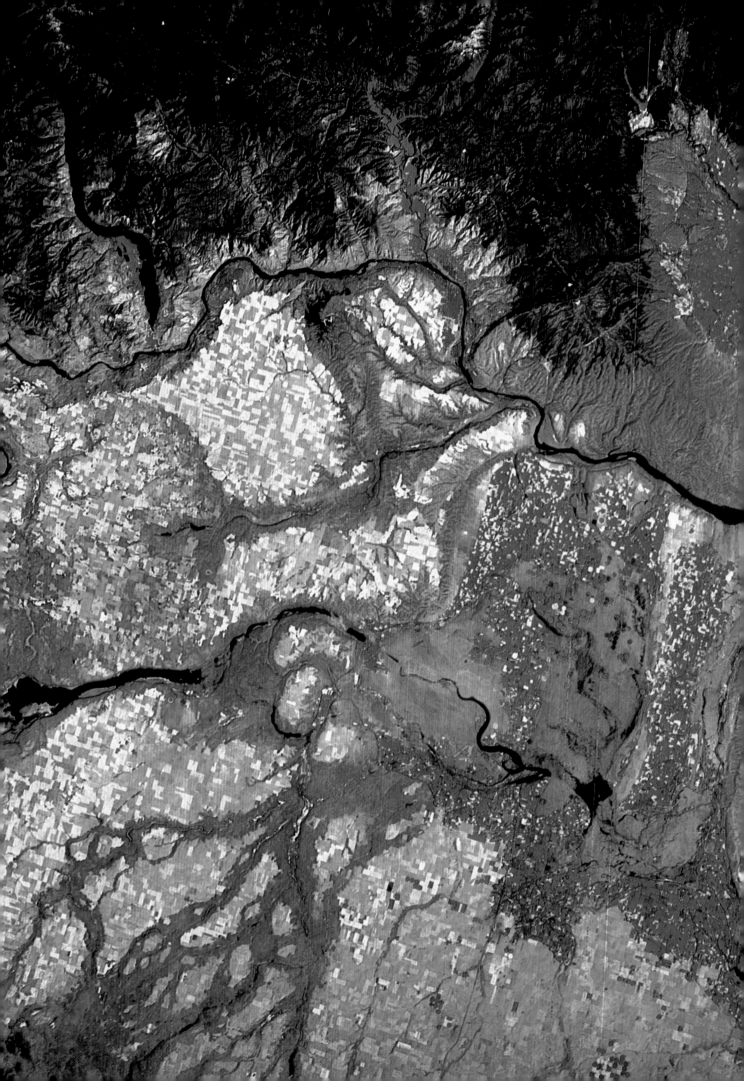

Left: This infra-red satellite picture was taken from Apollo 7 and shows part of India, Nepal, Tibet and the Himalayas. It shows a river watershed and several tributaries, as well as some cloud formations.

Below: Apollo 11, on top of its Saturn V launch rocket, carries astronauts Aldrin, Armstrong and Collins towards man's first landing on the Moon.

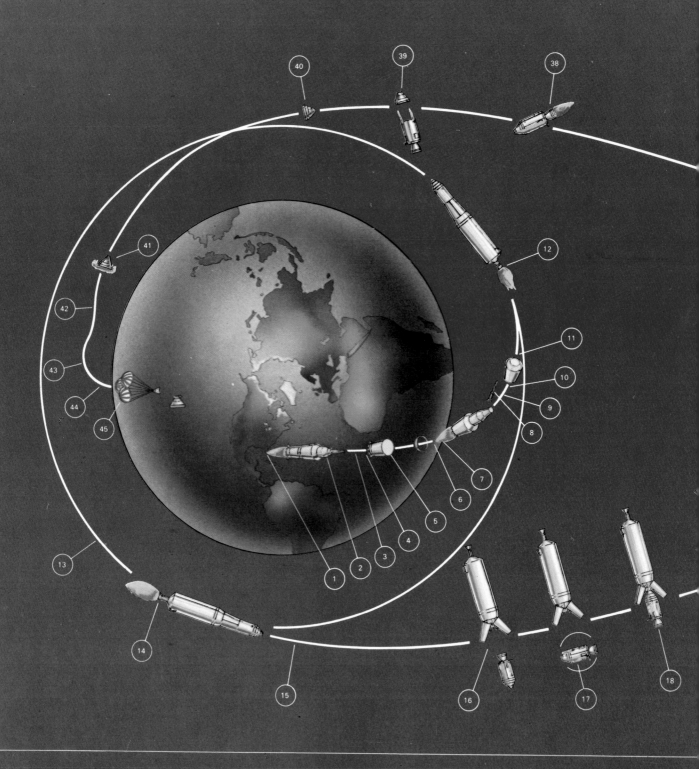

Apollo
Space
Mission

1 S-1C engines ignition	16 CSM separation from LM adapter	31 LM—RCS ignition
2 Lift-off	17 CSM 180° turnaround	32 Rendezvous
3 S-1C powered flight	18 CSM docking with LM/S-1VB	33 CSM/LM initial docking
4 S-1C engines cutoff	19 CSM/LM separation from S-1VB, S-1VB jettison	34 Transfer crew and equipment from LM to CSM
5 S-1C/S-11 separation, S-1C retro, S-11 ullage	20 SM engine ignition	35 CSM/LM separation and LM jettison
6 S-11 engines ignition	21 SM engine ignition	36 Transearth injection
7 S-1C/S-11 interstage jettison	22 SM engine firing lunar orbit insertion	37 SM engine ignition
8 Launch escape tower jettison	23 Lunar orbit insertion	38 SM engine ignition
9 S-11 powered flight	24 Begin lunar orbit evaluation	39 CM/SM separation
10 S-11 engines cutoff	25 Pilot transfer to LM	40 Orient CM for re-entry
11 S-11/S-1VB separation, S-11 retro, S-1VB ullage	26 CSM/LM separation	41 122,000m (400,000ft) altitude penetration
12 S-1VB engine ignition	27 LM descent engine ignition	42 Communication blackout period
13 Translunar injection 'GO' decision	28 LM descent	43 61,000m (200,000ft) altitude
14 S-1VB engine ignition	29 Touchdown	44 76,200m (250,000ft) altitude
15 Translunar injection	30 Lift-off	45 Deploy main chute at 3,000m (10,000ft)

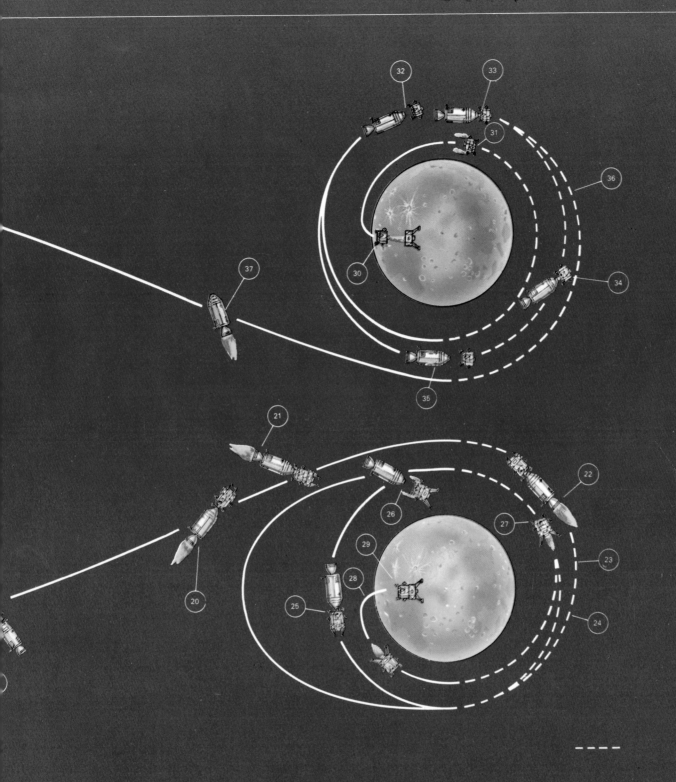

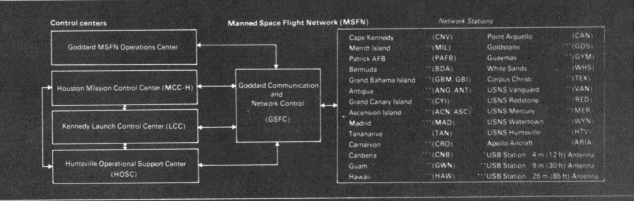

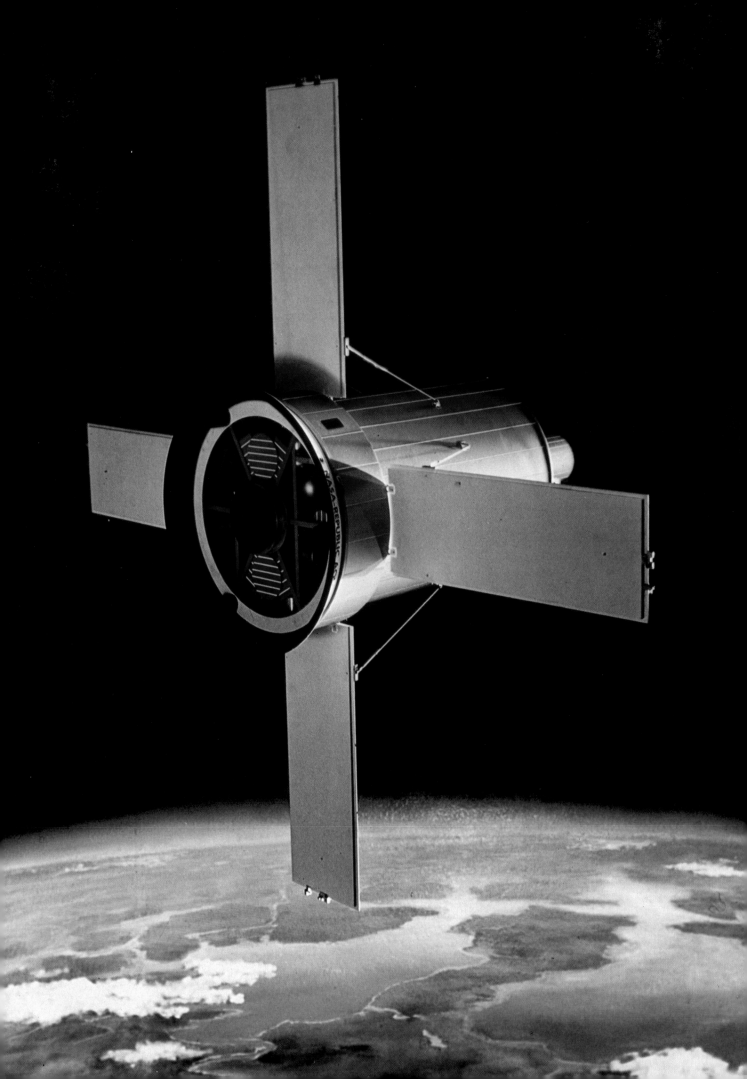

distinguished (with the exception of *No 1*) by their swivelling engines, which assisted in manoeuvring the vessels laterally and vertically.

Britain's first military airship, British Army Dirigible No. 1, was built by the Balloon Factory at Farnborough, Hants. Known unofficially as *Nulli Secundus*, this non-rigid had an envelope 37·19m (122ft) in length, 7·92m (26ft) in diameter, and with a capacity of 1,557m³ (55,000cu ft). No suitable engine was available, so Balloon Factory commanding office Colonel J. E. Capper obtained an Antoinette engine of about 50hp from Paris. This was installed along with the crew positions in a small car suspended beneath the envelope. *Nulli Secundus* flew for the first time on 10 September 1907, piloted by Colonel Capper and with the colourful Samuel F. Cody as engineer. A month later this airship flew from Farnborough to London, causing enormous excitement when it circled around St Paul's Cathedral.

Redesigned in the following year, *Nulli Secundus* acquired a triangular-section fabric-covered keel which made her a semi-rigid, and the addition of pointed ends to the envelope provided another 28·3m³ (1,000cu ft) of capacity. These changes improved maximum speed to 35km/h (22mph).

She was followed by the *Beta* (unofficially *Baby*), which measured 25·6m (84ft) in length and 7·52m (24ft 8in) in diameter, and had a capacity of 623m³ (22,000cu ft). The original pair of 8hp Buchet engines were later replaced by a single 25hp radial engine driving two propellers. *Beta* was also rebuilt, and when flown in May 1910 had a gas capacity of 934m³ (33,000 cu ft) and a British-built Green in-line engine developing 35hp. In this form *Beta* could carry a crew of three and had a powered endurance of about five hours. The first practical airship to serve with any of Britain's armed services, *Beta* was flown at night in June 1910 and in a summer of activity was reported to have flown some 1,609km (1,000 miles) without any serious problems.

Airship development in Italy

Italy, the only other European nation actively interested in the development of airships, started by building a number of non-rigids and semi-rigids for the army's Corps of Engineers. The most important of these was the Forlanini semi-rigid of some 3,680m³ (129,995cu ft) capacity, which first flew in 1909. Airships of this type were to serve with the Italian Army for some years, although they could hardly be classed as particularly successful experiments.

Airships in World War I

By building the airships which operated with the pioneering airline, Delag, the designers of the Zeppelin Company learned a great deal that was to prove of enormous value in the development of military airships for both the German Army and Navy.

While Delag was developing its airship services, a new design had appeared in Germany. A rigid airship designed by D. Johann Schütte and Heinrich Lanz, the Schütte-Lanz SL.1 made its first flight on 17 October 1911. Measuring 128m (420ft) in length and 17·98m (59ft) in diameter, the SL.1 had a capacity of about 19,822m³ (700,000 cu ft). Its pair of 270hp engines gave a maximum speed of about 56km/h (35mph). One of the SL.1's most advanced features was an improved streamline shape, which had an important influence on the development of the Zeppelin airships when Schütte-Lanz became a part of the Zeppelin Company. But a second revolutionary idea, the use of wood for the SL.1's rigid frame, did not survive the amalgamation of the companies.

There was one other significant airship company in Germany: the concern established to build the non-rigid airships designed by Major von Parseval. These non-rigids had pressure envelopes and were considered unsuitable for operational military use, but there remained a small number used for training.

Germany's principal military airships during World War I were the Zeppelins of the German Navy, which earned a fearsome reputation with their bombing raids on Britain. Though this was due more to the inadequacy of Britain's home defences than the accuracy and weight of the attacks, the Zeppelins were intensively developed, from the L3, with which Germany began the war, to the L70 and L71.

The L3 was 158m (518ft 4in) long, 14·78m (48ft 6in) in diameter and had a volume of 22,500m³ (794,575cu ft). Powered by three engines giving a total of 630hp, the L3 had a maximum speed of about 75km/h (46·5mph). Four years later, Zeppelin attacks on Britain ended after the loss of the L70 and *Fregatten Kapitän* Peter Strasser, Chief of the German Naval Airship Division. L70 fell to the de

Left: Model of the advanced orbiting solar observatory.
Previous page: Apollo 15's Lunar moving vehicle.

Havilland D.H.4 of Major Egbert Cadbury and Captain Robert Leckie on 5 August 1918. The L70 was 211m (692ft 3in) long and 23·9m (78ft 5in) in diameter, and had a volume of 62,180m³ (2,195,800cu ft). Its engines developed a total of 1,715hp and were reported to have propelled the airship through the air at 130km/h (81mph) during a trial flight.

Military airship development in Britain and France followed a different pattern, for neither nation envisaged these craft as war-winning strategic weapons. The two Allies proposed to use the airship mainly for naval patrol, a role in which it was to give valuable service, especially for Britain.

France, pioneer of the dirigible, used about 25 airships – 21 non-rigids and 4 semi-rigids – during World War I. The former were mainly of the Astra-Torres type, a design in which the three lobes of the envelope ran from stem to stem. Designed by a Spaniard, Torres Quevedo, they were built and

A Goodyear-built airship in service with the US Navy. More airships have been built by Goodyear than any other company in the world, totalling 303 in 1979.

developed by the French *Astra Société des Constructions Aéronautiques*. The internal rigging of the Astra-Torres passed within the folds of the lower lobes, considerably reducing the parasite drag generated by the rigging. The other French manufacturer worthy of mention was Clément-Bayard, which built a small number of non-rigid airships of the Lebaudy type. The majority of the French airships were used mainly for patrols over the Mediterranean Sea.

The RNAS, which became the main user of dirigibles among Britain's military services, began the war with a Parseval, an Astra-Torres and a Lebaudy. The RNAS urgently required a fleet of semi-rigids for coastal anti-submarine patrols, and an experimental vessel was built from the spare envelope of a Willows airship and the shortened fuselage and powerplant of a Royal Aircraft Factory B.E.2C aeroplane. It proved an immediate success and a whole family of related non-rigids – or Blimps as they became known – evolved from this first successful experiment. Designations for these types included Sea Scout (SS), Coastal (C) and North Sea (NS), and there were variants of each. Undoubtedly the most successful was the N.S. class, which had an envelope of Astra-Torres form, length of 79·86m (262ft), diameter of 17·37m (57ft) and a capacity of 10,194m³ (360,000cu ft). A completely enclosed gondola 10·67m (35ft) in length and 1·83m (6ft) high accommodated the crew of ten. The two 240hp engines were mounted in a completely separate car and gave a maximum speed of about 93km/h (58mph), making it an effective patrol aircraft.

The success of the Royal Navy's scouting airships lay not so much in their ability to carry out attacks as in their efficiency in tracking U-boats and calling up surface vessels to harass or destroy them.

The United States entered the war comparatively late and did not feel the same urgency about developing airships for a military role. During the war a number of blimps were built for the US Navy, mainly by the Goodyear Tire and Rubber Company, which continues to build airships to this day. The main US user of ligher-than-air craft in World War I was the US Army which deployed large numbers of kite balloons on the Western Front to observe enemy movements and direct artillery fire.

The only other combatant to make any significant use of airships was Italy, which employed a small number of non-rigids and semi-rigids on naval duties. Largest of the semi-rigids were the three built by Forlanini, which were 71·32m (234ft) long, 17·68m (58ft) in diameter, and with a volume of 11,330m³ (400,113cu ft).

Post-war development of the airship

Britain was left after the war with a small number of indifferent rigid airships, and with the R.33 and R.34 just about to enter service. Both were modelled on the German Zeppelin L.33 that was forced to land at Little Wigborough, Essex, on 24 September 1916. While R.33 is renowned mainly for her struggle to regain her base at Pulham, Norfolk, after being damaged and blown across the North Sea, R.34 has an important place in aviation history. Between 2 and 6 July 1919, this vessel carried out the first North Atlantic crossing by an airship, taking just over 108 hours. The return west–east crossing was completed between 9 and 13 July in just over 75 hours. The only other really successful British airship was the R.100.

Designed by the great inventor and engineer Sir Barnes Wallis, it had a very strong and light duralumin frame and was powered by six 650hp Rolls-Royce Condor IIIA engines. Capable of 130km/h (80mph) the R.100 flew to Canada and back in 1930, the return journey taking just 58 hours. Built at the same time as the R.100 was the R.101, which crashed and was destroyed in late 1930. After this disaster the R.100 was taken out of commission and scrapped without making a further flight.

Germany, forbidden by the Versailles Treaty to build aeroplanes, lost little time in starting up a post-war airship service. Delag was resuscitated and on 24 August 1919, a new purpose-built passenger-carrying airship, the *Bodensee*, started to operate between Friedrichshafen and Berlin. So successful was this service that a second vessel, the *Nordstern*, was called for. But before it could enter service the Allies put a stop to the operation.

At the end of World War I most of Germany's Zeppelin airships had been destroyed by their crews instead of being surrendered to the Allies in accordance with the Armistice agreement. America had been due to receive at least one of the Zeppelins and was seeking financial compensation. In Germany, meanwhile, the fortunes of the Zeppelin Company were at a low ebb. Count von Zeppelin had died in 1917, and direction of the company had passed to Dr Hugo Eckener, the Count's close associate from the early days of rigid airships. Eckener suggested to the Americans that his company should build them a new airship. That way America would get her airship, and Zeppelin would stay in the forefront of airship development. The resulting LZ.126 began flight tests in the summer of 1924 and, on 12 October 1924, was flown across the Atlantic from Friedrichshafen to Lakehurst, New Jersey, in a time of 79 hours.

The uneventful Atlantic crossing was to be typical of this airship's service life. Designated ZR-3 and named USS *Los Angeles* in US Navy service, she accumulated 5,368 flight hours before being retired in 1932. She was subsequently recommissioned and used occasionally before being scrapped altogether in 1939.

Zeppelin was to build two more superb vessels, the *Graf Zeppelin* and the *Hindenburg*. The former, designated officially LZ.127, was 236m (774·28ft) long and had a maximum diameter of 30·5m (100·07ft). Volume was 105,000m³ (3,708,020cu ft), of which 30,000m³ (1,059,434cu ft) was occupied by blaugas fuel and the remaining 75,000m³ (2,648,586cu ft) by the hydrogen lifting gas. Power was provided by five 530hp Maybach engines each mounted in a separate gondola. Luxury accommodation was provided for 20 passengers, and up to 12 tons of mail or other types of cargo could be carried.

First flown on 18 September 1928, the *Graf Zeppelin* made a round-the-world flight in 21 days, 7 hours and 34 minutes between 8 and 29 August 1929. She last flew during 1939 off the east coast of England, unsuccessfully attempting to monitor British radar installations. Before being scrapped finally in 1940, LZ.127 had completed 590 flights, including 140 Atlantic crossings, carried 13,100 passengers, and travelled 1,695,252km (1,053,383 miles) in 17,178 flying hours.

But even while the *Graf Zeppelin* was going from success to success, an even larger sister ship was being created by the Zeppelin Company. When completed in March 1936 the LZ.129 (or *Hindenburg*) was the world's largest rigid airship: 245m (803·81ft) in length, with a maximum diameter of 41m (134·51ft) and a volume of approximately 200,000m³ (7,062,895cu ft). Its four 1,000hp Daimler diesel engines gave a maximum speed of

Germany's superb airship *Hindenburg* explodes in flames while mooring at Lakehurst, New Jersey, on 6 May 1937. This disaster brought an end to the development of the airship as a long-range passenger-carrier.

130km/h (81mph), and this giant vessel could accommodate 75 passengers and a crew of 25. Entering service in the summer of 1936, the *Hindenburg* carried out a number of North and South Atlantic crossings. Then, on 6 May 1937, a tongue of flame broke out at the top of the envelope near the stern as the great ship was approaching the mooring mast at Lakehurst, New Jersey. Within a matter of seconds *Hidenburg* was a blazing mass of wreckage from which 62 of the 97 people on board miraculously escaped with their lives.

This disaster, coming on the heels of the R.101 tragedy and accidents to two other large dirigibles in US service, marked the end of German attempts to create a worldwide fleet of commercial transport airships. It also put a full stop to the furtherance of development work in this field by other nations, notably Italy and the United States.

The most successful of the large, non-rigid Italian airships built after the war was the N.1, later named *Norge*. In this ship a crew of 16, including the American Lincoln Ellsworth, Colonel Umberto Nobile and the Norwegian explorer Roald Amundsen, reached the North Pole on 12 May 1926. The entire flight, about 4,830km (3,000 miles) from Spitzbergen to Teller, Alaska, took approximately 71 hours.

A little more than two years later Umberto Nobile led a polar expedition of his own in a similar semi-rigid, the N.4 *Italia*. The *Italia* and her crew reached the north pole on 24 May 1928, only to crash into the pack ice about 282km (175 miles) north-east of Spitzbergen on the morning of 25 May. The recue of some of the survivors was an epic of courage in which many lives were lost, including that of Roald Amundsen.

The United States Army and Navy have at differing times been the world's major operators of military airships. The US Army's heyday lasted from 1921 to 1935, with the Navy continuing during World War II. In 1921 an agreement between the Army and Navy resulted in the former concentrating on the use of non-rigid and semi-rigid types for coastal and inland patrol, a division which lasted until the termination of US Army lighter-than-air operations in 1936.

As well as some notable rigid airships, the US Navy acquired a total of 241 non-rigids between 1917 and 1958. Some of these were excellent, fast vessels: the final examples of the N series, delivered between 1958 and 1960, had lengths of 122·83m (403ft), volumes of 42,929m³ (1,516,000cu ft), and maximum speeds of 145km/h (90mph). Of the

Navy's rigid airships, the most successful was the German-built ZR-3 USS *Los Angeles*. Built subsequently in America by the Goodyear-Zeppelin Corporation were the 239·27m (785ft) ZRS-4 *Akron* and ZRS-5 *Macon*. Fighter aircraft were carried inside and could be launched and recovered on a trapeze. Both of these airships were lost at sea: the *Akron* on 4 April 1933, with the loss of all but three of her crew; the *Macon* on 11 February 1935, with the loss of only two lives. This brought an end to the Navy's rigid airship programme.

Airships survive in the jet era
Although the USA no longer uses airships for military purposes, Goodyear has retained a fleet of four non-rigids for publicity purposes. Goodyear can claim to have built more airships than any other company in the world. Its 303rd, the *Mayflower*, was completed in 1978.

Hot-air airships have been flown by Cameron in Britain (1973) and Raven Industries in America (1975). The *Albatross*, which first flew in October 1975, was designed by a husband and wife team, Brian and Kathy Boland.

Non-rigid airships designed in recent years include the WDL-1, a 55m (180ft 5in) helium-filled airship built by WDL in Essen-Mülheim, West Germany, and first flown in August 1972. British non-rigids include the 23·16m (76ft) helium-filled *Santos-Dumont* constructed by Anthony Smith and first flown in May 1974, and also the 50·00m (164ft) Aerospace Developments AD-500 which flew in early 1979. Japan had several versions of the Fuji Model 500, small research airship, undergoing flight tests in 1976, and other examples were also being built.

Airships under construction or flying in the United States include the Hov-Air-Ship HX-1, a 5.79m (19ft) long remotely controlled experimental airship, and the Tucker Airship Company's TX-1 semi-rigid, which is 27.74m (91ft) long, and which was scheduled to make its first flight in the late summer of 1979.

Some of these craft have been built 'just for fun', while others are intended for roles which – in a period of aviation history when money and fuel are both in short supply – their designers believe can best be carried out by airships. These tasks include maritime patrol; cargo carriage to points of difficult access; aerial survey and photography; meteorological and pollution measurement; distribution of seeds, fertilizers and insecticides; logging, and finally, advertising.

Missiles and Space Exploration

Missiles

The year 1927 had seen the foundation of the *Verein für Raumschiffarte. (VfR)*, the German Society for Space Travel. For the rocket enthusiasts it was a time when small amateur groups could still make important contributions by building and firing rockets.

Leading members of the *VfR* included Hermann Oberth, Walter Hohmann, Guido von Pirquet, Max Valier, Rudolf Nebel, Klaus Riedel, Kurt Hainish and Willy Ley. A youth of 18 who joined in 1930 was to become a legend in his own time: Wernher von Braun.

By then the liquid-fuel rocket had been firmly established and in August 1931 the *VfR* launched a Repulsor rocket to an altitude of 1,006m (3,300ft) which floated back to the ground by parachute. The experiments at the *Raketenflugplatz* began to attract international attention, but Germany was in the grips of an economic depression and membership of the *VfR* fell significantly. Money ran short and the Berlin authorities – quite understandably – objected to rockets being fired within the city limits.

It was clear that further progress would have to depend on gaining some kind of government support. Accordingly, in the summer of 1932, Nebel and von Braun set up a demonstration for the German Army near Kummersdorf. It was an appropriate moment. Hitler was a 'rising star' (he came to power in 1933) and the Army recognized that rockets were outside the scope of the Versailles Treaty which forebade the manufacture of aircraft in Germany. By the end of the year work at the *Raketenflugplatz* had ended and von Braun had received an invitation to conduct experimental work for his doctor's thesis on combustion at the Army's proving ground.

Other early experiments

By that time the *VfR* had fired the enthusiasm of people in other countries and March 1930 saw the foundation of the American Interplanetary Society (later the American *Rocket* Society), which carried out a wide range of experimental work with liquid-fuel rockets before World War II. The British Interplanetary Society, barred from making rocket experiments under the Explosives Act of 1875, nevertheless made a number of significant theoretical contributions to rocket flight, including the first engineering concept of a vehicle for landing on the Moon. The Society, founded in October 1933, is still active today.

The V-2

With the resurgence of German militarism, huge sums of money became available to Werner von Braun's research team, from both the German Army and the Luftwaffe. This allowed work to begin in 1935 on a large experimental rocket establishment near the village of Peenemünde on the Baltic coast.

The major project was to be a large artillery rocket for which the designation A-4 had already been chosen. To achieve it, features of the design would be tested on a small-scale A-5 rocket. Launched at Griefswalder Oie in mid-1938, without the guidance system, the rockets reached heights of 12,875m (42,240ft). Over the next two years different control techniques were tried and some A-5s, launched on inclined trajectories to achieve maximum range, were made to fly along radio guide beams.

Thus the concept of the A-4 rocket was born, designed to carry a 1,000kg (2,205lb) warhead over a distance of 275km (170 miles). This involved major advances in almost every department of rocket engineering. The A-4, which weighed more than 12,500kg (27,557lb), did not, however, achieve immediate success at Peenemünde. The first example, set upright on its fins on a small launch table, failed to lift off as the thrust of its engine gradually faded; it toppled over and exploded. The second A-4, held stable by its gyro-controlled exhaust vanes, flew perfectly for more than 45 seconds, then began to oscillate sideways, and broke apart in the air.

After necessary strengthening modifications had been made, the third A-4 made a perfect ballistic flight, reaching a maximum height of 85km (53 miles) before splashing into the Baltic some 190km (118 miles) from the launch pad. The date was 3 October 1942 – World War II had already been in progress for three years. In the eyes of the German High Command, the rocket had now become something far more important than a mere artillery weapon. Before adequate trials could be carried out, Hitler ordered the A-4 into large-scale production.

Mobile A-4 batteries were organized and rockets were deployed at coastal sites. The A-4 attack on London began on 8 September 1944, from a site near the Hauge in Holland; and at last the Goebbels propaganda ministry was able to reveal the existence of the V-2-Retaliation Weapon No 2 – against which there was no available defence.

Some 4,320 V-2s were launched between 6 September 1944 and 27 March 1945. Of these some 1,120 were directed against London and the London Civil Defence area, killing 2,511 people and seriously injuring nearly 6,000 more.

A multiple rocket launcher, as used by the Mongols during the mid-to-late 13th century, probably represented the very first use of unguided missiles in warfare. Britain's William Congreve played a significant role in developing the simple rocket as a war weapon.

When the war in Europe ended, the full extent of the V-2 operation was at last open to inspection, including the extensive underground factory in the Kohnstein Hills, near Nordhausen, where V-1 flying bombs and V-2 rockets had been made on a production line basis. The principals at Peenemünde, in the face of the Russian advance, had fled westward, preferring to place themselves in the hands of the Western Allies. General Dornberger and von Braun surrendered to the welcoming and open arms of the US 7th Army on 2 May 1945.

Wernher von Braun's interrogation report showed the depth of progress that had been achieved by the Peenemünde team, and that it had kept alive the spaceflight ideal despite the pressures of the military establishment. In order to improve the range of the V-2, test rockets had been fitted with swept-back wings to see if they would glide in the upper atmosphere. One had flown with some degree of success in 1944, and drawings existed for a version in which a pressure cabin replaced the warhead so that a man could fly in space. Even more ambitious was the A-9/A-10 project in which an improved A-4, with dart-like wings, was to be launched in flight from the nose of a huge liquid-fuel booster to fly some 5,000km (3,100 miles).

Although the scheme never got beyond the design stage in 1942, it was a clear indication of the shape of things to come.

Initial post-war developments
At the end of the war, Germany stood head and shoulders above all others in the technology of rockets and guided weapons, and major efforts were made by America and Russia to round up as many rocket specialists as could be found. Many of the principals, including von Braun and a large part of his Peenemünde team, were encouraged to continue their work in the United States. Others were taken to Russia and put to work in isolation from mainstream Soviet technology, for extra secrecy.

In the meantime, work in America, spurred by the pioneering efforts of Robert H. Goddard (who died in August 1945), had led to a research group at the California Institute of Technology to developing in 1944 to 1945 a small rocket called WAC-Corporal. It was powered by a nitric acid/aniline engine of 680kg (1,500lb) thrust. Launched from a tower by a solid-fuel booster, the rocket sent a payload of instruments to an altitude of 70km (43·56 miles).

Surface-to-air missiles
One of the first operational missiles to employ storable liquids was America's MIM-3A Nike-Ajax, the ramp-launched anti-aircraft missile which took off with the assistance of a three-finned solid-fuel booster. The missile itself had triangular cruciform wings and steerable nose vanes and was powered by a nitric acid/aniline motor. Guidance was by radio command. One radar tracked the target aircraft and another tracked the missile. A computer continuously integrated the two plots and caused steering signals to be sent to the missile, vectoring it towards the target. When fairly close, the missile became self-homing, responding to ground-radar reflections from the target.

Its Russian counterpart, (SA-2, NATO code-name Guideline – Russian missiles are identified by their NATO code names throughout) burned nitric acid and a hydro-carbon fuel and was also guided by automatic radio command. Evidence of the effectiveness of this weapon came on 1 May 1960, when it brought down Lt Francis Gary Powers in his high-flying U-2 reconnaissance aircraft.

Other anti-aircraft weapons followed in quick succession. In America the MIM-14A Nike-Hercules gave more height and range, as did the LIM-49A Nike-Zeus, which had a small nuclear warhead. Bomarc, an aircraft-like missile with cropped delta wings, had twin underslung ramjets and a rocket booster.

Man-portable guided weapons have also made their mark for the self-defence of army units against low-level air attack. America's tube-mounted Redeye, launched from the shoulder, was designed to home automatically on the engine heat of a departing aircraft. Russia's equivalent – the SA-7 Grail – is believed to have a slant range of 2·9 to 4·0km (1·8 to 2·5 miles). Britain has produced the Blowpipe, a similar weapon but which the operator steers along his line of sight by radio signals.

The post-war years also saw a major growth in arms exports as the leading powers manoeuvred on the chessboard of world influence.

In the Yom Kippur War of 1973, Israeli pilots faced Soviet anti-aircraft weapons deployed in depth. While batteries of Sa-2 Guideline missiles engaged their aircraft at long range, SA-3 Goa took on low-flying targets. A relatively new anti-aircraft missile, the SA-6, was fired from triple mounts on tracked vehicles which advanced with the Egyptian infantry, giving them effective cover. Of 104 Israeli aircraft lost in 1973, 100 are said to have fallen to missiles.

Tactical missiles

Battlefield weapons include large, unguided and spin-stabilized rockets with conventional high explosive, chemical or nuclear warheads launched from tracked or wheeled chassis. American examples are the MGR-1B Honest John, MGR-3A Little John and MGM-52B/C Lance. Russia has produced a whole family of such weapons known to the West as Free Rocket Over Ground (FROG). More complex missiles of medium range – some able to attack near battlefield areas – have also been developed by both countries.

Early missiles were beam-riders, centring themselves on radio beams aligned with the target; others were guided by radio command. Such systems, however, are vulnerable to electronic counter-measures and they were rapidly superseded by inertial guidance systems with pre-set target instructions which could not be 'jammed'. The MGM-13B Mace – an American battlefield missile in the form of a ramp-launched pilotless aircraft – had a Goodyear Atran terrain/map-comparison guidance system, a forerunner of the highly advanced terrain-comparison systems used in America's latest hedge-hopping cruise missiles.

The German V-2 of World War II was the first strategic intercontinental ballistic missile. Here it is on its launcher being elevated to the firing position.

France too has developed its own surface-to-surface battlefield missile, Pluton. This highly effective solid-fuel weapon has inertial guidance and operates from the AMX-30 tracked launcher.

Nor has the infantryman been neglected. To meet the challenge of the new battle tanks, a whole range of anti-tank missiles have become operational with the world's armies. The MGM-51A Shillelagh – developed for the US Army – was launched from a combination gun/launcher carried by a variety of vehicles, including the General Sheridan lightweight, armoured reconnaissance vehicle and the M60A1E2 medium tank. The missile itself had flip-out fins and was steered by deflecting the exhaust of the sustainer rocket according to a series of signals from a microwave guidance system. The operator kept his optical sight fixed on the target and the missile responded automatically.

Another significant US anti-tank weapon is the MGM-71A Tube-launched, Optically sighted, Wire-guided (TOW). The missile and its tripod-mounted launcher can be operated by two men. Its effective range is 1,830m (6,000ft). In Vietnam, TOW was mounted in launch pods carried by Huey helicopters. All the 'gunner' had to do was hold a target such as a supply truck in the cross-hairs of his stabilized sight. The pilot could do limited evasive manoeuvres after missile launch and the weapon was so effective that it was quite capable of scoring a bullseye at a range of 3,000m (9,840ft).

When Egyptian troops stormed across the Suez Canal on 6 October 1973, special units engaged Israeli tanks with Russian-built Sagger missiles, which they unpacked from suitcase-like containers. With this weapon, the infantryman attaches the warhead to the missile and unfolds the guidance fins. A launch rail is set up on the container's lid and Sagger mounted on it. Wire connections can then be run out to a firing unit which the operator uses to launch and steer the missile in conjunction with an optical sight and a joystick. With this and other weapons some 200 Israeli tanks were destroyed in one day.

Air-to-air missiles

During World War II, Germany expended much effort on air-to-air guided weapons. The 90·7kg (200lb) Henschel Hs298 was actually put into production in 1945 just as the war was ending. It took the form of a small swept-wing monoplane powered by a dual-thrust solid-fuel motor, with fast- and slow-burning propellants for boost acceleration and sustained flight, one inside the other. Its maximum speed was about Mach 0·8. Launched at a distance of up to 1·6km (1 mile) from the target, it was radio-controlled and had a warhead which was detonated by a proximity fuse.

A more advanced weapon was the Ruhrstahl X-4, which had a crude form of self-homing and a maximum range of some 3·2km (2 miles). Like some other German guided weapons, it was impervious to radio jamming since control signals were transmitted through fine, 0·2mm (0·008in) insulated wires which unwound from bobbins on the wings and connected it to the launch aircraft.

The missile, which had cruciform wings, was spin-rotating and controlled by solenoid-operated spoilers on the tail fins. In normal flight these vibrated at 5 cycles per second but, in order to change the flight path, were made to operate asymmetrically by control signals from the launch aircraft. The smaller X-7 variant was designed by the same team for air-launching against ground targets.

The lessons learnt in Germany were quickly applied by the United States, the USSR, Britain and France. Early examples of American air-to-air missiles were the AIM-4E/F Falcon, which armed the F-101, F-102 and F-106 interceptor/fighters, and the AIM-7E Sparrow IIIB, which became standard armament on the F-4 Phantom. Both employ semi-active radar homing and depend on continuous-wave radar beams being directed from the launch aircraft and they respond to the reflections from the

target. Sidewinder, on the other hand, had an Infra-Red (IR) seeker which steered the missile towards any heat source of a certain intensity, such as an engine or its exhaust.

The Soviet Union replied with a large solid-fuel missile (NATO code-name Alkali) which armed all-weather versions of the MiG-19. Like Sparrow and Falcon, it homed on the reflections from the target of the interceptor's radar signals.

Russian designers have generally tended to develop two versions of their air-to-air missiles, one with semi-active radar homing and the other with an IR homing head. Both versions are then flown on the same interceptor to enhance the probability of a kill. Two versions of Anab, for example, are carried by the Yak-28P, Su-9 and Su-11 all-weather interceptors. The Tu-28P can accommodate four of the larger AA-5 Ash missiles, two with radar homing and two with IR.

Britain, too, was well to the fore with missile armament during the 1950s and 1960s. Firestreak, an IR-homing pursuit course weapon, became operational on Lightning and Sea Vixen interceptors. It was followed by the IR-guided Red Top, which could attack from virtually any direction by virtue of a lead-collision IR guidance system.

One of the most promising of the new breed of air-to-air weapons is Skyflash, which has been adapted in Britain from the well tried Sparrow airframe. Using a new and advanced semi-active radar seeker, it is designed to engage targets from sea level to high altitudes and can 'snap down' to intercept low-flying aircraft. It is claimed to be particularly effective in electronic-warfare environments where the enemy is operating 'jammers' and other devices.

Air-to-surface missiles

Germany played a leading role in developing anti-shipping guided missiles in World War II when glide-bombs were used against Allied merchant shipping. The Hs 293, which took the form of a small monoplane, had an underslung rocket motor which gave a thrust of 590kg (1,300lb) for 12 seconds. The maximum speed was about 600km/h (375mph).

Released at 8 to 9·7km (5 to 6 miles) from the target, the Hs 293 was radio-controlled from the launch aircraft, where the pilot visually tracked a flare in the tail of the missile. Experiments were also made with a crude television system to give the crew of the launch aircraft a missile-eye view of the approaching target. There was even a scheme to control such relatively long-range glide missiles by wire transmission to prevent the Allies jamming the

control signals. An enlarged version, developed in 1944, was the Hs 295.

The Hs294 was a more advanced weapon, designed to attack warships in a spectacular manner. After being visually guided by radio from the launch aircraft, the missile entered the sea within some 45/7m (150ft) of the target ship, shedding the wings and rocket motor to complete the attack under water. It was then exploded below the waterline either by the action of a proximity fuse or automatically at a predetermined depth.

Air-to-surface (tactical)

The need to pick off point targets in Vietnam – especially bridges defended by anti-aircraft missiles, and single vehicles carrying supplies from the North – led the Americans to introduce the so-called 'smart' bomb. Paveway was a conventional bomb fitted with a laser seeker and extra control surfaces. A laser beam aimed at the target by a supporting aircraft meant the bomb, released from an attack aircraft such as a Phantom, could steer itself towards the bright spot of laser energy reflected from the target. This allowed it to score a direct hit. From such devices have come a whole range of precision guided missiles which allow launch aircraft to leave the scene immediately the weapon has been released.

Another example of precision guidance is found in the AGM-65A Maverick. It has a small television camera which focuses on the target, steering the missile on to it with lethal effect.

A European example of a precision tactical missile is Martel, which exists in two versions. One, built in Britain, is TV-guided; the other, built in France,

Russia's SA-5 Gammon, a surface-to-air missile which is claimed to be equally effective against high-flying aircraft and rocket-powered missiles, is shown on its transporter.

homes on to radar emissions.

So effective are such techniques that TV-guided Mavericks have been launched against ground targets from radio-controlled pilotless drones, removing the human element entirely from the attack zone.

Air-to-surface (strategic)

Efforts to enhance the ability of strategic bombers to reach their targets led to air-launched 'stand-off' missiles which fly on as the bomber turns away. Among the first of such weapons was the AGM-28B Hound Dog carried by Boeing B-52Gs and Hs of the US Air Force's Strategic Air Command. Powered by an underslung turbojet, this 4,355kg (9,600lb) winged missile had a range of some 1,110km (690 miles).

Blue Steel, the British equivalent, was carried by Victor B.2s and Vulcan B.2s of RAF Bomber Command. It was powered by a Stentor two-chamber rocket engine fuelled by kerosene and High Test Peroxide. One chamber was for cruising flight, the other for a high-speed dash. The missile was 10·7m (35ft) long, had a maximum diameter of 127cm (50in) and a wing span of 3·96m (13ft). It was particularly effective in allowing bombers to 'hedge-hop' and attack at low level.

Russia is also prominent in this league. As early as 1961 a stand-off missile, about 15m (50ft) long and resembling a swept-wing turbojet fighter, made its appearance beneath the fuselage of a Tu-95 Bear B. It was given the NATO code-name Kangaroo.

Designed to attack peripheral targets within the NATO area, it was guided by radio from the launch aircraft.

Smaller winged missiles were developed mainly for anti-shipping duties. Two turbojet missiles which bear the NATO code-name Kennel are carried by the Tu-16 Badger B on underwing pylons. A small radar mounted above the engine air intake in the nose allows the weapon to home on to its target automatically. A land-based version for coastal defence is called Samlet.

A more recent air-launched weapon is the rocket-powered Kelt, two of which can be carried under the wings of the Tu-16 Badger G. Yet another winged missile for anti-shipping use is Kipper, which is belly-mounted on the Tu-16 Badger C. It resembles a swept-wing aircraft with an underslung turbojet.

Intercontinental ballistic missiles

By the end of 1947 the first launchings were being made of improved V-2 rockets built by Soviet engineers on Soviet soil. A secret launch base was used at Kapustin Yar (literally 'cabbage crag'), south-east of Stalingrad (now Volgograd).

The Russian version of the German missile had an RD-101 rocket engine. The same basic vehicle was later to appear as a geophysical rocket, the V-2-A. Experience with this vehicle led to the first Soviet short-range ballistic missile, known to the West as the SS-3 Shyster and to the Russians as Pobeda (Victory). The missile first appeared in a Red Square military parade in 1957, but US radar stations in Turkey had long since detected its early trials and listened in to its telemetry, radio signals which carry coded data.

The engine, developed between 1952 and 1953, was the RD-103 which burned kerosene and liquid oxygen to produce a vacuum thrust of some 50,000kg (110,230lb). America's equivalent was the US Army's Redstone, developed by the von Braun team at Redstone Arsenal, Alabama.

Next to appear in Russia was the SS-4 Sandal, powered by a four-chamber RD-214 engine of 74,000kg (163,140lb) vacuum thrust. Developed between 1955 and 1957, this engine had variable thrust in flight, achieved by adjusting the supply of propellants to the gas generator of the turbo-pumps. It was this missile which featured in the Cuban missile crisis of 1962, when Nikita Khruschev tried to put pressure on the United States by stationing offensive missiles on its 'doorstep'. Reconnaissance photographs of Cuba showed that concrete had been poured for launch pads with support vehicles

and missiles drawn up around the sites. The United States was within range of a missile capable of carrying a nuclear warhead over 1,770km (1,100 miles). Operational flexibility had been greatly improved compared with earlier missiles because Sandal employed storable liquids – nitric acid and kerosene – and could thus be held ready, fully fuelled, for long periods.

Although the Cuban sites were dismantled at the end of the confrontation, and the missiles returned to the Soviet Union, the episode proved to be the final justification for the build-up of missile forces in the United States. Already, Thor intermediate-range missiles stood ready at four RAF bases in eastern England pending the development of the big Intercontinental Ballistic Missiles (ICBMs) which could reach Soviet targets from launch sites in the United States. These rockets, fuelled with liquid oxygen and kerosene, had a range of some 2,775km (1,725 miles), putting Moscow into the front line of any nuclear exchange. Thor, instead of emplying exhaust vanes to steer, had an inertial guidance system which issued commands to the rocket motor itself; the engine was gimbal-mounted and thus the thrust direction was changed for course corrections.

In the meantime the Russians had built a second major test centre north of Tyuratam in Kazakhstan. The big intercontinental rocket was rolled ponderously out to the launch pad for its first firing. In 1957 came the triumphant (if terse) announcement from the *Tass* news agency: ... a super long-distance international ballistic multistage rocket flew at ... unprecedented altitude ... and landed in the target area.

It was many years before the West was to be acquainted with the design, although the broad features were known to Western intelligence. At lift-off, no fewer than 20 main thrust chambers fired, as well as 12 smaller swivel-mounted engines for control. Sixteen of the main chambers were in four strap-on boosters which were thrown off sideways as the vehicle left the atmosphere. The central core of the rocket, powered by four main chambers, drove the vehicle to its final cut-off velocity as four Vernier motors responded to commands from the inertial guidance system. The warhead separated successfully and continued to its target on a ballistic trajectory.

The engine of the central core was the RD-107 of 96,000kg (211,650lb) vacuum thrust. Each of the four boosters had an RD-108 engine of 102,000kg (224,870lb) vacuum thrust. All of them burned liquid oxygen and kerosene. Launch pads for these huge rockets were elaborate and extremely vulnerable to nuclear attack. At the test centre they comprised a large reinforced-concrete platform linked with the preparation building by a standard-gauge railway. Rockets assembled horizontally on a transporter-erector were transferred to the launch pad by diesel locomotive. After being elevated into a vertical position on the pad, they were fuelled from rail-mounted tankers moved up on adjacent track.

America's reply to Korolev's ICBM was the $1\frac{1}{2}$-stage Atlas, which broke new ground by having thin but pressurized stainless-steel tanks which withstood the thrust loads of the rocket engines without heavy reinforcement. Fuel (kerosene) and oxidant (liquid oxygen) were kept in separate compartments.

The original Atlas tank, 18·29m (60ft) long and 3·05m (10ft) diameter had no internal framework and comprised the entire airframe from propulsion bay to nosecone. At the base was a single gimbal-mounted rocket engine with two small swivel-mounted engines outboard for roll control and Vernier adjustment of velocity. In a jettisonable skirt – into which the main sustainer engine projected – were two thrust chambers drawing their propellants from the same tanks.

All engines burned at lift-off, producing a total thrust exceeding 163,295kg (360,000lb). The outboard engines were jettisoned with the skirt after 145 seconds of flight. The sustainer accelerated Atlas to a velocity of about 24,140km/h (15,000mph), when it was shut off by the guidance system. Launch tests began at Cape Canaveral, Florida, as the Russians were preparing to launch their first ICBM at Tyuratam – but early success was to be denied. In June 1957, a test missile blew up soon after take-off and the following September another went off course and disintegrated completely in the atmosphere.

Only the booster engines were live in these initial tests. But when Atlas was launched with all engines firing, on 2 August 1958, success was complete; the rocket splashed into the Atlantic some 4,000km (2,500 miles) downrange. Although testing was far from complete, the US Defence Department lost no time in ordering the missile into production and Atlas D became operational on 'soft' sites in the United States in 1959.

In the meantime, a two-stage ICBM of more conventional design had been tested in January with a dummy top stage. The rocket – Titan I – employed the same propellants as Atlas and the engine gave a lift-off thrust of some 136,080kg (300,000lb). The

vulnerability of missiles on open sites to nuclear attack, however, had already made clear the need for protective measures. Efforts to improve the reaction time of Atlas D led to above-ground 'coffin sites' at Vandenberg Air Force Base (AFB), California, and at Warren AFB, Wyoming. These offered a degree of protection from blast effects. The missiles were stored horizontally and fuelled behind protective walls before being elevated into a vertical launch position.

Titan I was stored in a deep underground silo. When the launch command was given, the silo doors opened and the rocket was raised to the surface by an elevator. Like Atlas, however, Titan I suffered the disadvantage of having to have liquid oxygen loaded on board immediately before launch; this added precious minutes to its reaction time, which was about 15 minutes. An improved version of Atlas, the F model, was tested in 1961. This too was designed for protection in an underground silo.

In the Soviet Union, the two-stage ICBM known to the West as SS-7 Saddler, became operational in 1961. Unlike some other missiles of the period, it was omitted from Red Square military parades. Evidence from reconnaissance satellites showed it to be a two-stage missile over 30·5m (100ft) tall and some 3m (9·84ft) in diameter. It was thought to employ storable liquid propellants.

The Russians were then discovered to be installing missiles in underground silos from which they could be launched direct, the exhaust flame being diverted through vents on two sides of the silo. This rocket was thought capable of delivering a 5-megaton (MT) warhead over a range of more than 10,460km (6,500 miles). Another two-stage rocket of similar range, the SS-8 Sasin, became operational in 1963.

Strenuous efforts were being made in the United States to develop ICBMs which could be launched from inside their silos. By 1963, the two-stage Titan II was in service. This used fuel and oxidant which (as in some contemporary Russian missiles) ignited on contact. This big ICBM stands 31·39m (103ft) high, has a range of more than 8,047km (5,000 miles) and carries a warhead of over 5-MT yield. With propellants stored in the missile, Titan II has a reaction time of one minute from its fully hardened underground silo. Fifty-four of them are still operational at air force bases in Arizona, Kansas and Arkansas.

Russia replied to the new Titan with another two-stage rocket, also employing storable propellants and silo-launched. This is the ICBM known in the West as the SS-9 Scarp, which entered service in

1965 and has since appeared in four different versions. The so-called Mod 1 version carries a single re-entry vehicle of about 20-MT yield; Mod 3 has been used to test Fractional Orbit Bombardment System (FOBS) vehicles.

The missile build-up of the 1960s, however, was only beginning. The need to achieve faster-reacting missiles which could be held at instant readiness underground, or deployed in mobile launch systems, led to solid-fuel rockets of high performance. At the same time, advances in inertial guidance and reductions in the size and weight of thermonuclear warheads allowed ICBMs of high destructive power to be designed very much smaller than Atlas and Titan.

In the United States, the Minuteman I ICBM was officially declared operational in December 1962, only four years after the programme began. Eleven months later, Minuteman Wing I at Malmstrom AFB, Montana, and Wing II at Ellsworth AFB, South Dakota, were fully operational and missiles were being installed at Minot AFB, North Dakota. No fewer than 1,000 silos were built for Minuteman ICBMs, which lie hidden below ground in the wheat and cattle country of North and South Dakota, Wyoming, Montana and Missouri. Each flight of ten missiles is controlled from a metal capsule located 15m (50ft) underground and which is manned 24 hours a day, on a rota basis, by pairs of Strategic Air Command officers.

The original three-stage missiles have now been replaced by new models of increased range, accuracy and warhead power. The latest version, Minuteman III, carries three Multiple, Independently targetable Re-entry Vehicles (MIRVs). The MIRVs are carried in a post-boost 'bus' and are discharged against different targets according to instructions from a computer in the missile.

The Soviet counterpart of Minuteman I is the silo-launched SS-13 Savage, which entered service in 1968. It is believed to carry a single warhead of between 1- and 2-MT yield or three multiple Re-entry Vehicles (MRVs) which explode in the same general area as each other but which are not individually guided towards specific targets.

Soviet designers developed the SS-13 on a modular basis. The second and third stages were used in a tracked transporter-erector (Scamp) to achieve a mobile system which can be driven to different parts of the country and operated from points of conceal-

A Polaris A-3 submarine-launched missile breaks the surface of the sea after leaving its under-water carrier before streaking away to a pre-programmed target.

ment, among trees or in mountain passes for example. Some have been deployed near the border with China. A longer-range development of the same launch technique is Scrooge, which carries a larger missile (possibly also derived from the SS-13) in a long tubular container. The SS-11 Sego liquid-propellant rocket, however, although in the same general class as the SS-13, had greater performance and was produced in greater numbers.

France joined the 'ballistic club' in August 1971, when the first squadron of nine two-stage, solid-fuel *Sol-Sol Balistique Stratégique* (SSBS) missiles became operational in silos on the Plateau d'Albion in Haute Provence, Southern France. These S1 missiles could launch a 150-kT warhead a distance of 3,000km (1,864 miles). Two years later a second squadron of nine S2 missiles became operational. These have a range of 3,150km (1,957 miles). In 1979 to 1980, these early missiles are to be replaced by the slightly smaller but more effective S3, which is designed to carry a 1·2-MT thermonuclear warhead. Manufactured at the French Atomic Centre at Pierrelatte, warheads of this type have been tested amid environmentalists' protests at the *Centre d'Expérimentation du Pacifique*, Fangataufa Atoll, near Mururoa.

China, too, began to build ballistic missiles in the 1960s. The mid-70s saw the introduction of multi-stage ICBMs of limited range – the CSS-3. Installed in underground launch cells in north-west China they are capable of reaching Moscow and the European part of the USSR. A modification of this rocket is believed to be responsible for launching China's largest observation satellites which eject data capsules for recovery, e.g. China 4 on 26 November 1975 and China 7 on 7 December 1976.

The latest development, the CSSX-4 – a true ICBM with a range of some 11,000km (6,835 miles) – brings the United States within range of thermonuclear warheads. It is expected to be operational about 1980 in limited numbers.

Submarine-launched Ballistic Missiles

The increasing vulnerability of land-based ICBMs to thermonuclear attack led to the super-powers installing ballistic missiles in submarines. America began with Polaris A-1, which first became operational aboard the Fleet Ballistic Missile (FBM) submarine USS *George Washington* on 15 November 1960, with a full complement of 16 missiles. By 4 October 1967, the forty-first FBM submarine, USS *Will Rogers*, had set out on patrol.

Thirty-three of these vessels were assigned to the Atlantic Fleet and eight to the Pacific, target selection and assignment being under the control of the Joint Chiefs of Staff. The first five vessels, which originally carried Polaris A-1, were re-fitted to carry the longer-range Polaris A-3; most of the 616-class vessels have since been converted to accommodate the larger-diameter Poseidon C-3 missile.

Broadly, the development history of these two-stage sea-going missiles reflects the increasing demand for more range, payload and accuracy. The Polaris A-2, slightly longer than the original, led to the A-3 which was a big advance on both earlier models. The Navy-industry team described it as 'an 85 per cent new missile', with 60 per cent more range and only a minimal increase in overall size. In fact, it had the same diameter of 137cm (54in) and was just 30·5cm (12in) longer. The maximum effective range was about 4,635km (2,880 miles).

Britain abandoned the Blue Streak Intermediate-Range Ballistic Missile (IRBM) in April 1960 in favour of the American Skybolt air-launched ballistic missile. Skybolt, however, was subsequently cancelled by the Americans, and by the Nassau agreement of December 1962 Britain was to buy Polaris missiles, without warheads. By the end of 1969 the base at Faslane on the Clyde in Scotland had been built and all four of the British-produced Polaris submarines had been commissioned.

They have been named HMS *Resolution* (launched September 1966), HMS *Renown* (February 1967), HMS *Repulse* (November 1967) and HMS *Revenge* (March 1968). Each carries 16 Polaris A-3s with warheads of British design and manufacture. Like their American contemporaries, each submarine thus has more firepower at its disposal than the total of conventional explosives used in World War II.

Russian submarine-launched missiles

Russia, too, was in the race and, after gaining experience with large and primitive solid-propellant missiles carried in both diesel- and nuclear-powered submarines (in two and three vertical launch tubes extending from the base of the hull through an extended bridge fin), much improved missiles began to appear.

Russian Yankee-class nuclear submarines have 16 SS-N-6 Sawfly missiles apiece in the pressure hull. The two-stage Sawfly has appeared in three versions: Mod 1 has a single re-entry vehicle of about 1-MT yield; Mod 2 has improved performance; and Mod 3 has three MRVs. Late models have a range of some 3,000km (1,865 miles).

Greater surprises were in store when high-resolution photographs of Soviet submarine yards were secured by US Air Force reconnaissance satellites. At Severodvinsk they uncovered the new Delta-class submarines which carry the SS-N-8 ballistic missile.

For a submarine-launched missile, SS-N-8 performance is outstanding. Test firings carried out from a submarine in the Barents Sea in October 1974 resulted in high-accuracy impacts in the central Pacific, suggesting that the United States could be attacked from as far away as the North Sea.

The SS-N-8 is believed to obtain a mid-course navigational fix from a stellar-inertial guidance system. It is thought to employ a storable liquid propellant of the type which the Americans dismissed as too dangerous in their own FBM submarines in the face of counter-launched missiles, torpedoes and nuclear depth charges.

America's counterpart of the Delta 2 submarine is the Trident, which is planned to carry no fewer than 24 missiles with ranges of between 7,400–7,800km (4,600–4,850 miles). Each UGM-93A Trident I missile can carry eight MIRV warheads of 100-kT yield. Associated with Trident is the world's most advanced re-entry vehicle, the General Electric Mk 500 Evader. This is the first Manoeuvrable Re-entry Vehicle (MARV), which can make abrupt directional changes in flight in order to confuse anti-ballistic missile defences.

The Chinese also have begun to develop their own ballistic missiles for launching from nuclear powered submarines.

France, too, has a fleet of FBM nuclear-powered submarines, each of which carries 16 two-stage solid-fuel *Mer-Sol Balistique Stratégique* (MSBS) missiles of French design and manufacture. The M1 and M2 versions carry a 500-kT warhead with ranges of 2,593km (1,610 miles) and 3,148km (1,955 miles) respectively. The M1 was installed in the submarines *Le Redoutable* and *Le Terrible*, and the M2 in *Le Foudroyant*. M20 missiles with 1-MT thermonuclear warheads are carried in *L'Indemptable*. The fifth and sixth French missile submarines, *Le Tonnant* and *L'Inflexible*, to be commissioned in 1979 and 1983, will carry the larger M4 missile with six or seven multiple re-entry vehicles, each having a 150-kT yield.

Anti-ballistic missiles

The 'missile race' caused the super-Powers to seek ways of protecting vital targets from the worst effects of a nuclear attack. Since most ICBMs have multiple warheads and/or eject decoys, this is no easy task and a determined attack would surely saturate any practical defence system.

In the United States the Safeguard Anti-Ballistic Missile (ABM) system was developed largely as a means of protecting Minuteman ICBM bases. The first sites were earmarked for Malmstrom AFB, Montana, and Grand Forks AFB, North Dakota, each complex having 100 missiles – 30 long-range LIM-49A Spartan and 70 short-range Sprint. (However, see below.)

Controlled by a Missile Site Radar (MSR) the three-stage Spartan is silo-launched to intercept missile re-entry vehicles above the atmosphere. The third stage is ignited by ground command at a time which depends on the precise path of the incoming warhead as worked out by fast computers on the ground. Interception, carried out automatically by radar command, ends with the detonation of a thermonuclear warhead.

The two-stage Sprint was meant to catch ICBM warheads which had eluded Spartan *after* they had entered the atmosphere. In consequence it has a very high acceleration and is launched from its underground cell by a separate charge, the first stage igniting in mid-air. Rapid course changes after launch are brought about by a thrust vector control system by fluid injection into a single propulsion nozzle, and then during second stage flight by aerodynamic controls.

Associated with Safeguard are Perimeter Acquisition Radar (PAR) located at strategic points around the United States which provide the MSRs with preliminary tracking data. Supporting data comes from radars external to the United States and US Air Force early warning satellites which can detect the launch of ICBMs and Submarine-Launched Ballistic Missiles (SLBMs) from geo-stationary orbit by sensing energy emitted in the rocket's exhaust.

Similar developments have, of course, taken place in the Soviet Union. Four sites of 16 ABM-1 Galosh anti-ballistic missiles defend Moscow. They operate in conjunction with two Try Add radar sites which contain target tracking and guidance/interception radars. Advance warning of missile attack depends on large Hen House phased-array radars in peripheral areas of the country and Dog House acquisiton radars near the capital.

A new SH-4 exo-atmosphere ABM and a Sprint-type high-acceleration ABM have been tested together with more advanced radars. Advance warning satellites have also reached the test stage.

Space Exploration

Russia shook the world on 4 October 1957, by launching the world's first artificial satellite. This signal event in world history had been made possible by the big intercontinental ballistic missile developed by the team under Sergei P. Korolev. The satellite itself, called *Sputnik 1*, weighed 83·6kg (184lb). It took the form of a polished aluminium sphere of 0·58m (1·9ft) diameter with long whip antennae. The distinctive 'bleeps' of its radio transmitter, picked up by radio stations around the globe, signified the dawn of a new age.

America's position in the field of rocketry was seriously challenged, and President Eisenhower felt the need to re-assure the American public, on television and radio, that the nation's defences were sound. In Moscow there was unsuppressed glee. Enthusiastically, *Pravda* commented that 'The launching of the first satellite proved that . . . the possibility of flying to the Moon is reasonable and justified.'

In the United States efforts were redoubled to bring to the launch pad at Cape Canaveral the three-stage Vanguard rocket which had previously been announced as part of America's contribution to the International Geophysical Year (1957–58). The objective was the launching of a 9kg (19·8lb) scientific satellite. The first rocket with all stages 'live' was ready on the launch pad on 6 December 1957. Within its nose cap was a test satellite of 16·3cm (6·4in) diameter, which weighed just 1·47kg (3·25lb).

However, when Vanguard's engine was ignited it failed to develop sufficient thrust, and instead of blasting off into space the rocket toppled over and erupted in a sea of flames. The tiny satellite, thrown clear of the inferno, was still bleeping.

Almost immediately the Eisenhower administration agreed to allow the US Army Ballistic Missile Agency to develop their Jupiter C rocket as a satellite launcher. The concept had already been worked out by rocket scientist Dr Wernher von Braun and his associates, and it took a remarkably short time to bring it to the pad.

On 1 February 1958, the rocket – Juno I – lifted off from Cape Canaveral to put into orbit America's first artificial satellite, *Explorer 1*. The satellite swung round the earth in a wide, elliptical orbit and the data it sent back led to the discovery that the Earth is girdled by electrically-charged particles from the Sun, trapped by the Earth's magnetic field (ie, the Van Allen radiation belts, named after the scientist who devised the experiment).

Soon afterwards a Vanguard rocket succeeded in placing a 1·47kg (3·25lb) test satellite into orbit, which Nikita Khruschev promptly dubbed America's 'grapefruit' satellite. Further salt was rubbed into the wound when, two months later, Russia launched *Sputnik 3*, a big cone-shaped geophysical laboratory nearly 100 times as heavy as *Explorer 1*.

By then a major re-think of space policy was taking place in America, leading to the National Aeronautics and Space Act which was passed on 29 July 1958. This opened the way for the formation, from the nucleus of the existing National Advisory Committee for Aeronautics (NACA), of an important new body, the National Aeronautics and Space Administration (NASA), inaugurated on 1 October 1958. All space-related functions were to be transferred to NASA over the next four years, and the civilian space agency was thus able to acquire the services of the Army Ballistic Missile Agency and von Braun's concepts for large (Saturn) launch vehicles. In March 1960 von Braun was appointed director of NASA's Marshall Space Flight Center at Huntsville, and additional funds were given to support the Saturn rocket programme.

On 12 April 1961 Russian cosmonaut Yuri Gagarin became one of the immortals in aviation history, orbiting the Earth in the Vostok 1 spacecraft.

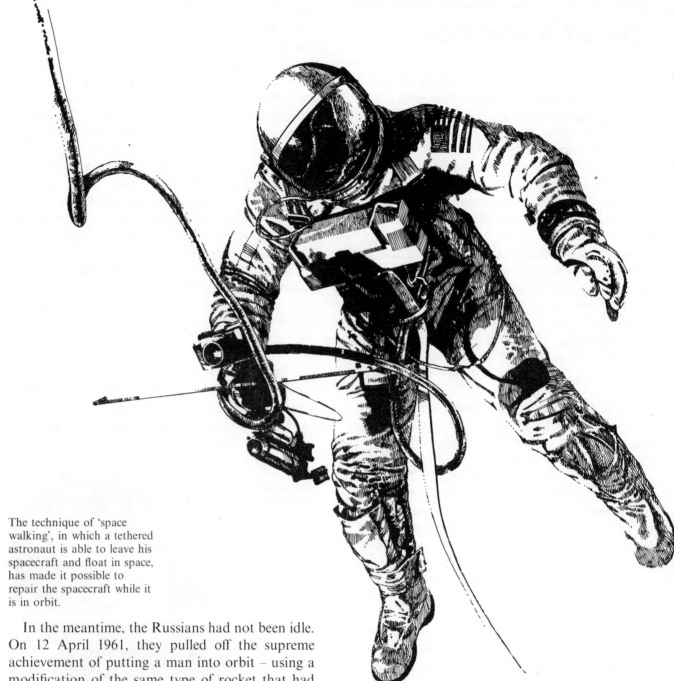

The technique of 'space walking', in which a tethered astronaut is able to leave his spacecraft and float in space, has made it possible to repair the spacecraft while it is in orbit.

In the meantime, the Russians had not been idle. On 12 April 1961, they pulled off the supreme achievement of putting a man into orbit – using a modification of the same type of rocket that had launched the first sputniks. That man was Yuri Gargarin. He travelled in a ball-like re-entry capsule, attached to a service module which kept the craft supplied with air (oxygen and nitrogen at about 1 atmosphere pressure). The cone-shaped service module had small thrusters for changing the craft's attitude in space and a retro-rocket which, fired at a backward angle, braked it for re-entry. Only the cosmonaut's capsule returned to Earth; the service module, which separated before re-entry, was burned up by friction in the atmosphere.

Gagarin made one circuit of the Earth in 108 minutes, proving that a human being could survive in space at least for relatively short periods, despite the high accelerations experienced at take-off and re-entry, a period of weightlessness and exposure to radiation while in orbit.

America was still some way from a similar achievement, having restricted initial tests to ballistic lobs over the Atlantic. On 5 May 1961, Alan Shepard, in the Mercury capsule *Freedom 7*, was launched by a Redstone rocket to reach a height of 187km (116 miles) and travel a maximum distance of 478km (297 miles). Just over two months later Virgil 'Gus' Grissom made a similar flight in the capsule

Liberty Bell splashing down 488km (303 miles) downrange. Although his capsule became waterlogged and sank, he was rescued in good shape by helicopter. (Grissom was later to lose his life with Ed White and Roger Chaffee in the Apollo spacecraft fire at Cape Canaveral on 27 January 1967).

These early American experiments had one advantage over those carried out by Russia. The astronauts themselves used hand controls for changing the attitude of their spacecraft in flight, providing valuable experience for America's first attempt to place a man into orbit. The opportunity fell to John Glenn, who lifted off from Cape Canaveral on 20 February 1962, aboard the capsule *Friendship 7* on the nose of a modified Atlas ICBM. Glenn completed three orbits of the Earth in a flight which lasted 4 hours 55 minutes. Although the Russian Gherman Titov had already made a spaceflight lasting over 25 hours, Glenn's flight was a milestone in American space exploration history.

More American and Russian manned flights followed in swift succession. Valentina Tereshkova, the first spacewoman, made a spaceflight of nearly 71 hours in June 1963. Russia launched a three-man spacecraft, *Voskhod 1*, in October 1964, and less than five months later Russia's Alexei Leonov 'space-walked' from the orbiting *Voskhod 2*.

America followed this with the two-man Gemini spacecraft, from which Ed White spacewalked in June 1965, manoeuvring in space with the help of a hand-held 'gas-gun'. In *Gemini 7* Frank Borman and James Lovell remained aloft for nearly two weeks, during which they flew close to *Gemini 6*, manned by Walter Schirra and Tom Stafford. Manned spacecraft docked with Agena 'targets' in orbit, and astronauts perfected the art of spacewalking.

Events were now moving swiftly towards a major American challenge. On 25 May 1961, President Kennedy told a joint session of Congress, 'I believe that this nation should commit itself to achieving the goal, before the decade is out, of landing a man on the Moon and returning him safely to Earth . . .' So Project Apollo was born.

Unmanned Moon probes

Before men could land on the Moon it was necessary to send unmanned probes to explore its surface. Russia succeeded first in crash-landing an object (*Lunik 2*) on the Moon, in 1959. In the same year they sent *Lunik 3* around the Moon to photograph the hidden side. Another major achievement was the landing of *Luna 9* in February 1966, which sent the first television pictures from the Moon's surface.

America sent Ranger spacecraft, which photographed the Moon continuously until they crash-landed on it. The three-legged Surveyor craft examined the soil by means of a mechanical scoop operated under remote control from Earth. Another American spacecraft, Lunar Orbiter, photographed large areas of the Moon from orbit. These robot forays proved that the Moon's surface was safe for a manned landing, and allowed landing sites to be selected for the forthcoming Apollo missions.

Men around the Moon

At last – in the face of circumlunar flights by Russian unmanned Zond spacecraft – NASA gave the go-ahead to launch three men into lunar orbit aboard *Apollo 8*. During Christmas 1968 Frank Borman, James Lovell and Willian Anders gave television viewers on Earth their first sight of the Moon's bleak surface from a distance of 112·6km (70 miles).

This was the turning point in the 'Moon Race'. After a final dress rehearsal by the *Apollo 10* astronauts, a manned landing was at last in sight.

After separating from the *Apollo 11* command ship in lunar orbit, Neil Armstrong and Edwin Aldrin began their descent onto the Moon's Sea of Tranquillity, skirting a boulder field to put their Lunar Module safely on to the surface. The date was 20 July 1969.

A large part of the world, linked by television, watched the misty image of Neil Armstrong slowly descend the ladder to jump lightly on to the virgin moondust. He was soon joined by Aldrin. The two men, quickly adjusted to low gravity conditions. They set up scientific instruments and obtained 21·8kg (48lb) of moon samples.

At the end of their nearly 22-hour stay, they blasted off in the ascent stage of the Lunar Module, leaving the descent stage behind. A metal plaque records their visit: 'Here, men from planet Earth first set foot upon the Moon, July 1969 AD. We came in peace for all mankind.'

By the end of 1972, five more Apollo craft had landed and 12 men had left footprints on the Moon.

More than 380kg (837·7lb) of lunar rock and soil were brought to Earth by the six successful missions, ranging from loose soil and surface rocks to core samples from *mare* ('lunar seas') and highlands. The mare rocks turned out to be relatively rich in titanium and iron, and the highland rocks had a greater abundance of aluminium. A typical *Apollo 11* soil sample contained: oxygen (40 per cent), silicon (19·2 per cent), iron (14·3 per cent), calcium (8·0 per cent), titanium (5·9 per cent), aluminium

(5·6 per cent) and magnesium (4·5 per cent).

Of the rocks taken from the first four landing sites, none was older than 4,200 million years nor younger than 3,100 million years. The soils were older than the rocks, and dated back 4,500 or 4,600 million years. The age of the crater Copernicus, dated by material found in its 'rays', appeared to be only about 850 million years.

A total of 166 man-hours had been spent in surface exploration at six different sites, and the astronauts had travelled more than 96km (59·6 miles). Five Apollo Lunar Surface Experiment Packages – science stations powered by nuclear energy – continued to send data to Earth years after the astronauts had departed. A total of 60 major scientific experiments in geology, geophysics, geo-chemistry and astrophysics were left behind on the Moon. Another 34 were conducted in lunar orbit from the command ships.

In the light of this tremendous achievement, one must inevitably ask what had happened to the Russian challenge. In 1968–69 cosmonauts seemed to be on the brink of making circumlunar flights. By linking up separate spacecraft modules in orbit, they might even – eventually – have landed on the Moon. The immense success of the Apollo programme, however, put Soviet plans for manned moonflight into abeyance, and effort was concentrated on perfecting automatic Moon samplers, able to land, collect soil and rock specimens, and bring them back to Earth. Even as the *Apollo 11* astronauts were preparing to land, a robot spacecraft, *Luna 15*, was being manoeuvred in lunar orbit under control from a Russian ground station. There was speculation that this was an attempt to obtain a Moon sample and fly it to Earth in a final bid to up-stage Apollo. In the event, the probe crash-landed.

In September 1970 *Luna 16* soft-landed, drilled 35cm (13·8in) into the Moon's surface, transferred a 100 gramme (3·5oz) core sample to a capsule and blasted it back to Russia. Two months later *Luna 17* succeeded in landing a Lunokhod roving vehicle, which the Russians steered over the lunar surface by remote radio control.

Interplanetary probes

Similar means of robot exploration were developed in the United States. With the 'Moon Race' won, Congress looked with disfavour on more manned missions as new priorities and the unpopular war in Vietnam began to engage its attention. Any long-term prospects of astronauts pressing on to Mars were abandoned.

Instead, research was directed to developing an increasing number of ingenious automatic vehicles to range deeper into the Solar System. Built at a fraction of the cost of manned spacecraft, they could be sent to places when men dare not go: deep into the crushing atmosphere of Venus, for example, and through the intense radiation belts of Jupiter.

In 1962 America had already flown *Mariner 2* past Venus and probed the planet with instruments which indicated surface temperatures above the melting point of lead. Russia then parachuted instrument capsules through the planet's thick carbon dioxide atmosphere to find that its surface pressure was 90 to 100 times that on Earth. There was strong competition between Russia and America to probe the secrets of Mars, with fly-by and orbital missions. The US *Mariner 9* spacecraft became the first artificial satellite of the Red Planet on 13 November 1971, as a dust storm raged on Mars' surface. When the dust storm cleared *Mariner 9*, scanning from orbit, discovered huge volcanoes, canyons and features like dried-up river beds.

At the same time Russia was attempting to land instrument capsules. The first was destroyed on impact; the second came down in the region between Electris and Phaethontis (about 48 deg S, 158 deg W), but unfortunately just as it began to send a picture.

In 1974 *Mariner 10* flew past Venus on its way to Mercury, showing for the first time that the tiny planet nearest the Sun had a moon-like surface of craters, mountains and valleys.

Pioneers 10 and *11* swung around the giant planet Jupiter before departing on different paths. The first is due to leave the Solar System in 1987; the second transmitted pictures of the ringed planet Saturn in 1979. Calculations show that *Pioneer 10* will reach the environs of the star Aldebaran in the constellation of Taurus after some 1,700,000 years. Both craft carry pictorial greetings to any extra-terrestrial intelligence that may find them in the remote future, giving information on the senders, their home planet known to them as Earth, and its position in the Galaxy.

In 1973 Russia again launched a spacecraft to Mars, using the powerful Proton D-I-e rocket and spacecraft of a new generation – *Mars 4, 5, 6* and *7*. The results were disappointing. One arrived in orbit around Mars; the others flew past, one doing so unintentionally when its braking engine failed to fire. However, the biggest setback was the loss of the two landing capsules released by the *Mars 6* and *7* mothercraft as they approached the planet.

Despite these costly setbacks, Russia pressed on and, in November 1975 two large automatic spacecraft succeeded both in landing capsules on Venus and going into orbit. The capsule of *Venera 9* plunged into the planet's atmosphere at 10·7km/sec (6·6 miles/sec) and at an angle of 20 degrees. When speed had dropped to 250m/sec (820ft/sec) the probe threw off clamshell protective covers and deployed a parachute. After it had floated down to a height of 50km (31 miles) from the surface, the parachute jettisoned and the capsule was slowed in the thickening atmosphere by a disc-like drag brake attached to the body. It hit the surface at 7–8m/sec (21–24ft/sec), the landing shock being absorbed by a ring of crushable metal. The capsule transmitted

data (via its orbiting mothercraft) for 53 minutes, including the first picture to show surface conditions on Venus.

Clearly visible was a level surface strewn with sharp-edged rocks 30 to 40cm (12 to 16in) across, which might have come from volcanoes. Atmospheric pressure at the surface was about 90 atmospheres, and the temperature 485°C (905°F).

A model of NASA's Viking spacecraft, two of which landed on Mars on 20 July and 3 September 1976, a journey of nearly 740 million km (460 m miles).

The capsule released by *Venera 10* came down some 2,200km (1,367 miles) from the first landing site, transmitting for 65 minutes and sending a picture showing outcrops of rock among rock debris. There the atmosphere was 92 atmospheres, which suggested that the craft had dropped into a lower lying region.

Two years later America sent two Viking spacecraft into orbit around Mars, on 19 June and 7 August 1976 respectively. The first released its landing craft onto the dry plains of Chryse Planitia on 20 July, seven years to the day since men first walked on the Moon. Photographs received on Earth showed a scattering of rocks; the surface had a red-brown colouring and the sky was a pinkish hue. The thin carbon dioxide atmosphere had a pressure of 7·69 millibars.

On 3 September 1976, the lander of *Viking 2* made its touchdown in Utopia Planitia some 200km (125 miles) west of the crater *Mie*, which has a diameter of about 100km (62 miles). It is believed that some of the rocks in the pictures came from this impact crater.

In a highly successful series of experiments, both robot Landers scooped up Martian soil and transferred samples to an onboard biology laboratory for automatic analysis. The results revealed that the main constituents of Mars' soil were iron, calcium, silicon, titanium and aluminium. But whether the reactions of treated soil samples are evidence of chemical or biological activity remain an open question. The results indicated a surprising amount of water and oxygen in the soil, but there was no certain evidence of organic material.

Meanwhile, the Viking Orbiters had made extensive photographic surveys of the surface, using two high-resolution cameras which provided spectacular new details of the ancient river beds, the equatorial canyon and the polar caps. Temperature measurements revealed that, contrary to expectations, the latter consist mainly of water ice and not dry ice (frozen co^2).

Earth satellites

The results of deep space astronomy conducted from unmanned Earth satellites and manned space stations have been no less exciting. Orbiting above the Earth's hazy and turbulent atmosphere, they have opened up a 'new' universe to which we were practically blind – a universe of fantastic spectral range extending far beyond visible light and radio waves to X-rays, ultraviolet and gamma rays. Astronomers have begun to make fundamental

discoveries about supernovae (exploding stars), pulsars (pulsating stars), neutrons (superdense stars), 'black holes' and quasars (quasi-stellar objects). The advantages of having observation platforms outside Earth's atmosphere are incalculable; radio astronomers, who listen to the Universe rather than looking at it through an ordinary telescope, are accumulating evidence confirming the 'Big Bang' theory of the origin of the Universe: reception of radiation from space in various parts of the electromagnetic spectrum is much easier without the interference of the atmosphere. Astronomers and particle physicists are helping to confirm each others' work, and are on the trail at last of a Unified Field Theory of energy, in an explosion of scientific knowledge to which the satellite is a priceless contributor.

Not all the work is in the interests of pure science. Meteorological satellites keep us informed from day to day of the world's changing weather conditions and give advance warning of destructive storms like hurricanes and typhoons. Early examples of such satellites were Tiros and Nimbus (US) and Meteor (USSR). They send television pictures of cloud, ice and snow cover on the day and night sides of the Earth, and also provide data on the thermal energy reflected and emitted by the Earth and its atmosphere, important for the greater understanding of meteorological conditions.

At the same time the super-powers lost no time in exploiting the military potential of space – especially by the use of photo-reconnaissance satellites and satellites able to provide advance warning of missile attack.

Earth satellites have also become 'radio stars', to assist the precision navigation of ships and aircraft in all weathers, and networks of military satellites have been established to maintain reliable, fade-free communications with naval vessels. Elint (electronic intelligence) missions are also carried out from orbit, with satellites 'listening in' to defence radars and communications, logging their codes and frequencies.

After America, for some obscure reason, had shelved development of an anti-satellite rocket (saint), the Russians began testing 'killer' satellites within the Cosmos programme which fragmented close to orbiting targets. Other military test satellites were observed to eject re-entry vehicles at the end of a single orbit. These were the so-called Fractional Orbit Bombardment System (FOBS) vehicles which, launched in anger, could attack the United States through the 'back door' of the South Pole.

Nonetheless, enormous efforts have been made by

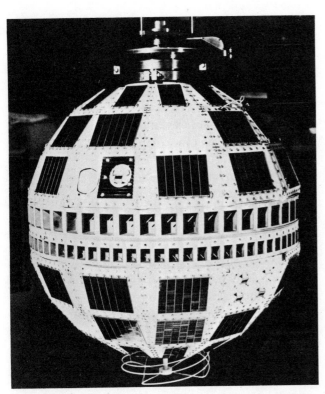

Communications satellite Telstar 1, put in geostationary orbit in July 1962, gave a first taste of international television.

the super-powers to apply the fruits of space technology to human need. More and more people are becoming involved in Space in their ordinary work – geologists, oilmen, doctors, farmers, ecologists, teachers and others. And although few of them are likely ever to see a rocket blast off from a launch centre, they will be called upon to use their skills, via space, to help solve some of mankind's most terrifying problems – poverty and famine, overpopulation, illiteracy, pollution, and perhaps even the energy crisis.

It all began with a space age communications explosion sparked by a technical memorandum which the 'space prophet' Arthur C. Clarke placed before the Council of the British Interplanetary Society (BIS) in 1945. That brief document, headed: 'The Space Station: Its Radio Applications' heralded the greatest revolution in global communications of all time. It described how three satellites, spaced at equal distances apart in 24-hour 'stationary' orbit high above the equator, could serve to spread communications around the world, and showed how a single satellite could broadcast directly to several regions simultaneously.

Clarke wrote: 'No communications development which can be imagined will render the chain obsolete, and since it fills what will eventually be an urgent need, its economic value will be enormous.'

Although Clarke and the BIS did all they could to promote the scheme, it was left to America to put the 24-hour satellite into commercial practice with *Early Bird* built by the Hughes Aircraft Company in 1965. Today's communications explosion is the result. Matching the Earth's rotation above the Atlantic, Pacific and Indian Oceans are big, drum-shaped satellites of the Intelsat network, any one of which can relay up to 6,000 telephone calls, or 12 colour television programmes or a combination of the two.

By 1975 the chain linked North and South America, Europe and Africa, and girdled the Far East via 111 antennae at 88 Earth stations in 64 countries. (By the use of land-lines from the existing Earth stations, the system actually reaches more than 100 countries on a full time basis).

Since 1965 transoceanic telephone traffic has grown from an estimated three million calls a year to more than 50 million in 1974, and it is predicted that calls will increase to 200 million by 1980. Indeed, because of satellites, a three-minute telephone call between New York and London is 55 per cent cheaper today than when *Early Bird* entered service. And satellites are the only way to send television around the globe.

The big test of the social applications of space technology began in 1974 in a programme co-ordinated by the US Department of Education and Welfare. The powerful ATS-6 'umbrella' satellite was placed in geostationary orbit just west of the Galapagos Islands in the Pacific to transmit educational and medical services directly to low-cost television receivers in scores of isolated communities in the United States.

Teachers participating in the experiment were able to see instructors, pictures and charts 'bounced' to them from the satellite across hundreds of miles, and to ask questions as if they were in the same lecture hall. The satellite allowed doctors in city hospitals to 'visit' patients in remote regions of Alaska, monitor their medical condition and advise locums of specialized medical treatments. Medical data, including electro-cardiograms and X-rays, were sent to doctors over the space link.

The same versatile platform was later moved eastward, by operating a small rocket motor aboard the satellite, to take up a position some 35,880km (22,295 miles) above Lake Victoria in East Africa. From this lofty vantage point, it broadcast directly to children and adults in some 5,000 towns and villages in seven states of India.

All the time space investment is producing new opportunities for social advance. 'If I had to chose one spacecraft, one Space Age Development, to save the world,' said NASA Administrator Dr James C. Fletcher, 'I'd pick ERTS and the satellites which I believe will be evolved from it later in this decade.' He was speaking about the first of two observation satellites – later re-named Landsat – which take special photos of the Earth. Rocketed into polar orbits some 925km (575 miles) high, they circle the globe every 103 minutes. Every 18 days they view the same spot anywhere in the world at the same local time of day.

The 952kg (2,100lb) Landsats carry television cameras as well as radiometric scanners, to obtain data in various spectral ranges of visible light (red, blue and green) and infrared, which show up hidden features on the ground. The data obtained relates to crop species, crop quality, moisture content of the soil, and soil fertility over wide areas. This in turn helps to find the best places to develop new land for farming and the best time to plant and harvest crops for maximum yield.

Satellite observation also promises to achieve an early prediction of the total harvest, so that food supplies are known in advance and can be set against future need. An early discovery was that crops and forest areas damaged by blight or insects showed up blue-black in the satellite photos and healthy vegetation pink or red.

India is now moving into the field of space technology with determination. Having had a scientific satellite launched by the Russians in April 1975, India is now building her own satellites and launch vehicles for meteorological study and storm warning, Earth resource observation and help with agriculture. If, for instance, the onset of the monsoon rains can be predicted by satellites – information which is crucial to the transplantation of rice – millions of dollars a year could be saved. And in these regions the saving of a rice crop is not just a financial gain; it could mean the difference between life and death.

In retrospect, it is that sort of data, streaming back to us from spacecraft sent to explore the Moon, Mars and other planets, that has been most valuable. It has prepared the way for a spectacular and much needed rediscovery of our own planet. And, what is even more remarkable, these benefits are coming from only a small share of the one cent of the tax dollar which currently supports the space programme in the United States.

Space stations

The desire to gather information on Earth's natural resources and expand other useful areas of science and technology, led to the first manned space stations. After teething troubles with test models, the Russians launched a series of 18·5-tonne (18·2-ton) Salyut orbital laboratories. Cosmonauts docked with the stations at intervals in Soyuz ferries to carry out various research tasks, which ranged from photographing the earth's resources to making experiments in biology, medicine and industrial processing under conditions of weightlessness. Telescopes were carried to observe the Sun and the stars from above the atmosphere.

The much larger Skylab space station launched by America on 14 May 1973, carried out a wide range of similar experiments which included use of the Apollo Telescope Mount to increase knowledge of the Sun and its influence on earth's environment. Metals were melted in a small electric furnace, and experiments made in the growth of crystals of a kind which could be of value to the electronics industry. Photographs of the comet Khoutek were taken by astronauts.

Skylab was made from the S-IVB stage of a Saturn rocket. Fitted out with two-storey accommodation, it included a large workshop area for experiments and, on the 'lower floor', a wardroom and living area including a combined washing and toilet compartment and separate sleeping compartments.

The station was damaged during the launch – accomplished by a two-stage *Saturn V* – but the first two boarding parties of three men each were able to carry out running repairs. The workshop's external meteoroid shield and one of the large solar 'wings' had been torn off by air pressure as the rocket passed through the atmosphere; the other 'wing' was held fast by a piece of torn metal. Not only was the first boarding party able to erect a temporary sunshade over part of the workshop to reduce the temperature, but the astronauts were able to climb out of the station to cut away the obstruction and release the jammed solar 'wing'. The second team brought a larger sunshield which they erected without great difficulty. In the end Skylab exceeded all expectations, with three highly successful astronaut missions lasting 28, 56 and 84 days respectively.

Not long after, Russia and America pooled their resources in a joint space experiment which entailed astronauts and cosmonauts training in each others' countries. This was the celebrated Apollo-Soyuz Test Project (ASTP).

The mission got underway in July 1975, when a three-man Apollo spacecraft docked with a two-man Soyuz 225km (140 miles) above the Earth. Both craft had been modified to make their life support and docking systems compatible. The crews visited one another and carried out joint experiments before separating their ships and landing in their respective recovery areas, proving the feasibility of space rescue missions and of joint operation of space stations. If nations can now learn to co-operate in other important ways, the dream of aviation as an instrument of peace may yet come true.

The space shuttle

Meanwhile, a revolutionary new space vehicle which promises to bring down the cost of space travel was taking shape in America. Unlike conventional rockets, which fall to destruction each time a spacecraft is launched, this winged craft – blasting off like a rocket and flying home like an aeroplane – can be re-used 100 times or more. Called the Space Shuttle, it consists of the winged Orbiter, about the size of a DC-9 airliner, a large external tank and two powerful solid rocket boosters. The Orbiter has a crew of three – pilot, co-pilot and mission specialist.

Launched vertically from the pad like a normal rocket, it rides on the large external tank which supplies the ascent propellants (liquid oxygen and liquid hydrogen). The two solid fuel rockets, mounted on the sides of the tank, assist during the lift-off and jettison when the craft has climbed some 45km (28 miles). They are intended to be recovered after

parachuting into the sea. Some of them will be used again after being cleaned and re-filled with propellant.

The big tank is jettisoned just before the 'spaceplane' arrives in orbit under the thrust of small manoeuvre engines. This is the only part which does not return to Earth, as it burns up in the atmosphere during re-entry.

On certain missions the Orbiter will carry the European Spacelab, an entirely new concept in manned space stations. Unlike Salyut and Skylab, which had to be abandoned in space along with their costly instruments, Spacelab – which remains fixed inside the spaceplane's cargo bay – will return to earth after each mission. Up to four specialists can ride as payload into orbit with the astronaut flight crew for missions lasting 7 to 30 days.

To return from orbit, the spaceplane is turned to a backward angle so that its manoeuvre engines can serve as retro-rockets. It descends through the atmosphere at a high angle of attack, and is protected from the high temperatures of frictional heating by special surface insulation. At the end of its flight the Orbiter will fly across Florida at high speed, to circle round and land without power on a 4·8km (3 mile) runway at the Kennedy Space Center.

The Shuttle is the key to a major new effort in manned spaceflight for the 1980's. It could lead to large space stations built up piece by piece from modules carried up in the cargo bay.

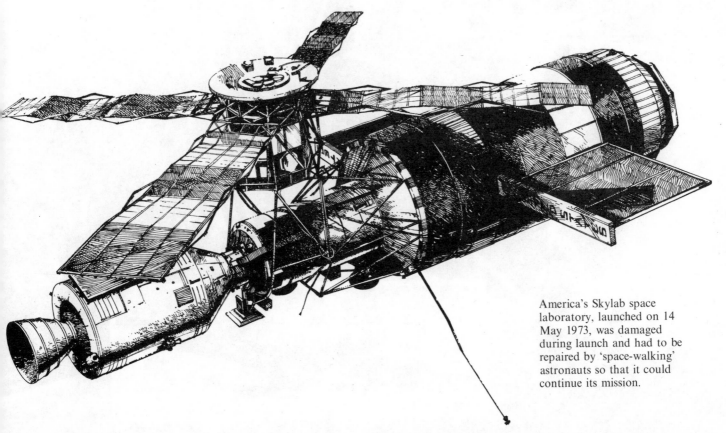

America's Skylab space laboratory, launched on 14 May 1973, was damaged during launch and had to be repaired by 'space-walking' astronauts so that it could continue its mission.

Capt. J. ALCOCK. D.S.C.

ST. JOHN'S

LUNCHEON
TO THE
WINNER
OF THE
DAILY MAIL
£10,000
PRIZE

20, 19

Facts, Feats and Records

The TRANS-ATLANTIC AIR RACE 1919

LT. A. WHITTEN BROWN
CLIFDEN

On 13 January 1908 at Issy Les Moulineaux in France Henry Farman flew the first circular flight in Europe, thus winning the 'Grand Prix d'Aviation' donated by two French benefactors. He covered a distance of 1km (·625 miles) in a modified Voisin biplane and remained airborne for 1 minute 28 seconds. Though a mere hop by modern standards this was the most impressive flight that had been made in Europe to that date. Yet in America the Wright Brothers, before they temporarily gave up flying in 1905, had already covered 40·25km (25 miles) in one flight and stayed up more than half an hour at a time; and they could turn and fly figure eights with ease. In Europe no one knew or believed this until, in 1908, the Wrights returned to flying, and gave their first demonstrations in public. Wilbur took a Flyer to France, while Orville stayed home to fly their new two-seat Flyer for the US Army.

The Wrights' achievement

Wilbur's first flight in Europe was a two-minute circuit of Le Mans race-course on 8 August 1908. He astonished spectators with the degree of control he had over his machine. 'One of the most exciting spectacles in the history of applied science,' wrote one. 'Good grief!' exclaimed another, 'We are beaten . . . we don't exist.' When he finished his French demonstrations at Auvours on the last day of 1908 Wilbur Wright had made more than 100 flights, with flying-time of more than 25 hours; he had taken up passengers on more than 60 occasions; had attained an altitude of 110m (360ft); and had made one flight of more than 2 hours.

Orville's flights in America began impressively, but ended tragically on 17 September with structural failure of his Flyer while aloft.

Louis Blériot

One of the French experimenters most inspired by watching Wilbur's demonstrations was Louis Blériot. His 1909 Type XI monoplane was his first really successful design, and in it he made a cross-country flight of 42km (26 miles) from Etampes to Chevilly on 13 July. On 25 July he made the first aerial crossing of the English Channel. This flight won him the £1,000 prize offered by the London *Daily Mail* newspaper to the first person to achieve this feat as well as attracting extraordinary publicity around the world. It brought home to people for the first time that the aeroplane might in time become a formidable weapon of war. Blériot's flight was also a feat of great courage: the 27 minutes it took was about as long as his frail 25hp 3-cylinder Anzani

engine had ever run without stopping. He had no compass or real means of navigation for the misty crossing and he was still suffering the effects of an injury received in an earlier crash. (His arrival on the cliffs of Dover was in truth another crash-landing.) The feat earned him a fortune. In addition to the £1,000 newspaper prize he soon had orders for more than 100 of his Type XI monoplanes. The design of the XI was interesting in that it incorporated many features that we have since come to take for granted in aeroplane design: monoplane wings, tractor propeller directly attached to the engine crankshaft, hinged tilting stick and rudder pedal controls, and a covered-in fuselage structure. (The Wright Flyers by contrast had none of these features.)

Reims air show

The first flying meeting took place soon after Blériot's Channel crossing, between 22 and 29 August at Reims in France. It was sponsored by local champagne interests and 38 aircraft were entered. Of these, 23 actually got airborne to make 120 flights. Speed records were established by Blériot and the American Glenn Curtiss of between 75 and 77km/h (47 to 48mph); an Antoinette monoplane climbed to 155m (508ft); and the distance prize was won by Henry Farman, who covered 180km (122·5 miles) in 3 hours 5 minutes. Among the many impressed visitors was Lloyd George, who commented: 'Flying machines are no longer toys and dreams . . . the possibilities of this new system of locomotion are infinite.' However, as much as the many fine flights made at Reims, most visitors were impressed by the general appalling unreliability of the engines.

1909–1910

On 7 September 1909 the first pilot of a powered aeroplane was killed and on 22 September the second, and on 6 December, the third. Since the beginning of the 19th century five men and one woman had been killed in parachuting accidents; two experimenters in glider crashes; and a considerable number in balloons. There were no less than 32 aviation fatalities in 1910. That year, however, also saw more than 20 aviation meetings held, in Europe, the USA and Egypt; the first night-flying accomplished, in the Argentine; and the first lady to qualify as a pilot, Baroness de la Roche. The Hon C. S. Rolls, the wealthy sportsman who was one of the founders of Rolls-Royce cars, made the first out-and-return non-stop Channel crossing on 2 June 1910 and lost his life on 12 July during a

Two aircraft seen in flight at the Reims Air Meeting of 1909. The Voisin biplane (*top*) was virtually a 'dead-end' design, but the Blériot XII (*below*) pointed towards future monoplanes.

meeting at Bournemouth when his French-built Wright Flyer broke up in the air. On 17 August a French-American pilot, J. B. Moisant, crossed the Channel carrying his mechanic as a passenger.

The Peruvian pilot, Georges Chavez, flying a Blériot monoplane, made the first aerial crossing of the Alps, through the Simplon pass, although he lost his life at the end of the attempt while trying to land at Domodossola. Curtiss biplanes made several remarkable flights in the USA; from Albany to New York City; and New York to Philadelphia and back when radio was used for the first time. Eugene Ely took off in a Curtiss biplane from a platform built on the USS *Birmingham* – another first in aviation. Glenn Curtiss himself made the first formal bombing tests, dropping dummies onto flagged buoys which marked the outline of a warship.

1911

By 1911 aviation had become an enormously popular public spectacle, and the first air races were held that year. The first non-stop flight from London to Paris was made on 12 April in a Blériot by Pierre Prier, who flew the 402km (250 miles) in a little less than 4 hours. There were races (in stages) from Paris to Rome and Madrid and a return trip to Paris via London, Brussels and Amiens. In America Calbraith P. Rodgers made the first coast-to-coast flight in a Wright biplane. The 6,437km (4,000 miles) took him 82 flying hours in 82 stages over a period of 49 days. He had 19 crashes en route and was followed all the way by a special train containing spares as well as his wife and mother. He had hoped to win a $50,000 prize for the journey offered by William Randolph Hearst, but arrived outside the specified 30-day limit. Another American pilot flew a Curtiss 90 miles over water – from Key West to Havana, Cuba. Curtiss flew the first practical seaplane in 1911. And Orville Wright made a nine-minute flight in a glider which remained a soaring world record for ten years. Near the end of the year a *Concours Militaire* was held at Reims, for the display of aeroplanes that might be used in war. And one *was* used, for the first time, when an Italian pilot, in a Blériot, made an hour-long flight on 22 October 1912 to observe Turkish positions near Tripoli. The first air mail flights were made. The first parachutist to drop from an aeroplane was Captain Albert Berry in the USA, and American Harriet Quimby became the first woman to fly the Channel. In England Frank McClean flew a Short float-plane through Tower Bridge and was promptly in trouble with the police (as has been every other pilot to try it since) when he landed on the Thames by Westminster.

1912–1918

During 1912 and 1913 the world airspeed record was raised no less than ten times by two French pilots, Jules Vedrines and Maurice Prévost, flying very advanced Deperdussin monocoque racers. The first such record set was Vedrine's 145km/h (90mph) on 13 January 1912. By 29 September next year Prévost had achieved 203km/h (126mph).

Two outstanding flights of 1913 were the first air crossing of the Mediterranean, by Roland Garros in a Morane-Saulnier monoplane. He flew 729km (453 miles) from southern France to Tunisia in just under 8 hours. The first flight from France to Egypt was made, in stages, by Vedrines in a Blériot. The Norwegian pilot Tryggve Gran was the first man to fly the North Sea, in a Blériot, on 30 July 1914.

Civil flying generally ceased in Europe during World War I. But South America was far from the conflict and two remarkable record flights were made there. The Argentinean Army pilot Teniente Candelaria made the first crossing of the Andes from east to west on 13 April 1918. He flew a Morane and reached an altitude of about 3,960m (13,000ft).

1919

Substantial prize money had already been put up before the war for the first air crossing of the Atlantic Ocean. As soon as the war was over many began to plan for the attempt. Newfoundland, the easternmost province of Canada, had never seen an aeroplane, but, in the spring and summer of 1919, it became a familiar dateline for press reports of preparations for the transatlantic crossing. When the two British aviators John Alcock and Arthur Whitten Brown arrived as latecomers in the race. they were hopefully asked to pay $25,000 for the temporary lease of one small meadow when land could be bought there for 35 cents an acre! The prize for the first successful crossing was £10,000 (then $50,000), put up by the *Daily Mail* in 1913 for the first flight between England and America to be completed within 72 hours.

Meanwhile the US Navy planned an ambitious Atlantic crossing with 3 huge Curtiss flying-boats aided by a string of destroyers stationed every 80·5km (50 miles) across the ocean. Two of the flying-boats were wrecked during the attempt, but the third, the NC-4, left Newfoundland on 8 May. She landed twice to refuel, in the Azores and in Portugal, and finally arrived in England on 31 May – the first Atlantic crossing by any aircraft. It was a formidable achievement, with 6,315km (3,925 miles) covered in 57 hours 16 minutes flying time at an average airspeed of 126·8km/h (78·8mph).

News of the US Navy's attempt stampeded two landplane crews also in Newfoundland into making ill-advised and hasty starts from there on 18 May. First off were Harry Hawker and MacKenzie Grieve in a Sopwith biplane named, appropriately, the *Atlantic*. An hour later followed Frederick Raynham and William Morgan (who had lost a leg in the war) in a Martinsyde. They crashed on take-off, but survived. Nothing was heard of Hawker and Grieve for a week; then a Danish tramp steamer which had no radio signalled to a semaphore station in the Hebrides that she had picked up the Sopwith's crew who had ditched alongside her after experiencing trouble with the aircraft's engine.

Alcock and Brown left Newfoundland on 14 June. They were flying a converted twin-engine Vickers Vimy bomber fitted with extra fuel tanks. It was so overloaded that spectators watching its take-off run were convinced it would crash. Even when airborne and headed out to sea they were still struggling for altitude. Later, flying higher but in cloud, they lost control of the aircraft and it fell into a spin from which it was only recovered just above the waves. They also became badly iced-up, with rime clogging the engine air intakes. Brown crawled out between the biplane wings to each engine in turn to clear them. Another hazard awaited them even after they had arrived safely over Ireland: the green field on which they chose to land was in fact a soft peat bog, and the Vimy went over on its nose in a crash-landing. Nevertheless, Alcock and Brown made the first ever non-stop Atlantic crossing, flying 3,041km (1,890 miles) in 16 hours 28 minutes. Subsequently, both were knighted by King George V for the feat. Sir John Alcock did not have long to enjoy his fame for he lost his life on 19 December that year while flying through extremely bad weather over France.

In July 1919 the Atlantic was crossed for the first time by an airship, the British R-34. It set out from Scotland on 2 July, captained by Major G. H. Scott of the RAF and carrying Lieutenant Commander Zachary Lansdowne of the US Navy as an observer.

A Sikorsky heavy-lift helicopter (*right*) illustrates the typical use of flying-crane helicopters, able to lift bulky loads over difficult terrain to sites inaccessible to other transport.

The Boeing Model 247 prototype (*below*) was not only an advanced concept when introduced in 1933, but it must be regarded as the great-grandfather of the Model 747 'jumbo jet'. The BAC (Folland) Gnat (*right*) has been used worldwide for precision aerobatic displays by the RAF's Red Arrows. In 1980 they will be flying the new British Aerospace Hawk trainer.

The Avro Lancaster bomber (*below*), operated by the RAF's Battle of Britain Flight, has acquired a typical dorsal turret since this picture was taken. Lockheed's S-3A Viking anti-submarine warfare aircraft (*top right*) combines the hunter and killer roles. An F-104 Starfighter in Luftwaffe service (*below right*) is used to test zero-length launch with a booster-rocket.

Boeing's B-17 Flying Fortress (*below*) is typical of the four-engine bombers developed at the beginning of World War II. The Handley Page Victor (*right*) is an example of the new generation of turbine-powered V-bombers which Britain introduced in the 1950s to maintain a nuclear deterrent policy.

The R-34 also carried the first translatlantic aerial stowaway, a rigger who, disgruntled at being left off the crew roster, had hidden among the gas-bags. The airship battled headwinds, and over Newfoundland met thunderstorms that threatened to wrest control of it from the crew, who radioed for help from the US Navy. The crew recovered control and made their way unaided to Long Island, where they landed at Montauk (with just two hours' fuel remaining) on 6 July. The R-34 was the first big airship – 202·69m (665ft) long – to be seen in the United States and attracted huge crowds during its short stay. Just before midnight on 9 July it took off, circled the Chrysler Building with its searchlights playing, and set off on the return journey. It arrived back in England in 75 hours 3 minutes, helped by good tailwinds.

A Vimy was used later that year when two Australians, brothers Ross and Keith Smith, made the first flight from England to Australia, in just under 28 days, from 12 November to 10 December. Their flight from Hounslow to Darwin covered 18,175km (11,294 miles). It earned them knighthoods and a £10,000 prize donated by the Australian government. There is another and more tragic coincidence with Alcock and Brown's flight: Sir Ross Smith was killed in an accident not long afterwards, near Brooklands in England on 13 April 1924 while flying a Vickers Viking – the same type in which Sir John Alcock had been killed.

1920–1925

A Vimy was also the chosen mount of Pierre van Ryneveld and Christopher Quintin Brand when they set off from Brooklands for South Africa on 4 February 1920. When they crashed at Wadi Halfa the South African government provided them with another Vimy in which to continue. This in turn they crashed at Bulawayo on 6 March and it was replaced by a DH-9. They arrived at Wynberg Aerodrome, Cape Town, on 20 March . . . the first men to fly from England to South Africa. They too received knighthoods and £5,000 prize-money.

In the United States, the first coast-to-coast crossing in a single day was accomplished by James H. Doolittle in a DH-4. He flew from the east coast of Florida to San Diego, California, on 4 September 1922 with one refuelling stop in Texas. His flying time was 21 hours 19 minutes over a distance of

3,480km (2,163 miles). (Doolittle, then an Army lieutenant, was to go on to greater fame as a racing pilot and later still organizer and leader of the first US bombing raid on Tokyo during World War II.)

On 2–3 May 1923 two other US Army aviators, Lieutenants O. G. Kelly and J. A. MacReady, flew a Fokker T.2 monoplane from Long Island to San Diego for the first non-stop coast-to-coast flight. They covered 4,050km (2,516 miles) in 26 hours 50 minutes. A little later they established a new world endurance record for aeroplanes by staying aloft in the Fokker for 36 hours 5 minutes over a 4,052km (2,518 miles) measured course. The record was beaten on 27–28 August the same year when two other US Army aviators, Captain L. H. Smith and Lieutenant J. P. Richter, kept a DH-4B aloft for 37 hours 15 minutes using in-flight refuelling – their DH-4 was flight-refuelled 15 times from another DH-4.

In 1924 the US Army mounted an even more elaborate exercise than that of the US Navy's transatlantic flight of 1919 – the first around-the-world flight. Four Douglas World Cruisers, open-cockpit single-engined biplanes, were used. Of these two finished the flight – flying 42,398km (26,345 miles) in 363 hours 7 minutes of flying time.

This American project was not the first attempt to fly around the world. Sir Ross Smith had been killed testing a Vickers amphibian preparing for such an attempt in 1922. An Englishman, Norman Macmillan, using a DH-9 landplane and then a Fairey III-C seaplane, had got as far as the Bay of Bengal in 1922, before the Fairey was wrecked in a storm. A second attempt was made in 1923, but was abandoned when the supply yacht was arrested in San Pedro, California, for breaches of the US prohibition laws. A French expedition was mounted in 1923 with 5 aircraft and 14 personnel, but they did not get far. There were three other attempts proceeding at the same time as the successful American one: that of Squadron Leader Stuart MacLaren of the RAF, who actually saw the Douglas World Cruisers fly overhead while he was on the water in Burma waiting out a storm; that of Lieutenant Antonio Locatelli of the Italian Air Service, who flew with the Americans on one leg, but got no further than Iceland; and that of Major Pedro Zanni of Argentina, who crashed his first aeroplane in French Indo-China in August and his spare in Japan in May 1925.

There is a point in all this that is insufficiently emphasized by most historians. Most great flying feats and record flights become achievable due to

The vast size of aircraft carriers tends to diminish the size of the aircraft which operate from them. Here pilots restore the size of McDonnell F-4 Phantoms to their true perspective.

technical advances and that if whoever *did* first achieve them had *not* succeeded, some other aviator, not far behind him, surely would have.

Charles Lindbergh

It was thus, though it is none the less praiseworthy for that, with the flight that is perhaps the greatest single feat in flying history: Charles Lindbergh's first solo Atlantic crossing, from New York to Paris, on 20–21 May 1927. Technically, it was made possible by the development of the Wright Whirlwind engine which possessed a reliability, power:weight ratio and fuel efficiency of a high order. The financial spice for such a flight was a $25,000 prize put up in 1919 by Raymond Orteig. Before Lindbergh's flight six men had already died in the many attempts on the Orteig prize. France's two greatest surviving fighter aces of World War I, René Fonck (who survived) and Charles Nungesser (who vanished in mid-Atlantic with his navigator), were among the failed contenders for the prize.

Lindbergh chose to make his flight solo in a single-engined plane in contrast to several of his rivals who had chosen trimotors or twins with multi-man crews. Lindbergh's decision was based on the consideration that the weight he would save by flying alone on one engine would be more useful as extra fuel capacity. He commissioned a modified version of a Ryan monoplane mail-carrier, named *Spirit of St Louis*. He designed and controlled the modifications (principally to increase its fuel capacity) himself. Throughout his preparations he doubted that he would be ready in time to cross the Atlantic before his competitors and kept an alternate set of charts of the Pacific.

His take-off roll, which began at 07.52 on 20 May 1927, was one of the great cliff-hangers of all time. The earth was soft with rain, and spectators followed in cars with fire extinguishers convinced they would be needed since his plane was accelerating so slowly. Lindbergh just staggered over the wires at the end of the field, and set off at low altitude northeastwards. Lindbergh battled ice, an involuntary spin in mid-ocean (as had Alcock and Brown) and the craving to sleep. When he arrived over Le Bourget airfield near Paris in the darkness of his second night aloft, at 22.24 on 21 May, traffic jammed every access road from all directions and such a wave of humanity poured over the field that he could hardly find space to land without injuring someone. He had been airborne 33 hours 39 minutes (after having hardly had any sleep the previous night) and had flown 5,810km (3,610 miles) non-

Charles Lindbergh (*above*) achieved the first solo crossing of the North Atlantic from West to East, New York to Paris, in May 1927, in the Ryan NP monoplane *Spirit of St Louis*. This achievement created worldwide interest in civil aviation. Amelia Earhart (*right*), the first woman to emulate Lindbergh's feat by flying from Newfoundland to Londonderry in 1932, subsequently made the first solo flight from Hawaii to California in 1935, but disappeared during a round-the-world flight attempt in 1937.

stop, averaging 173km/h (107·5mph). His aircraft, which hangs today in the national Air & Space Museum in Washington, had a performance that was extraordinary then and is still remarkable now. With a 14m (46ft) span, 975·5kg (2,150lb) empty weight and 237hp engine, it could carry 1,700 litres (450 US gallons) of fuel, for a range of about 7,000km (4,350 miles). Its useful load was thus considerably greater than its empty weight. It offered Lindbergh (the only pilot who ever flew it) no direct forward visibility, for the huge fuel tank was where the windshield should have been. He could only peer obliquely forward through side panels.

Perhaps the major significance of Lindbergh's feat was that it captured the imagination of many ordinary people, who had never given aviation much attention, and made them aware of the possibilities of powered heavier-than-air flight.

was used by Lady Heath to make the first solo flight by a woman from South Africa to London (12 February-17 May 1928). At about the same time another woman pilot was setting out in the opposite direction. Lady Bailey, flying a Cirrus II Moth, left London on 9 March 1928 and returned on 16 January 1929 having in that period flown round Africa, crashed her Moth and replaced it with another.

Charles Kingsford Smith with a crew of 3 made the first full crossing of the Pacific by air between 31 May and 9 June 1928 in a trimotor Fokker F.VIIB-3M, *Southern Cross*. Flying from Oakland to Brisbane via Honolulu and Fiji, they covered 11,890km (7,389 miles) in 83 hours 38 minutes flying time. Kingsford Smith captained the same Fokker for the first air crossing of the Tasman Sea. He flew from Sydney to Christchurch in 14 hours 25 minutes on 10–11 September of that year.

Two RAF pilots, G. Jones Williams and N. H. Jenkins, made the first non-stop flight from England to India, in a Fairey Longrange Monoplane, between 24–26 April 1929. They successfully flew 6,647km (4,130 miles) in 50 hours 37 minutes – that is more than 2 days and nights without once having to land.

Graf Zeppelin

The first, and only, airship flight around the world, and one of the most romantic journeys ever, was made by the *Graf Zeppelin*, captained by Dr Hugo Eckener, between 8–29 August 1929. It was only the second world flight by any aircraft. German airships were noted for their comfort, good food and fine wines, and it is reported that, with the clocks on board beind advanced 1 hour every 7, the 20 or so passengers, some of whom had paid $9,000 for the trip, complained they were being fed too often!

Commander Byrd

The first flight over the South Pole took place on 28–29 November 1929 in a Ford 4-AT Trimotor commanded by Commander R. E. Byrd of the US Navy and flown by Bernt Balchen with two other crew members. The flight, from and returning to the expedition's base at 'Little America' on the edge of Antarctica, took 18 hours 59 minutes. It is interesting that no less than 10 aircraft were in Antarctica with various expeditions that summer. The first aerial expedition to the South Pole (a British one) had been planned as early as 1920.

Byrd, with Floyd Bennett as co-pilot, had earlier made the first-ever flight over the North Pole, out of and back to Spitsbergen, on 9 May 1926 in a Fokker

1927–1929

In the same year as Lindbergh's flight, on 28–29 June, two US Army Lieutenants, Albert F. Hegenburger and Lester J. Maitland, in a Fokker C-2 trimotor, made the first non-stop flight from the USA westwards to Hawaii. They flew 3,874km (2,407 miles), from Oakland to Honolulu, in 25 hours 50 minutes.

The South Atlantic was also first flown non-stop that year, on 14–15 October, when Dieudonné Costes and Joseph Le Brix of France flew a Breguet XIX from Senegal to Brazil covering 3,420km (2,125 miles) in 19 hours 50 minutes.

The first solo flight between England and Australia was made by H. J. L. 'Bert' Hinkler in a prototype Avro 581 Avian (7–22 February 1928). An Avian III

F.VIIA-3M. He had been greatly assisted on this occasion by the Norwegian explorer Roald Amundsen, who had been the first man ever to reach the South Pole.

1930s

Long-distance and over-ocean records continued to be set throughout the 1930s. Clyde Pangborn and Hugh Herndon, in a Bellanca, made the first non-stop flight from Japan to the USA between 3–5 October 1931. The flight, from Tokyo to Washington state, lasted 41 hours 13 minutes. The great American woman pilot, Amelia Earhart, made the first female solo crossing of the North Atlantic from Newfoundland to Londonderry, in a Lockheed Vega monoplane during 20–21 May 1932. James Mollison made the first solo east–west Atlantic crossing, from Dublin to New Brunswick on 18–19 August 1932 in a de Havilland Puss Moth. Mollison had set an Australia-to-London speed record in 1931 in a Moth, and from London to Cape Town in 1932, both times using the same aeroplane. In 1933 Mollison made the first solo east–west crossing of the South Atlantic, in the course of a marathon flight which took him from Lympne, England to Natal in Brazil in a time of 3 days 10 hours 8 minutes.

Mollison was married to the British aviator heroine Amy Johnson, who in 1930 had made an impressive solo flight – the first by a woman – from England to Australia. She flew a de Havilland D.H.60G Gipsy Moth. Jim and Amy Mollison flew the North Atlantic together during 22–24 July 1933 in a twin-engined D.H. Dragon from Pendine in Wales to an unpleasant crash-landing at Bridgeport, Connecticut.

A world long-distance record, and the first-ever non-stop flight from England to South Africa, was made by Squadron Leader O. R. Gayford and Flight Lieutenant G. E. Nicholetts in a Fairey Long-range Monoplane between 6–8 February 1933. They flew 8,595km (5,431 miles) in 57 hours 25 minutes.

The first solo flight around the world was made by the one-eyed American Pilot Wiley Post between 15–22 July 1933. Piloting a Lockheed Vega, *Winnie Mae*, he flew 25,099km (15,596 miles) in 7 days 18 hours 49 minutes.

Sir Charles Kingsford Smith, accompanied by P. G. Taylor, made the first flight from Australia to the USA between 22 October–4 November 1934, flying a Lockheed Altair. Amelia Earhart used a Lockheed Vega monoplane to make the first solo flight from Hawaii to the USA on 11–12 January 1935. Amelia Earhart's solo flight lasted 18 hours 16 minutes.

Jean Batten of New Zealand made the first solo South Atlantic crossing by a woman on 13 November 1935 flying a Percival Gull. She used the same aircraft to make the first flight from Britain to New Zealand by a woman, in the record time of 11 days 45 minutes (5–16 October 1936).

Other events of that time foretold the way aviation was headed. The prototype Bf 109 first flew in September 1935; the first Hurricane on 6 November; and the first Spitfire on 5 March 1936. A Messerschmitt racer called the 109R, flown by Fritz Wendel, set a world speed record of 781·97km/h (486mph) on 29 April 1939. This remained the world record for piston-engined aircraft for 30 years. The first successful helicopter, the Focke-Wulf Fw61V1, made its first flight on 6 June 1936. Frank Whittle ran his first gas turbine aero engine on 12 April 1937. The first jet aircraft to fly was the Heinkel He 178, on 27 August 1939. The first British jet, the Gloster E.28/39, did not make its maiden flight till 15 May 1941. Germany also flew the first rocket-propelled aircraft, a modified glider named the *Ente* piloted by F. Stamer. The flight lasted a minute, on 11 June 1928.

Post-war achievements

A rocket-powered aircraft, air-launched from under the fuselage of a bomber, made the first manned supersonic flight. The Bell XS-1 was flown by Captain Yeager on 14 October 1947. Rocket aircraft remain those which have flown fastest and highest, even within the atmosphere. A North American X-15A-2 research aircraft, air-launched from under the wing of a B-52 jet bomber and piloted by W. J. Knight, reached a speed of Mach 6·72 (7,297km/h/4,534mph) on 3 October 1967. The highest altitude achieved during the X-15 programme was 107,860m (354,200ft), reached by J. A. Walker on 23 August 1963.

The first non-stop flight around the world was made by a piston-engined Boeing B-29 Superfortress in March 1949, captained by James Gallagher and manned by a crew of 18. The *Lucky Lady II* flew about 36,225km (22,500 miles) at an average speed of 378·35km/h (235mph), and was refuelled in the air four times: over the Azores, Saudi Arabia, the Philippines and Hawaii. Three 8-jet Boeing B-52 bombers flew around the world in formation and non-stop in 1957, covering 39,147km (24,325 miles) in 45 hours 19 minutes – less than half the time the B-29 had taken only 8 years earlier. The B-52s were refuelled four times, and followed a route that took them over the Atlantic to Morocco, to Dharan, across India and Ceylon, and home to California via Guam. Each aircraft carried a crew of 9 instead of the usual 6, and one tail gunner set a record of a kind as the first man to circumnavigate the world facing backwards.

The world has been circumnavigated since World War II by innumerable pilots, including at least one in a homebuilt aeroplane. Among the women who have flown around the world is Miss Sheila Scott who achieved her feat in 33 days 3 hours, between 18 May–20 June 1966, in a single-engined 260hp Comanche. Her other records include flying from London to New York in 24 hours 48 minutes on 4–5 May 1969; and from London to South Africa in 3 days 2 hours 14 minutes, between 6–9 July 1967. The world has been circumnavigated over the North and South Poles by, among others, Max Conrad flying solo in a light twin aircraft. This pilot has made a number of remarkable long distance solo flights. He holds a world class record for distance, at 11,211·83km (6,967 miles) in a single-engined Comanche. At the age of 62 he flew solo from Cape Town to St Petersburg, Florida, in a Twin Comanche.

The more important world records since World War II have generally been competed for by air forces and governments, with private individuals

Significant Air Races

Race	Date	Place/Course	Winner	Aircraft	Speed/time
First Gordon Bennett aeroplane race	29 October 1910	Belmont Park, N Y /100km (62·12 miles)	Claude Grahame-White	100 hp Gnome-Blériot	1hr 1m n 4·74sec
Circuit of Europe	18 June–7 July 1911	Paris–Reims–Liège–Spa–Liège–Verloo–Utrecht–Breda–Brussels–Roubaix–Calais–Dover–Shoreham–London–Shoreham Dover–Calais–Amiens–Paris	Lieutenant de Vaisseau Conneau	Blériot	
Daily Mail Circuit of Britain	22–27 July 1911	Brooklands–Hendon–Harrogate–Newcastle–Edinburgh–Stirling–Glasgow–Carlisle–Manchester–Bristol–Exeter–Salisbury Plain–Brighton–Brooklands/1,625km (1,010 miles)	Lieutenant de Vaisseau Conneau	Blériot XI	22hr 28min
Daily Mail First Aerial Derby	8 June 1912	Hendon–Kempton Park–Esher–Purley–Purfleet–Epping–High Barnet–Hendon/130km (81 miles)	T. O. M. Sopwith	Blériot	Av. 94km/h (85·5mph) 1hr 23min 8·4sec
First Kings Cup Air Race	8–9 September 1922	Croydon–Glasgow–Croydon	Captain F L Barnard	D H -4A	198·9km/h (123·6mph) 6hr 32min 50sec
First Pulitzer Trophy Race	1920	Mitchel Field, Long Island, N.Y./closed-circuit	Lieutenant Corliss C. Moseley	638hp Verville-Packard	251·87km/h (156·5mph)
Last Pulitzer Trophy Race	1925	Mitchel Field, Long Island, N.Y./closed-circuit	Lieutenant Cyrus K. Bettis	620hp Curtiss R3C-1	400·72km/h (248·99mph)
First Thompson Trophy Race	August 1930	Chicago/closed-circuit	Speed Holman	450hp Laird Solution	324·9km/h (201·9mph)
Last Thompson Trophy Race	1949	? Cleveland	Cleland	4,000hp F2G-1 Corsair	639·04km/h (397·07mph)
First Bendix Trophy Race	4 September 1931	Burbank–Cleveland	James Doolittle	535hp Laird Super Solution	358·94km/h (223·03mph) 11hr 16min 10sec
Last Bendix Trophy Race	1951	Rosamond Dry Lake, California–Cleveland	Joe DeBona	2,270hp P-51C	756·63km/h (470·136mph) 4hr 16min 17sec
MacRobertson Race	20 23 October 1934	England–Australia/18,329km (11,333 miles)	Charles W. A. Scott/Tom Campbell Black	D H -88 Comet	255·7km/h (158·9mph) 70hr 54min 18sec
Schlesinger Air Race	29 September–1 October 1936	Portsmouth–Johannesburg/9,897km (6,150 miles)	C W A Scott Giles Guthrie (only finishers)	Vega Gull	186·7km/h (116mph) 2 days 4hr 56min
Circuit of the Alps	25 July 1937	Zurich International Flying Meeting/125 miles	Oberst Karle Franke	Bf-109V13	387km/h (240·9mph)
London–Christchurch (NZ) Air Race	8–9 October 1953	London–Christchurch, New Zealand/19,623km (12,193 miles)	Flight Lieutenants R. L. E. Burton and D. H. Gannon	English Electric Canberra	795·887km/h (494·48mph) 23hr 51min 11sec
Daily Mail Transatlantic Air Race	4–11 May 1969	London–New York	Lieutenant Commanders Brian Davies and Peter Goddard, RN	McDonnel-Douglas F-4K	4hr 36min 30·4sec flying time

lacking the financial resources to embark on such ventures. A number of memorable world records have, however, been set in recent years. The Lockheed test pilot Darryl Greenamyer, in a privately sponsored project, flying his much-modified Grumman Bearcat Special *Conquest One*, took the world airspeed record for piston-engined aircraft, achieving 776·449km/h (483·041mph) on 16 August 1969.

James R. Bede, flying a sailplane which he had converted to motor power, gained the world record for distance around a closed circuit for piston-powered aeroplanes, when he covered 17,083km (10,615 miles) between 7–10 November 1969 in 70 hours 15 minutes.

In September 1975 a Lockheed SR-71A aircraft, en route to participate in the Farnborough Air Show, set a record of 1 hour 54 minutes 56·4 seconds for the New York to London sector of its flight – a distance of 5,584km (3,470 miles). It returned to Los Angeles a distance of 9,085km (5,645 miles), in just 3 hours, 47 minutes 35·8 seconds. The same type of aircraft holds the world airspeed record at 3,523km/h (2,189mph), and the (sustained) altitude record of 26,213m (86,000ft). During 1–3 May 1976, a Pan American Boeing 747 SP, commanded by Capt. Walter H. Mullikin, established a new airliner round-the-world speed record of 809·24km/h (502·84mph), the journey occupying 1 day 22hr 50 sec. From 10 December 1977 to 16 March 1978, Russian cosmonauts Yuri Romanenko and Georgi Grechko established a world record, ratified by the FAI, for the amount of time spent in space, amounting to 96 days 10 hr 0 min 20 sec. Since that time the record has been broken twice, both as yet unratified, first by cosmonauts Kovalyonok and Ivanchenko with almost five months aboard Salyut 6, and cosmonauts Lyakhov and Ryumin with 6 months spent in Earth orbit. On 25 January 1978 Philip Clark of Britain achieved a new world distance record in a hot air balloon of 564·47km (350·7 miles). From 12–17 August 1978 Ben Abruzzo, Maxie Anderson and Larry Newman achieved the world's first balloon crossing of the North Atlantic in the American Yost HB-72 helium-filled balloon. A new world speed record for helicopters of 368·4km/h (228·9mph), established over a 15/25km course near Moscow, was achieved by Gourgnen Karapetyan, flying a Mil A-10 on 21 September 1978. On 12 June 1979, a date very close to the 70th anniversary of Blériot's first Channel crossing, the English Channel was crossed for the first time by a man-powered aircraft, Dr Paul MacReady's *Gossamer Albatross*, piloted by Bryan Allen.

Air Crimes

Crimes connected with aviation fall into four main categories: theft, smuggling, sabotage and hijacking. They may be categorized as follows:

Theft
Unless the villain is a pickpocket, or one prepared to snatch hand-luggage from the overhead rack, theft rarely occurs *within* an aircraft. It is in the terminal and its environs that theft flourishes. In the terminal itself the loss of luggage is commonplace, either before departure, after arrival, or at some transit airport. There have been enormous losses by bullion robberies from airport warehouses, as well as pilfering from stacks of cargo awaiting shipment, which offer tempting opportunities for amateur and professional alike.

Terminal car parks – where automobiles are left for extended periods – are also targets for thieves, who know that in many instances the theft may not be reported for two or more weeks after the event. The only safeguard against theft, whatever form it takes, lies in constant and increased vigilance by police, security patrols and by the public generally.

Smuggling
Smuggling, in some form or other, whether it be watches, drugs, diamonds, currency, or any other dutiable or forbidden items, has flourished since the beginning of international air travel, and customs officers at airports throughout the world remain constantly alert for such attempts. Hiding places for would-be smuggled items frequently display imagination – inside hats and babies' nappies, hidden in hollow heels and wooden legs, and concealed around the figures of stout men and women, who become surprisingly thin when examined by customs officials.

Some smugglers are prepared to gamble with their lives. For example, some attempt to smuggle drugs by swallowing liquid cannabis contained in rubber contraceptives. Although smuggling by swallowing is an old practice, this idea is comparatively new – only about two years old, in fact. Although this idea originated in America, customs investigators in Europe are concerned because the idea is fast spreading on their side of the Atlantic. Some cannabis smugglers give themselves away by refusing a cup of tea or drink of water when being quizzed by investigators. The drugs are usually swallowed in

India just before a flight, and nothing else must be swallowed until they are passed naturally – usually in the United States. Liquid swallowed could bring about an embarrassing situation, or worse still, cause death by perforation of the contraceptive. Light aircraft have also been used frequently for smuggling cocaine and marijuana from South America, Mexico, Jamaica and elsewhere into the United States and Canada.

Another comparatively new form of smuggling is that of human beings – mainly illegal immigrants – into the United States and into Britain. Many succeed with the connivance of staff already employed at international airports. Many attempts have been very ambitious. On one occasion a chartered airliner arrived at Heathrow, London, from the East; its passengers, all carrying musical instruments, were ushered through customs, gaining entry as an orchestra. Two weeks later the airliner returned to take them back, but not one member of the 'orchestra' was seen again.

Sabotage

This may be regarded as the most cowardly of all crimes, especially when innocent men, women and children lose their lives in the explosion or subsequent crash, as so often happens. Saboteurs fall into three categories: those with strong political or ideological motives; those who threaten sabotage in the hope of gain – either in cash or the release of persons from custody; and demented, compulsive murderers.

From their very nature, aircraft are particularly vulnerable to sabotage, and almost any concessions can be wrung from the operating companies.

The important subject of sabotage, and how to combat it, was discussed at a conference organized by the international Civil Aviation Organization (ICAO) in the summer of 1970. This conference, held in Montreal, revealed that there had been 14 sabotage or armed-attack cases during the 18 months prior to the end of June 1970. A six-point manifesto was issued, in which one item called for 'concerted action' in suppressing such criminal acts. Today, although acts of sabotage appear to be less frequent, it is possible that some aircraft whose disappearance is marked down as 'cause unknown', could be traced to an act of sabotage.

Hijacking

Of all modern aviation crimes, 'hijacking' is the most recent. Isolated incidents of aerial piracy have been brought to the notice of the public since 1930, but the first official statement concerning this crime came from the Tokyo Convention of September 1963, held to discuss 'Offences Committed on Board Aircraft'. The Convention stated that '. . . if a person on board unlawfully or by force or threat of force seized an aircraft, or wrongly interfered with the control of it in flight, all parties to the Convention should have an obligation to take all appropriate measures to restore control of the aircraft to its lawful commander or to preserve his control of it'.

While outlawing the crime, the Tokyo Convention proposals were considered by most to be inadequate. It discussed the release of passengers, crew and aircraft, but laid down no penalties for the hijacker himself.

By 1963 – the year of the Tokyo Convention – the number of hijack incidents had fluctuated through the years as follows: 1947, 1; 1948, 7; 1949, 6; 1951, 1; 1952, 2; 1953 and 1956, 1 each; 1958 and 1959, 7 each; 1960, 8; 1961, 11; 1962 and 1963, 3 each. Then, as if in contempt of the Tokyo Convention, the number of incidents began to grow from 7 in 1967 to a fantastic 92 in 1969. The years 1968 to 1972 were those of the 'Havana hijackings', which were to become so frequent that they finally lost their newsworthiness and continued virtually unnoticed by all but the airlines and the unfortunate passengers involved. These Cuban hijackings involved the flying of aircraft from the USA to Havana. Not all were successful, as this table shows:

1968　35 hijackings, of which 29 were to Cuba (including 3 failures)
1969　92 hijackings, of which 58 were to Cuba (including 12 failures)
1970　85 hijackings, of which 36 were to Cuba (including 3 failures)
1971　61 hijackings, of which 21 were to Cuba (including 7 failures)
1972　62 hijackings, of which 11 were to Cuba (including 2 failures).

Originally the Cuban authorities allowed the passengers to return home only after making them wait for another aircraft to be flown out specially. Before the hijacked aircraft could be released, landing and other charges had to be met. This was changed in February 1969 and passengers were afforded a virtual 'turn around' at Havana Airport for the return flight.

The most spectacular of the many hijacks to Cuba at that time was that of a US Boeing 747 on a New York to Puerto Rico flight, which was diverted to Havana with 378 passengers. The aircraft was actually inspected at the airfield by Cuba's leader, Fidel Castro, who admitted that he had never had the opportunity to see a Jumbo jet before.

In 1970 President Nixon spoke out against hijacking saying: 'Piracy is not a new challenge for the community of nations. Most countries, including the United States, found effective means of dealing with piracy on the high seas a century and a half ago. We can, and we will, deal effectively with piracy in the skies today.' This was the first time that a leader of one of the world's great powers had made specific reference to hijacking and offered recommendations to combat it.

Hijackings to other parts of the world continued but were becoming rarer. Nevertheless, in October 1969, a most unusual hijacking occurred. Ex-US Marine Raffaele Minichiello boarded a TWA Boeing 707 at Los Angeles and when airborne, demanded to be flown to Rome, backing up his request with a rifle. The Boeing was refuelled at Denver, New York and Bangor (Maine), then took off for Italy. After a 17-hour non-stop flight of 11,263km (7,000 miles) the aircraft touched down in Rome. The hijacker escaped but was later arrested, without resistance, in a country church.

In March 1970, seven Japanese extremist left-wing students hijacked a Boeing 707 on a Tokyo to Fukuoka flight and demanded to be flown to Pyongyang, North Korea. The pilot was ordered by radio to land at Kimbo, Seoul, where elaborate attempts were made to delude the students that they had actually landed at Pyongyang. Western signs were concealed, a large 'Welcome to North Korea' poster covered the fascia of the airport building, and the few soldiers to be seen were in North Korean uniforms. Nevertheless the hijackers were suspicious and the aircraft finally took off for Pyongyang. On arrival the North Korean authorities were furious at having 'such men forced on us'. The aircraft was released immediately and flown back to Tokyo.

The ICAO Montreal Conference of 1970 proposed sanctions against those countries which aided or sheltered hijackers. This was a great advance on the insipid proposals of the Tokyo Convention of seven years earlier, yet three months later the most expensive hijack in history took place. On 6 September 1970, a TWA Boeing 707 on a Frankfurt to New York flight with 145 passengers and a crew of 10, and a Swissair DC-8 flying from Zurich to New York, were both hijacked and flown to a desert airfield north of Amman, Jordan. On the same day a Pan Am 747 on an Amsterdam to New York flight with 158 passengers and 18 crew was also hijacked,

flown to Cairo, and blown up on the following day. On 9 September a British BAC VC-10 with 105 passengers and 10 crew flying from Bombay to London, was hijacked and flown to the desert airfield near Amman. After more than 300 passengers and crew had been released the three aircraft were blown up, at a loss of 50 million dollars.

Following this serious incident, a number of governments and airlines considered following Israel's practice of using armed guards and closed-circuit television to enable pilots to keep an eye on passengers. The adoption of the El Al system, whereby incapacitating gas could be pumped into

The first recorded incident of hijacking occurred in 1930, followed by two in 1931. They were all politically motivated and were connected with the Peruvian revolution of the time. There was then a cessation of such incidents until 1947.

1947
The pilot of a Romanian aircraft was shot dead after refusing to fly it to Turkey. It was the first fatality in hijacking.

1948 7 July
A Cathay Pacific Consolidated Catalina on flight Macao–Hong Kong crashed into the Canton River delta after its pilot had been shot by an armed passenger. Of the 23 passengers, 10 were killed and 12 listed as missing. This was the world's first non-political hijack.

1964 7 May
San Francisco control tower received the message, 'We are being shot'. The aircraft, a Fairchild F-27, crashed, and in the wreckage was found a revolver and six empty cartridges. Forty-four occupants were killed.

1969 29 August
A TWA Boeing 707 was hijacked over southern Italy by two Arabs as the aircraft was on a New York to Lydda flight. The pilot was forced to land at Damascus and the aircraft was evacuated and destroyed on the ground. The hijacking was a reprisal for a US sale of 50 Phantom jet fighters to Israel.

13 September
An Ethiopian Airlines Douglas DC-6 flying from Addis Ababa to Djibouti was seized by three hijackers. One was shot and wounded by a security officer on board, the other two were arrested.

31 October
A TWA Boeing 707 was hijacked over California and diverted half-way across the world.

1970 17 March
As an EAL DC-9 approached Boston Airport, an armed passenger ordered the pilot to fly east. Both pilot and co-pilot were shot whilst struggling with the hijacker, who was overcome and shot. The co-pilot died on the way to hospital.

30 March
The hijacking of a Japanese Boeing 707 on a flight from Tokyo to Fukuoka led to an elaborate hoax in South Korea.

1 July
While flying from Sao Paulo to Buenos Aires, an attempt to hijack a Sud-Aviation Caravelle of Cruzeiro was made by four persons. The pilot was shot and injured. The Brazilian Air Force intervened and forced the aircraft to return to Rio where, on landing, police immobilized the aircraft by machine-gunning the tyres. Using a smoke and tear-gas screen, they then forced their way into the aircraft and captured the hijackers.

6–9 September
Three airliners were hijacked and an attempt made on a fourth in the most spectacular hijacking of all time. Eventually four airliners were destroyed by terrorists.

1971 30 March
A Philippine airliner flying Manila to Davac was hijacked by five armed Filipino students, and the aircraft diverted to China. The authorities made the hijackers apologize to passengers.

23 July
An armed man seized a TWA airliner bound for Chicago, forcing it to return to La Guardia, New York. He then held a stewardess as hostage and ordered a TWA maintenance man to drive them to Kennedy Airport. The hijacker ordered an aircraft to fly him to Italy, but was shot dead by FBI marksmen.

the aircraft's air-conditioning system, was also considered. Magnetometers were immediately installed at airports to check passengers' clothing and hand-luggage for metal objects, a system that has definitely prevented many attempts at both hijacking and sabotage.

The year 1971 saw an appreciable decrease in hijackings, more than half of the attempts being foiled. The trend continued into 1972 due to an overall tightening of security throughout the world. During a three-month period at Heathrow Airport, London, checks revealed 19 pistols, shotguns and rifles, 300 rounds of ammunition, 160 replicas of toy firearms plus 400 knives, swords, axes, bayonets and open razors. By 1973 airport officers had been granted stronger powers for search, arrest and detention than afforded to any other peace-time corps.

Nevertheless, hijackings still persist. In September 1976, a TWA Boeing 727 was diverted to France, the pilot being forced to fly at 609 metres (2,000ft) over London whilst leaflets printed in English were dropped, and over Paris where French leaflets were similarly scattered. They called for independence for the Croatians of Yugoslavia. When the hijackers surrendered on arrival in Paris, their dynamite and detonators were found to be made of modelling clay.

Significant Post War Hijackings

25 November
A Northwest Airlines aircraft was seized by a passenger known as 'D. B. Cooper' prior to touchdown at Seattle. The passengers were allowed to disembark, and the hijacker then collected 200,000 dollars ransom and a parachute. The aircraft was ordered to fly to Reno, but on arrival the hijacker was missing, having parachuted out *en route*.

1972 2 February
Five members of an ultra-militant wing of the Popular Front for Liberation of Palestine (PFLP) hijacked a Lufthansa Boeing 747. After releasing the 182 passengers, the hijackers offered the lives of five crew members for five Palestinian commandos held by Germany. A ransom of five million dollars was subsequently paid by Lufthansa.

9 May
Three Palestine Guerillas threatened to blow up a Sabena Boeing 707 with 95 occupants unless Israel freed some 100 Palestinian guerillas. The hijackers were overcome by 15 'technicians' with guns concealed in their boots.

30 May
Three Japanese walked into the passenger terminal at Lydda, Israel, pulled rifles from their luggage and fired indiscriminately, killing 20 people and wounding 50. One attacker committed suicide by throwing himself on his own grenade, one was caught, and the other escaped.

9 June
Ten Czechs (seven men, three women) hijacked an aircraft of the Czechoslovakian State Airline, seeking asylum in West Germany. The pilot was killed in the struggle. The aircraft, on a Marienbad to Prague flight, landed in Weiden.

2 July
An Asian tried to hijack a Pan Am flight to Hanoi, but was overpowered and shot dead by a passenger. His 'bombs' were two lemons wrapped in tinfoil.

12 July
A National Airlines Boeing 727 was hijacked while making its final approach to Kennedy Airport, New York, with 113 passengers and crew of five. It was forced to return to Philadelphia, and the passengers were held on board in stifling heat for nine hours. A ransom of 600,000 dollars was paid, then the two remaining crew members (the pilot and two others being injured) and four stewardesses were transferred to a new aircraft which flew to Lake Jackson, Texas, where the hijackers surrendered.

18 August
A hijacker cycled through a hole in the airport fence at Reno, Navada and forced a United Airlines Boeing 727 to fly to Vancouver, Canada, where he demanded 15 gold bars and 2 million dollars in cash. The gold was handed over and the aircraft then flew on to Seattle, Washington, where the cash was being gathered by the F.B.I. The hijacker was shot later by FBI men in Tacoma.

8 December
Six hijackers (four men, two women) died in a mid-air battle with security guards aboard an Ethiopian Airlines Boeing 720B as a grenade tore a hole in the side of the aircraft. Nine other people were treated in hospital and released.

1973 30 May
A Colombian Lockheed Electra with 80 passengers was hijacked by two armed men who ordered the pilot to head for Cuba and demanded the release of all political prisoners in Columbia and 200,000 dollars ransom for the passengers. The Colombian government rejected the demand, and the aeroplane finally landed at Buenos Aires on 5 June after the longest hijack flight on record.

25 November
A KLM Boeing Jumbo on flight from Amsterdam to Tokyo with 280 passengers was hijacked at Beirut by three armed Arabs. After being diverted to Damascus for refuelling, permission was sought to land at Cairo, Baghdad, Kuwait, Bahrain and Aden. All refused, and the aircraft finally landed at Dubai, where the hijackers surrendered.

1974 24 January
Three men and a woman hijacked a helicopter in Eire and forced the pilot to fly them to Strabane at gun point. They tossed out two milk-churns filled with explosive. No damage was done, but it was the first 'air raid' on Northern Ireland since World War II.

15 September
A Boeing 727 informed radio control at Da Nang that a hijacker was on board. The aircraft tried to land at Phan-Rang, but overshot the runway and exploded on touching the ground, killing 67 passengers and the crew of 9.

21 November
Four Palestinian guerrillas forced a British Airways BAC VC-10 to fly to Tunis, where the 40 occupants were held hostage until 25 November as ransom for the release of 15 Palestinians serving sentences for hijacking. A compromise was reached: the passengers were released in exchange for seven Palestinians. The guerrillas then finally surrendered to the Tunisian authorities.

6 June
A helicopter pilot was forced at knife-point to sweep down into a prison yard near Jackson, Michigan, to free a convict. A waiting car took him to safety.

1976 27 June
Three men and a woman hijacked an Air France A300B with 247 passengers and 12 crew aboard soon after take-off from Athens. The pilot was forced to land at Entebbe, Uganda. The hijackers were joined by six more Palestinians and asked for the release of 'freedom fighters' imprisoned in various countries. A successful rescue operation was made by the Israeli government using three Lockheed Hercules C-130 transport aircraft. The passengers and crew were flown to Israel after a brief refuelling stop in Nairobi, Kenya.

12 September
A TWA Boeing 727 on a flight from New York to Chicago was hijacked; 52 passengers and 7 crew were held hostage at Charles de Gaulle Airport, Paris. A Boeing 707 was forced to accompany the hijacked aircraft, whose pilot was inexperienced in trans-Atlantic travel. A 'leaflet raid' followed before the hijackers gave themselves up in Paris after negotiations.

1978
Eight hijacking attempts were made in the United States during the year, the highest number since screening of passengers and carry-on baggage was initiated in 1972. Of great significance was the fact that in none of these cases had the hijackers managed to board the aircraft with weapons or explosives. In each case their claim to be armed turned out to be a hoax.

1979 21 June
Serbian Nikola Kavaja, on his way to stand trial for conspiracy to assassinate, took control of an American Airlines Boeing 727 on an internal flight from New York. Threatening to destroy the aircraft by means of explosives attached to his body, Kavaja ordered the aircraft first to Chicago, where the 127 passengers on board were released. He then directed the return of the 727 to New York, where he transferred to an American Airlines 707, demanding to be flown to South Africa. He decided subsequently to be flown to Shannon, Eire, where he surrendered to police.

Air Disasters

From the day that man made his first tentative attempts to become airborne in a heavier-than-air machine, it was inevitable that the first fatal accident should soon follow. That doubtful honour fell to Lieutenant Thomas Selfridge, who was killed while flying as a passenger with Orville Wright, when their Wright biplane crashed at Fort Myer on 17 September 1908. He thus became the world's first powered aeroplane occupant to be killed whilst flying.

The increasing number of fatalities is allied to the incredible progress of aviation in general. They have grown from the first, isolated accidents to large and then even larger accidents. However, a comparison of passenger miles travelled per accident reveals that flying has become progressively safer over the years. Accident statistics relating to regular British air transport show that between August 1919, when the first regular air transport service was inaugurated, and the end of 1933, the number of aircraft miles flown per accident resulting in death or injury to occupants increased from 574,535km (357,000 miles) during the period 1919 to 1924, to 3,759,400km (2,336,000 miles) during the period 1930 to 1933. A typical 'serious accident' of 1933 occurred when an Imperial Airways' aircraft flying from Cologne to London caught fire and crashed at Diksmuide, the crew of 3 and all of its 12 passengers being killed.

A large proportion of early accidents was attributed to the inability of some aircraft to cope with sudden stresses. A report of 1912 stated that most mishaps were caused by 'frames broken by sudden deflecting forces thrown upon them from the rotating machinery due to an unusually quick change in the direction of the axis of rotation of the driving mechanism. More than half of the casualties caused by men falling to the ground in aeroplanes are traceable to this instability, known as the "gyroscopic effect".'

In World War I, 'an unusually quick change in direction' often made the difference between life or death, and this resulted in the development of far more robust and handier machines. By then the greatest single contributory cause of fatal accidents was that of 'stalling'. This occurs when flying speed is lost, and the smooth airflow over the upper surface of a wing (aerofoil surface) breaks up, causing a breakdown of the low-pressure area above the wing, an increase in drag and a sudden loss of lift. The wing is said to have stalled. The effect is for the nose to drop and the aircraft to lose height. Unless the aircraft has sufficient altitude to regain flying speed in its dive, enabling the pilot to regain control, it will crash.

Types of accidents

Aircraft accidents today, other than those caused by sabotage or air piracy (hijacking), fall into 23 categories:

Airframe failure; Aquaplaning/Hydroplaning; Bird strike/ingestion; Cargo breaking loose; Collision with high ground; Collision with water; Crew incapacitation; Door, windows opening/falling off in flight; Electrical systems failure/malfunction; Failure of power units; Flying control system malfunctions; Fuel contamination; Fuel exhaustion, starvation, mismanagement; Hail damage; Ice/snow accretion on airframe/engine; Inflight fire/smoke; Instruments incorrectly set, misread, failure, malfunction; Lightning strike; Major powerplant disruption/loss of propeller in flight; Mid-air collision with another aircraft; Over-running/Veering off runway; Third-party accidents; Tyre burst after retraction.

Despite all these possibilities, and no matter by what yardstick the safety of airline operations is measured, air travel is very safe. Jet operations alone – comprising some 90 per cent of all flying done today – have improved tenfold since 1959, when jet transport was established. A rate of one jet aircraft loss in every 50,000 hours was recorded in that first 'jet year'. By 1976 it had improved to about one in 560,000 hours flown. Assuming an average sector length of three hours (a 'sector' is the average time between a take-off and a landing), and that each passenger averages two sectors, a passenger should have to make 84,500 journeys before he is likely to be involved in an accident. Even flight crews on long-haul operations, whose time in the air is much greater than that of the passengers, would have to live to be about 400 years old before they were likely to be involved in a fatal accident.

The last decade has shown a remarkable improvement in safety standards. In 1966 4,480 million km (2,784 million miles) were flown as against 7,550 million km (4,691 million miles) in 1975. In 1966 there were 31 fatal accidents and 1,001 passengers killed, representing 0·33 fatal accidents for every 100,000 aircraft hours flown, or 0·44 fatalities per million passenger-km. In 1975 there were 21 accidents and 473 passengers killed, representing the related factors of 0·17 and 0·07 respectively.

Major Air Disasters

Note: Only those with 100 or more fatalities are shown. Obviously, as aircraft developed, so the seating capacity increased, and individual disasters grew larger by comparison. The first disaster with a 100+ fatality occurred in 1956

Date	Aircraft	Location	Fatalities Total	Pass	Crew
1956					
30.6	Lockheed L-1049	Arizona, USA	128	64	6
	Douglas DC-7			53	5
1960					
16.12	Lockheed L-1049	New York, USA	133	39	5
	Douglas DC-8			77	7
				(Others, 5)	
1962					
4.3	Douglas DC-7C	Cameroons	111	101	10
16.3	Lockheed L-1049	Between Guam and Philippines	107	96	11
3.6	Boeing 707–328	France	130	122	8
22.6	Boeing 707	Guadeloupe	112	102	10
1963					
1.2	Vickers Viscount 754	Turkey	104	11	3
	Douglas C-47				3
				(Others, 87)	
3.6	Douglas DC-7C	Alaska	101	95	6
29.11	Douglas DC-8F	Canada	118	111	7
1965					
20.5	Boeing 720B	Egypt	119	114	5
1966					
24.1	Boeing 707–437	France	117	106	11
4.2	Boeing 727	Japan	133	126	7
24.12	Canadair CL-44	South Vietnam	111	0	4
				(Others, 107)	
1967					
19.4	Bristol Britannia 313	Cyprus	126	117	9
11.11	Ilyushin Il-18	Russia	130	n.a.	n.a.
1968					
20.4	Boeing 707–344C	S.W. Africa	123	111	12
1969					
16.3	Douglas DC-9	Venezuela	154	74	10
				(Others, 70)	
20.3	Ilyushin Il-18	Egypt	100	93	7
1970					
15.2	Douglas DC-9	Argentina	102	97	5
3.7	de Havilland Comet 4	Spain	112	105	7

Date	Aircraft	Location	Fatalities Total	Pass	Crew
1970					
5.7	Douglas DC-8	Canada	109	100	9
9.8	Lockheed Electra L-188A	Peru	101	99	0
				(Others, 2)	

Date	Aircraft	Location	Fatalities Total	Pass	Crew
1971					
30.7	Boeing 727	Japan	162	155	7
4.9	Boeing 727	Alaska	111	104	7
1972					
7.1	Sud-Aviation Caravelle	Spain	104	98	6
14.3	Sud-Aviation Caravelle	Muscat & Oman	112	106	6
5.5	Douglas DC-8	Sicily	115	108	7
18.5	Antonov An-10	Russia	108	n.a.	n.a.
18.6	Hawker Siddeley Trident	England	118	109	9
14.8	Ilyushin Il-62	Germany	156	148	8
2.10	Ilyushin Il-18	Russia	100	n.a.	n.a.
13.10	Ilyushin Il-62	Russia	174	164	10
3.12	Convair CV-990A	Tenerife	155	148	7
1973					
22.1	Boeing 707	Nigeria	176	170	6
21.2	Boeing 727	Sinai Desert	106	n.a.	n.a.
10.4	Vickers Vanguard	Switzerland	108	104	4
11.7	Boeing 707	France	123	116	7
22.12	Sud-Aviation Caravelle	Morocco	106	99	7
1974					
5.3	Douglas DC-10	France	345	334	11
22.4	Boeing 707	Bali	107	96	11
4.12	Douglas DC-8	Sri Lanka	191	182	9
1975					
24.6	Boeing 727	New York	133	107	6
3.8	Boeing 707	Morocco	188	181	7
19.8	Ilyushin Il-62	Nr. Damascus	126	115	11
1976					
5.3	Ilyushin Il-18	Russia	127	109	11
				(Others, 7)	
10.9	Douglas DC-9	Yugoslavia	176	108	5
	Hawker Siddeley Trident			54	9
13.10	Boeing 707	Bolivia	100+	0	2
				(Others, 100+)	
1977					
27.3	Boeing 747 (KLM)	Tenerife	561	229	15
	Boeing 747 (PanAm)			308	9
19.11	Boeing 727	Funchal, Madeira	129	123	6
1978					
1.1	Boeing 747	Bombay, India	213	190	23
25.9	Boeing 727	San Diego, USA	136	129	7
15.11	Douglas DC-8	Katunayake, Sri Lanka	182	174	8
23.12	Douglas DC-9	Palermo, Sicily	108	103	5
1979					
25.5	Douglas DC-10	Chicago, USA	272	259	13
11.8	Tupolev Tu-134	Donetsk, Russia (two in collision)	173	163	10

Wreckage of a Brazilian airliner, Orly airport, Paris, in July 1973.

The human factor

Such has the standard of design and manufacture improved that today something like 80·5 per cent of all accidents are the result of human error. Broken down into percentages, these comprise: crew 62 per cent; operational (procedures etc) 15 per cent; maintenance 3·5 per cent; non-operational sabotage 6·5 per cent; material or system failure 8 per cent; and weather 4·5 per cent.

The Lovelace Foundation's Department of Aerospace and Environmental Medicine studied a number of approach and landing accidents in an attempt to identify common factors. Of the 197 critical elements catalogued, 117 were grouped under 'man', 12 under 'machine' and 68 under 'environment'. This is further shown by an analysis of jet losses during two significant periods.

Aircraft position when

accident occurred	1962–1971	1972–1975
Take-off and initial climb	20%	18·25%
Climb	13%	2%
Cruise	4%	6%
Descent	9%	10%
Approach	35%	36·5%
Landing	12%	18·25%
On ground (during maintenance, etc.)	4%	8%
Test and training	3%	—
Missed approach	—	1%

The International Air Transport Association (IATA) has identified so-called 'human factors' as a major causal element in accidents. The term 'human factors' is used to describe events where the flight crew would seem to have failed to act as they should have done, thereby contributing to an accident. Intensive work on this problem over a period of years culminated in the 20th IATA Technical Conference, held in Istanbul in November 1975. This Conference led to a number of recommendations which, it is believed, will play an important role in reducing the number of 'human factors'-related accidents.

International co-operation

Some countries insist that certain occurrences are reported regularly and investigated thoroughly. By this it is hoped to prevent recurring problems in both machinery and operating practices. These are, of necessity, 'after the event' actions.

Today there are a number of exchange schemes where airlines share their experiences freely, albeit on a confidential basis. IATA, America's Air Transport Association (ATA) and the UK Flight Safety Committee all have similar schemes. These provide a forum where any airline operator who has experienced an occurrence from which he has learned a lesson, or which has caused him to change his procedures or drills, can report to the others participating in the scheme.

Most countries have some form of Flight Safety Committee, a group which includes representatives of their country's airlines, manufacturers, insurers and ministries. IATA's Regional Technical Offices spend a great deal of their time persuading governments to install and maintain modern navigation and landing aids. It is well known that a very large majority of accidents occur in the descent, approach and landing phase, and by providing pilots with these facilities at all airports, the occurrence of this type of accident should in fact be substantially reduced.

Whilst IATA basically consists of airlines, ICAO is comprised of governments and well over 100 are represented. The constitution of ICAO is the Convention on International Civil Aviation, drawn up by a conference in Chicago in November/December 1944, and to which each ICAO member state is a party.

One of its chief activities is standardization, the drafting of international standards and procedures covering the whole field of aviation, including airworthiness, search and rescue, accident inquiry, and so on.

IATA maintains an incident exchange scheme known as the 'Safety Information Exchange'. This provides for exchange between IATA members (and more recently non-IATA airlines) of anonymous reports on accidents and serious incidents – particularly the latter. There is a great deal to be learned from incidents, that is to say, unusual or alarming in-flight events which did not lead to an accident, but nevertheless taught some safety lesson. Incidents occur much more frequently than accidents, and the lessons they teach are often just as important as those learned from accidents themselves. The incident exchange scheme therefore immensely broadens the foundation for future preventive action.

'The little black box'

One important mechanism that has helped investigators to reach their conclusions more quickly and accurately than before is the tell-tale device popularly known as 'the little black box', introduced as long ago as 1924.

Unfortunately, the title is very much a misnomer

as, for the most part, flight recorders, as they are more properly known, are painted in brilliant colours such as fluorescent scarlet or orange, to make them easier to locate amid aircraft wreckage. The vital information concerning time, speed, pitch attitude, altitude, magnetic heading and vertical acceleration, plus other significant data, is produced through a sophisticated programme of digital flight data recording. As many as 125 channels a second from 60 separate sources can be recorded.

Equally important is the introduction of machines that prevent mid-air crashes. Systems currently in use are basically of two types; those used by the Air Traffic Control system on the ground, to indicate to the controller that he has a conflict, or those in the aircraft that interrogate all the aircraft nearby to establish if collision potential exists. In the former case the Air Traffic controller would pass on the necessary avoidance instructions, or, as in the B–CAS System, the information could also be carried on board the aircraft and displayed to the pilot. In the latter case, it is necessary to have compatible equipment in *every* aircraft. This has to be set against a common time base which needs to be incredibly accurate. The 'on board' systems tell the pilot to turn, climb or dive to avoid the conflict.

The search for improved safety standards never ceases. It is partly because of this search that ICAO came into being, in an attempt to achieve cooperation and coordination wherever commercial aircraft fly. The success which international civil aviation has achieved in the past two decades has already proved that nations *can* work together effectively for the public good . . . and do.

Aviation Law

'Air law' may be defined as a functional and interdisciplinary branch of the legal system. It deals with the rights and duties of different subjects with respect to the airspace and the aeronautical use thereof, and with respect to aircraft and its operations. It falls both in the category of *international law* (i.e. rules created by and applicable to sovereign states in their mutual relations) and *national legislation* (i.e. legal rules issued by an appropriate authority of a state and enforced by the judicial and other authorities of that state). The national legislation comprises private law, administrative law, labour law, criminal law, etc. Due to the international character of aviation and the need for world-wide uniformity of legal regulation, the national legislation is to a large extent modelled on or derived from international agreements and other sources of international law.

The following are just a few examples of the spectrum of problems governed by air law on national or international level: licensing of the flight personnel; rules of the air; registration and nationality markings of the aircraft; certification of airworthiness of aircraft; operation of the aircraft; establishment, equipment and operation of airports; aeronautical communications; search and rescue in case of aircraft in distress; investigation of accidents; economic regulation and licensing of operators; rules on contract of carriage in respect of passengers, cargo and mail, including problems of liability and insurance; protection of aviation against unlawful acts (hijacking, sabotage); protection of the environment against noise, sonic boom and pollution; and so on.

The 'aircraft' is internationally defined as 'any machine that can derive support in the atmosphere from the reactions of the air, other than the reactions of the air against earth's surface. A hovercraft derives its support from the reactions of the air against the earth's surface. Consequently it is not an 'aircraft', and is not governed by the international air law conventions or by the national air legislation.

The Roman law maxim '*cuius est solum eius est usque ad coelum*' (who owns the land, owns it up to the skies), generally considered as historically the first rule of air law, is hardly more than a clarification of the scope of the proprietary rights of the landowner, and was never intended to regulate aviation. The law cannot develop before the situation it is supposed to regulate.

The law-maker and the law-enforcer closely followed the aviators from their earliest exploits. The historians seem to agree that the first real legal regulation of aviation is contained in an ordinance of a certain M. Lenoir – a *lieutenant de police* in Paris – which prohibited balloon flights without special permit from 23 April 1784 – some five months after the first manned flight of a hot-air balloon over Paris. In January 1822 the New York Supreme Court issued the first recorded judicial decision in an aviation case – a balloonist was found liable for damage caused on the plaintiff's land by a forced landing of the balloon (incidentally, the damage was caused mostly by a crowd of curious onlookers, who rushed to the place of landing and damaged the plaintiff's fences).

In the first years of this century most developed countries had some form of legal regulations for the safety in air navigation, customs regulations applicable to foreign aviators, penalties and jurisprudence. In practice, if not by law, many states clearly started to assert their sovereign rights in the airspace above their territories. The first international incident occurred in August 1904, when Russian borders guards shot down a German balloon allegedly overflying Russian territory.

The cardinal disputed question at the outset of development of air law was the legal status of the airspace above national territory. At the beginning of the 20th century three basic theories were advocated:

Freedom of the air: by analogy with the regime of the high seas, it was argued that the airspace was a common property of mankind and not subject to appropriation or claims of sovereignty

Complete and exclusive sovereignty: according to this theory the airspace above the territory of a state was to be considered an integral part of that state's sovereign territory.

Mixed theory: this theory recognized the 'freedom of the air' while safeguarding certain rights of the state in its national airspace.

Whatever merits these theories may have had, it was the practice of states to determine the sovereign character of the airspace above its territory. There was an abortive attempt, in Paris in 1910, to prepare an international convention on air navigation. However, World War I interrupted further diplomatic negotiations, aircraft proved to be a formidable new weapon and states resorted to unilateral – and surprisingly uniform – practice. All international boundaries were closed to all aircraft of the belligerents, and all neutral states strongly protested against any violation of their airspace, and in numerous cases had the intruding aircraft shot down. Thus, by 1918, the international practice recognized unequivocally the legal principle of state sovereignty in the airspace.

The milestones in the development of public international air law are the following conventions:

International Convention on Air Navigation (Paris, 1919): This Convention recognized that every state has complete and exclusive sovereignty over the airspace above its territory, and provided for the freedom of innocent passage of civil aircraft of other contracting states over the state territory. It also established the International Commission for Aerial Navigation (ICAN) as an international organization, the function of which was to exercise the legislative, administrative and judicial functions set out under the Paris Convention.

Ibero-American Convention on Air Navigation (Madrid, 1926): Essentially identical with the Paris Convention, the Madrid Convention was to Group Spain and Latin American countries into an organization similar to ICAN, but with equal voting power for all states.

Pan-American Convention (Havana, 1929): Accepted by 16 states at the 6th Pan-American Conference, this was also modelled on the Paris Convention, and included detailed technical legislation as well as some rules of private air law.

Convention on International Civil Aviation (Chicago, 1944): 135 states are presently Parties to this Convention (generally known as 'Chicago Convention'), which is the backbone of the current public international air law. Part I of the Convention sets out the rules applicable to air navigation. The Convention recognizes that each state has complete and exclusive sovereignty over the airspace above its territory (land areas and territorial waters adjacent thereto). Aircraft of other contracting states are entitled to make non-scheduled flights into or over the territory of a state and make stops for non-traffic purposes without the necessity of obtaining prior permission, but subject to certain limitations. Consequently, the Convention does not recognize (unlike the Paris Con-

vention) the right of 'innocent passage' as such. No scheduled international services may be operated over or into the territory of a state except with special permission or other authorization of that state. Such authorization is usually contained in bilateral agreements on air traffic services, in which states contractually agree on a mutual exchange of commercial rights for the operation of their airlines. There are presently over 2,000 such bilateral agreements.

The Chicago Conference in 1944 also adopted the **International Air Services Transit Agreement** (generally known as 'two freedoms' agreement), in which contracting states agreed to grant to other contracting states, in respect of scheduled international air services, the privilege to fly across their territory without landing and the privilege to land for non-traffic purposes (e.g., refuelling, maintenance, but not to take on or put down passengers, mail or cargo). This agreement is presently in force for 92 states.

The **International Air Transport Agreement** (also known as 'five freedoms' agreement) was also adopted by the Chicago Conference. In addition to the above-mentioned 'two freedoms' this also grants the commercial 'freedoms' (privilege to put down passengers, mail and cargo originating from the territory of the state of registry of the aircraft, the privilege to take on passengers, etc. destined for such territory and, finally, to take on or put down passengers, etc. coming from or destined to the territory of any contracting state). Only 12 states are parties to this agreement, and states prefer to negotiate the granting of the traffic rights on a bilateral basis.

The Chicago Convention deals in detail with many technical and economic aspects of international civil aviation, and in Part II establishes the International Civil Aviation Organization (ICAO) and sets its constitutional rules.

The International Civil Aviation Organization (ICAO), created by the Chicago Convention, came into being on 4 April 1947. It is a specialized agency in the United Nations system and its membership – 135 states – makes it practically universal. The Headquarters of ICAO is in Montreal, Canada, and Regional Offices are located in Paris (European Office), Dakar (African Office), Bangkok (Far East and Pacific Office), Cairo (Middle East and Eastern African Office), Mexico City (North American and Caribbean Office) and Lima (South American office).

The aims and objectives of the Organization are to develop the principles and techniques of international air navigation, and to foster the planning and development of international air transport to ensure the safe and orderly growth of international civil aviation throughout the world, encourage the arts of aircraft design and operation for peaceful purposes, promote safety of flight in international air navigation and to promote generally the development of all aspects of international civil aeronautics.

The supreme body of ICAO is the Assembly, which meets ever three years. The executive body is the Council, a permanent body composed of 30 states elected by the

Assembly. The Air Navigation Commission is a permanent body of 15 technical experts appointed by the Council from among persons nominated by contracting states. The Air Transport Committee dealing with economic matters is composed of Representatives on the Council. The Legal Committee of ICAO was established by the Assembly, and is open to participation by all states. The President of the ICAO Council is elected by the Council, and is an international civil servant; he has no vote in the Council. The chief executive officer is the Secretary General who is appointed by the Council.

From the legal point of view, the essential function of ICAO is the formulation and approval of international standards and recommended practices in the field of civil aviation. These standards are approved by the Council and designated as Annexes to the Chicago Convention. All contracting states undertake to collaborate in securing the highest practicable degree of uniformity in regulations, standards, procedures, etc., but may file a difference if they find it impracticable to comply in all respects with any international standard or procedure. So far the Council of ICAO has approved 17 Annexes to the Chicago Convention containing international standards and recommended practices.

The International Air Transport Association (IATA) is a non-governmental organization of airlines, and was created in 1945 to succeed the International Air Traffic Association of 1919. It was incorporated in Canada by Act of Parliament and has the status of a corporation, to which Canadian law attributes an international character. IATA is sometimes compared with an international cartel of airlines – particularly because of its rate-making functions. However, since in many states the airlines are subject to anti-trust legislation, and since the tariffs and rates are subject to governmental approval, IATA should not legally be considered as a business cartel.

Apart from technical and economic regulation of the airline industry. IATA performs important functions in the legal field by establishing uniform air transport documents and general conditions of carriage which are accepted uniformly in the tariffs of a great majority of world airlines.

The international character of civil aviation necessitates a high degree of uniformity of legal regulations. The unification of law is best achieved by multilateral international agreements, the rules of which are transformed into the national legislation of contracting states. Since the establishment of ICAO, all agreements in this field have been prepared by the ICAO Legal Committee and adopted by Diplomatic Conferences convened and held under the auspices of ICAO. The following air law conventions have been adopted so far.

Warsaw Convention of the Unification of Certain Rules Relating to International Carriage by Air (1929). This Convention, to which 108 states are parties, is the most widely accepted unification of private law. It unifies the rules on documentation in the carriage of passengers, baggage and cargo (passenger ticket, baggage check and air waybill). Secondly, it unifies the regime of liability of the carrier. The carrier is liable on the basis of fault, but there is a presumption of fault and the burden of proof is reversed. The extent of liability of the carrier is limited to 125,000 Poincaré francs (less than US $10,000) in respect of passengers, and to 250 francs (about US $20) per kilogramme of baggage or cargo. Finally, the Convention unifies the question of jurisdiction by defining the courts before which an action may be brought.

The Convention was amended in 1955 by **The Hague Protocol**, which simplified the documentation and doubled the limit of liability in respect of passengers. In 1961 a Supplementary Convention was adopted at Guadalajara, which extends the applicability of the Warsaw Convention to the 'actual' (as distinct from 'contractual') carrier. In 1966 the Warsaw Convention was *de facto* amended by a private agreement of air carriers ('**Montreal Agreement**, 1966') with respect to carriage to, from or via the territory of the USA. For such carriage the carrier is absolutely liable with respect to passengers, and the limit of liability is $75,000.

The **Guatemala City Protocol** of 1971 (not yet in force) further simplified the documentation requirements with respect to passengers, introduced the principle of absolute liability and increased the limit of liability to about US $100,000.

The **Montreal Protocols** of 1975 (not yet in force) would furthermore replace the gold value clause (Poincaré franc) in the system of the Warsaw Convention by the Special Drawing Rights of the International Monetary Fund.

Geneva Convention on the International Recognition of Rights in Aircraft (1948). The purpose of this Convention, to which 38 states are parties, is to ensure protection of rights in aircraft (such as the right of creditors), when the aircraft and its equipment are situated in another country or where the registration of the aircraft is changed from one country to another.

Rome Convention on Damage Caused by Foreign Aircraft to Third Parties on the Surface (1952). There are 28 state parties to this Convention, which introduces the regime of absolute liability for damage caused by aircraft on the surface. The liability is limited and the amount varies with the weight of the aircraft. The Legal Committee of ICAO is now studying a possible amendment of the Rome Convention in order to make it clear that it would deal only with 'physical impact' damage and not with damage caused by noise, vibrations or sonic boom. The system of the limits of liability will also be reviewed. The preparation of a new separate instrument on liability for damage caused by noise and sonic boom is under study in ICAO.

Under the auspices of ICAO three Conventions in the field of criminal law have been prepared, the last two of them dealing with criminal acts against the safety of civil aviation such as 'hijackings'. They are the **Tokyo Convention on Offences and Certain Other Acts Committed on Board Aircraft** (1963); the **Hague Convention for the Suppression of Unlawful Seizure of Aircraft** (1970) and the **Montreal Convention for the Suppression of Unlawful Acts Against the Safety of Civil Aviation** (1971).

Famous Names

Ader, Clément (1841–1925)
French pioneer, whose bat-winged *Éole* monoplane became, on 9 October 1890, the first man-carrying aeroplane to achieve a powered 'hop' into the air.

Alcock, Sir John (1892–1919)
Who, with Lieutenant Arthur Whitten-Brown (later Sir Arthur) as navigator, achieved the first non-stop crossing of the North Atlantic in a Vickers Vimy bomber, 14 to 15 June 1919.

Aldrin, Edwin (1930–)
Astronaut and crew member of the *Apollo XI* spacecraft which carried the first men to land on the Moon's surface, on 21 July 1969.

Amundsen, Roald (1872–1928)
Renowned explorer, a member of the crew of 16 of the Italian airship *Norge* which flew over the North Pole on 12 May 1926.

Archdeacon, Ernest (1863–1957)
Whose prize awards did much to foster European aviation, was the first passenger to be carried in an aeroplane in Europe, on 29 May 1908.

d'Arlandes Marquis
First aerial passenger in history, who accompanied Pilâtre de Rozier in a Montgolfier hot-air balloon which flew over Paris on 21 November 1783.

Armstrong, Neil (1930–)
Commander of the *Apollo XI* spacecraft which carried the first men to the Moon, and the first man to step on the lunar surface on 21 July 1969.

Bacon, Roger (1214–1292)
English scientist-monk who, so far as is known, was the first person to write in a scientific manner about the possibility of manned flight. His *Secrets of Art and Nature*, in which he first put forward such ideas, was written about 1250.

Bader, Sir Douglas (1910–)
Famous British fighter ace of World War II, who achieved his 22 credited victories while flying with two artificial legs. His legs were amputated following a pre-war flying accident.

Balbo, General Italo (1896–1940)
Italian Air Minister, who led two mass flights of Savoia-Marchetti flying-boats across the South and North Atlantic in 1931 and 1933 respectively.

Balchen, Bernt (1899–1973)
Pilot of the Ford Trimotor *Floyd Bennett*, which made the first flight over the south Pole on 29 November 1929.

Baldwin, F. W. (Casey)
First British subject (a Canadian) to fly a heavier-than-air machine. This flight was made at Hammondsport, N.Y., 12 March 1908, in an AEA Red Wing.

Ball, Albert (1896–1917)
First great British fighter ace of World War I; was credited with his 47th victory on 6 May 1917. He died on the following day when his aircraft crashed after following a German single-seater scout aircraft into dense cloud, but the cause of his death was never known.

Banks, F. Rodwell (1898–)
Of the Ethyl Export Corporation, whose specialized knowledge of aviation fuels played an important part in the development of the Rolls-Royce racing engines and, subsequently, of the British gas turbines.

Barber, Horatio (1875–)
Pilot of a Valkyrie monoplane which, on 4 July 1911, carried the first recorded consignment of air-freight between Shoreham and Hove in Sussex.

Barnard, F. L.
Commodore of Britain's early Instone Airline, who won the first King's Cup Air Race on 8–9 September 1922. Captain Barnard was flying Sir Samuel Instone's de Havilland D.H.4A *City of York*.

Batten, Jean (1909–)
Pioneer woman pilot from New Zealand, whose exploits included the first solo air crossing of the South Atlantic by a woman, 11–13 November 1935.

Beachey, Lincoln
Daring pioneer pilot in America, first became famous for an airship flight around the Washington Monument on 13 June 1906.

Béchereau, Louis
Responsible primarily for the design of the streamlined, circular, monocoque wooden fuselage of the highly successful French Deperdussin monoplanes of 1912/1913.

Bell, Alexander Graham (1847–1912)
Famous inventor born in Edinburgh, was associated with Glenn Curtiss and Thomas E. Selfridge in Bell's Aerial Experiment Association which, from 1903 to 1905, carried out successful aeroplane experiments in America, following the work of the Wright brothers.

Bennett, D. C. T. (1910–)
Piloted the upper, Short S.20 *Mercury* seaplane, component of the Short-Mayo composite, from Foynes, Ireland to Montreal, Canada on 21 to 22 July 1938, the first commercial use of composite aeroplanes.

Bennett, Floyd (1890–1928)
Pilot of the Fokker F.VIIA/3M monoplane *Josephine Ford* which, on 9 May 1926, became the first aeroplane to fly over the North Pole.

Birkigt, Marc (1878–1953)
Swiss engineer who in 1904 founded with Damien Mateu the Hispano-Suiza Motor Company. During World War I the company turned to production of aero-engines. Another famous product of the company was the 20mm Hispano gun.

Bishop, R. E. (1903–)
Designer of the de Havilland D.H. 106 Comet 1, which on 2 May 1952, operated the world's first passenger service by a turbo-jet-powered aircraft.

Bishop, William A. ('Billy') (1894–1956)
Canadian ace pilot of World War I, credited with 72 confirmed victories. Awarded Britain's highest military honour, the Victoria Cross, he rose subsequently to the rank of Honorary Air Marshal in the Royal Canadian Air Force.

Black, Tom Campbell
Co-pilot of the de Havilland D.H.88 Comet which won the 1934 MacRobertson race from Mildenhall, England, to Melbourne, Australia, in a time of 71 hours.

Blériot, Louis (1872–1936)
French aviator pioneer who, flying a Blériot XI monoplane of his own design, made the first crossing of the English Channel in an aeroplane, on 25 July 1909.

Boeing, William (1881–1956)
Co-pilot with Eddie Hubbard of the Boeing C-700 seaplane which made the first international air mail flight from Canada to America, on 3 March 1919, he was the founder of the now famous Boeing Company.

Boothman, John N. (1901–1957)
A member of the Royal Air Force High Speed Flight of 1931, and pilot of the Supermarine S.6B racing seaplane which won the Schneider Trophy outright for Great Britain on 13 September 1931.

Brabazon of Tara, Lord (Moore-Brabazon, J. T. C.) (1884–1964)
First Englishman to make an officially recognized aeroplane flight in England, on 30 April 1909, and holder of the first Aviator's Certificate in Great Britain, awarded by the Aero Club on 8 March 1910.

Braun, Dr Wernher von (1912–1977)
German rocket pioneer, under whose supervision was developed the world's first ballistic rocket missile, the V-2 (A-4), first launched successfully on 3 October 1942. He was subsequently to become one of America's leading space scientists, actively involved in the Apollo project, which put the first man on the moon.

Breguet, Louis (1880–1955)
French pioneer of rotary-wing flight. In association with Professor Richet he de-signed and built a helicopter which, on 29 September 1907, was the first to lift a man from the ground. This was not a controlled flight, the aircraft being stabilized by four men with poles.

LeBris, J. M. (1808–1872)
A French sea captain who studied bird flight on his ocean voyages. In 1856 he completed construction of a glider which was launched from a horse-driven cart. A second aircraft, built in 1868, was launched frequently, with ballast instead of a human pilot.

Butler, Frank Hedges (1856–1928)
Aware of the encourage-ment which early aviation activities in Europe were re-ceiving from the Aéro-Club de France, was the first to suggest the formation of an Aero Club in Britain. This was duly registered on 29 October 1901.

Byrd, Richard E. (1888–1957)
American explorer, first man to fly over both North and South Poles, on 9 May 1926, and 29 November 1929, respectively.

Cabral, Sacadura
Who, together with Gago Coutinho, achieved the first aeroplane crossing of the South Atlantic in 1922. Their arrival in Brazil, on 16 June, marked the end of án epic three-month journey involving three aircraft.

Camm, Sir Sydney (1893–1966)
Famous British aircraft de-signer, who joined the H. G. Hawker Engineering Com-pany in the early 1920s. Knighted in 1953 for his ser-vices to aviation, his name will always be associated in aviation history with the Hawker Hurricane of World War II, but as early as 1924 he had designed the ultra-light cygnet, his first aircraft built for HG Hawker En-gineering Company.

Capper, J. E. (1861–1955)
Officer Commanding the Balloon Sections of the British Army in 1903 and pilot of the first British Army airship, *Nulli Secundus*, which flew for the first time on 10 September 1907.

Castoldi, Dr Ing Mario
Famous aircraft designer, whose work is associated with the Italian company Aeronautica Macchi. He is known universally for the superb racing seaplanes which he designed to com-pete in the Schneider Trophy Contests.

Cavendish, Henry (1731–1810)
British scientist who isolated the gas hydrogen in 1766. The discovery of this light-weight gas, which Cavendish called Phlogiston, or inflam-mable air, led to the develop-ment of practical lighter-than-air craft.

Cayley, Sir George (1773–1857)
British inventor, now regar-ded universally as the 'Father of Aerial Navi-gation'. As early as 1799 he engraved on one side of a silver disc – now in London's Science Museum – a dia-gram showing the forces of lift, drag and thrust.

Chadwick, Roy (1893–1947)
Famous aircraft designer of the British A. V. Roe Com-pany, renowned for the design of the Avro Lancaster bomber which played a sig-nificant role in World War II.

Chanute, Octave (1832–1910)
American railway engineer, who became one of the three great pioneer designers of gliders. He began to build hang gliders in 1896, and helped the Wright brothers in their early research.

Charles, Professor J. A. C. (1746–1823)
Designed and built the first hydrogen-filled balloon. This made the first ascent, unmanned, of a hydrogen balloon on 27 August 1783. He was also one of the first men to be carried aloft in such a balloon on 1 Decem-ber 1783.

Chavez, Georges (1887–1910)
Peruvian pioneer pilot, made the first aeroplane flight over the Alps, in a Blériot monoplane, on 23 September 1910. Success turned to disaster, for Chavez was killed when his aircraft crashed when land-ing at Domodossola, Italy.

Chichester, Sir Francis (1901–1971)
British aviator, sailor and navigator extraordinary, achieved the first solo flight in a seaplane between New Zealand and Japan in 1931.

Cierva, Juan de la (1886–1936)
Spanish designer of rotary-wing aircraft, whose C-4 Autogiro flew successfully for the first time at Getafe, Spain, on 9 January 1923.

Coanda, Henri (1885–1972)
Romanian aviator engineer who, in 1910, built the world's first jet-propelled aeroplane. This was not powered by a turbojet, but by what today would be called a ducted-fan. He subsequently became a significant aircraft designer with the British & Colonial Aeroplane Com-pany (later Bristol).

Cobham, Sir Alan (1894–1973)

Gained early renown for a number of long distance flights, including a 8,047m (5,000 mile) circuit of Europe in 1921, London to Cape Town and return in 1926 and a 37,015m (23,000 mile) flight around Africa in 1927. Sir Alan later developed in-flight refuelling.

Cochrane, Jacqueline
Who, flying a North American F-86 Sabre fighter aircraft on 18 May 1953, became the first woman in the world to pilot an aeroplane at supersonic speed.

Cody, Samuel Franklin
(1861–1913)
Born an American, made the first officially recognized aeroplane flight in Britain, in the *British Army Aeroplane No. 1*, which he designed and built. This flight of 424m (1,390ft) was made on 16 October 1908. Cody later became a British citizen.

Conrad, Max
American pilot who made numerous record long-distance flights in light aircraft, including Casablanca–Los Angeles, non-stop in 58·5 hours. Between 1928 and 1974 he logged 52,929 hours' flying time, a world record.

Cornu, Paul
Pioneer designer of rotary-winged aircraft, achieved the first free flight of a man-carrying helicopter on 13 November 1907. This was in a twin-rotor aircraft of his own design, flown at Lisieux, France.

Costes, Dieudonné
With Joseph le Brix, achieved the first non-stop crossing of the South Atlantic flying the Breguet XIX *Nungesser-Coli* on 14 to 15 October 1927.

Coutelle, Capitaine J.-M.-J. (1748–1835)
Commander of the French Company of Aérostiers in the Artillery Service who, on 26 June 1794, carried out military reconnaissance from the tethered observation balloon *Entreprenant* at the battle of Fleurus – the first operational use of an aircraft in warfare.

Curtiss, Glenn H.
(1878–1930)
American pioneer designer and aircraft builder. He made the first official public flight in America of more than 1km (0·62 miles) on 4 July 1908. His Curtiss Aeroplane & Motor Company was to become famous for the design and construction of naval aircraft.

Dassault, Marcel
French aircraft designer who, under the original name of Marcel Bloch, was responsible for the design of a number of bomber aircraft which equipped the French Air Force during the inter-war period and at the beginning of World War II. During the war his brother Paul, a General in the French Army, adopted the Resistance code-name of d'Assault. After the war, in 1946, both brothers adopted the family name Dassault, and Marcel rebuilt his aircraft company under this name. From this company has come the most advanced fighter, bomber and business-jet aircraft to be produced by the post-war French aircraft industry.

Delagrange, Léon
(1873–1910)
French pioneer pilot who accomplished the first aeroplane flight in Italy in May 1908, and in which country he flew the world's first woman aeroplane passenger on 8 July 1908. On 17 September 1908, he flew non-stop for 30min 27sec at Issy-les Moulineaux, then the longest duration recorded in Europe.

Doolittle, James H.
(1896–)
Renowned American pilot, who has contributed much to aviation. He made the first one-day trans-continental crossing of the United States in 1922, won the 1925 Schneider Contest at Baltimore, made the first instrument landing in fog in 1929 and became a national hero when he led a raid on Tokyo in April 1942.

Dornier, Dr Claudius
(1884–1969)
German pioneer designer and aircraft builder, first employed in the development of large all-metal flying-boats for Count von Zeppelin. This early work culminated in the evolution of some superb flying-boats in the between-wars period, some of which remained in service until 1970.

Douhet, Giulio (1869–1930)
Italian exponent of air power whose outspoken criticism of Italy's conduct of World War I earned him a court-martial. Subsequently restored to command, his book *Command of the Air* is a classic argument for strategic bombing.

Dunne, J. W. (1875–1949)
British pioneer designer of tailless aircraft. His then-advanced designs were aimed at producing an automatically-stable aircraft. His first really successful aircraft flew in 1910.

Durouf, Jules
French balloonist, pilot of the first balloon to ascend from besieged Paris, on 23 September 1870, to initiate the world's first 'airline' and 'airmail' services.

Earhart, Amelia
(1898–1937)
American pioneer pilot, the first woman to achieve a solo North Atlantic flight in May 1932. Subsequently she made the first solo flight from Hawaii to California in 1935. She disappeared over the Pacific during a round-the-world attempt in 1937.

Eckener, Dr Hugo
(1868–1954)
Early associate of Count Von Zeppelin. After the death of the latter he headed the Zeppelin Company and was responsible for development and construction of the LZ.126 (later *Los Angeles*), *Graf Zeppelin* and *Hindenburg*, before loss of the latter ended construction of civil airships.

Ellehammer, Jacob C. H.
(1871–1946)
Danish inventor who designed and built an early aeroplane with which he achieved 'hopping' flights in 1906. It was powered by a 20hp, three-cylinder, radial air-cooled engine of his own design.

Ely, Eugene
American pilot with the Curtiss Aeroplane & Motor Company who achieved the first aeroplane take-off from a ship, the *USS Birmingham*, on 14 November 1910. On 18 January 1911 he piloted the first aeroplane to land on a ship, the *USS Pennsylvania*, anchored in San Francisco Bay.

Esnault-Pelterie, Robert
(1881–1957)
French aircraft designer and founder of the R.E.P. Aircraft Company, his developed R.E.P. 2 *bis* monoplane flying successfully on 15 February 1909. He is better remembered as a pioneer researcher into the possibility of space flight.

Fabre, Henri (b. 1883)
French pioneer of the sea-

plane, whose Hydravion became the first aircraft to take off from water on 28 March 1910. Fabre was to become well-known as a designer and constructor of seaplane floats.

Farman, Henry
(1874–1958)
British-born pilot and aircraft builder who became a naturalized French citizen in 1937, and was the first Briton to establish an internationally ratified world distance record. A distance of 771m (2,530ft), it was achieved in a Voisin aircraft on 26 October 1907.

Flynn, Reverend J.
(1880–1951)
Founder of the Australian Flying Doctor Service which was inaugurated on 15 May 1928.

Fokker, Anthony H. G.
(1890–1939)
Aviation pioneer, pilot and founder of the company which today bears his name in the Netherlands. His name is perpetuated in aviation history not only for World War I fighter aircraft, but for a superb family of transport aircraft which have contributed much to civil aviation.

Gagarin, Yuri (1934–1968)
Russian cosmonaut who was the first man to orbit the Earth in spaceflight, in the *Vostok I* spacecraft, on 12 April 1961.

Garnerin, André Jacques
(1769–1823)
A Frenchman, recorded the first successful parachute descent from a vehicle in flight on 22 October 1797. He descended from a balloon which was at a height of about 915m (3,002ft).

Garros, Roland
(1888–1918)
Made the first air crossing of the Mediterranean on 13

September 1913 flying a Morane-Saulnier monoplane. This French pilot has another niche in aviation history, recorded in World War I, when he used a crude method of firing a machinegun through the propeller disc to make head-on attack possible.

Giffard, Henri (1825–1882)
French pioneer airship builder, whose steam-engine powered vessel, in which he flew from Paris to Trappes on 24 September 1852, is regarded as the world's first successful airship.

Glenn, John Jr (1921–   )
American astronaut who, on 20 February 1962, became the first American to orbit the Earth. His three-orbit flight in the Mercury capsule *Friendship 7* marked the start of US attempts to catch up on Russia's lead in space technology.

Goddard, Robert H.
(1882–1945)
American pioneer of rocket propulsion for space flight. His research and development of high energy liquids led to the launch of a liquid-propellant rocket of his own design, the world's first, on 16 March 1926.

Grahame-White, Claude
(1879–1959)
Pioneer British pilot and competitor in many of the early aviation races; developer of the aerodrome at Hendon, north London, which became the venue of many important aviation meetings, and is now the site of the Royal Air Force Museum.

Green, Charles (1785–1870)
British balloonist and showman who, by 1854, had made well over 500 ascents. His *Royal Vauxhall* balloon, first flown in August 1836, could carry 12 persons when

the envelope was filled with coal gas, and as many as 28 when pure hydrogen was used for lift.

Guidoni, A.
One of the significant names in Italian aviation history, he pioneered the naval torpedo-carrying aeroplane. His first successful launch of a 340kg (750lb) torpedo was made in February 1914.

de Gusmão, Bartolomeu
(1686–1727)
Luso-Brazilian priest who, on 8 August 1709, demonstrated before King John V of Portugal and his court a practical model of a hot-air balloon which rose to a height of about 3·6m (11ft).

Hamel, Gustav (1889–1914)
Pioneer aviator, and pilot of the Blériot monoplane which, on 9 September 1911, carried the first official air mail in Britain, between Hendon and Windsor.

Handley Page, Sir
Frederick (1885–1962)
British aviation pioneer, founder of Handley Page Ltd in 1909 and builder of Britain's first strategic bomber aircraft in World War I. His name will be perpetuated in aviation history for the Handley Page slot, an aerodynamic feature which has been contributing to aircraft safety since it was first demonstrated in 1920.

Hargrave, Lawrence
(1850–1915)
Australian pioneer of kite design, whose development and perfection of the box-kite in 1893 was an important contribution to the construction of early European aircraft.

de Havilland, Sir Geoffrey
(1882–1965)
One of the best-known British aviation pioneers, whose career extended from

the very beginning of powered flight to the advent of the jet age. His name is linked in aviation history with the D.H.60 Moth that his company produced in 1925, and which fostered the private flying movement around the world, leading to the extensive current use of light planes, both for business and for pleasure.

Hawker, Harry G.
(1889–1921)
Australian pioneer pilot, recipient of a £1,000 consolation prize in the *Daily Mail* Hydro-Aeroplane Trial of August 1913. He was a founder of the H. G. Hawker Engineering Company at Kingston-upon-Thames, Surrey, and his name was perpetuated in Hawker Siddeley Aviation.

Heinkel, Professor Ernst
(1888–1958)
German pioneer aircraft designer. Amongst a whole series of superb aircraft, he is likely to be best remembered for the Heinkel He 178, the world's first turbojet-powered aircraft, which flew for the first time on 27 August 1939.

Henson, William Samuel
(1805–1888)
Early British disciple of Sir George Cayley, he designed and patented a heavier-than-air craft, the *Aerial Steam Carriage*, in 1843. Although never built in full-size form, it incorporated many advanced ideas.

Hill, G. T. R. (1895–   )
British designer of tailless aircraft who, like J. W. Dunne, sought to evolve an extremely stable aircraft. Most advanced was the Pterodactyl Mk V, built by the Westland Aircraft Company, which flew in 1932.

Hinkler, H. J. L. ('Bert')
(1892–1933)

Australian pilot who made the first solo flight from England to Australia, between 7 and 22 February 1928. Later, between 27 October and 7 December 1931, he made the first solo flight in a light aircraft, a de Havilland Puss Moth, from New York to London.

Hughes, Howard
(1905–1976)
American billionaire and movie magnate, in 1935 built an advanced racing aircraft designated H-1, with which, on 13 September 1935, he established a new landplane speed record of 567·026 km/h (353·388 mph). He was later to build the Hughes H-2 *Hercules*, a 180-ton giant flying-boat which he flew once only, on 2 November 1947.

Immelmann, Max
(1890–1916)
Renowned German pilot of World War I, whose name is always linked with those of Oswald Boelcke and Manfred von Richthofen. This legendary trio developed German fighter tactics. Immelmann died after achieving 15 victories in air combat.

Jatho, Karl (1873–1933)
German pioneer designer and constructor, whose powered aeroplane made a number of short 'hops' in 1903, including one of about 18·3 m (60 ft) recorded on 18 August 1903.

Johnson, Amy (1903–1941)
British airwoman who became famous by achieving the first female solo flight from England to Australia between 5 to 24 May 1930 flying a de Havilland Gipsy Moth.

Junkers, Professor Hugo
(1859–1935)
German aircraft designer and pioneer of all-metal aircraft construction, whose

Junkers J.I, the world's first all-metal cantilever monoplane, flew for the first time on 12 December 1915. Built of iron and steel, it had a Mercedes engine of 120 hp.

Kingsford-Smith, Sir Charles (1899–1935)
Australian pioneer pilot whose famous flights include the first trans-Pacific crossing, between 31 May and 9 June 1928; the first crossing of the Tasman Sea, 10/11 September 1928; and the first aeroplane flight from Australia to the United States between 22 October and 4 November 1934.

Knight, W. J.
Pilot of the North American X-15A-2 rocket-powered research aircraft which, on 3 October 1967, he flew at a speed of 7,297 km/h (4,534 mph), making the X-15 the fastest aeroplane ever flown.

Krebs, A. C.
Associate of Charles Renard, whose electric-motor-powered airship, *La France*, flew for the first time on 9 August 1884.

Langley, Samuel Pierpont
(1834–1906)
American railway surveyor and civil engineer, who demonstrated successful steam-powered model aeroplanes in 1896. A full-size man-carrying aircraft was developed from these in 1903, but two attempts to launch the *Aerodrome*, as it was called, proved wholly unsuccessful.

Latécouère, Pierre
(1883–1943)
French aircraft manufacturer and founder of the Lignes Aériennes Latécoère, which he established to forge an air route between France and South America. On 12 May 1930 a Latécoère 28 floatplane *Comte de la Vaux*

first completed the trans-Atlantic route which had been Pierre Latécoère's dream.

Latham, Hubert
(1883–1912)
French pioneer pilot, renowned for his two unsuccessful attempts to beat Louis Blériot in achieving the first crossing of the English Channel in a heavier-than-air craft.

Levavasseur, Léon
(1863–1922)
French pioneer aviation engineer and designer of the graceful Antoinette monoplanes and Antoinette engines which contributed importantly to the early development of aviation.

Lilienthal, Otto
(1848–1896)
German pioneer glider-builder and pilot, one of the most important names in aviation history. Before his death on 10 August 1896 – the result of a flying accident the previous day – he had made thousands of flights and had recorded meticulously the results of his theoretical and practical research work.

Lindbergh, Charles
(1902–1974)
American airmail pilot who, on 21 May 1927, landed at Paris after a 33½-hour flight non-stop from New York. This historic flight was the first solo non-stop flight across the North Atlantic, and brought everlasting fame to Lindbergh.

Link, Edward Albert
American inventor of a ground-based flight trainer which could reproduce the behaviour and response to control of an aeroplane in flight. His first trainer, sold in 1929, had led to the complex flight simulators used in training pilots today.

Lippisch, Professor Alexander (1894–   )
German aircraft designer responsible for much research into advanced aerofoils. He designed the DFS 194 which, powered by a 272 kg (600 lb) thrust Walter rocket, was the first successful liquid-rocket-powered aircraft, flown in 1940 by Heini Dittmar, and he contributed to the development of tailless and delta aircraft.

McCurdy, John A. D.
Canadian engineer who made the first successful controlled flight in a heavier-than-air machine (the AEA Silver Dart) in the British Commonwealth by a British subject in Baddeck, Novia Scotia, 23 February 1909.

Manly, Charles M.
(1876–1927)
Associate of American Samuel Pierpont Langley, and designer of the remarkably advanced five-cylinder radial air-cooled petrol engine which powered Langley's *Aerodrome*.

Maxim, Sir Hiram Stevens
(1840–1916)
American who, while resident in England, designed and built a giant flying-machine of 31·70 m (104 ft) wing span. Powered by two 180 hp steam-engines of his own design, and launched along a special railway track in 1894, it lifted at a speed of 64 km/h (40 mph) but broke its restraining guard rails and came to a halt.

Mayo, R. H. (1890–1957)
Technical general manager of British Imperial Airways, who proposed a piggy-back concept to help achieve a North Atlantic mail service. The resulting Mayo Composite, a modified Short 'C Class' flying boat carrying a specially-built four-engined seaplane, first separated on 6 February 1938.

Mermoz, Jean
Pilot of the French Lignes Aériennes Latécoère and later Cie Générale Aéropostale, renowned for his pioneering flights across the South Atlantic. He disappeared without a trace on 7 December 1936 during his 24th trans-Atlantic flight.

Messerschmitt, Willy
(1898–1978)
Famous German aircraft designer who as early as 1921 collaborated in the design of a successful glider. It was followed by other gliders and a series of lightplanes before his creation of the Bf 108 Taifun in 1934, Messerschmitt's first all-metal aircraft. Then came what must be regarded as his masterpiece, the Bf 109 fighter, often claimed as the most famous German aircraft of all time. It was to record distinguished service with the Luftwaffe throughout the war, and remained in service with foreign air arms for almost 20 years.

Mitchell, R. J. (1895–1937)
British aircraft designer of the Supermarine Aviation Works, Southampton, Hants, responsible for the design of the S.4, S.5, S.6, S.6A and S.6B racing seaplanes which gained the Schneider Trophy for Britain. These aircraft had an important influence on Mitchell's Spitfire design.

Mitchell, William E. (**'Billy'**) (1880–1936)
Controversial exponent of air power, and commander of all US Army fighter squadrons at the Western Front in World War I. Court-martialled in 1926 for his outspoken views, he resigned his commission. In 1946, ten years after his death, he was posthumously awarded the Congressional Medal of Honour, and so vindicating his beliefs.

Mollison, J. A. (1905–1959)
British long-distance pilot, first to make a solo east–west crossing of the North Atlantic, on 18/19 August 1932. He was also the first man to fly solo east–west across the South Atlantic and the first man to cross both North and South Atlantic.

Montgolfier, Etienne (1745–1799) and **Joseph** (1740–1810)
French makers of the world's first hot-air balloon capable of lifting a man, and whose balloon, launched on 21 November 1783, was the first to carry men in free flight.

Mozhaiski, Alexander F. (1825–1890)
Russian designer of an odd-looking monoplane which, powered by two steam-engines, achieved a short 'hop' of 20 to 30m (66 to 98ft) in 1884 after take-off down an inclined ramp.

Nesterov, Lieutenant
Pilot of the Imperial Russian Army who, on 27 August 1913, was the first pilot in the world to execute the manoeuvre known as a loop, flying a Nieuport Type IV monoplane.

Nobile, Umberto (1885–1978)
Italian designer of the semi-rigid airship *Norge* in which he, together with Lincoln Ellsworth and Roald Amundsen, flew over the North Pole on 12 May 1926. Nobile was later to build the *Italia*, which was lost after flying over the Pole on 24 May 1928.

Oberth, Hermann (1894–   )
Austrian pioneer of the theory of rocket power and the problems of interplanetary navigation. His thesis *The Way to Space*

Travel is regarded as a basic classic of the theory of astronautics.

Ohain, Pabst von (1911–   )
German pioneer of jet propulsion who, financed by Ernst Heinkel, demonstrated his first turbojet engine of about 250kg (551lb) thrust in September 1937. His experiments were in no way influenced by the work of Frank Whittle in Britain.

Opel, Fritz von (1899–1971)
German tycoon of the motor industry, who financed development of several rocket-powered aircraft. Most successful was the Opel-Sander-Hatry Rak. 1, first flown by von Opel on 30 September 1929.

Parke, Wilfred
British naval pilot, was the first to demonstrate the method of recovery from a spin on 25 August 1912. Prior to that time spinning aircraft had almost invariably crashed.

Paulhan, Louis (1883–1963)
French pioneer pilot and winner of the *Daily Mail* £10,000 prize for the first flight from London to Manchester. Paulham, flying a Farman biplane, became involved in a race with Claude Grahame-White, who was equally anxious to win the prize, but Paulhan was the victor, landing at Didsbury at 05.32am on 28 April 1910.

Pégoud, Adolphe (1889–1915)
The first pilot in the world to demonstrate an aircraft in sustained inverted flight, at Buc, France, on 21 September 1913. Inversion was achieved by a half-loop and after which another half-loop restored his aircraft to an even keel, for the roll

manoeuvre was not then known.

Pénaud, Alphonse (1856–1880)
French inventor who built and flew a successful rubber-driven model aeroplane in 1871.

Phillips, Horatio (1845–1924)
Little-known British designer who patented, in 1884, a wing section which contributed to the design of modern aerofoil sections. His multiplane of 1907 had some 200 narrow-chord wings of venetian-blind form and was reported to have flown for a short distance, but this claim is unsubstantiated.

Pilcher, Percy (1866–1899)
British pioneer glider builder and pilot, inspired and given practical help by the German Otto Lilienthal. Pilcher, like his mentor, died when one of his gliders crashed, on 2 October 1899, the first Englishman to lose his life in a heavier-than-air craft.

Platz, Reinhold
German aircraft designer, who was responsible for the basic design of the DrI triplane and D VII biplane fighter aircraft built during World War I by the German Fokker aircraft factory and also designed the post-war Fokker airliner.

Porte, John C.
A Member of the British Royal Naval Air Service, who developed from the Curtiss *America* flying boats – which he had assisted to design – a superb series of RNAS flying boats during World War I, known as Felixstowe 'boats.

Post, Wiley (1899–1935)
Renowned American long-distance pilot who, flying the

Lockheed Vega monoplane, *Winnie Mae*, made the first solo round-the-world flight in July 1933. Subsequently killed on another flight.

Prévost, Maurice
French pioneer pilot and winner of the first Schneider Trophy Contest, held at Monaco on 16 April 1913. His low average speed of 73·63km/h (45·75mph) was due to his having to complete an extra lap to comply with the rules of the contest.

Reitsch, Hanna (1912–1979)
Famous German airwoman, and the first woman in the world to pilot a helicopter. She demonstrated the Focke-Achgelis Fw61 twin-rotor helicopter in the Deutschland-Halle in Berlin in 1938.

Renard, Charles
(1847–1905)
French airship pioneer who, with the assistance of A. C. Krebs, designed and built the first practical airship. Powered by an electric motor, it reached a speed of 19km/h (12mph) when first flown on 9 August 1884.

Roe, Sir Alliott Verdon
(1877–1958)
British pioneer aircraft designer, constructor and pilot, and founder of the famous aircraft company which bore his name. On 13 July 1909 he flew his Roe 1 triplane to become the first Briton to fly an all-British aeroplane.

Rolls, The Hon. Charles S.
(1877–1910)
Associate of Henry Royce in the founding of Rolls-Royce Limited, accomplished the first double crossing of the English Channel in a French-built Wright biplane on 2 June 1910. He was killed on 12 July 1910 when his aircraft crashed at Bournemouth.

Royce, Sir Henry
(1863–1933)
Co-founder with C. S. Rolls of Rolls-Royce Limited, he authorized development of the Buzzard in-line engine as a power plant for the British Schneider Trophy contender of 1929. This led to the evolution of the Rolls-Royce Griffon engine.

Rozier, François Pilâtre de
(1754–1785)
First man in the world to be carried aloft in a tethered Montgolfier balloon on 15 October 1783; first to make a free flight, in company with the Marquis D'Arlandes, on 21 November 1783; and the first to die in a ballooning accident on 15 June 1785.

Santos-Dumont, Alberto
(1873–1932)
Brazilian aviation pioneer resident in Paris, was first to fly round the Eiffel Tower in his No. 6 dirigible, and the first to achieve a manned, powered and sustained aeroplane flight in Europe in his 14 *bis* tail-first biplane on 12 November 1906.

Schneider, Jacques
(1879–1928)
Son of a French armaments manufacturer, in 1912 he offered a trophy for international competition to speed the development of waterborne aircraft. The resulting Schneider Trophy Contests had significant influence on the design of high-speed aircraft.

Scott, Charles W. A.
(1903–   )
Pilot of the de Havilland D.H.88 Comet who, together with Tom Campbell Black, won the 1934 Mac-Robertson England-Australia air race. Charles Scott had, in 1931, made two record-breaking solo flights, England to Australia and Australia to England, in a de Havilland Gipsy Moth.

Scott, Sheila
British record-breaking woman pilot of the post World War II years. Between 18 May and 20 June 1966 she completed a solo round-the-world flight – the longest solo flight by any pilot of any nationality, and the first round-the-world flight by a British pilot.

Seguin, Laurent and Louis
French designers of the Gnome rotary engine which, from 1908, was the most important of early aircraft engines. The performance of the Gnome rotaries was surpassed by new designs evolved during World War I.

Shepard, Alan B.
(1923–   )
First American to travel in space, was launched in a sub-orbital trajectory in the Mercury capsule *Freedom 7* on 5 May 1961. In his 486km (302 mile) flight he attained a height of 185km (115 miles).

Short, Eustace (1875–1932), **Horace** (1872–1917) **and Oswald** (1883–1970)
British pioneers who founded the nation's first aircraft factory. By early 1909 they were building their first six Short-Wright biplanes under licence. The company exists to this day, under the title of Short Brothers Ltd, and is still designing and building aeroplanes.

Sikorsky, Igor A.
(1881–1972)
Russian pioneer aircraft designer and builder, creator of the world's first four-engined aircraft, the *Bolshoi*, which flew on 13 May 1913. Later becoming a naturalized American citizen, he concentrated on the development of rotary-winged aircraft, his VS-300, which was first flown on 14 September 1939, being developed into what is now regarded as the world's very

first practical single-rotor helicopter.

Smith, Sir Keith
(1890–1955) **and Sir Ross**
(1892–1922)
Australian brothers who, between 12 November and 10 December 1919, achieved the first flight from England to Australia in a Vickers Vimy bomber. The brothers won a prize of £10,000 from the Australian Government, and both were knighted.

Smith-Barry, R. R.
Founder of the School of Special Flying at Gosport, Hampshire, in July 1917, he was the originator of formal pilot instruction.

Sohn, Clem (d. 1937)
American 'bird-man' who devised webbed 'wing' and 'tail' which enabled him to manoeuvre in semi-gliding 'flight' after jumping from an aircraft, landing finally by parachute. He was killed on 25 April 1937 when his parachute failed to open fully.

Sopwith, Sir Thomas
(1888–   )
British pioneer pilot and founder of the Sopwith Aviation Company at Kingston-upon-Thames, Surrey; later a director and subsequently Chairman of the Hawker Siddeley Group.

Sperry, Lawrence B.
(1892–1923)
American designer of gyro-stabilizing mechanisms to control an aircraft in flight. His earliest device was fitted to a Curtiss flying boat and this, providing longitudinal and lateral stability was tested successfully in 1913.

Stringfellow, John
(1799–1883)
British associate of W. S. Henson in the construction of the 6·10m (20ft) span, Aerial Steam Carriage.

Stringfellow was responsible in particular for its steam-engine power plant.

Sukhoi, Pavel Osipovich (1895–1975)
Contemporary of Andrei Tupolev at the TsAGI (Central Aero and Hydro-dynamic Institute), he has been concerned primarily with the design of fighter aircraft. The greatest success of his design bureau has come in the years since 1953, with the development of a family of jet fighters, the most interesting feature of which is the steady and econ-omic improvement of the type, extending from the Su-7 of 1956 to the latest Su-19 twin-engined variable-geometry fighter-bomber which has the NATO code name *Fencer*.

Tank, Kurt
A contemporary of Ger-many's Willy Messerschmitt, Kurt Tank achieved similar fame as an aircraft designer with the Focke-Wulf com-pany. His first significant design was the Fw 56 Stösser trainer of 1933, but he is remembered especially for the Fw 190, regarded as the most advanced fighter air-craft in the world when it entered active service in 1941, and had its first Major use during the 'Channel-dash' of the German battle-ships *Gneisenau* and *Scharnhorst* and cruiser *Prinz Eugen*, on 12–13 February 1942. Also famous, in a quite differing military role, was the Fw 200 Condor adopted for maritime patrol.

Temple, Félix du (1823–1890)
French naval officer who de-signed and built a powered monoplane aircraft which, piloted by an unknown young sailor in 1874, gained the distinction of being the first manned aircraft to ach-ieve a brief 'hop' after

launch down a ramp.

Tereshkova, Valentina (b. 1937)
Russian cosmonaut, the first woman to travel in space, in *Vostok 6*, 16 to 18 June 1963, completing 48 orbits of the Earth before landing.

Thomas, George Holt (1869–1929)
British founder of Aircraft Transport and Travel Ltd., first British airline to be registered (on 5 October 1916). On 25 August 1919 this company inaugurated the world's first daily sche-duled international commer-cial airline service.

Trenchard, Sir Hugh (1873–1956)
British soldier turned av-iator, regarded as the 'Father of the Royal Air Force', who fought strenu-ously to keep it an indepen-dent service. He was to become Lord Trenchard, Marshal of the Royal Air Force.

Trippe, Juan T. (1899–)
Airline pioneer and founder of Pan American World Air-ways System in 1927. Before then Trippe had organized Long Island Airways with seven World War I aero-planes. For many years Pan American dominated American Civil aviation.

Tsiolkovsky, Konstantin (1857–1935)
Russia's 'father of astro-nautics' who, as early as 1898, suggested the use of liquid propellants to power space rockets. He first dis-cussed the principle of reac-tion, proved theoretically that a rocket would work in a vacuum, and made the first serious calculations relating to interplanetary travel, and anticipated the thrust-vector system of control, the prob-lem of zero gravity, multi-stage rockets, etc.

Tupolev, Andrei Nikolaevich (1888–1972)
Most famous internation-ally of all Soviet aircraft de-signers, he helped to found the TsAGI (Central Aero and Hydrodynamic Insti-tute) in 1918, becoming the head of its aircraft design department in 1920. His design and development of the Tu-104 jet-powered air-liner earned him the Lenin Prize and the Gold Medal of the FAI, and in 1970 he was made an Honorary Fellow of the Royal Aeronautical Society. Apart from the cre-ation of some important bomber and civil transport aircraft, the Tupolev design bureau's Tu-144 was the world's first supersonic air-liner to fly.

Turnbull, W. R.
Canadian engineer who tested aerofoils in a wind tunnel he built in 1902. He may have been the first to recommend dihedral to in-crease stability. He deve-loped the first controllable pitch propellers between 1916 and 1927.

Twiss, L. P. (1921–)
Pilot of the Fairey Delta 2 research aircraft which, on 10 March 1956, set the first world speed record estab-lished at over 1,609km/h (1,000mph). His average speed for the two runs over the selected course was 1,822km/h (1,132mph).

da Vinci, Leonardo (1452–1579)
Italian genius born in 1452 who is renowned in aviation history for drawings which show his early designs for a rotary-wing aircraft and a parachute. He believed from his observations of bird flight that man would be able to achieve flight by his own muscle power.

Voisin, Charles (1882–1912) **and Gabriel** (1886–)

French pioneer aircraft buil-ders. Flying a boxkite glider towed by a launch, Gabriel recorded the first flight from water by a manned aircraft on 6 June 1905.

Walden, Henry W.
Designer and builder of the first successful monoplane aircraft to fly in America. This was the Walden III which, powered by a 22hp Anzani engine, made its first flight on 9 December 1909.

Wallis, Sir Barnes N. (1887–1979)
British aviation engineer and designer, designer of the rigid airship R.100, origi-nator of geodetic construc-tion, inventor of the skip bomb which breached the German Moehne, Edar and Sorpe dams, and of the earthquake bombs which helped to bring allied victory in Europe. Post World War II he developed the variable-geometry wing.

Watt, Sir Robert Watson (1892–1973)
Leader of the British re-search team which, in the three years prior to the out-break of World War II, de-signed and created ground and airborne radar systems for the detection and lo-cation of enemy aircraft.

Wenham, F. H. (1824–1908)
His classic paper *Aerial Locomotion* was read to the Aeronautical Society in 1866. One of the leading theorists of his day, he also built the world's first wind-tunnel in conjunction with John Browning.

Whittle, Sir Frank (1907–)
British designer of the air-craft gas turbine engine who, while a Cadet at RAF Cran-well in 1928, published a thesis entitled '*Speculation*' which outlined the basic equ-

ations of thermodynamics for such a power plant. It was not until 1935 that he was able to begin development, and he ran the first aircraft turbo-jet engine in the world on 12 April 1937. This later powered the Gloucester Meteor.

Willows, E. T. (–1926)
British designer of some small, but successful semi-rigid airships, the first of which flew in 1905. He patented the design of swivelling propellers to simplify control of an airship in flight.

Wright, Orville (1871–1948) and Wilbur (1867–1912)
The proud names of the two American brothers renowned in aviation history for designing, building and flying the first man-carrying aeroplane in the world to achieve powered, controlled and sustained flight, on 17 December 1903.

Yakovlev, Alexander (1906–   )
One of the younger generation of Soviet aircraft designers, he completed his first military design in 1939. This had the designation BB 22, but became better known as the Yak-4. His first important military aircraft was the Yak-1 fighter, seen at the October Day flypast in 1940, and which with developments was built in huge numbers (about 37,000). Since those days the Yakovlev design bureau has been responsible for aircraft ranging from basic trainers, the important Yak-42 civil transport, and military aircraft which include the VTOL combat aircraft (*Forger*) first seen aboard the Soviet carrier/cruiser *Kiev* in the summer of 1976.

Yeager, Charles (1923–   )
United States Air Force pilot who, on 14 October 1947 piloted the Bell XS-1 rocket-powered research aircraft at a speed of 1,078km/h (670mph) (an equivalent of Mach 1·015) at a height of 12,800m (42,000ft). His was the first supersonic flight in history.

Zeppelin, Count Ferdinand von (1838–1917)
Founder of the German Zeppelin Company at Friedrichshafen on Lake Constance. His first rigid airship, the LZ.I, flew for the first time on 2 July 1900, carrying a total of five people.

Chronology

1709
8 August
Bartolomeu de Gusmão demonstrated the world's first model hot-air balloon.

1783
25 April
First flight of a Montgolfier hot-air balloon capable of lifting a man.
4 June
First public demonstration by the Montgolfiers of a small hot-air balloon.
19 September
A Montgolfier balloon carried a sheep, a duck and a cockerel into the air, the first animals to fly.
15 October
First man carried in tethered flight in a Montgolfier hot-air balloon.
21 November
First men carried in free flight in a Montgolfier hot-air balloon, travelling about $5\frac{1}{2}$ miles ($8\frac{1}{2}$km).
1 December
First men carried in free flight in a hydrogen balloon.

1784
4 June
First woman carried in free flight in a Montgolfier hot-air balloon.

1785
7 January
First crossing of the English Channel by balloon.

1793
9 January
First free flight of a manned hydrogen balloon in America.

1794
26 June
Observation balloon used at battle of Fleurus, the first use of an aircraft in war.

1797
22 October
First successful parachute descent from a balloon made by the Frenchman A. J. Garnerin.

1849
22 August
First air raid by balloons: the Austrians launched pilotless balloons against Venice, armed with time-fuse-controlled bombs.

1852
24 September
First flight of the world's first powered and manned airship.

1861
1 October
American Army's Balloon Corps was formed.

1866
12 January
Foundation of the Aeronautical society of Great Britain (later the Royal Aeronautical Society).

1870
23 September
First balloon ascent from besieged Paris at the height of the Franco-Prussian War.

1884
9 August
The dirigible *La France* made the world's first flight of a powered, manned and controlled airship.

1890
9 October
Clément Ader's bat-winged *Éole* monoplane made a brief hop.

1896
10 August
Death of Otto Lilienthal as the result of a gliding accident.

1898
In this year the Aero Club de France was founded.

1899
2 October
Death of Percy Pilcher as the result of a gliding accident.

1900
2 July
Count von Zeppelin's first rigid airship first flew.

1901
19 October
Alberto Santos-Dumont flew his No. 6 dirigible round the Eiffel Tower, Paris.
29 October
Foundation of the British Aero Club (later Royal).

1903
17 December
Wright brothers achieve world's first powered, controlled and sustained flight by a man-carrying aeroplane.

1905
12 October
Foundation of the Fédération Aéronatique Internationale (FAI), authority for air record certification.

1906
12 November
Alberto Santos-Dumont makes first powered and sustained flight by a man-carrying aeroplane in Europe.

1907
10 September
First flight of *Nulli Secundus*, first British army airship.
13 November
Paul Cornu's helicopter made world's first free flight by a manned rotary-wing aircraft.

1908
16 October
First officially recognized aeroplane flight in Britain made by S. F. Cody.

1909
14 May
First flight of over 1·6km (1 mile) in Britain recorded by S. F. Cody.
25 July
First aeroplane flight across the English Channel made by Louis Blériot in his Type XI monoplane.
25 July
First official aeroplane flight in Russia, made by Van den Schkrouff flying a Voisin biplane at Odessa.
29 July
First official aeroplane flight in Sweden, made by French pilot Legagneux flying a Voisin biplane.
2 August
The US Army bought its first aeroplane from the Wright brothers.
22 August
World's first international aviation meeting held at Reims, France.
30 October
First official aeroplane flight in Romania made by Louis Blériot flying a Blériot monoplane.
9 December
First official aeroplane flight in Australia. Colin Defries flew an imported Wright

biplane over the Victoria Park Racecourse at Sydney.

1910
8 March
Frenchwoman Mme la Baronne de Laroche became the first certificated woman pilot in the world.
10 March
First night flights in an aeroplane made by Frenchman Emil Aubrun.
28 March
First successful powered seaplane, built and flown by Henri Fabre, takes off from water.
2 June
First 'there and back' double crossing of the English Channel by an aeroplane, flown by the Hon C. S. Rolls.
27 August
First transmission of radio messages between an aeroplane and a ground station made by James McCurdy in America.
11 September
First crossing of the Irish Sea, from Wales to Ireland (Eire) made by Robert Loraine in a Farman biplane.
23 September
Georges Chavez made the first aeroplane flight over the Alps in a Blériot monoplane.
14 November
A Curtiss biplane flown by American pilot Eugene Ely was the first aircraft to take off from a ship.

1911
18 January
Eugene Ely, flying a Curtiss biplane, recorded the first landing of an aeroplane on a ship.
18 February
World's first official air mail flight at Allahabad, India, flown by Frenchman Henri Pequet.
12 April
Pierre Prier, flying a Blériot monoplane, made the first non-stop flight from London to Paris.

18 June
The Circuit of Europe, the first real international air race, started from Paris.
1 July
First flight of the US Navy's first aeroplane, a Curtiss A-1 Hydro-aeroplane.
22 July
First 'Round Britain' air race started from Brooklands, Surrey.
17 September–5 November
First coast-to-coast flight across the United States made by Calbraith P. Rodgers in a Wright biplane.
19 September
First official air mail flight in Great Britain, made by Gustav Hamel in a Blériot monoplane, which carried air mail between Hendon and Windsor.
23 September
First official air mail flight in the United States, made by Earl L. Ovington flying a Blériot monoplane.
22 October
First use of an aeroplane in war, Italian Capitano Piazza making a reconnaissance of Turkish positions in his Blériot monoplane.

1912
13 April
The British Royal Flying Corps was constituted by Royal Warrant.
12 November
A Curtiss A-1 Hydro-aeroplane, flown by Lt Ellyson, was launched by a compressed-air catapult.

1913
16 April
First Schneider Trophy Contest was won at Monaco by Maurice Prevost.
13 May
First flight of a four-engine aeroplane, the *Bolshoi* biplane designed by Igor Sikorsky, at St Petersburg, Russia.
23 September
First flight across the Mediterranean Sea made by

Roland Garros in a Morane-Saulnier monoplane.

1914
1 January
Inaugural date of the world's first scheduled service by an aeroplane, the Benoist Company's St Petersburg to Tampa service in Florida, USA.
1 August
Beginning of World War I.
30 August
Bombs were dropped on Paris, the first to be dropped from an aeroplane on a capital city.

1915
19 January
First air raid on Britain by Zeppelin airships.
3 March
NACA, National Advisory Committee for Aeronautics, predecessor of NASA, established in America.
31 May
Zeppelin LZ.38, an airship of the German Army, made the first Zeppelin attack on London, causing panic amongst the population.
12 December
The Junkers J.1, the world's first all-metal cantilever monoplane, flew for the first time.

1916
15 March
First use of US Army aircraft in military operations, in support of a punitive expedition into Mexico.
31 May
A British naval seaplane spotted for the British fleet at the Battle of Jutland, the first use of an aeroplane in a major fleet battle.
12 September
The Hewitt-Sperry biplane, in effect the world's first radio-guided flying-bomb, was tested in America.
5 October
First British airline company, Aircraft Transport and Travel Ltd, registered by George Holt Thomas.

1917
20 May
First German submarine (U-36) sunk as the direct result of action by an aircraft.
2 August
A Sopwith Pup, flown by Sqdn Cdr E. H. Dunning, made the first landing of an aeroplane on a ship under way.

1918
11 March
The first regular and scheduled international air mail service in the world began, between Vienna and Kiev.
1 April
British Royal Air Force formed.
13 April
First flight across the Andes, from Argentina to Chile, made by Lt Luis C. Candelaria.
24 June
First official air mail flight in Canada, from Montreal to Toronto, by a Curtiss JN-4, flown by Capt B. A. Peck, RAF.
12 August
First regular air mail service inaugurated in America, with a New York to Washington route.

1919
3 March
First American international air mail service established between Seattle and Victoria, British Columbia.
8–31 May
First crossing of the North Atlantic by air achieved by Lt Cdr A. C. Read of the US Navy, flying a Navy/Curtiss NC-4 flying-boat.
14–15 June
First non-stop crossing of the North Atlantic by air, achieved by Capt John Alcock and Lt Arthur Whitten Brown in a Vickers Vimy bomber.
2–6 July
First crossing of the North Atlantic by an airship, the

British R.34 made a return flight between 9–13 July.
7 August
First flight across the Canadian Rocky Mountains, from Vancouver to Calgary, made by Capt E. C. Hoy.
25 August
World's first scheduled daily international commercial airline service began, Aircraft Transport and Travel's London to Paris service.
1 November
America's first scheduled international commercial air service began, between Key West, Florida, and Havana, Cuba, by Aeromarine West Indies Airways.
12 November–10 December
First flight from England to Australia completed by Australian brothers Keith and Ross Smith.

1920
4 February–20 March
First flight from England to South Africa made by Lt-Col P. van Ryneveld and Sqdn Ldr C. Q. Brand, using three aircraft.
7–17 October
First trans-Canada flight, from Halifax, N.S., to Vancouver, B.C., by aircraft and airmen of the Canadian Air Force.
16 November.
QANTAS (Queensland and Northern Territory Aerial Services), Australia's famous international airline, was first registered.

1921
22 February
Inaugural flight of America's first coast-to-coast air mail service, between San Francisco and New York.
5 December
Australia's first regular scheduled air services inaugurated by West Australian Airways.

1922
4 September

First coast-to-coast crossing of the US in one day achieved by Lt J. H. Doolittle flying a de Havilland D.H.4B.
2 November
First scheduled service operated by Australia's QANTAS, between Charleville and Cloncurry.

1923
9 January
Spaniard Juan de la Cierva's first successful rotating-wing aircraft, the C-4 *Autogiro*, made its first flight.
2–3 May
First non-stop air crossing of the US accomplished by Lts O. G. Kelly and J. A. Macready flying a Fokker T-2 monoplane.
27 June
First successful in-flight refuelling of an aeroplane, made during the setting of a world endurance record by the US Army Air Service.
4 September
The *Shenandoah*, America's first helium-filled rigid airship, made its first flight.

1924
1 April
Establishment of Imperial Airways, Britain's first national airline.
6 April–28 September
Two out of four Douglas DWC floatplanes, flown by pilots of the US Army Air Service, made the first successful round-the-world flight.
1 July
Inauguration of America's first regular transcontinental air mail service, under control of the US Post Office.

1926
16 March
American R. H. Goddard launched the world's first liquid-propellant rocket.

1927
20–21 May
First non-stop solo crossing of the North Atlantic, from

New York to Paris, achieved by Charles Lindbergh in the Ryan monoplane *Spirit of St Louis*.
28–29 June
First non-stop flight between America and Hawaii, made by Lts A. F. Hegenberger and L. J. Maitland in the Fokker C-2 monoplane *Bird of Paradise*.
14–15 October
First non-stop aeroplane crossing of the South Atlantic, from Saint-Louis, Senegal, to Port Natal, Brazil, made by Capt D. Costes and Lt-Cdr J. Le Brix in the Breguet XIX *Nungesser-Coli*.

1928
7–22 February
Sqdn Ldr H. J. L. Hinkler made the first solo flight from England to Australia.
15 May
Inauguration date of the Australian Flying Doctor Service.
31 May–9 June
First flight across the Pacific Ocean, from San Francisco to Brisbane, made by Capt C. Kingsford Smith and C. T. P. Ulm, in the Fokker F.VIIB/3m *Southern Cross*.
11 June
First flight of the world's first rocket-powered aircraft in Germany.
8 October
The first cinema film was projected in an aircraft in flight.

1929
30 March
First commercial air route between London and Karachi, India, inaugurated by Imperial Airways, with passengers carried by rail on Basle-Genoa stage.
24–26 April
First non-stop flight from England to India, made by Sqdn Ldr A. G. Jones Williams and Flt-Lt N. H. Jenkins in a Fairey Long-range Monoplane.
8–29 August

First airship flight around the world, made by the German *Graf Zeppelin* commanded by Dr Hugo Eckener.

28–29 November
First aeroplane flight over the South Pole, made in a Ford Trimotor, the *Floyd Bennett*, piloted by Bernt Balchen.

1930
5–14 May
The first solo flight from England to Australia was made by a woman pilot, England's renowned Amy Johnson flying a de Havilland Gipsy Moth.

15 May
An aircraft of Boeing Air Transport carried the first airline stewardess, Ellen Church, on a commercial flight.

1931
13 March
First successful firing of a solid-fuel research rocket in Europe, by Karl Poggensee near Berlin.

13 September
Schneider Trophy won outright for Britain, by a Supermarine S.6B seaplane flown by Flt-Lt J. N. Boothman.

3–5 October
First non-stop flight from Japan to America made by Clyde Pangborn and Hugh Herndon in a Bellanca aircraft.

1932
20–21 May
Amelia Earhart achieved the first solo crossing of the North Atlantic by a woman.

1933
6–8 February
First non-stop flight from England to South Africa, made by Sqdn Ldr O. R. Gayford and Flt-Lt G. E. Nicholetts flying a Fairey Long-range Monoplane.

3 April
First aeroplane flights over Mount Everest, a Westland

PV-3 flown by the Marquess of Clydesdale and a Westland Wallace flown by Flt-Lt D. F. McIntyre.

15–22 July
American pilot Wiley Post, flying the Lockheed Vega *Winnie Mae*, made the first solo round-the-world flight.

31 December
First flight of Russia's Polikarpov I-16, first monoplane fighter with enclosed cockpit and retractable landing gear to enter squadron service.

1934
8–9 August
First non-stop flight from Canada to Britain by J. R. Ayling and L. G. Reid, from Wasaga Beach, Ont., to Heston.

20 October
Start of the MacRobertson Air Race from England to Australia.

22 October–4 November
First flight from Australia to the United States.

8 December
Inauguration of the first regular weekly air mail service between England and Australia.

1935
11–12 January
First solo flight by a woman from Honolulu, Hawaii, to America.

13 April
Inauguration of the first through passenger service by air from England to Australia.

6 November
First flight of the Hawker Hurricane, Britain's first monoplane fighter with enclosed cockpit, retractable landing gear and armament of eight machine-guns.

22 November
First scheduled air mail flight across the Pacific Ocean, San Francisco to Manila, Philippines.

1936
March
First Spitfire flight.

26 June
The Focke-Wulf Fw 61 twin-rotor helicopter, the first entirely successful powered rotary-wing aircraft in the world, made its first flight.

1937
12 April
Frank Whittle in Britain successfully ran the world's first aircraft gas turbine engine.

6 May
The German airship *Hindenburg* destroyed by fire in America.

1938
26 June
The first through flying-boat service from England to Australia was operated by Short C Class 'boats of Imperial Airways.

20–21 July
First commercial aeroplane flight across the North Atlantic, flown by the *Mercury* upper component of the Short-Mayo Composite.

1939
28 June
Inauguration of Pan American's weekly New York to Southampton service, flown with Boeing Model 314 flying boats.

5 August
Imperial Airways began an experimental Southampton to Montreal and New York mail service with Short C Class flying boats, which were refuelled in flight.

27 August
The German Heinkel He 178, the world's first aircraft to fly solely on the power of a gas turbine engine, made its first flight.

1 September
Beginning of World War II.

1941
5 April
First flight of the Heinkel He 280, the world's first aircraft designed specifically as a

turbojet-powered fighter.

15 May
The first British aircraft powered by a turbojet, the Gloster Whittle E.28/39, flew for the first time at Cranwell, Lincolnshire.

20 May
The Luftwaffe's largest airborne assault of the war was launched against the island of Crete.

2 October
The German Messerschmitt Me 163, the world's first rocket-powered combat aircraft, was flown at a speed of 1,004km/h (624mph) during early tests.

1942
7–9 May
The Battle of the Coral Sea, the first naval battle in history which was fought entirely by carrier-based aircraft.

3 October
First successful launching of a German V-2 (A-4) ballistic rocket.

1943
9 September
The Italian battleship *Roma* was hit and sunk by a German Rurstahl/Kramer X-1 radio-guided armour-piercing bomb.

1944
12 July
Britain's first jet fighter, the Gloster Meteor Mk 1, began to enter service with the Royal Air Force.

3 October
Germany's first jet fighter, the Messerschmitt Me 262, began to enter operational service.

1945
14 March
First operational use of the British *Grand Slam* bomb, weighing 9·980kg (22,000lb).

6 August
First military use of an atomic bomb, when a nuclear device was exploded over the Japanese city of

Hiroshima.

3 December

A de Havilland Vampire I, used for deck-landing trials aboard HMS *Ocean*, was the first jet fighter to operate from the deck of an aircraft carrier.

1946

24 July

First manned experimental use of an ejection seat: the 'guinea pig', Bernard Lynch, being ejected at 515km/h (320mph).

1947

14 October

The Bell XS-1 rocket-powered research aircraft became the first aircraft in the world to exceed the speed of sound, piloted by Charles Yeager, USAF.

1948

26 June

Beginning of the Berlin Airlift, the city being provisioned by allied aircraft until 12 May 1949.

1950

25 June

Beginning of the Korean War.

29 July

A Vickers Viscount made the world's first scheduled passenger service to be flown by a turboprop-powered airliner.

1952

2 May

The world's first regular passenger service flown by a turbojet-powered airliner, the de Havilland Comet 1.

1953

18 May

An American airwoman Jacqueline Cochrane, flying a North American F-86 Sabre, became the first woman in the world to fly at supersonic speed.

1954

15 July

First flight of America's first turbojet-powered commercial aircraft, the Boeing 707.

1956

21 May

The world's first airborne hydrogen bomb was dropped on Bikini Atoll in the Pacific Ocean.

1957

4 October

First man-made satellite, *Sputnik 1*, launched into Earth orbit by Russia.

19 December

First transatlantic passenger service flown by a turbo-prop-powered airliner.

1958

4 October

First transatlantic passenger service flown by turbojet-powered airliner.

1961

12 April

Russian spacecraft *Vostok 1* launched into Earth orbit carrying Russian cosmonaut Yuri Gagarin, the first man to travel in space.

1963

16 June

Russian cosmonaut Valentina Tereshkova, launched into Earth orbit in *Vostok 6*, became the first woman to travel in space.

1966

3 February

Russia's *Luna 9* spacecraft soft-landed on the Moon's surface and transmitted the first pictures of the lunar scene.

20 June

British airwoman Sheila Scott landed at Heathrow after setting a new woman's record of 33 days 3 mins for a round-the-world flight.

1967

3 October

North American X-15A-2 rocket-powered research aircraft was flown at a speed of 7,279km/h (4,534mph) by W. J. Knight.

1968

31 December

Russia's Tupolev Tu-144 prototype made its first flight, the first supersonic commercial transport aircraft to fly.

1969

9 February

First flight of the world's first wide-body turbofan-powered airliner, the Boeing Model 747.

2 March

The 001 prototype of the Anglo-French Concorde supersonic transport aircraft made its first flight at Toulouse, France.

1 April

The Hawker Siddeley Harrier, the world's first V/STOL close support aircraft, entered squadron service.

21 July

Neil Armstrong, US astronaut, became the first man to step on the Moon's surface, following a moon landing in the Sea of Tranquility.

1970

12 September

Launch of Russia's unmanned spacecraft *Luna 16*, which landed on the Moon, collected soil samples, took off from the Moon under remote control and was recovered successfully on 24 September.

1971

31 July

A Boeing Lunar Roving Vehicle was driven on the Moon's surface for the first time.

1974

14 August

First flight of a Panavia MRCA (now Tornado) multi-role combat aircraft prototype.

23 December

First flight of a Rockwell International B-1 strategic bomber prototype.

1976

21 January

Concorde SST introduced into international commercial airline service simultaneously by Air France and British Airways.

20 July

Touchdown on Mars of the first Viking Lander.

1977

7-25 February

Russia's two-man Soyuz 24 docked with the Salyut 5 spacelab during a flight lasting 424hr 48min.

26 September

Laker Airways inaugurated the cheap fare transatlantic Skytrain.

1978

12-17 August

First crossing of the North Atlantic by a balloon, the American Yost HB-72 *Double Eagle II*, crewed by Ben L. Abruzzo, Maxie L. Anderson and Larry M. Newman.

1979

12 June

Dr Paul MacReady's man-powered aircraft, *Gossamer Albatross*, pedalled/piloted across the English Channel by Bryan Allen.

Glossary

Absolute ceiling
The maximum altitude above sea-level at which a heavier-than-air craft can maintain level flight.

Accumulator
A reservoir to supply a fluid under pressure.

ADF
Automatic direction finder. An automated system of radio direction finding.

Aerobatics
Manoeuvres, other than those necessary for normal flight, carried out voluntarily by an aircraft's pilot.

Aerodrome
An area intended for the take-off and landing of aircraft, and including generally the associated buildings and equipment for the operation of aircraft.

Aerodynamics
The science dealing with air in motion and the reactions of a body moving in air.

Aerodyne
A heavier-than-air aircraft.

Aerofoil
A suitably shaped structure which, when propelled through the air, generates lift.

Aeronautics
Dealing with flight within the Earth's atmosphere.

Aeroplane
A fixed-wing powered aerodyne.

Aerostat
A lighter-than-air aircraft.

Afterburner
A thrust-augmenting addition to a turbojet or turbofan.

AI
Airborne interception. An airborne radar system evolved to assist in the interception of hostile aircraft.

Aileron
Movable control surface hinged to aft edge of each wing, usually adjacent to wingtip, to control roll.

Airbrake
An aerodynamic surface which can be deployed in flight to increase drag.

Aircraft
Vehicles, both lighter- and heavier-than-air, which fly in the atmosphere.

Airfield
See aerodrome.

Airfoil
See aerofoil.

Airframe
An aircraft's structure.

Airlift
The carriage of personnel or supplies by air, normally not for financial gain.

Airplane
U.S. usage: aeroplane (*q.v.*).

Airscrew
See propeller.

Airship
A steerable lighter-than-air craft.

Airstrip
A natural surface adapted – often in an unimproved state – for the operation of aircraft. Located usually in a forward area, such strips are sometimes reinforced by aluminium planks or steel netting.

Aldis lamp
Signalling device, projecting a narrow beam of light, visible only at the point to which it is aimed by a sight.

Altimeter
A height-indicating instrument.

Altitude
Height above ground level.

Amphibian
A heavier-than-air craft which can operate from land or water.

Analog computer
A computing device, based on the principle of measuring as opposed to counting.

Angle of attack
The angle at which an aerofoil surface meets the air.

Angle of incidence
The angle at which an aerofoil surface is attached to a fuselage in relation to the longitudinal axis.

Anhedral
The angle which an aerofoil surface makes to the fuselage when its tip is lower than the root.

Area rule
A formula for shaping an aircraft for minimum drag in supersonic flight.

Artificial horizon
A gyro-stabilized flight instrument which shows graphically pitching and rolling movements of the aircraft, to aid the pilot to maintain desired altitude.

ASI
Air-speed indicator, records aircraft speed relative to the surrounding air.

Aspect ratio
The ratio of the span to the mean chord of an aerofoil surface.

Astrodome
Transparent dome on fuselage upper surface to permit celestial navigation.

Astronaut
Crew member of a spacecraft.

Astronautics
Dealing with flight in space.

ASW
Anti-submarine warfare.

Automatic pilot
A gyro-stabilized system to maintain an aircraft in steady flight.

Autopilot
See automatic pilot.

Autorotation
The continued automatic rotation of a rotary wing by means of air flowing through it due to forward or downward motion of a gyroplane.

APU
Auxiliary power unit carried within an aircraft for such tasks as main engine starting, ground air conditioning, and the provision of electric, hydraulic or pneumatic power.

Ballast
Carried in an aerostat to permit variation of lift.

Ballistic missile
One which becomes an unguided free-falling body in the unpowered stage of its flight, and subject to ballistic reactions.

Ballonnet
An air-filled compartment in the envelope of an airship, used to maintain a desired gas pressure at varying altitudes.

Balloon
An unsteerable lighter-than-air craft.

Biplane
An aircraft with two superimposed wings.

Blimp
Small non-rigid, occasionally semi-rigid, airship.

Boundary layer
The thin stratum of air in contact with the skin of the airframe.

Buffet
Irregular oscillations imposed upon an aircraft's structure by turbulent airflow.

Bungee
A tensioning device, usually of rubber cord.

Cabin
An enclosed compartment for the crew and/or passengers of an aircraft.

Camber
The curvature of an aerofoil surface from leading-edge to trailing-edge, usually the upper or lower surfaces, or a mean line between them.

Canard
An aircraft which has its tailplane ahead of the wing.

Cantilever
A beam or other structural member which is supported at one end only.

CAT
Clear air turbulence. Invisible, and sometimes violent, whirlpools of air, created usually by the shearing action between jet streams or other high velocity winds, and a larger mass of slow moving air.

Ceiling
Maximum operating height of an aircraft, exceeded by the absolute ceiling.

Centre of gravity
(CG), the point of a structure at which the combined total weight acts.

Centre of pressure
The point at which the lift of

an aerofoil surface is considered to act.

Centre-section
The central portion of a wing.

Chaff
Narrow metallic foil strips dropped by an aircraft to confuse an enemy's radar system. Known also by its original British code name Window.

Chord
The distance from the leading- to trailing-edge of an aerofoil.

Cockpit
An open compartment in an aircraft's fuselage to accommodate a pilot, other crew member, or passenger.

Constant-speed propeller
One in which the pitch angle of the blades is changed constantly in flight to maintain maximum engine efficiency.

Cosmonaut
Term used by the Soviet Union to describe a crew member of a spacecraft.

Cowling
Usually the fairing which encloses a power plant.

Delta wing
One which, viewed in plan, has the appearance of an isosceles triangle: the apex leads, the wind trailing-edge forming the flat base of the triangle.

Digital computer
A computing device, based on the principle of counting as opposed to measuring.

Dihedral
The angle which an aerofoil surface makes with the fuselage when the tip is higher than the root.

Dirigible
Steerable aerostat, airship.

Ditch
To set down a landplane on water, due to emergency conditions.

Dive brake
A drag-inducing surface which can be deployed to reduce the speed of an aircraft in a dive.

DME
Distance measuring equipment, a navigational aid which interrogates a selected grand station to give a digital read-out of distance.

Dope
A cellulose-base paint used to waterproof and/or render taut the fabric covering an aircraft.

Drag
The resistance of a body to its passage through the air, creating a force parallel to the airstream.

Drag chute
A heavy-duty parachute attached to an aircraft's structure which can be deployed to reduce its landing run.

Drift
The lateral movement of an aircraft away from its track.

Drone
A pilotless remotely controlled aircraft, known also as an RPV (*q.v.*).

Drop tank
An auxiliary tank, generally containing fuel, which can be jettisoned if required.

ECM
Electronic countermeasures. Airborne electronic equipment designed to neutralize or reduce the effectiveness of an enemy's radar or other devices which employ or are affected by electromagnetic radiations.

Elevator
Movable control surface

hinged to the aft edge of the tailplane (stabilizer) to permit control about the aircraft's lateral axis.

Elevons
Control surfaces which on a tailless aircraft combine the duties of ailerons and elevators.

Elint
Electronic intelligence, regarding the location, volume, direction and, especially, the signature of an enemy's electronic devices.

Envelope
The gas container of a lighter-than-air craft, or of air in a hot-air balloon.

Exhaust-gas analyser
An instrument to determine and indicate the ratio of fuel to air fed into an engine.

FAA
Fleet Air Arm (UK).

FAC
Forward air controller. An officer operating in a forward position with a land force, and with suitable radio equipment to call in close-support aircraft.

Fairing
An addition to a structure to reduce its drag.

Fillet
A faired surface to maintain a smooth airflow at the angular junction of two components, for instance between wing and fuselage.

Fin
Fixed vertical surface of tail unit, to provide stability about the aircraft's vertical and longitudinal axes.

Flap
Usually a hinged wing trailing-edge surface which can be lowered partially to increase lift, or fully to increase drag.

Flaperon
Control surface which can serve the dual functions of a flap or aileron.

Flight deck
Term for a separate crew compartment of a cabin aircraft.

Flight envelope
The permitted boundary of an aircraft's performance.

Flight plan
A written statement of the route and procedures to be adopted for other than local flights.

Flight simulator
Permits the practice of flight operations from a ground-based training device. Some are for general flight routines, but many are configured for training in the operation of a specific aircraft.

FLIR
Forward-looking infra-red, consists of an electronic system which allows a pilot to see surface features on a cathode ray tube under no-visibility conditions.

Floatplane
An alternative name for a seaplane (*q.v.*) as opposed to a flying-boat (*q.v.*).

Flutter
An unstable oscillation of an aerofoil surface.

Flying-boat
A heavier-than-air craft which operates from water, and in which the hull or body is the main flotation surface.

Flying wires
Bracing wires of a non-cantilever wing which take the load of the wing in flight.

Fully-feathering propeller
One in which, in the event of engine failure, each blade can

be turned edge on to the airstream. This prevents windmilling and minimizes drag.

Fuselage
The body of an aircraft.

GCA
Ground controlled approach. Landing procedure when a pilot receives radio instructions for his approach and landing from a ground-based controller who monitors the operation by radar.

Glider
An unpowered heavier-than-air craft. The same is used sometimes to include both the sailplane and the much smaller hang-glider.

Gosport tube
Flexible speaking tube used for communication between instructor and pupil in early training aircraft.

Ground effect
The effect upon an aircraft of the cushion of air compressed against the ground by a helicopter's rotor or a low-flying fixed-wing aircraft.

Gyroplane (gyrocopter, 'Autogiro')
An aircraft with an unpowered rotary wing.

Harmonize
To align a fighter aircraft's guns and gunsight to produce a desired pattern of fire at a given range, or to match all aircraft controls correctly in manoeuvres.

Helicopter
An aircraft with a powered rotary wing.

Helium
A lifting gas for lighter-than-air craft which is not inflammable.

High-wing monoplane
One in which the wing is

mounted high on the fuselage.

Horizontally-opposed engine
In which two banks of in-line cylinders are mounted horizontally opposite to each other on a common crankcase.

HUD
Head-up display, which projects performance and/or attack information on to an aircraft's windscreen for constant appraisal by the pilot.

Hull
The fuselage or body of a flying-boat.

HVAR
High velocity aircraft rocket.

Hydrogen
The lightest known lifting gas, for lighter-than-air craft; unfortunately highly inflammable.

IAS
Indicated air-speed, from an ASI, which progressively reads lower as altitude increases and air density decreases.

ICBM
Intercontinental ballistic missile (see ballistic missile).

Icing
The process of atmospheric moisture freezing upon the external surfaces of an aircraft.

IFF
Identification, friend or foe; an electronic interrogation device.

ILS
Instrument landing system, to permit safe landing in conditions of bad visibility.

ILS marker
A radio marker beacon of an instrument landing system.

Inertia starter
Starting device of many piston-engines, the inertia of a hand- or electrically-driven flywheel being suddenly connected to the engine by a clutch.

In-line engine
In which cylinders are one behind another.

INS
Inertial navigation system. This relies upon the ability of three highly sensitive accelerometers to record, via a computer, the complex accelerations of an aircraft about its three axes. From this information the computer can continually integrate the aircraft's linear displacement from the beginning of the flight, pinpointing the position of the aircraft at all times.

Interference drag
Drag induced by conflicting airstreams over adjacent streamlined structures.

ISA
International Standard Atmosphere. An agreed standard of sea-level pressure and temperature (1,013·2 millibars at 15°C) to facilitate the comparison of aircraft performance figures.

JATO
'Jet'-assisted take-off. The augmentation of an aircraft's normal thrust for take-off, by the use of rockets.

Kinetic heating
The heating of an aircraft's structure as a result of air friction.

Kite
A tethered heavier-than-air craft sustained in the air by lift resulting from airspeed in a wind.

Landing weight
The maximum weight at

which an aircraft is permitted to land.

Landing wires
Bracing wires of a non-cantilever wing which support the wing when it is not in flight.

Landplane
A heavier-than-air-aircraft which can operate from land surfaces only.

Leigh-light
An airborne searchlight used for ASW operations in World War II.

Lift
The force generated by an aerofoil, (such as an aircraft wing) acting at right angles to the airstream which flows past it.

Longeron
A primary longitudinal member of a fuselage structure.

Loran
Long-range navigation, a navigational aid.

Low-wing monoplane
One with the wings attached at or near the bottom of the fuselage.

Mach number
The speed of a body expressed as a decimal fraction of the speed of sound in the medium in which that body is moving. Hence, Mach 0·75 represents three-quarters of the speed of sound. The speed of sound in dry air at 0°C (32°F) is about 331m (1,087ft) per second (1,193km/h; 741mph). Named after the Austrian physicist Ernst Mach.

Mean chord
The area of an aerofoil divided by its span.

Mid-wing monoplane
One with the wings attached mid-way on the fuselage.

Minimum control speed
The lowest certificated speed at which an aircraft can be flown and controlled.

Monocoque
Aircraft structure in which the skin carries the primary stresses.

Monoplane
A fixed-wing aircraft with one set of wings.

NACA
National Advisory Committee of Aeronautics (US, predecessor of NASA).

NASA
National Aeronautics and Space Administration, US airspace research body.

NATO
North Atlantic Treaty Organization.

Navigation lights
Mounted on the exterior of an aircraft so that its dimensions, altitude and position can be seen at night.

Omni
Omnidirectional radio range, a navigational device giving bearing from the transmitter, irrespective of the location or heading of the aircraft.

Ornithopter
A flapping-wing aircraft.

Parachute
A collapsible device which, when deployed, is used to retard the descent through the air of a falling body. Worn originally by aircrew for escape from a stricken aircraft, the parachute has since been adopted for paratroopers (parachute equipped troops), and for such uses as the dropping of cargo and supplies, recovery of spacecraft, and to reduce the landing run of high-speed aircraft. 'Sky-diving' has also become a sport recently.

Parasite drag
Induced by structural components which make no contribution to lift.

Parasol monoplane
One in which the wing is mounted above the fuselage.

Pitch
The movement of an aircraft about its lateral axis, i.e. that which extends from wingtip to wingtip, resulting in up and down movements of the tail. Also, the angular setting of rotary blades, as in a propeller aeroplane or helicopter.

Pitot head
Comprising two tubes, one to measure impact pressure, the other static pressure. They are used in conjunction with an airspeed indicator which registers airspeed from the pressure difference between the two tubes.

Planing
The condition when a flying-boat's hull or seaplane's floats are supported by hydrodynamic forces.

Pressurization
A condition in an aircraft's cabin, cockpit or a compartment, where the air is maintained at a pressure higher than that of the surrounding air. This is to compensate for the lower pressure at high altitude, permitting the normal respiratory and circulatory functions of crew and passengers at heights abnormal to the body.

Port
To the left of, or on the left-hand side of an aircraft.

Propeller
Rotating blades of aerofoil section, engine driven, to create thrust for the propulsion of an aircraft.

Pulse jet engine
Essentially an aerodynamic

duct with a series of spring-loaded inlet valves at the forward end. Works on intermittent combustion principle and will run in a static condition.

Pusher propeller
A propeller mounted aft of an engine.

RAAF
Royal Australian Air Force.

Radar
A method of using radio waves for location and detection of a variety of objects, and for navigational purposes.

Radial engine
In which the radially disposed cylinders are fixed and the crankshaft rotates.

RAE
Royal Aircraft Establishment (UK).

RAF
Royal Air Force. (Also, previously, Royal Aircraft Factory.)

Ramjet engine
Essentially an aerodynamic duct in which fuel is burned to produce a high-velocity propulsive jet. It requires acceleration to high speed before any thrust is produced.

RATO
Rocket-assisted take-off. Describes the augmentation of an aircraft's normal thrust for take-off, by the use of a rocket motor.

RCAF
Royal Canadian Air Force; now known as the CAF.

RDF
Radio direction finding. A method of obtaining by radio the bearing of an aircraft to two or three ground stations to determine its position.

Reversible-pitch propeller
One in which the blades can be turned to provide reverse thrust to reduce an aircraft's landing run.

RFC
Royal Flying Corps (UK, predecessor of RAF).

Rib
A component of an aerofoil structure, shaped to locate the covering or skin of such structure in the desired aerofoil section.

RNAS
Royal Naval Air Service (UK, predecessor of FAA).

RNZAF
Royal New Zealand Air Force.

Rocket engine
One which consists simply of an injector, combustion chambers and exhaust nozzle, and which carries with it a liquid or solid fuel and an oxidizer, allowing it to operate outside the Earth's atmosphere.

Roll
The movement of an aircraft about its longitudinal axis, i.e. that which extends through the centre-line of the fuselage, representing a wing-over rolling movement.

Rotary engine
In which radially disposed cylinders rotate around a fixed crankshaft.

Rotor
A rotary-wing assembly, comprising the rotor blades and rotor hub, of an autogyro or helicopter.

RPV
Remotely piloted vehicle (see Drone).

Rudder
Movable control surface hinged at aft edge of fin to permit directional control.

SAAF
South African Air Force.

Sailplane
An unpowered heavier-than-air craft designed for soaring flight.

Scanner
The antenna of a radar installation which moves (scans) to pick up signal echoes.

Scarff ring
Movable mounting for a machine-gun, usually in the aft cockpit of a two-seat combat aircraft.

Seaplane
A heavier-than-air craft which operates from water, and in which floats provide the flotation surface. The derived name of a float plane is also in common use.

Semi-monocoque
Aircraft structure in which frames, formers and longerons reinforce the skin in carrying the load stresses.

Skin
The external covering of an aircraft's structure.

Skin friction
Drag induced by skin surfaces.

Slat
An auxiliary aerofoil surface, mounted forward of a main aerofoil.

Slot
The gap between a slat or flap and the main aerofoil surface.

Spacecraft
A vehicle for travel in space – that is outside the Earth's atmosphere – which may be manned or unmanned.

Spar
A principal spanwise structural member of an aerofoil.

Specific fuel consumption (sfc)
The amount of fuel required per hour by an engine per unit power.

Spoilers
Drag-inducing surfaces used for both lateral control and for lift dumping to steepen descent, improve braking and reduce landing run.

Stall
Describes the condition of an aerofoil surface when its angle of attack becomes too great and the smooth airflow over the upper surface of the aerofoil, breaks down, destroying lift.

Starboard
To the right of, or on the right-hand side of an aircraft.

Step
A break in the smooth undersurface of a flying-boat hull or seaplane float to resist suction and facilitate planing (q.v.).

STOL
Short take-off and landing.

Streamline
To shape a structure or component so as to minimize aerodynamic drag.

Stringer
An auxiliary, lightweight structural member, the primary use of which is to establish the desired external form of a structure before covering or skinning.

Strut
Structural member in compression.

Subsonic
Flight at speeds below the speed of sound.

Supercharger
A form of compressor to force more air or fuel/air mixture into the cylinders of a piston-engine (to supercharge) than can be induced by the pistons at the prevailing atmospheric pressure.

Supersonic
Flights at speeds above the speed of sound.

Swept wing
In which the angle between the wing quarter-chord and the fuselage centre-line is less than 90 degrees.

Tabs
Small auxiliary surfaces which are used to trim an aircraft or move a control surface.

Tailplane
Horizontal aerofoil surface of tail unit, fixed or pivoted, to provide longitudinal stability or serve as a primary control surface. Also known as stabilizer.

Thickness/chord ratio
The thickness of an aerofoil expressed as a ratio of the wing chord.

Thrust
Force generated by propeller or jet efflux which propels an aircraft through the air. For stable forward flight the forces of thrust and drag must be equal.

Tractor propeller
A propeller mounted forward of an engine.

Transonic
Relating to the phenomena arising when an aircraft passes from subsonic to supersonic speed.

Triplane
A fixed-wing aircraft with three superimposed wings.

Turbofan
A gas turbine in which air is ducted from the tip of the fan blades to bypass the tur-

bine. This air, added to the jet efflux, gives high propulsive efficiency.

Turbojet
A gas turbine engine which produces a high velocity jet efflux.

Turboprop
A gas turbine which, through reduction gearing, also drives a conventional propeller.

Turboshaft
A gas turbine which, through a gearbox, drives a power take-off shaft. Used most commonly for rotary-wing aircraft.

USAAC
United States Army Air Corps (US, predecessor of USAAF).

USAAF
United States Army Air Forces (US, predecessor of USAF).

USAAS
United States Army Air Service (US, predecessor of USAAC).

USAF
United States Air Force.

USMC
United States Marine Corps.

USN
United States Navy.

Variable-geometry wing
One which can be swept or spread in flight to provide optimum performance for take-off/landing, cruising and high-speed flight.

Variable-pitch propeller
One in which the pitch angle of the blades can be changed in flight to give optimum performance in relation to airspeed.

Vee-engine
In which two banks of in-line cylinders are mounted at an angle on a common crankcase.

V/STOL
Vertical and/or short take-off and landing.

VTOL
Vertical take-off and landing.

Wing warping
Early method of lateral control in which a flexible wing was twisted, or warped, to provide control similar to that of an aileron.

Yaw
The movement of an aircraft about its vertical axis, to make the nose move left or right.

Air Museums

The major international air shows of the world are those held in alternate years at Paris, France (1981 and 'odd' years) and Farnborough, England (1980 and 'even' years). The Paris Show is organized by the Union Syndicale des Industries Aeronautiques et Spatiales, and is normally held during the first week in June. This has been staged at Le Bourget Airport, Paris, for many years, although it may be transferred to a new site in a few years' time. The Farnborough Show is arranged by the Society of British Aerospace Companies, and has been held at its present site since shortly after World War II. The Farnborough display usually takes place during the first week in September.

Other important trade shows are held at Hanover in the Federal German Republic – this takes place during the first week in May in the same year as Farnborough shows – and at Iruma, near Tokyo in Japan, during October.

Surprisingly the USA does not have a regular showpiece to display the products of its aerospace industry, but there are two specialist shows held annually. These are the annual conventions of the National Business Aircraft Association – the location of this varies but it is held in late September – and of the Experimental Aircraft Association – which is held at Oshkosh, Wisconsin, during the first week in August. The latter is a gathering of amateur aircraft constructors from all over the United States and beyond. Other regular events in the USA are the National Air Races, held at Stead Field, Reno, Nevada, during September, and the Confederate Air Force annual air show, a unique gathering of surviving World War II aircraft types, held in October, at Harlingen, Texas.

There are many major aviation museums in countries all over the world. Much effort has been put into preserving and restoring significant historical aircraft, especially in the last ten years, a period which has seen the growth of many new museums in Europe, the USA and elsewhere. Although some of this work is comparatively recent, it is fortunate that much of the world's early aviation heritage had already been preserved for posterity – for example, 6 original Wright biplanes still exist in museum collections, together with over 30 Blériot monoplanes, and – from a more modern era – more than 70 Supermarine Spitfires.

The tabulation below includes all major museums that have aircraft on display and are open to the public, with the location of each, the number of aircraft actually on display, and some indication of the particular specialization of each.

ARGENTINA

Museo Nacional de Aeronautica, Aeroparque Airport, Buenos Aires.
Twenty aircraft, relating to the history of the Argentine Air Force and aircraft industry, from a Comper Swift of 1930 to an I.A.35 Huanquero of 1957.

AUSTRALIA

Camden Museum of Aviation, P.O. Box 72, Kogarah, New South Wales.
Twenty-two aircraft, almost all of military types, from a D.H. Moth of 1927 to two Sea Venom jet fighters of 1955.
Moorabbin Air Museum, P.O. Box 242, Mentone, Victoria. (At Moorabbin airfield.)
Ten aircraft of both military and civilian types, dating from a D.H. Moth of 1929 to a Gloster Meteor jet fighter of 1951.
Warbirds Aviation Museum, Mildura Airport, Muldura, Victoria.
Ten military aircraft.

BELGIUM

Musée de l'Armée et d'Histoire Militaire, Palais du Cinquantenaire, Brussels.
Sixty aircraft relating to the history of Belgian military and civil aviation.

BRAZIL

Air Museum, Campo dos Afonsos, Rio de Janeiro.
Ten aircraft, connected with the history of Brazilian military aviation.

CANADA

Calgary Centennial Planetarium, Mewata Park, Calgary, Alberta.
Six relatively modern aircraft are displayed here.
Canadian National Aeronautical Collection, Rockliffe Airfield, Ottawa, Ontario.
Well presented display with an excellent and varied collection with forty exhibits dating from 1918 to the present day, ranging from a Sopwith Camel to a Vickers Viscount. Products of the Canadian aircraft industry and aircraft which have served Canadian airlines and the Armed Forces are comprehensively covered. About four aircraft each at Ottawa International Airport, the National War Museum, and the Museum of Science and Technology, all in Ottawa.
Canadian Warplane Heritage, Hamilton Civic Airport, Hamilton, Ont.
Fairey Firefly, Vought Corsair, B-25 Mitchell and about six other World War II aircraft, all in flying condition or being restored to it.

CZECHOSLOVAKIA

Letecka Expozice Vojenskeho Muzea, Prague-Kbely.
Forty aircraft, including gliders, light aircraft, fighters, bombers and helicopters, mostly dating from World War II or later.
National Technical Museum, Kostelni Str. 42, Prague 7.
Thirteen aircraft, dating from a 1910 Blériot, to a Mraz Sokol light aircraft of 1947.

DENMARK

Egeskov Museum, DK-5772, Kvaerndrup. (South of Odense.)
Thirteen aircraft, mostly light aircraft and gliders owned by the Royal Danish Aero Club.

FRANCE

Musée de l'Air, Aeroport du Bourget, Paris; and 8 rue des Vertugadins, Chalais Meudon, Hauts-de-Seine.
Two separate exhibitions, with a total of over sixty aircraft on display, together giving a comprehensive view of French aviation history, with extra exhibits from other countries.

GERMAN FEDERAL REPUBLIC

Deutschesmuseum, 800-München (Munich).
Fifteen aircraft, covering the history of German aviation from an early Lilienthal glider to a Dornier DO 31 vertical take-off transport.
Hubschraubermuseum (Helicopter Museum), Bückeburg. (Near Minden.)
Seventeen helicopters of German, British, French and US manufacture.
Luftwaffen Museum, 2082-Uetersen, Holstein. (Near Hamburg.)
Twenty aircraft, mostly related to the post-1955 history of the Luftwaffe, but also including a Messerschmitt Bf109 and Heinkel He 111.

HOLLAND

Aviodome, National Aeronautical Museum, Schiphol Airport, Amsterdam.
Seventeen aircraft, covering the history of Dutch aviation, from a replica Fokker Spin (Spider) of 1911 to a Fokker S.14 jet trainer of 1951.
Luchtmacht Museum, Soesterburg Air Force Base. (Between Utrecht and Amersfoort.)
Twelve aircraft connected with the history of the Royal Netherlands Air Force, mostly of World War II or more modern design.

INDIA

Indian Air Museum, Palam Airfield, New Delhi.
A collection of seventeen aircraft relating to the history of the Indian Air Force, from a unique Westland Wapiti of 1935 to a Folland Gnat jet fighter.

ITALY

Museo Aeronautico Caproni di Taliedo, Via Durini 24, Milano.
Thirty-two aircraft, mostly of Italian (and many of Caproni) design. Oldest is a Caproni Ca.6 of 1910, newest a Bernardi Aeroscooter of 1961.
Museo Nazionale della Scienza e della Technica, Via San Vittore 21, Milano.
Twenty aircraft, mostly of Italian design, dating from 1910 to 1950.

NEW ZEALAND

Museum of Transport and Technology, Western Springs, Auckland 2.
Twenty aircraft, both military and civilian types, from the Pearse monoplane of 1904 to a de Havilland Vampire of 1952.

POLAND

Army Museum, Warsaw.
A display of nine military aircraft ranging from a Yak-9 of 1942 to a MiG-15 of 1954.
Museum of Aeronautics and Astronautics, Kracow.
Thirty-three aircraft, ranging from a replica Blériot XI to a Yakovlev Yak-23 jet fighter.

ROMANIA

Central Military Museum, Bucharest.
Eleven aircraft of Romanian, Italian, Russian and Canadian design, from a replica Vuia monoplane of 1906 to a MiG-15 jet fighter.

SOUTH AFRICA

South African National Museum of War History, Erlswold Way, Saxonwold, Johannesburg.
Nine aircraft are displayed, all being British or German types used in World War II.

SWEDEN

Air Force Flymuseer, F13M Malmslätt, near Linköping.
Over forty aircraft are held at this Royal Swedish Air Force base, including a wide variety of Swedish, British German, Italian and American designs relating to the history of the Air Force.

SWITZERLAND

Verkehrshaus, Lidostrasse 5, CH-6006 Lucerne.

Thirty aircraft, dating from the Dufaux biplane of 1910 to modern jets such as a Swissair Convair 990 airliner, covering the complete spectrum of the history of the Swiss aircraft industry.

THAILAND

Royal Thai Air Force Museum, Don Muang Air Force Base, Bangkok.

Twenty-eight aircraft of types used by the Royal Thai Air Force, from a Boeing P-12 of 1931 to a North American F-86 Sabre of 1953.

UNITED KINGDOM

Battle of Britain Museum, Aerodrome Road, Hendon, London.

Contains a collection, to which additions are being made, of British, German and Italian aircraft involved in the Battle of Britain.

Cosford Aerospace Museum, Royal Air Force Cosford, Shropshire. (Near Wolverhampton.)

Twenty-five aircraft, most of which are World War II or modern types, ranging from a Spitfire to a TSR-2.

Fleet Air Arm Museum, RNAS Yeovilton, Ilchester, Somerset.

Forty-five aircraft connected with the history of the Fleet Air Arm, mainly being World War II or modern types, and also including the first British Concorde.

Historic Aircraft Museum, Aviation Way, Rochford, Southend-on-Sea, Essex. (Adjacent to Southend Airport.)

Thirty aircraft, mostly of World War II or more modern design, but also including some pre-war D.H. Moths.

Imperial War Museum, Lambeth Road, London.

Fourteen aircraft, including missiles. Another thirty-six, ranging from a 1917 R.E.8 to a Concorde, are at Duxford Aerodrome, Cambridgeshire, together with over twenty of the Duxford Aviation Society's mainly civil collection.

Annual flying display.

Lincolnshire Aviation Museum, Old Station Yard, Tattershall, Lincolnshire.

Ten aircraft, dating from a Mignet Flying Flea of 1936 to a Vampire T.11 jet trainer of 1953.

Mosquito Aircraft Museum, Salisbury Hall, London Colney, Hertfordshire.

Ten aircraft, all of de Havilland manufacture, of which the most famous is the prototype Mosquito.

Newark Air Museum, Winthorpe Airfield, Newark, Nottinghamshire.

Twelve aircraft relating to aviation history in Nottinghamshire and Lincolnshire.

Royal Air Force Museum, Aerodrome Road, Hendon, London.

Forty aircraft, all related to the history of the Royal Air Force and the British aircraft industry, ranging in age from a Blériot XI of 1910 to a Lightning supersonic jet fighter.

Royal Scottish Museum (Museum of Flight), East Fortune Airfield, near Haddington, Lothian.

Fifteen aircraft, related to the history of aviation in Scotland, dating from a D.H. Dragon Rapide of 1935 to modern jets.

Science Museum, Exhibition Road, South Kensington, London.

Twenty-five aircraft depicting milestones in aviation technology, including types as varied as a Cody biplane of 1912 and a Rolls-Royce 'Flying Bedstead' vertical take-off test rig of 1954.

Shuttleworth Collection, Old Warden Aerodrome, near Biggleswade, Bedfordshire.

Twenty-five aircraft, mostly in flying condition, including all airworthy aircraft in the UK over fifty years old, and examples of most RAF training types from the Avro 504 to the Hunting Provost. Regular monthly flying displays in summer.

Skyfame Museum, Staverton Airport, Gloucestershire. (Between Gloucester and Cheltenham.)

Fifteen aircraft, mostly World War II and modern types, including the unique Saro A.1 jet fighter flying-boat and a rare Hawker Tempest.

Strathallan Collection, Strathallan Airfield, Auchterarder, Tayside.

Fifteen aircraft, mostly still in flying condition, and the majority being World War II types. Included are a Lancaster and a Mosquito. Annual flying display, usually held in July.

Torbay Aviation Museum, Higher Blagdon, Paignton, Torbay, Devon.

Twenty aircraft, mostly modern types, including light aircraft, jet fighters and helicopters.

UNITED STATES OF AMERICA

Aerospace Park, 413 West Mercury Boulevard, Hampton, Virginia 23666.

Ten aircraft, all American helicopters or jet fighters, except for one Hawker Siddeley XV-6A Kestrel VTOL prototype.

Air Force Museum, Wright-Patterson Air Force Base, Fairborn, Ohio 45433 (near Dayton).

A truly massive collection of 115 aircraft, concentrating on USAF aircraft types, but also including European and Japanese types, and ranging from a Wright Brothers Model B to a prototype North American SB-70 Valkyrie supersonic bomber.

Airpower Museum, Antique Airfield, Blakesburg, Iowa

52536. (Between Blakesburg and Ottumwa.)

Ten aircraft, mostly classic and vintage light types dating from the 1930–40 era, and mostly in flying condition.

Bradley Air Museum, Bradley International Airport, P.O. Box 44, Hebron, Connecticut 06248. (North of Hartford, near Windsor Locks, and North of Route 20).

Thirty aircraft, of which the majority are relatively modern types, ranging from a Boeing B-17G to a Douglas C-133B.

Confederate Air Force, Harlingen Industrial Airport, Harlingen, Texas 78550.

A museum of flying aircraft, largely devoted to US designs of World War II, but also including European World War II types and post-war US designs, the latter in a static exhibition. Usually about forty aircraft may be seen, although often airworthy aircraft may be involved in displays elsewhere in the USA. Annual air display normally held early in October.

Experimental Aircraft Association Air Museum, 11311 West Forest Home Avenue, Franklin, Wisconsin 53132. (Franklin is a suburb of Milwaukee.)

An impressive collection of ninety-five aircraft tracing the development of the light aircraft, both home-built and factory made, featuring both historic and modern designs, and displays of constructional techniques.

Florence Air and Missile Museum, Florence Municipal Airport, Florence, South Carolina 29501.

Fifteen modern US military aircraft are displayed, together with a collection of missiles.

Flying Circus Aerodrome, Route 17, Bealton, Virginia 22712. (Between Warrenton and Fredricksburg.)

Twelve flying aircraft ranging from a replica Sopwith Pup to a Stampe SV.4 and Stearman. Weekend flying displays in summer.

Flying Tiger Air Museum, P.O. Box 113, Paris, Texas 75460. (On Highway 82, at airport.)

Twelve aircraft, mostly in flying condition and concentrating on World War II types. Offers flying training to would-be owner-pilots of Mustangs and similar types.

Frederick C. Crawford Auto-Aviation Museum, 10825 East Boulevard, Cleveland, Ohio 44106.

Primarily an automobile museum, with eight aircraft on display ranging from a Curtiss A-1 of 1910 to a P-51K Mustang of 1944.

Glenn H. Curtiss Museum, Hammondsport, New York 14840. (North-West of Elmira, North of Route 17.) This museum houses eight aircraft, mostly of Curtiss design.

Greenfield Village and Henry Ford Museum, Dearborn, Michigan 48121.

Twenty aircraft, the oldest being a Blériot XI. Most date from the inter-war period.

Harold Warp Pioneer Village, Minden, Nebraska 68959. (South of Route 80 at Kearney.)

Fifteen aircraft, mostly of the 1910–1930 period, including a Curtiss Jenny and a Pitcairn autogiro.

Hill Country Transportation Museum, 15060 Foothill Avenue, Morgan Hill, California 95037. (South of San Jose.)

Thirty aircraft, mostly vintage and classic light aircraft types.

History and Traditions Museum, Lackland Air Force Base, San Antonio, Texas 78236.

Thirty aircraft, all relevant to the history of the USAF.

Movieland of the Air, Orange County Airport, Santa Ana, California 92707.

Thirty aircraft, many of which have featured in a variety of aviation films in the hands of Frank Tallman and remain in flying condition.

Museum of Flight, Seattle Centre, 305 Harrison Street, Seattle, Washington 98109.

Twelve gliders and powered aircraft, from a replica Wright 1902 glider to a prototype Northrop YF-5A jet fighter.

Museum of Science and Industry, Lake Shore Drive and 57th Street, Jackson Park, Chicago, Illinois 60637.

Ten aircraft ranging from a 1917 Curtiss Jenny to a 1940 Supermarine Spitfire set against the background of a major science and technology museum.

National Air and Space Museum, Smithsonian Institution, Washington D.C., 20560.

Sixty-five aircraft, including many famous and historic examples, from the original Wright Flyer to the North American X-15 rocket-powered research design of 1956, are on display in the new museum on the Mall in Washington, opened on 4 July 1976. In addition, more aircraft from the reserve collection may be viewed at the storage facility at Old Silver Hill Road, Silver Hill, Maryland.

Old Rhinebeck Aerodrome, P.O. Box 57, Route 1, Rhinebeck, New York 12572. (North of Poughkeepsie).

Twenty-five aircraft, mostly in flying condition, and representing the early days of aviation, with many original or replica World War I designs. Regular flying displays are held on Sundays in the summer.

Pate Museum of Transportation, P.O. Box 711, Highway 377, Fort Worth, Texas 76101. (South-West of city, between Benbrook and Cresson.)

Twelve modern US designs displayed as part of a larger exhibition.

Pima County Air Museum, Old Vail Road, Tucson, Arizona 85731. (Adjacent to the South side of Davis-Monthan Air Force Base.)

Fifty aircraft, mostly of US military origin.

Planes of Fame, Chino Airport, Chino, California 91710. (East of Los Angeles, to South of the Pomona Freeway, Route 60.)

Thirty aircraft varying in age from a rare Boeing P-12E fighter of 1932 to a Lockheed T-33A of 1953.

San Diego Aerospace Museum, 1649 El Prado, Balboa Park, San Diego, California 92101.

Thirty-five aircraft varying from a replica Wright Flyer to various jet fighters of the 1950s.

SST Aviation Exhibit Center, Kissimmee, Florida 32749.

Twelve aircraft, mostly light civilian types, in an exhibition dominated by the mock-up of the abandoned Boeing 2707 supersonic transport.

Strategic Aerospace Museum, 2510 Clay Street, Bellevue, Nebraska 68005. (Located at Offutt Air Force Base, South of Omaha.)

Thirty aircraft, including a collection of strategic bomber types loaned by the USAF Museum. A Convair B-36 is undoubtedly the most impressive exhibit.

USAF Armament Museum, Eglin Air Force Base, Valparaiso, Florida 32542.

Ten complete aircraft shown as part of an exhibition of aircraft armament dating from World War I until now.

US Army Aviation Museum, Fort Rucker, Alabama 36360. (North of Ozark, North-West of Dothan.)

Sixty aircraft, almost all of which have been associated with US Army Aviation, ranging from a Piper L-4 Cub to a Lockheed AP-2E Neptune.

US Army Transportation Museum, Fort Eustis, Virginia 23604. (North of Newport News.)

Twenty aircraft and helicopters relating to US Army aviation.

US Marine Corps Museum, Quantico Marine Corps Base, Quantico, Virginia 22134.

There is an aviation annex to the main military museum at this base, located at the Marine Corps Air Station – this contains twenty-five aircraft related to US Marine Corps history.

US Naval Aviation Museum, Naval Air Station Pensacola, Pensacola, Florida 32508.

Sixty aircraft related to the history of US Naval aviation.

USS Alabama Battleship Commission, Battleship Parkway, Mobile, Alabama 36601.

Nine aircraft, mostly of World War II origin and including a rare Vought OS2U Kingfisher.

Victory Air Museum, Polidori Airfield, Gilmer Road, Mundelein, Illinois 60060. (North of Chicago.)

Twenty military aircraft, mostly of World War II vintage.

Wings of Yesterday, P.O. Box 130, Santa Fe, New Mexico 87501.

Twenty aircraft, mostly in flying condition, and ranging in variety from a Stinson Detroiter tri-motor to a rare A-36 Mustang.

URUGUAY
Air Museum, Avenida de 31 Marzo, Montevideo.

Seventeen aircraft, dating from an Escafet 2 of 1910 to a Lockheed F-80 Shooting Star jet fighter.

400

Acknowledgments

The publishers would like to thank the following
individuals and organizations for their kind permission
to reproduce the photographs in this book:

Air Portraits Colour Library 42–43, 46–47, 117 above,
118–119, 352; Basil Arkell 243, 244 above; Aspect
Picture Library 272 below; Bildarchiv Preussicher
Kulturbesitz 283, 293 above and below, 299; The
Boeing Company 346, 350–351; British Aerospace
(H.S. Manchester/MARS, London) 138–139; British
Aircraft Corporation 269; Colorific (Mark
Meyer/Contact) 274–275, (Mary Fisher) 280–281;
Contact Press Images (Dirk Halstead 1979) 2–3; W.G.
Davis 24; Mary Evans Picture Library 33 right, 41,
296, 302 above; Flight International 83, 131, 135;
James Gilbert 48 below, 120, 160–161; Goodyear
Aerospace Company 302–303 below; Professor Clive
Hart 14, 33 left; F. Haubner 218; Hawker-Siddeley
Aviation Ltd. 192; Angelo Hornak 19, 22; Leslie Hunt
113 below, 348; Imperial War Museum, London
44–45, 78, 84, 123; Keystone Press Agency Ltd. 36–37,
363; Lockhead Aircraft Corporation 189 above,
270–271; Mansell Collection 240; Musee de L'Air 79;
National Maritime Museum 96; Photri 8, 100–101,
305, (Jack Novak) 307, 312, 341 below right;
Popperfoto 310–311, 340–341 above, 340–341 centre,
341 above right; John Rigby 347; Herbert Rittmeyer
4–5, 6–7; Science Museum, London 17 left and right;
Scottish Tourist Board 227; South African Airways
188 above; Spectrum Colour Library 231 below, 246;
Tony Stone Associates Ltd. 272 above; John Stroud
163, 171, 181, 183, 196, 201, 202, 209, 216, 219, 341
centre right; Suddeutcher Verlag 273, 292, 297, 298,
301, 340 centre left, 354, 355; John Taylor 21, 48
above, 68–69, 72, 73, 106, 107, 113 above, 114–116,
117 below, 134, 136–137, 176–177, 185–187, (Boeing
Company) 188–189 below, 198–199, 225, 226, 229,
230, 231 above, 232, 241, 244 below, 254–255, 265,
266–268, 322–323, 325, 331, 335, 337, (Smithsonian
Institute) 340 above left, (Canadair) 340 above centre,
(U.S. Airforce) 340 below left, 340 bottom, 343, 345,
349, 351 right; Michael J.H. Taylor 10–11, 62–63,
220–221, (Textron's Bell Helicopter Company) 238–239,
262–263, 290–291, 318–319; Radio Times Hulton
Picture Library 9, 295; U.S. Army 129; U.S. Navy 314;
Westland Helicopters Ltd. 228; Zefa Picture Library
(U.K.) Ltd. 1, 306.

VEGA GULL

FLYING WING

MAYO COMPOSITE

SARO LONDON

MAGISTER

WHITLEY

MILES R. R. TRAINER

HURRICANE

PERCIVAL Q6

LUTON MAJOR

WICKO

LYSANDER

SWORDFISH

MOTH MINOR

HORNET MOTH

CYGNET

SEA-FOX

BLENHEIM

ENSIGN

TIGER MOTH

SKUA